The Art and Science of Military Deception

For a listing of recent titles in the
Artech House Intelligence and Information Operations Series,
turn to the back of this book.

The Art and Science of Military Deception

Hy Rothstein
Barton Whaley
Editors

ARTECH
HOUSE
BOSTON | LONDON
artechhouse.com

Library of Congress Cataloging-in-Publication Data
A catalog record for this book is available from the U.S. Library of Congress.

British Library Cataloguing in Publication Data
A catalog record for this book is available from the British Library.

ISBN-13: 978-1-60807-551-5

Cover design by Vicki Kane

The two cover photos represent efforts to camouflage the Lockheed Burbank Aircraft Plant during World War II. The top photo is the airfield before being camouflaged. The second photo shows the plant after camouflage. The goal was to keep it from being found from the air and bombed.

10 9 8 7 6 5 4 3 2 1

To the United States Special Operations Forces: for your courage, patience and persistence in defending people against threats to their liberty and happiness and for pursuing justice, anywhere and anytime, we humbly dedicate this book to you.

All royalties for this book will go to the Special Operations Warrior Foundation which provides college scholarships for the surviving children of fallen Special Operations Forces, family services and educational counseling and support to wounded Special Operations warriors.

In Memoriam

Dr. Bart Whaley

Consummate gentleman, scholar, and the last great master of stratagem

Contents

Acknowledgments

This book is part of a research project supported by the Office of the Secretary of Defense. We would like to thank Ben Riley for his insights and encouragement and for always thinking "big." We also want to thank Glenn Fogg and Nancy Ann Budden who continuously ensure that fresh ideas are presented to the appropriate authorities in Washington for action.

The editors wish to thank Rebecca Lorentz. This work would not have been possible without her extraordinary assistance. Finally, this manuscript was significantly improved by the sharp eyes of Cindy Rothstein.

Introduction

I did send for thee to tutor thee
In stratagems of war.
—Shakespeare, Henry VI (c.1591), Pt. I, Act IV, Scene 5

Deception among people in a civilized society is something to be loathed even though it seems to be part of human nature. However, deception in war is a virtue. Stratagems aid commanders in hiding both intentions and capabilities from the enemy. They simultaneously reveal a false reality designed to guide the enemy to take actions that are detrimental to his cause. When properly designed and executed, stratagems reduce the horrific costs of war.

Yet the importance that U.S. military senior leaders have historically place on deception is uneven.[1] During World War II, the U.S. Army slowly and begrudgingly learned to use deception to great effect. General Eisenhower, perhaps more than anyone else, recognized the critical role of deception in seizing the Normandy beachhead before the Allied Forces drove towards Berlin. In 1947, as Chief of Staff of the Army and reflecting on his wartime experience, he wrote a memo to his Director of Plans and Operations saying, "...no major operations should be undertaken without planning and executing deception measures."[2]

However, Eisenhower also recognized the fragile place deception held in American military thinking. In the same memo he stated, "As time goes on... there is a danger that these two means [psychological warfare and cover and deception] may in the future not be considered adequately in our planning."[3] Again, the future President of the United States was reflecting of his wartime experience.

General Eisenhower considered it essential that the War Department "take those steps that are necessary to keep alive the arts of... cover and deception." He also directed the War Department to maintain "a nucleus of personnel capable in handling these arts in case an emergency arises." His expressed purpose in this guidance about deception was "the furtherance of national security."[4]

Eisenhower's concerns about the durability of deception in American military thinking seems odd considering the significant contributions of successful stratagems during World War II. Military officers, if nothing else, should be propelled by results, and the modest stratagem investments during the Second World War brought exceptional results.[5] Nevertheless, Eisenhower's insights and concerns regarding the future of deception in the U.S.

1. James Monroe, "Deception: Theory and Practice" (M.S. Thesis, Naval Postgraduate School, 2012), 3.
2. See Dwight D. Eisenhower, *The Papers of Dwight David Eisenhower: The Chief of Staff*, ed. Louis Galambos, Vol. VIII, Baltimore: Johns Hopkins Press, 1978, 1763; Monroe, 2.
3. Ibid.
4. Ibid.
5. Barton Whaley makes this point in "The one percent solution: costs and benefits of military deception," John Arquilla and Douglas Borer, eds., *Information Strategy and Warfare: A Guide to Theory and Practice,* New York: Routledge, 2007.

military were prescient. The U.S. military failed to act on General Eisenhower's instructions and continues to fall short on maintaining the necessary capabilities to competently plan and conduct deception in support of higher-level military operations. This failure places the United States at a disadvantage in the rough-and-tumble arena of international politics.

To be sure, deception has played an important role in American military history. During the American Revolution, General Washington continuously deceived the British about his capabilities and intentions, ultimately denying General Cornwallis the use of all of his forces and thus setting the stage for the final showdown at Yorktown. During the Civil War, General Johnston held Union forces in check with little more than decoy cannons made of painted tree trunks. During the Second World War, and following the lead of the British, deception reached a high point in both the war in the Pacific and in Europe. This is exemplified in the Battle of Midway and Operation Overlord. Even in Vietnam, the Military Assistance Command Vietnam—Studies and Observation Group (MACVSOG) made excellent use of deception, though it was never exploited at a theater level. During Operation Desert Storm, General Schwarzkopf deceived the Iraqi leadership into believing the coalition's strike into Kuwait would come from the sea rather than through the uncharted desert—a maneuver now referred to as the *left hook*.

This evidence of deception proficiency can lead defense experts to believe that concerns regarding U.S. military deception know-how are unwarranted. Yet self-deception can be as lethal as deception. Any careful and informed look into contemporary military operations validates General Eisenhower's concerns expressed in his 1947 memo. The careful and programmed use of stratagem in military operations is, at best, episodic. Why this is so is both astonishing and elusive. For some Americans, deception is sufficiently odious, even evil, and is unacceptable under any circumstances. In the U.S. military, where firepower, maneuver, speed, agility, and superb logistics almost always result in a perceived force advantage, the style of warfare that logically follows is of an attrition form that is aimed at the cumulative destruction of the enemy. Given these conditions, the habitual and systematic use of deception seems pointless. The United States seems to assume that the instrument of deception is a tool reserved for the weak or desperate. While these two broad explanations—the odious, unacceptable nature of deception and its incompatible features relative to the American way of war—have merit, they do not adequately account for America's halfhearted appreciation of deception.

The United States lacks a sufficient intellectual, ethical, and institutional framework to understand, plan, and execute sophisticated and high-level deception operations against its enemies.[6] In other words, the absence of both national policy and bureaucratic processes hamper the use of deception as an instrument of national policy. Recreating the capability to use deception requires a few key prerequisites: policy, apparatus, operators, and practice.

National policy legitimizes the use of deception as an instrument of statecraft. It describes the parameters for its use and shows the commitment that permeates a bureaucracy. Policy is the ultimate authorization that provides stable and continuous direction.

A *permanent apparatus,* at all appropriate levels, to contemplate, plan, control, and terminate deception is necessary in order to ensure that operations are integrated with and are mutually supporting of other activities. A permanent apparatus is much more likely to cultivate institutional knowledge and ensure compliance with national policies than with informal or temporary offices.

6. Walter Jajko, "Deception: Appeal for Acceptance; Discourse on Doctrine; Preface to Planning," *Comparative Strategy,* Vol. 21, 2002, p. 353.

Operators who have a penchant for the craft of deception are obviously indispensable. These individuals must be selected based on a combination of attributes that include intellect, imagination, operational art, and knowledge of the target. History has shown that individuals who excel in this craft are often eccentric and unorthodox. Leaders must not only tolerate unorthodoxy, but must ensure that there is a direct, responsive and unchallengeable relationship between leader and deceiver that transcends a normal chain of command. This special relationship will also make certain that deception operators are intimately familiar with the objectives and thinking of their leadership.

Finally, the *prerequisites of policy, apparatus, and operators* must be nurtured by realistic and persistent practice. Wartime is too late to learn how to conduct deception. Exercises are key for developing the skills for inventing, integrating, controlling, and terminating deception operations. Furthermore, exercises provide a mechanism for policy makers, commanders and operators to become accustomed to engaging in deception. They are a synchronization and socialization tool for both wartime and peacetime deception operations.[7]

These deception prerequisites seem rather obvious, straightforward, and uncomplicated. In fact, the U.S. military excels in similar enterprises that require the same prerequisites. Therefore, the military ought to be fully capable of planning and executing high level deception operations.

The purpose of this book is to set in motion a renaissance for using deception as an instrument of statecraft. Its contents have been carefully culled from the vast body of literature on deception. The various sections are designed to cumulatively provide sufficient breadth and depth on the subject in order to satisfy the novice as well as the expert. Additionally, sections and many of the individual pieces are preceded with expert commentary that is designed to illuminate the importance of the section or individual article, or to provide interesting background information to help establish context. The readings appear in their original format where practical. Properly understood, the contents of this book provide the reader with sufficient knowledge to pursue General Eisenhower's vision for the proper role of deception in support of the national interest.

7. Experts have identified various frameworks for deception that include both policy issues and processes. The ideas expressed above reflect aspects of many of these frameworks, but are mostly drawn from Walter Jajko's article previously cited.

CHAPTER 1

Beware of the Dog[1]

Roald Dahl

This short fictional piece can be an appropriate trigger for any discussion of deception basics.

Roald (pronounced RU-all) Dahl was born in Wales to Norwegian parents. Although he'd greatly exaggerated the autobiographical elements in this, his first short story, he had been an RAF fighter pilot in World War II when in 1940 he unromantically ran out of fuel and crash-landed his antiquated Gloster Gladiator biplane fighter in the Libyan Desert. Having been moderately injured in the face, Dahl recovered over the next few weeks in British military hospitals. He was then transferred to duty in the USA, where, for the remainder of the war, he worked with William Stephenson's secret British Security Coordination (BSC).

Dahl's simple story, substantially elaborated, became the basis of three major fictional tales. First came an episode in English spy thriller writer Len Deighton's best-selling "The Ipcress File" (1962) in which a British intelligence officer is kidnapped and awakens in a prison cell that he is led to believe is behind the Iron Curtain in Hungary. CIA Director Allen Dulles included that chapter in his "Great Spy Stories from Fiction" (1969).

The second version was the rather competent M-G-M thriller movie starring James Garner, "36 Hours" (1964). In that film, a U.S. Army major (Garner) who knows the where and when secrets of the forthcoming D-Day invasion is rendered unconscious and kidnapped by the Nazis. He awakens in a simulated American hospital in a fake post-war "Allied Occupied Germany" where the disguised Germans plan to get the secrets from him. A single discrepancy—a nick on his chin from shaving that he'd given himself only 36 hours earlier alerts him to the truth that he'd been in a coma for mere hours, not months.

This plot was recycled for the third time in 1989 for an American TNT-TV drama titled "Breaking Point," in which an American agent is captured by the Nazis who attempt to pry the D-Day secrets from him. When torture fails, they try the above type of deceptive charade.

Dahl is most widely remembered as the author of imaginatively bizarre children's stories, including "Charlie and the Chocolate Factory" (1964); and his rough marriage (1953–83) to American movie star Patricia Neal.

1. Roald Dahl, "Beware of the Dog," *Harper's,* October 1944, pp. 436–442. Reprinted with permission of the Estate of Roald Dahl.

Wing Commander Roald Dahl of the R.A.F., a veteran of the Greek and Syrian campaigns, has been writing for American magazines since his arrival in this country in 1942.

BEWARE OF THE DOG

A Story

ROALD DAHL

DOWN below there was only a vast white undulating sea of cloud. Above there was the sun, and the sun was white like the clouds, because it is never yellow when one looks at it from high in the air.

He was still flying the Spitfire. His right hand was on the stick, and he was working the rudder bar with his left leg alone. It was quite easy. The machine was flying well, and he knew what he was doing.

"Everything is fine," he thought. "I'm doing all right. I'm doing nicely. I know my way home. I'll be there in half an hour. When I land I shall taxi in and switch off my engine and I shall say, 'Help me to get out, will you?' I shall make my voice sound ordinary and natural and none of them will take any notice. Then I shall say, 'Someone help me to get out. I can't do it alone because I've lost one of my legs.' They'll all laugh and think that I'm joking, and I shall say, 'All right, come and have a look, you unbelieving bastards.' Then Yorky will climb up onto the wing and look inside. He'll probably be sick because of all the blood and the mess. I shall laugh and say 'For God's sake, help me out.' "

He glanced down again at his right leg. There was not much of it left. The cannon shell had taken him on the thigh, just above the knee, and now there was nothing but a great mess and a lot of blood. But there was no pain. When he looked down, he felt as though he were seeing something that did not belong to him. It had nothing to do with him. It was just a mess which happened to be there in the cockpit; something strange and unusual and rather interesting. It was like finding a dead cat on the sofa.

He really felt fine, and because he still felt fine, he felt excited and unafraid.

"I won't even bother to call up on the radio for the blood-wagon," he thought. "It isn't necessary. And when I land I'll sit there quite normally and say, 'Some of you fellows come and help me out, will you, because I've lost one of my legs.' That will be funny. I'll laugh a little while I'm saying it; I'll say it calmly and slowly, and they'll think I'm joking. When Yorky comes up onto the wing and gets sick, I'll say, 'Yorky, you old son of a bitch, have you fixed my car yet?' Then when I get out I'll make my report, and later I'll go up to London. I'll take that half-bottle of whisky with me and I'll give it to Bluey. We'll sit in her room and drink it. I'll get the water out of the bathroom tap. I won't say much until it's time to go to bed, then I'll say, 'Bluey, I've got a surprise for you. I lost a leg today. But I don't mind so long as you don't. It doesn't even hurt.' We'll go everywhere in cars. I always hated walking, except when I walked down the Street of the Coppersmiths in Bagdad, but I could go in a ricksha. I could go home and chop wood, but the head always flies off the axe. Hot water, that's what it needs; put it in the bath and make the handle swell. I chopped lots of wood last time I went home, and I put the axe in the bath. . . ."

Then he saw the sun shining on the engine cowling of his machine. He saw the rivets in the metal, and he remembered where he was. He realized that he was no longer feeling good; that he was sick and giddy. His head kept falling forward onto his chest because his neck seemed no longer to have any strength. But he knew that he was flying the Spitfire, and he could feel the handle of the stick between the fingers of his right hand.

"I'm going to pass out," he thought. "Any moment now I'm going to pass out."

He looked at his altimeter. Twenty-one thousand. To test himself he tried to read the hundreds as well as the thousands. Twenty-one thousand and what? As he looked the dial became blurred, and he could not even see the needle. He knew then that he must bail out; that there was not a second to lose, otherwise he would become unconscious. Quickly, frantically, he tried to slide back the hood with his left hand, but he had not the strength. For a second he took his right hand off the stick, and with both hands he managed to push it back. The rush of cold air on his face seemed to help. He had a moment of great clearness, and his actions became orderly and precise. That is what happens with a good pilot. He took some quick deep breaths from his oxygen mask, and as he did so, he looked out over the side of the cockpit. Down below there was only a vast white sea of cloud, and he realized that he did not know where he was.

"It'll be the Channel," he thought. "I'm sure to fall in the drink."

He throttled back, pulled off his helmet, undid his straps, and pushed the stick hard over to the left. The Spitfire dipped its port wing and turned smoothly over onto its back. The pilot fell out.

As he fell he opened his eyes, because he knew that he must not pass out before he had pulled the cord. On one side he saw the sun; on the other he saw the whiteness of the clouds; and as he fell, as he somersaulted in the air, the white clouds chased the sun and the sun chased the clouds. They chased each other in a small circle; they ran faster and faster, and there was the sun and the clouds and the clouds and the sun, and the clouds came nearer until suddenly there was no longer any sun but only a great whiteness. The whole world was white, and there was nothing in it. It was so white that sometimes it looked black, and after a time it was either white or black, but mostly it was white. He watched it as it turned from white to black, and then back to white again, and the white stayed for a long time, but the black lasted for only a few seconds. He got into the habit of going to sleep during the white periods, and of waking up just in time to see the world when it was black. But the black was very quick. Sometimes it was only a flash, like someone switching off the light and switching it on again at once, and so whenever it was white, he dozed off.

One day, when it was white, he put out a hand and he touched something. He took it between his fingers and crumpled it. For a time he lay there, idly letting the tips of his fingers play with the thing which they had touched. Then slowly he opened his eyes, looked down at his hand, and saw that he was holding something which was white. It was the edge of a sheet. He knew it was a sheet because he could see the texture of the material and the stitchings on the hem. He screwed up his eyes, and opened them again quickly. This time he saw the room. He saw the bed in which he was lying; he saw the gray walls and the door and the green curtains over the window. There were some roses on the table by his bed.

Then he saw the basin on the table near the roses. It was a white enamel basin, and beside it there was a small medicine glass.

"This is a hospital," he thought. "I am in a hospital." But he could remember nothing. He lay back on his pillow, looking at the ceiling and wondering what had happened. He was gazing at the smooth grayness of the ceiling which was so clean and gray, and then suddenly he saw a fly walking upon it. The sight of this fly, the suddenness of seeing this small black speck on a sea of gray, brushed the surface of his brain, and quickly, in that second, he remembered everything. He remembered the Spitfire and he remembered the altimeter showing twenty-one thousand feet. He remembered the pushing back of the hood with both hands and he remembered the bailing out. He remembered his leg.

It seemed all right now. He looked down at the end of the bed, but he could not tell. He put one hand underneath the bedclothes and felt for his knees. He found one of them, but when he felt for the other his hand touched something which was soft and covered in bandages.

Just then the door opened and a nurse came in.

"Hallo," she said. "So you've woken up at last."

She was not good-looking, but she was large and clean. She was between thirty and forty and she had fair hair. More than that he did not notice.

"Where am I?"

"You're a lucky fellow. You landed in a wood near the beach. You're in Brighton. They brought you in two days ago, and now you're all fixed up. You look fine."

"I've lost a leg," he said.

"That's nothing. We'll get you another one. Now you must go to sleep. The doctor will be coming to see you in about an hour." She picked up the basin and the medicine glass and went out.

But he did not sleep. He wanted to keep his eyes open because he was frightened that if he shut them again everything would go away. He lay looking at the ceiling. The fly was still there. It was very energetic. It would run forward very fast for a few inches, then it would stop. Then it would run forward again, stop, run forward, stop, and every now and then it would take off and buzz around viciously in small circles. It always landed back in the same place on the ceiling and started running and stopping all over again. He watched it for so long that after a while it was no longer a fly, but only a black speck upon a sea of gray, and he was still watching it when the nurse opened the door and stood aside while the doctor came in. He was an army doctor, a major, and he had some last-war ribbons on his chest. He was bald and small, but he had a cheerful face and kind eyes.

"Well, well," he said. "So you've decided to wake up at last. How are you feeling?"

"I feel all right."

"That's the stuff. You'll be up and about in no time."

The doctor took his wrist to feel his pulse.

"By the way," he said, "some of the lads from your squadron were ringing up and asking about you. They wanted to come along and see you, but I said that they'd better wait a day or two. Told them you were all right, and that they could come and see you a little later on. Just lie quiet and take it easy for a bit. Got something to read?" He glanced at the table with the roses. "No. Well, nurse will look after you. She'll get you anything you want." With that he waved his hand and went out, followed by the large clean nurse.

When they had gone, he lay back and looked at the ceiling again. The fly was still there and as he lay watching it he heard the noise of an airplane in the distance. He lay listening to the sound of its engines. It was a long way away. "I wonder what it is," he thought. "Let me see if I can place it." Suddenly he jerked his head sharply to one side. Anyone who has been bombed can tell the noise of a Junkers 88. He can tell most other German bombers, for that matter, but especially a Junkers 88. The engines seem to sing a duet. There is a deep vibrating bass voice and with it there is a high-pitched tenor. It is the singing of the tenor which makes the sound of a Ju. 88 something which one cannot mistake.

He lay listening to the noise, and he felt quite certain about what it was. But where were the sirens, and where the guns? That German pilot certainly had a nerve coming near Brighton alone in daylight.

The aircraft was always far away, and soon the noise faded into the distance. Later on there was another. This one, too, was far away, but there was the same deep undulating bass and the high singing tenor, and there was no mistaking it. He had heard that noise every day during the Battle.

He was puzzled. There was a bell on the table by the bed. He reached out his hand and rang it. He heard the noise of footsteps down the corridor, and the nurse came in.

"Nurse, what were those airplanes?"

"I'm sure I don't know. I didn't hear them. Probably fighters or bombers. I expect they were returning from France. Why, what's the matter?"

"They were Ju. 88's. I'm sure they were Ju. 88's. I know the sound of the engines. There were two of them. What were they doing over here?"

The nurse came up to the side of his bed and began to straighten out the sheets and tuck them in under the mattress.

"Gracious me, what things you imagine. You mustn't worry about a thing like that. Would you like me to get you something to read?"

"No, thank you."

She patted his pillow and brushed back the hair from his forehead with her hand.

"They never come over in daylight any longer. You know that. They were probably Lancasters or Flying Fortresses."

"Nurse."

"Yes."

"Could I have a cigarette?"

"Why certainly you can."

She went out and came back again almost at once with a packet of Players and some matches. She handed one to him and when he had put it in his mouth, she struck a match and lit it.

"If you want me again," she said, "just ring the bell," and she went out.

Once toward evening he heard the noise of another aircraft. It was far away, but even so he knew that it was a single-engined machine. But he could not place it. It was going fast; he could tell that. But it wasn't a Spit and it wasn't a Hurricane. It did not sound like an American engine either. They make more noise. He did not know what it was, and it worried him greatly. "Perhaps I am very ill," he thought. "Perhaps I am imagining things. Perhaps I am a little delirious. I simply do not know what to think."

That evening the nurse came in with a basin of hot water and began to wash him.

"Well," she said, "I hope you don't think that we're being bombed."

She had taken off his pajama top and was soaping his right arm with a flannel. He did not answer.

She rinsed the flannel in the water, rubbed more soap on it, and began to wash his chest.

"You're looking fine this evening," she said. "They operated on you as soon as you came in. They did a marvelous job. You'll be all right. I've got a brother in the R.A.F.," she added. "Flying bombers."

He said, "I went to school in Brighton."

She looked up quickly. "Well, that's fine," she said. "I expect you'll know some people in the town."

"Yes," he said, "I know quite a few."

She had finished washing his chest and arms, and now she turned back the bedclothes, so that his left leg was uncovered. She did it in such a way that his bandaged stump remained under the sheets. She undid the cord of his pajama trousers and took them off. There was no trouble because they had cut off the right trouser leg, so that it could not interfere with the bandages. She began to wash his left leg and the rest of his body. This was the first time that he had had a bed bath, and he was embarrassed. She laid a towel under his leg, and she was washing his foot with the flannel. She said, "This wretched soap won't lather at all. It's the water. It's as hard as nails."

He said, "None of the soap is very good now, and of course with hard water it's hopeless." As he said it he remembered something. He remembered the baths which he used to take at school in Brighton, in the long stone-floored bathroom which had four baths in a room. He remembered how the water was so soft that you had to take a shower afterwards to get all the soap off your body, and he remembered how the foam used to float on the surface of the water, so that you could not see your legs underneath. He remembered that sometimes they were given calcium tablets because the school doctor used to say that soft water was bad for the teeth.

"In Brighton," he said, "the water isn't . . ."

He did not finish the sentence. Something had occurred to him; something so fantastic and absurd that for a moment he felt like telling the nurse about it and having a good laugh.

She looked up. "The water isn't what?" she said.

"Nothing," he answered. "I was dreaming."

She rinsed the flannel in the basin, wiped the soap off his leg, and dried him with a towel.

"It's nice to be washed," he said. "I feel better." He was feeling his face with his hand. "I need a shave."

"We'll do that tomorrow," she said. "Perhaps you can do it yourself then."

That night he could not sleep. He lay awake thinking of the Junkers 88's and of the hardness of the water. He could think of nothing else. "They *were* Ju. 88's," he said. "I know they were. And yet it isn't possible, because they wouldn't be flying around so low over here in broad daylight. I know it is true, and yet I know it is impossible. Perhaps I am ill. Perhaps I am behaving like a fool and don't know what I am doing or saying. Perhaps I am delirious." For a long time he lay awake thinking these things, and once he sat up in bed and said aloud, "I will prove I'm not crazy. I will make a little speech about something complicated and intellectual. I will talk about what to do with Germany after the war." But before he had time to begin, he was asleep.

HE WOKE just as the first light of day was showing through the slit in the curtains over the window. The room was still dark, but he could tell that it was already beginning to get light outside. He lay looking at the gray light which was showing through the slit in the curtain, and as he lay there he remembered the day before. He remembered the Junkers 88's and the hardness of the water; he remembered the large pleasant nurse and the kind doctor, and now the small grain of doubt took root in his mind and it began to grow.

He looked around the room. The nurse had taken the roses out the night before, and there was nothing except the table with a packet of cigarettes, a box of matches, and an ash tray. Otherwise it was bare. It was no longer warm or friendly. It was not even comfortable. It was cold and empty and very quiet.

Slowly the grain of doubt grew, and with it came fear, a light, dancing fear that warned but did not frighten; the kind of fear that one gets not because one is afraid, but because one feels that there is something wrong. Quickly the doubt and the fear grew so that he became restless and angry, and when he touched his forehead with his hand he found that it was damp with sweat. He knew then that he must do something; that he must find some way of proving to himself that he was either right or wrong, and he looked up and saw again the window and the green curtains. From where he lay, that window was right in front of him, but it was fully ten yards away. Somehow he must reach it and look out. The idea became an obsession with him, and soon he could think of nothing except the window. But what about his leg? He put his hand underneath the bedclothes and felt the thick bandaged stump which was all that was left on the right-hand side. It seemed all right. It didn't hurt. But it would not be easy.

He sat up. Then he pushed the bedclothes aside and put his left leg on the floor. Slowly, carefully, he swung his body over until he had both hands on the floor as well; and then he was out of bed, kneeling on the carpet. He looked at the stump. It was very short and thick, covered with bandages. It was beginning to hurt and he could feel it throbbing. Suddenly he wanted to collapse, lie down on the carpet and do nothing, but he knew that he must go on.

With two arms and one leg, he crawled over toward the window. He would reach forward as far as he could with his arms, then he would give a little jump and slide his left leg along after them. Each time he did it, it jarred his wound so that he gave a soft grunt of pain, but he continued to crawl across the floor on two hands and one knee. When he got to the window he reached up, and one at a time he placed both hands on the sill. Slowly he raised himself up until he was standing on his left leg. Then quickly he pushed aside the curtains and looked out.

He saw a small house with a gray tiled roof standing alone beside a narrow lane, and immediately behind it there was a plowed field. In front of the house there was an untidy garden, and there was a green hedge separating the garden from the lane. He was looking at the hedge when he saw the sign. It was just a piece of board nailed to the top of a short pole, and because the hedge had not been trimmed for a long time, the branches had grown out around the sign so that it seemed almost as though it had been placed in the middle of the hedge. There was something written on the board with white paint, and he pressed his head against the glass of the window, trying to read what it said. The first letter was a G, he could see that. The second was an A, and the third was an R. One after another he managed to see what they were. There were three words, and slowly he spelled the letters out aloud to himself as he managed to read them. G-A-R-D-E A-U C-H-I-E-N. *Garde au chien.* That is what it said.

He stood there balancing on one leg and holding tightly to the edges of the window sill with his hands, staring at the sign and at the whitewashed lettering of the words. For a moment he could think of nothing at all. He stood there looking at the sign, repeating the words over and over to himself, and then slowly he began to realize the full meaning of the thing. He looked up at the cottage and at the plowed field. He looked at the small orchard on the left of the cottage and he looked at the green countryside beyond. "So this is France," he said. "I am in France."

Now the throbbing in his right thigh was very great. It felt as though someone were pounding the end of his stump with a hammer, and suddenly the pain became so intense that it affected his head and for a moment he thought he was going to fall. Quickly he knelt down again, crawled back to the bed, and hoisted himself in. He pulled the bedclothes over himself and lay back on the pillow, exhausted. He could still think of nothing at all except the small sign by the hedge, and the plowed field and the orchard. It was the words on the sign that he could not forget.

It was some time before the nurse came in. She came carrying a basin of hot water and she said, "Good morning, how are you today?"

He said, "Good morning, nurse."

The pain was still great under the bandages, but he did not wish to tell this woman anything. He looked at her as she busied herself with getting the washing things ready. He looked at her more carefully now. Her hair was very fair. She was tall and big-boned, and her face seemed pleasant. But there was something a little uneasy about her eyes. They were never still. They never looked at anything for more than a moment and they moved too quickly from one place to another in the room. There was something about her movements also. They were too sharp and nervous to go well with the casual manner in which she spoke.

She set down the basin, took off his pajama top, and began to wash him.

"Did you sleep well?"

"Yes."

"Good," she said. She was washing his arms and his chest.

"I believe there's someone coming down to see you from the Air Ministry after breakfast," she went on. "They want a report or something. I expect you know all about it. How you got shot down and all that. I won't let him stay long, so don't worry."

He did not answer. She finished washing him, and gave him a toothbrush and some tooth powder. He brushed his teeth, rinsed his mouth, and spat the water out into the basin.

Later she brought him his breakfast on a tray, but he did not want to eat. He was still feeling weak and sick, and he wished only to lie still and think about what had happened. And there was a sentence running through his head. It was a sentence which Johnny, the intelligence officer of his squadron, always repeated to the pilots every day before they went out. He could see Johnny now, leaning against the wall of the dispersal hut with his pipe in his hand, saying, "And if they get you, don't forget, just your name, rank, and number. Nothing else. For God's sake, say nothing else."

"There you are," she said as she put the tray on his lap. "I've got you an egg. Can you manage all right?"

"Yes."

She stood beside the bed. "Are you feeling all right?"

"Yes."

"Good. If you want another egg I might be able to get you one."

"This is all right."

"Well, just ring the bell if you want any more." And she went out.

He had just finished eating when the nurse came in again.

She said, "Wing Commander Roberts is here. I've told him that he can only stay for a few minutes."

She beckoned with her hand and the Wing Commander came in.

"Sorry to bother you like this," he said.

He was an ordinary R.A.F. officer, dressed in a uniform which was a little shabby, and he wore wings and a D.F.C. He was fairly tall and thin with plenty of black hair. His teeth, which were irregular and widely spaced, stuck out a little even when he closed his mouth. As he spoke he took a printed form and a pencil from his pocket, and he pulled up a chair and sat down.

"How are you feeling?"

There was no answer.

"Tough luck about your leg. I know how you feel. I hear you put up a fine show before they got you."

The man in the bed was lying quite still, watching the man in the chair.

The man in the chair said, "Well, let's get this stuff over. I'm afraid you'll have to answer a few questions so that I can fill in this combat report. Let me see now, first of all, what was your squadron?"

The man in the bed did not move. He looked straight at the Wing Commander and he said, "My name is Peter Williamson. My rank is squadron leader and my number is nine seven two four five seven."

Section I:
The Ethics of Political-Military Deception

MAN: Uh, is this a game of chance?
W. C. Fields: Not the way I play it.
—My Little Chickadee (movie, 1940)

CHAPTER 2

Intelligence Ethics[1]

R.V. Jones

Dr. Reginald Victor ("R.V.") Jones was responsible for coordinating scientific intelligence during World War II, and helped with the development of radar, breaking the secrets of German Beam navigation, preparing for D-Day and helping Britain deal with the V1 Flying Bombs and the V2 Rockets. After the war Jones was professor of natural philosophy at the University of Aberdeen (1946–1981).

1. R. V. Jones, *Reflections on Intelligence*, London: Heinemann, 1989, 35–56. Published by William Heinemann. Reprinted with permission of The Random House Group Limited.

Intelligence Ethics

When resources on a national scale can be applied in secrecy, as they have to be in intelligence, the power of public scrutiny cannot always be brought to bear against their prospective misuse. In the absence of such scrutiny, and particularly because of the opportunities for malpractice that intelligence offers, the only safeguard is a firm sense of ethics among its operators. This chapter therefore discusses some of the ethical problems that are involved.

Respect for Allies

Since intelligence often involves prying into another country's secrets, the question arises of the extent to which this can be justified. In my days in intelligence we observed a rule handed down to us that we should not spy on allies. The tradition was so strong that Churchill himself ordered that there should be no espionage against Russia once the German attack in 1941 had brought the Russians into alliance with us. And even after the alliance had dissolved in the Cold War, we have seen how Lord De L'Isle ruled that we should refrain from any reconnaissance against the *Sverdlovsk* at the Coronation Review of 1953.

Such a ruling, however, evidently had little effect at the later visit of the warship *Ordzhinikidze* which brought Khrushchev and Bulganin to Portsmouth in 1956, for it resulted in the affair in which Commander Crabb lost his life while attempting an underwater reconnaissance: and on the same visit the rooms to be occupied by Khrushchev and Bulganin were 'bugged', according to Peter Wright's account in *Spycatcher*.[1] It could, of course, be pleaded that this discourtesy was merely a reciprocation for the similar measures the Russians had long taken against us. But Peter Wright also relates how he and his colleagues 'bugged' the French Embassy between 1960 and 1963. It was clever work which resulted in our reading the French high-grade cipher coming in and out of the French Embassy in London. 'Every move made by the French during our abortive attempt to enter the Common Market was monitored.'[2] It is true that, as far as the Common Market was concerned, the French could no longer be regarded as our allies, and yet for me the episode jars.

The reason is that it brings back a memory of a very different episode in which we and the French were involved with ciphers. This was in 1945 when the British cryptographers at Bletchley were anxious to interrogate their German counterparts after the German surrender. We traced their flight from the wartime headquarters at Treuenbritzen and found that they were now in the relatively small zone of Germany that had been allocated for French control. After their wartime history the French were tending to re-establish their dignity whenever occasion arose, and this was one: they would not allow a mission of British and American cryptographers to enter their zone, and all representations, both military and diplomatic, had failed. I heard about the problem at Bletchley, and put it to my French colleague, Professor Yves Rocard, the Director of Research of the Free French Navy, whose contributions I described in *Most Secret War*.[3]

He said that he would see what could be done, and within a day he returned to my office to tell me that our cryptographers would now be welcome to go in, and to talk to the Germans. The French would not even ask to be present at the interrogations, and their only request was that if we discovered anything that affected the security of France, they would be grateful to know. Such a change had been brought about simply by the complete trust that the events of war had forged between Rocard and myself, and it shows how such a trust may prove much stronger than the wary relationships of diplomacy. With that memory, I for one could not have bugged the French Embassy.

Diplomatic Bags

A classic way of spying on embassies is, of course, the interception of diplomatic bags and the clandestine opening of the mail they contain. Much skill can be exerted in this process, as Peter Wright describes in *Spycatcher*, to avoid any sign that the contents have been read.[4] He also says that the simple application of Sellotape to the original sealing of the envelope makes it very difficult to open without giving the ultimate recipient any sign of tampering. During the war a colleague officer came to me with another simple suggestion for detecting whether our own confidential letters had been subject to unauthorized opening. This was simply to seal a letter in the ordinary way, and then to run it several times through a sewing machine. For if the letter were subsequently opened by cutting the thread it would be almost impossible to replace the enclosed sheets in such a way that all the holes would line up with those in the envelope so accurately that the ensemble could be re-stitched exactly as the original. He asked me whether I thought that it would work, and on receiving my encouragement he said he would send a stitched letter over to some of our security experts to see if they would agree. 'Just one more thing,' I suggested. 'Mark the cotton thread at intervals with a fluorescent dye, so that if they manage to replace the cotton we can identify that it is not what we used.' Here I banked on a human weakness that I had seen several times before (and many since) which beguiles clever minds to concentrate on the difficult parts of a challenge to such an extent that they overlook the simpler ones. After three days the envelope came back to us immaculately re-stitched with no visible sign of tampering, but including a complimentary card from our friends to show that they had opened it. They were furious when we told them that we were certain it had been opened, and that they had failed to spot the simple trick of fluorescent marking in concentrating, otherwise so successfully, on the very difficult task of re-stitching.

In such an instance the contest between the men trying to protect their country's secrets and the men who are trying to pry them open can be a battle of wits in which each side is aiming to be cleverer than the other, with all the thrust and parry of a mediaeval joust. And while such chivalry as Lord De L'Isle showed (see above, page 22) in the *Sverdlovsk* visit is rare, the contest tends to develop its own rules. Almost every power accepts that it is fair game to try and decrypt the signals of its potential opponents, even though there was the classic instance of the US Secretary of State, Henry Stimson, who in 1929 closed down the famous Black Chamber, the American cryptographic centre of World War I and afterwards, because, as he said, 'Gentlemen do not read each other's mail.' But while there may be something sneaky and furtive in reading a private letter, cryptography need have neither of these aspects and can be a highly intellectual exercise even though we may regret the diversion of so much intellect to such an uncreative end.

Open Skies

Much the same is true of both photographic and electronic reconnaissance. In pre-Sputnik days it could be argued that to fly a photographic aircraft over another country without its permission was an infringement of national sovereignty, and the Russians were prepared to shoot such aircraft down, as they showed in the U2 incident of 1960. This was after they had rejected President Eisenhower's offer of an 'open skies' policy in which the Americans and the Russians would have been free to fly photographic aircraft over one another's territory. In retrospect the Russian rejection looks all the more unreasonable because within a few years both sides were able to orbit reconnaissance satellites over each other's territory with impunity. What the Russian attitude would be if it should become readily possible to destroy or blind reconnaissance satellites is a matter for conjecture, especially in the face of the Strategic Defence Initiative. There was much to be said for Eisenhower's offer, if only because good intelligence can be a stabilizing influence in international affairs: it can contribute to both mutual deterrence and mutual reassurance.

The Espionage Convention

There was another noteworthy aspect of the U2 incident, for Eisenhower himself confirmed that he had authorized the U2 flights. He was criticized for making such an official acknowledgement, which was almost unprecedented in intelligence affairs, where the convention was that a government did not acknowledge its intelligence activities, and, for example, gave no support to any of its agents who had been caught in espionage. Notably, as William Colby records in *Honorable Men,*[5] it was Khrushchev's annoyance with Eisenhower's acceptance of responsibility for the U2 flight that led him to cancel the Paris Summit. But it was time that such a bogus convention was abandoned.

Curiously, although the pre-1914 era was the heyday of the convention, the European powers were surprisingly ready to make statements of their expenditure on their secret services. In 1912 the British Foreign Secretary, Edward Grey, easily obtained from his ambassadors the published expenditures in the countries to which they were respectively credited. The figures were published as a parliamentary paper[6] and they still make interesting reading:

Austria-Hungary	£62,500
France	£40,000
Germany	£80,387
Great Britain	£50,000
Italy	£120,800
Russia	£380,000 (+ £335,000 for secret police)

The preoccupation with secret service, both external and internal, is a Russian characteristic older than the 1917 Revolution. In fact in *The Craft of Intelligence*, Allen Dulles, who was Director of the Central Intelligence Agency from 1953 to 1961, traces the 'congenital suspicion' of the Russians at least as far back as the sixteenth century when, to resist the ravages of the Tartars and others, the inhabitants had to depend on the garrisons of walled stockades (kremlins) and cities in country that could easily be over-run. And since the Tartars sought to ascertain the strength of the defences in advance by sending agents in (as did Joshua at Jericho) the inhabitants developed a suspicion of strangers.

Secret Services

While there are relatively few problems of ethics in the operation of photographic, communications or electronic reconnaissance, it is otherwise with secret service. 'Oh, where hath our Intelligence been drunk, where hath it slept?' demanded King John, and knowing two of the traditional temptations – alcohol and sex – he might well ask. To these must be added money and disaffection with a native land, or greater affection for a foreign one. The problem for a secret service, or at least for one in a democracy which values decency, is how far the ends of national security can justify these means, nearly all of them disreputable, by which valuable information can sometimes be obtained.

The Russians, for example, have been ruthlessly unscrupulous in exploiting sexual weaknesses of both the hetero- and homo- varieties, not only by tempting employees of foreign powers to provide information in return for gratification but also by subsequent blackmail in threatening to bring such sexual indiscretions and treachery to the notice of an employee's superiors unless he (or she) continues to supply information. To any decent man such methods are thoroughly distasteful, even more so than the induction of an employee's treachery in return for money and material comforts; but the possibility that valuable information may thereby be obtained cannot be overlooked, although I for one would try to draw a borderline. This limit might be set at 'No sex, please, we're British!' or, rather shorter, at attempting to suborn an employee with money. I cannot conceive that a relation based upon mutual respect could be built up between the suborner and the suborned in such instances, and all my own experience of successes in intelligence has suggested that these have been based on some degree of mutual respect between the various elements in an intelligence chain.

But while I would have little respect for a man who attempted to sell his country's secrets for money, I would have to admit that information of value may be obtained in this way. For an example, the German cryptographic employee in 1932 to whom the French military intelligence gave the code-name 'Asche' (and whose real name may have been Hans-Thilo Schmidt) offered to sell them information connected with his employment in return for financial reward. When his offer was accepted, he provided among other items some operating instructions for the model of 'Enigma' machine then in use and tables of some current keys. These by themselves might not have been very helpful, but when the French gave them to the Polish cryptographers they greatly aided the latter's progress in solving the workings of Enigma. And, more recently, we saw the devastating Walker case in America. Not all such cases, though, end so brilliantly for either the accepter of the offer of treachery or its provider. Sometimes the offer is bogus, and ends in the accepter being duped; and sometimes the fear of being duped may lead the intended accepter to be suspicious and therefore to reject the offer, as appeared to be the case with Michael Bettany in 1983, whose treacherous overtures were treated so warily by the KGB that he was caught by his MI5 employers before his treachery could become effective.

The approach of Bettany to the KGB had some similarity to that of Oleg Penkovsky to the Americans in Moscow in 1960: to them his offer looked 'too good to be true', although it turned out to be for no personal reward but arose instead from a genuine disillusion with the Soviet system. Fortunately, although rejected by the Americans he established six months later a contact with the British, and a most valuable flow of information passed between him and his contact, Greville Wynn. Here mutual respect and friendship easily grew up between the two men.

That kind of respect was at its height of sublimity in World War II, among those of us in London who were working with agents in the many resistance networks in German-occupied territory. These agents were men and women who were risking themselves to help us and seeking no other reward than the liberation of their countries. On our side we were mainly amateurs who had been brought in to strengthen the previously slender organization, and one such was Jimmy Langley who so successfully headed the MI6 section responsible for helping our shot-down airmen and other service personnel to escape from the Continent. This effort was, of course, crucially dependent on specialized resistance chains whose sacrifices were such that Langley estimated that for each of the three thousand-odd men who were brought back one resistance worker had lost his or her life. No wonder, therefore, that some of us, including Langley, became completely committed to their support. In his book, *Fight Another Day*, he records the reaction of a 'professional' officer, Claude Dansey, the Vice-Chief of MI6, upon Langley's outburst at the directive that nothing could be done to help some workers in the famous Comet line who had been caught and condemned to death. 'Your trouble is, Jimmy,' said Dansey, 'that you love your agents.'[7] That cynical verdict would have applied to most of us.

Can Agents (and Others) Be Expendable?

Certainly we would not have countenanced any suggestion that one or more of our agents should be intentionally betrayed by another agent whom we aimed to ingratiate with the Germans as a double agent whose sympathies lay with them, and through whom we would subsequently feed false information about, for example, the coming landings in France. We were therefore concerned when the publication of Anthony Cave Brown's *Bodyguard of Lies* gave the impression in France that we had been prepared to do just that. One of the most gallant of French Resistance workers, the Vicomtesse de Clarens, wrote to me in 1981: '*Bodyguard of Lies* has stirred reactions and emotions, as you can well imagine, among those who have been led to think by Cave Brown that their lives and fates were considered of little value indeed when balanced with the dire necessity of avoiding the Germans finding out their Enigma was no longer secure.' Appealing to me to write to one of the Resistance journals, she continued: 'You will no doubt put things in their proper perspective, and publication in *Voix et Visages* could help to soothe some rather peeved feelings.' Faced with so much trust I took every possible step with old colleagues throughout the wartime intelligence organizations to check whether anyone could recall an instance where an agent had been intentionally betrayed, or which might be so interpreted. In reporting to *Voix et Visages* the firmly negative result, I could only console the French Resistance with the fact that the same author had said that Churchill himself had sacrificed both Coventry in 1940, to preserve the Ultra secret of our breaking Enigma, and also Bomber Command on the Nuremberg raid in 1944. He argued that the bogus agent who delivered such accurate information would be so entirely convincing that the Germans would believe the false information that he would subsequently provide about the coming D-Day landings. Neither the Coventry nor the Nuremberg claim has, to my knowledge, the slightest element of substance.

The Coventry story, though, for all its falsity has a widespread appeal because of the ethical problems that it illustrates for an operational commander and his intelligence officer. Let us suppose that the intelligence organization has established a source that can reveal the enemy's intentions, and so give notice of an impending attack in such detail that it can be parried by suitable preparation. The danger that then has to be faced is that this preparation may be so specific as to indicate to the enemy, either before or after his attack has been defeated, that information has been leaked, and that it has probably been leaked by a particular channel, for example, an insecure cipher or an agent in a key position. The enemy can then block the leak, and so valuable information on some future operation, or operations, will have been lost. Not only this, but the agent – if a human one – may have been caught and executed: an ungrateful reward for his work. Apart from this problem in ethics there are also the relative values to be assessed in the short-term advantage of acting on the information about an imminent threat and the long-term loss of information about later and possibly greater threats should the secret source be sacrificed.

Had Churchill in fact been faced with the Coventry dilemma, for example, a successful defence might have saved five hundred deaths in Coventry. But as the argument goes, if the necessary preparations could have led the Germans to realize that we were reading Enigma, and had they then changed the machine, some tens, or even hundreds, of thousands more lives might have been lost in the subsequent battles (the Atlantic, the Desert, and France) in which Enigma played a part. So Churchill must inevitably have had to decide on the sacrifice of Coventry. Actually, though, such decisions are rarely so clearcut: even in the hypothetical case of Coventry, for example, the Germans might have preferred to suspect that we had divined their intentions by some method other than breaking Enigma, such as observing the settings of their radio-beams as they were lined up in the afternoon prior to the bombing. Indeed, this is what they actually did when, later in the Blitz, they realized that we were somehow anticipating their nightly targets. Similarly, in the Battle of the Atlantic, where the Enigma information was invaluable, we were able to lead the Germans to think (as we had also done in 1914–18) that the information about U-boat positions came not from cryptography but from accurate direction-finding on the signals transmitted by the U-boats. The possibility of creating such an alibi may therefore ease the discomfort of making a painful decision; but it will always be a matter of judgment, and sometimes this judgment will have to be stern.

'Block 26'

As for whether it is ethical to risk sacrificing an agent as a result of using information he or she has provided, I am grateful never to have been confronted with the problem. But an instance has come to my notice where the agent himself took the decision for us, and it deserves to be placed on record. Pierre Julitte was an officer on the staff of General de Gaulle in London who volunteered to be parachuted back into France on intelligence missions. Captured on his third mission by the Gestapo in March 1943, he spent the next twenty-five months in prisons and concentration camps, including Buchenwald and Nordhausen (Dora). At the former, in early 1944 he and his fellow inmates of Block 26 were living in terrible conditions when they were set to work in a nearby factory making and assembling electronic components, gyroscopes and control equipment. With their knowledge of engineering they considered ways in which they might subtly sabotage the components so that whatever these were to be used for would fail in operation. Gradually they realized that the 'whatever' must be something new and important, and Julitte deduced that this could be 'a self-propelled projectile, navigating in space, subject to vibration and remote-controlled by radio'. A colleague returning from a stint in the works where the V2 (A4) rockets were being assembled confirmed Julitte's guess. And when they learned from the German radio that the V1 campaign had opened on 13 June 1944, they decided to try to warn the British and American Air Forces about the production of the V2.

One of Julitte's colleagues was to attempt to escape with a message to Julitte's sister in Paris and his cousin in Neuilly. The message gave instructions that the information be forwarded to London with the advice that their factory should be bombed – even though Julitte and his colleagues might be killed if this advice were acted upon. When I wrote *Most Secret War* I was quite ignorant of this heroic episode; but Julitte himself then sent me a copy of his own book *Block 26* in which he had recounted his story, though with all the personal names changed. This gave the book the atmosphere of a novel; but so many of the details rang true that I was convinced. Later I was able to meet M. Julitte and ask him why he had falsified the names. His characteristic reply was that some of his colleagues had not shown up so well, and that if he had changed only their names and given credit to those to whom it was truly due, a reader might deduce those who had behaved badly, and so he decided to falsify all names.

The one remaining item in his story that puzzled me was his precise statement that the factory had been heavily bombed in daylight on 24 August 1944. The factory had been destroyed and five hundred of his co-workers had been killed, as he himself might have been. I could, though, find no mention of such a raid in the Official Histories of either Bomber Command or the Air Defence of Great Britain, although other raids to inhibit the development of the V-weapons had been faithfully listed. Ultimately the Air Historical Section found the answer for me: the factory had indeed been attacked by 128 Flying Fortresses on 24 August 1944, but the target had merely been described in the records as 'an armaments factory at Weimar (Buchenwald)' and so the historians had failed to connect the factory with V2 production.[8] The air intelligence summary described the results of the subsequent photographic reconnaissance: 'Smoke from extensive fires obscures the greater part of the armaments factory but in the southern and eastern parts which are visible it can be seen that the USAAF attack on 24 August has caused severe damage to almost every building. The radio factory to the north has been completely gutted and some barrack huts in the concentration camp have been severely damaged.'[9] Through the lost production of V2s the lives of many must thereby have been saved in London and Antwerp who never knew what they owed to Julitte and his colleagues. Can anyone wonder that Langley and I 'loved our agents'? Their sacrifice stands in sublime contrast to the hideous conditions under which they made it.

Torture (and Reprisals)

The German treatment of captured agents raises the question of how far a captor is entitled to go in forcing a prisoner to disclose information, and in discouraging intelligence activities in occupied territory by reprisals on the civilian population. The German treatment of prisoners-of-war was, with some exceptions, correct: but their attitude towards civilians was harsh and, often, brutal. They followed the doctrine of Clausewitz, that terror was justified as a way to shorten war. They had employed it after Sedan in 1870, and again in 1914 in Belgium. Barbara Tuchman in *August 1914* gives the examples of 211 civilians shot at Andenne, 50 at Seilles, 400 at Tamines, 612 at Dinant, and uncounted numbers at Louvain.[10] In World War II they again used terrorism as a policy; the massacres at Oradour and Lidice were among the appalling results. While it may not be entirely fair to cite these examples as inhumanity specifically to intelligence agents, they do indicate the utter harshness of the German viewpoint during both wars. And the harshness undoubtedly had its effect on intelligence agents; of which there are two examples in *Most Secret War*. One was of the outstanding agent, Georges Lamarque, who gave himself up to the Germans when they surrounded the village from which they had detected his radio transmissions, sacrificing himself to protect the villagers who might otherwise have suffered the fate of those at Oradour. The other was of the agent Yves le Bitoux, who surrendered to the Gestapo at Tréguier in 1944 because he feared that the Germans would savage the town if he had succeeded in escaping. He later died in a concentration camp.[11]

Apart from the questions of the morality of reprisals and also of the extent to which they may be counterproductive by stimulating resistance rather than cowing it, we need to consider the morality of torture as a means of forcing a captive to reveal information. We in Britain eschewed it. From the interrogators whom I knew personally I am sure that this was on the grounds of humanity, although it was also doctrine with them that torture could produce misleading results because a person under torture is likely to tell his torturers what he thinks they would like to hear rather than what he truly knows. But we would have to admit that the Germans did achieve some successes by torturing captives; and we would also have to admit that the sadism from which we so firmly refrained has sometimes been exhibited in Britain.

I am personally grateful never to have been confronted with a problem of this type: suppose you have captured a terrorist who has planted a bomb which has not yet exploded but which you have reason to think he has left in some place where its explosion will cause casualties and damage. In an extreme case it might be an atomic bomb. He refuses to tell you its location. To what limits are you justified in going to make him talk?

I would hate to resort to torture, all the more so if he has shown himself to be a brave man; but the same fundamental respect for human life and dignity that makes torture so repulsive would also justify driving him to the point of death if you could thereby save the many more lives that would perish if the bomb exploded. And then, if you admit the argument for torture in such an instance, where do you draw the line? A rationalist answer might be: at the point where you are reasonably certain that the total of human suffering will be less if the torture succeeds in extracting the information required. But while this argument may have been valid for the overall saving of human lives by the atomic bomb on Hiroshima, it also underlies Clausewitz's justification of terror tactics, and we have seen the depths into which it can so easily slide.

If we admit that torture might in some circumstances be justified, we have to return to the responsibility of an intelligence service towards its agents in the field who risk capture and torture by the opposing security service. They may also risk death. My own experience was almost entirely in the circumstances of war, where such considerations were much simpler. In peace, there are not only agents at risk but also foreign relations if an operation goes wrong. Such problems have been described by Admiral Stansfield Turner, who was Director of the CIA from 1977 to 1980, in *Secrecy and Democracy*.[12] Coming into intelligence after a distinguished career in the US Navy, the Admiral looked afresh at all such problems, and his conclusions have much in common with those that I myself formed in wartime. This might be partly due to the fact that even in peacetime an intelligence service is in effect engaged in a war with its opposing security service. One difference, though, that makes peacetime working more difficult is that intelligence operations are then regarded far less sympathetically by the public, and there is much more time for post-mortems and recrimination if mistakes have been made.

Covert Action

This last factor has dogged the CIA over the last thirty years, all the more so because the agency has been involved not only with intelligence but with the more aggressive forms of covert action. In Britain in World War II there was originally a section in MI6 concerned with sabotage, but in 1940 this was hived off into a completely new organization, the Special Operations Executive. On the whole this was beneficial to both activities, since the intelligence organization – if it had an agent placed in a good position to gather information – aimed to keep him there, even sometimes to the extent of not using his information if this might lead to his being discovered, and therefore to his inability to provide more valuable information in the future, whereas SOE tended to regard its sabotage agents more as elements in an active army where casualties were to be expected, and where the results of their work would be much more immediately obvious to the enemy. Again, sabotage usually required greater numbers of personnel, and the organization had therefore to be larger: and so, although the same conflict between short- and long-term operations could arise in sabotage as well as in intelligence, there was much merit in separating the two activities as far as possible. They also attracted rather different types of officer – some who would be repelled by actions of the 'dirty tricks' variety would have less hesitation in working for intelligence.

Admiral Turner, though, takes a differing view. While accepting that separating covert action from intelligence might solve some problems, he argues that it would create others: 'The CIA's intelligence agents overseas are often the same people needed for covert action. It would be confusing, and at times dangerous, to have two agencies giving them orders and managing their activities. And if one agency did only covert action, what would it do during periods of slack demand?' This last consideration did not apply to us in war because there were continuous demands for both activities; but we certainly encountered problems in rivalries between the two types of organization operating in the same territory.

A further complication arises from the fact that an intelligence organization – if it is doing its job – will often be the first to realize, from its unique viewpoint, the vulnerability of an opponent, actual or potential. Therefore intelligence will know where and how the enemy can best be hindered by deception, sabotage, or – in war – overt military action. For example, I myself was sometimes in this position in World War II, as in the cases of 'Window' (or 'Chaff') and countermeasures to German radio bombing techniques: but it was clear that once I had made a technical or tactical suggestion for action the responsibility for its implementation belonged to the operational staff. Obviously action is facilitated if both intelligence and operations can be controlled under one organization; but an offsetting merit of separation is that an intelligence unit is more likely to be impartial in its assessment of the success of operations if it is independent of any attempt by the operational side to interpret the evidence regarding success or failure too favourably. For all these reasons I continue to stand for the independence of intelligence from operations, both covert and overt, to the highest possible level in government organization, although of course recognizing the need for the greatest possible understanding between intelligence and operational staffs.

Returning then to America, it can be seen how the responsibility of the CIA for both intelligence and covert operations has created many problems for the Agency. Its involvement in, and part responsibility for, the Bay of Pigs disaster in 1961, which attempted to overthrow Castro, led to the resignation of the Director, Allen Dulles. Richard Helms, who had been Director from 1966, was fired by President Nixon in 1973 because, according to his biographer Thomas Powers, he refused to provide a CIA cover-up for Nixon in the Watergate affair.[13] These grounds would have been quite implausible if the CIA had always kept itself apart from covert operations concerned with internal security. William Colby, who became Director in 1973, was in his turn dismissed in 1975 by President Ford largely because of public outcry at the CIA's involvement in assassinations, over which Colby himself had been scrupulously honest – too honest, many of his staff thought – in exposing the Agency's past mistakes. His departure was finally precipitated by Press and public indignation over his revelation to a Senate Committee of the existence of a dart gun and small quantities of virulent poisons that had come to light in an obscure store-room which had been overlooked in his previous survey, conscientious though this had been. Colby welcomed me in his office shortly after his dismissal. When I expressed my regret he nobly told me that he did not mind how much he himself suffered from the episode if by 'taking the rap' he could leave the Agency in a healthy condition. Finally, this story of the fates of CIA Directors might have had a further melancholy chapter over the arms for Iran affair, had not the death of William Casey early in 1987 intervened before the official inquiries into the affair could be completed – although Casey himself was convinced that he and the Agency would be vindicated.

Assassination

Helms himself observed that 'war corrupts and secret war corrupts secretly'.[14] So it is easy to slip towards the acceptance of assassination as one of its techniques. In 1940 there was a suggestion that there should be a special British clandestine operation to attack the aircrews of the German pathfinders who were causing us so much trouble in the Blitz. They were to be ambushed while being ferried by bus between their billets and the airfield at Vannes before taking off for an attack. This was vetoed by Sir Charles Portal, the Chief of Air Staff; but as the war progressed, feelings became less delicate. Later, at the time of the Suez crisis in 1956, there was even talk – it appears – of attempting to assassinate President Nasser.[15] In America, Allen Dulles, aware of the danger of the CIA slipping towards an acceptance of assassination, specifically forbade it when he was Director, and an embargo was included in presidential directives on three occasions. Nevertheless it was sometimes attempted, with disastrous results for the CIA when it came to light.

Torture and Duress

Torture, too, was officially barred – at least to the extent of a rule being formulated by Helms in 1955 under Dulles's directorship: 'You may not use electrical, chemical, or physical duress,' it read, although, as Powers then went on to point out, 'psychological duress was okay'.[16] We, in World War II, certainly tried to bluff prisoners-of-war into thinking we knew much more than we did in the hope that this would mislead them into giving away things about which we knew little or nothing. And it seemed not too unfair to 'fence' psychologically with a prisoner and thus to trap him into saying more than he had originally intended. But there had to be a limit such as, for example, not playing on a man's emotions by encouraging him to worry about the safety of his family. I take it that no such restraint is likely to be widely observed today.

'Character Assassination'

Short of killing a key individual on the opposing side, his value to them may be destroyed if his colleagues or countrymen can be misled into suspecting that he is a secret agent for your own side, or has some other motive for working against his side's interests. This mischievous technique is only likely to succeed when the individual in question is working in an organization prone to mutual suspicion; but in so far as it exploits a moral weakness in that organization, so 'hoisting the engineer with his own petard', it may be less repugnant than other measures.

Intervention

In parallel with the ethical problem of the humane limits of conduct towards individuals, there is the problem of the extent to which any one state can be justified in attempting to interfere in the internal affairs of another. The 'Irangate' affair and the clandestine mining of Nicaraguan harbours under CIA auspices are cases in point. In discussing such questions Stansfield Turner in *Secrecy and Democracy*[17] quotes John Stuart Mill's *A few words on non-intervention* (1859):

> The doctrine of non-intervention, to be a legitimate principle of morality, must be accepted by all governments. The despot must consent to be bound by it as well as the free state. Unless they do, the profession comes to this miserable issue – that the wrong side may help the wrong, but the right must not help the right.

Thus a Gresham's Law will tend to operate in international affairs, especially in any field of covert action where the operators think that their actions will be free from public scrutiny. All the more reason, therefore, to endorse Turner's specification for CIA personnel selection: 'The CIA needs people not only with skills, but with high moral standards, with the confidence to be independent, and with the desire to be innovative.'[18] He also stated that 'there is one overall test of the ethics of human intelligence activities. That is whether those approving them feel they could defend their decisions before the public if the actions became public. This guideline does not say that the overseers should approve actions only if the public would approve them if they knew of them. Rather it says that the overseers should be so convinced of the importance of the actions that they would accept any criticism that might develop if the covert actions did become public, and could construct a convincing defense of their decisions.'[19]

Internal Intelligence

Besides the ethical questions concerning how a nation and its officials and agents should restrain their activities in gaining information about another nation's secrets, or in attempting to interfere in its internal affairs, there are other questions that will arise when a nation's officials have to seek out activities within its own borders that could jeopardize its security. In Britain these are the province of MI5, while MI6 covers intelligence about other nations. Although this is a neat division of functions, MI5 and MI6 at times have joint interests. For example, when foreign agents operating outside Britain attempt to organize sub-agents for espionage, sabotage, terrorism, or whatever inside it, the agents abroad are, strictly, targets for MI6 and the sub-agents in Britain for MI5. For such problems, liaison between MI5 and MI6 has to be very close: in 1939–45 this was largely effected by having a special section inside MI6 – ironically its head was Kim Philby.

In America a rather similar separation of functions is made between the Central Intelligence Agency and the Federal Bureau of Investigation, although the separation there is now more on a territorial basis; the CIA being responsible for all intelligence and counter-intelligence activities abroad, and the FBI for all inside the United States. Much the same problems arise in America as in Britain respecting individual privacy; for example, in opening private correspondence, tapping telephone conversations, or breaking into private premises. According to Colby in *Honorable Men*,[20] the CIA in 1952 started a programme of opening selected letters in the mail between the United States and Russia in the hope of detecting undesirable activity. This was 'a direct violation of a criminal statute' and the practice was shelved in 1973. Admiral Turner records that the tapping of telephones and other forms of electronic communication in the United States have since 1978 been governed by the Foreign Intelligence Surveillance Act, which stipulates that if it has to be done it should be carried out using the least obtrusive technique that will do the job; every such operation should be certified as necessary before the Attorney General and reviewed by a special court of senior judges. Breaks into private premises presumably now require similar authorization, since the Watergate and Ellsberg (1971) cases brought such break-ins so embarrassingly to public notice.

Following disclosures by former MI5 officers about break-ins by MI5 in Britain, the Government has introduced a Bill to put such operations – hitherto illegal – on a legitimate basis and subject to the safeguard that each operation must be authorized by the Home Secretary and 'only when he was satisfied that the information was likely to be of substantial value and assistance to [the Security Service] . . . and . . . could not reasonably be obtained by other means'.[21] An independent commissioner would be appointed to review the issuing of warrants and to make an annual report to the Prime Minister. There appear to be few reasonable alternatives to these proposals in a democratic society.

Privacy

While any decent individual instinctively reacts against break-ins and less violent infringements of privacy, a state legitimately requests some details about each of its individual citizens for a wide range of social purposes such as taxation, educational planning, transport facilities, and potential for military and other forms of public service. It might therefore be asked why we set so much store by a right to privacy. Apart from the nightmare of a 'Big Brother' state, there is an instinctive dislike of surveillance, even parental surveillance; and there may well be an apprehension based deeply back in the evolutionary process arising from a feeling of vulnerability while executing bodily functions or in sickness, and of latent trouble from a stalking predator whose staring eyes betray his intentions. More rationally, I for one would have little objection to any authority having any information it wished about my actions – or even my thoughts – provided that I could be sure that it would not misinterpret the information to come to false conclusions about me.

Just as technology is tending to modify concepts of sovereignty (from a three-mile limit for territorial waters based on the range of a gun to a limit of 200 miles today, and the free movement of satellites over the territories of other nations, for example) so also it is tending to change the balance between the rights of individuals and the states of which they are members. The complex organization of a modern state needs to know more about its individuals for optimum functioning, and this is all the easier to achieve because of the technical advances in handling and storing information; and, unfortunately, it will be only too easy for zeal to replace judgment in the process.

Oversight

Some of the problems of internal intelligence, both in the United States and in Britain, have arisen from excessive zeal. Thirty years ago Percy Sillitoe's determination as head of MI5 to stop Britain from becoming a police state (see above, page 21) was all the safeguard that was needed; but today with the various exposures regarding both MI5 and the greater surveillance needed to protect against penetration, it is difficult to be so sure. The attempted remedy in America is 'oversight', where the activities of the intelligence and security services are overseen by committees, one from the Senate and the other from the House of Representatives. Much obviously depends on the selection of members for these committees, but on balance Admiral Turner, in whose time as Director of the CIA they first became effective, records a favourable impression. On the negative side, he found them restrictive: 'We tend to apply our new enthusiasm for oversight of the ethics of intelligence to micromanagement by the Congress of the development of new intelligence technologies . . . Another false economy that congressional oversight has fostered is the frugality in stockpiling intelligence collection systems . . . One other dangerous hindrance is that the CIA's research branch is gradually losing out to the large and clever military bureaucracy at the Pentagon . . . The espionage people also deserve better protection from the Congress and the White House.'[22] But on the positive side, he wrote, 'oversight, especially by Congress, can give helpful guidance to the CIA as to what is and what is not acceptable conduct in the pursuit of secrets . . . the better the oversight process is, the less concern there need be about concentrating too much authority in the hands of the DCI . . . If we want good intelligence in the long run, our only option is to make oversight work.'[23] At the same time Admiral Turner would fuse the Senate and Congressional oversight committees into one with a membership limited to reduce the possibility of leaks.[24]

While, though, Admiral Turner so firmly supports oversight, the record in Washington since it was introduced has not been one of unqualified success. It is now generally agreed that the CIA-sponsored mining of Nicaraguan harbours in 1983 was a mistake, and it was allowed to go ahead despite nominal oversight by both the Senate Intelligence Committee under Barry Goldwater and the Congressional Intelligence Committee under Edward Boland. The latter had been so disturbed by earlier developments that in August 1982 he had succeeded in moving that the CIA and the Defense Department should be prohibited from furnishing military equipment, training or support to anyone 'for the purpose of overthrowing the Government of Nicaragua'.[25] Robert Woodward, who with Carl Bernstein exposed the Watergate affair, has described this episode at length in *Veil: The Secret Wars of the CIA 1981–7*.[26] Formally, the members of the oversight committee were in a difficult position. The law which established them stipulates that they must be informed of major intelligence activities, but appears not to have given them a right of veto: moreover, the individual members had been sworn to secrecy, and so they could be in trouble if they made a public disclosure.

It is not easy to see a way out of this difficulty. No intelligence organization can surrender its executive responsibility to an oversight committee, no matter how able, experienced and responsible the individual members of the committee may be. Two reasons preclude any such procedure. The first is that an intelligence organization is constantly waging a war, and any operations in that war need a commander with full responsibility for their planning and execution. At times risks have to be taken, and a committee approach is likely to be inhibiting. Clive of India said that he had held a council of war but once, and had he heeded the advice of the council rather than his own judgment the British would never have been masters of India.

The second reason against the committee approach is allied to the first; it arises from the way in which intelligence appreciations have to be formed from the information that has been gained. Despite Napoleon's worry about the danger of 'making a picture of the enemy', this is what an intelligence organization has to do in presenting its conclusions for assimilation by those who have then to undertake executive action. Just as an artist has to present as faithful an impression as possible, according to his lights, with the economy of detail of a painted portrait as contrasted with a photograph, so an intelligence officer has to convey a portrait of the opponent constructed to give the truest possible impression from the limited amount of detailed information that will be available even to a good intelligence service. An intelligence report therefore has much in common with a work of art; and my experiences on the Joint Intelligence Committee (see above, page 28) suggest that committees are hardly more likely to produce good intelligence reports than they would be to paint good pictures. When information is sparse it permits a multitude of explanations, and committees can lose much time over indecisive argument as to which is the correct one.

This was also the experience of General Eugene Tighe, head of the Defense Intelligence Agency in the Pentagon from 1977 to 1979. In thirty-six years of intelligence work, 'he had seen administrations, Secretaries of Defense, and DCIs come and go, and the shape and tone of intelligence work change. But the real squabbles arose when they didn't have enough information. When US intelligence had a lot of good data, there was rarely a fight.'[27]

While not wishing to under-rate the second of the foregoing reasons, I believe the first is decisive in determining that an intelligence service must have freedom to act as it thinks best, and to that extent oversight must have limits. The example of the mining of Nicaraguan harbours may suggest that these limits were too loosely drawn, or that an enterprising DCI could too easily find a way round them. The same example may also suggest that it is dangerous for an intelligence organization to have such autonomy, even though nominally subject to oversight. But, once again, intelligence is waging a war and risks sometimes have to be run. Churchill's delighted soldiers in the Royal Scots Fusiliers in 1916 told how he, as their commanding officer, was visited in the front line by a general from some rear headquarters. 'If you would care, sir,' said Churchill, 'to step over the parapet, we could go for a walk in no-man's land.' 'Wouldn't that be dangerous?' asked the general. Churchill replied, 'Sir, this is a dangerous war!' And that will always be true of war fought by intelligence agencies too, where the lines may be covertly drawn.

As for whether we in Britain should adopt oversight of the intelligence services on the American pattern I would have said in the days of men like Percy Sillitoe that there was little need for it – at least as regards the danger of the services overstepping the limits of reasonable conduct. The oversight that was needed then was more in looking for inefficiencies in the system such as were all too evident in 1939 and, more recently, were the subject of the Franks inquiry over the Falklands. But moral standards have fallen from the days when an Englishman's word was his bond, as we have seen in other walks of public life, such as the City. It cannot be guaranteed that this is never likely to affect the conduct of intelligence affairs, with all the temptations that they may afford. Some measure of oversight is therefore desirable, if only to assure Parliament and the public that intelligence is being wisely and scrupulously conducted.

Successive prime ministers have strongly discouraged the discussion of intelligence affairs in Parliament, where obvious dangers could arise from disclosure. But, while it was easy to protect the intelligence services from scrutiny when these were small, it is less easy today when organizations like GCHQ are much larger. Besides the problems that thereby arise from trades unions, increases in numbers give rise to increased chances for treachery and for leaks that will sooner or later come to public notice and so lead to discussion in Parliament. There appears to be a body inside government itself that could provide oversight, the Ministerial Steering Committee on Intelligence. But whether this is the best body for the purpose, or whether a small inter-party body of senior politicians (and perhaps others of experience, discretion and judgment) might be better, would itself be a fitting matter for Parliamentary discussion.

Minimum Trespass

Two final points on ethics are worth making. The first is that, despite all the opportunities – and temptations – that it offers for malpractice, intelligence can be an honourable pursuit. Indeed, by improving the assessment by one nation of another it can, on occasion, contribute to international stability. On less happy occasions, of course, the children of light will be at a disadvantage unless they know enough of the ways of the world to forestall or counter any unworthy exploitation of those ways by the children of this world. There need be absolutely no dishonour in trying to ascertain what a potential or actual opponent is likely to attempt, be it by external armed threat or by internal subversion. The risk of disrepute will depend on the extent to which the individual intelligence officer or his organization departs from the norms of morality in uncovering an opponent's activities.

The second point follows from the first. One of the canons governing military or police action is the doctrine of minimum force, and a parallel canon should govern intelligence: it should be conducted with the minimum trespass against national and individual human rights. This canon applies to all forms of intelligence, both external and internal, that a civilized state may find it necessary to undertake.

Section II: Principles and Other Fundamentals

Introduction

It was beautiful and simple as all truly great swindles are.
— O. Henry, "The Octopus Marooned," The Gentle Grafter (1908)

He [Gregg Toland] showed me the inside of that bag of tricks, and, like all good magic, the secrets are ridiculously simple.
— Orson Welles, 1972

The Structure of Deception

The basic structure of deception is simple—one thing is hidden (dissimulated) while another is shown (simulated) in its place. Deception is the deliberate distortion of perceived reality. Operationally, it is done by changing the pattern of distinguishing characteristics of the thing (whatever object or event) detected by the sensory system of the target.

Reality is distorted, deceptively portrayed by both nature and man. Nature's deceptions are either without purpose—as with physical illusions such as a mirage or mirror images—or purposefully (and then, always to the advantage of the species involved as with all evolutionary bits of camouflage that have survival value). Humans' deceptions are also either without purpose—unintentional representations, or purposeful—with intent to deceive. Nature does this unconsciously; humans do it either unconsciously, as with self-deception and some deceptions of others, or consciously (and then, always to some personal advantage).

Every deception operation, whether of man or nature, is comprised of only two basic parts: dissimulation and simulation. In 1597, English philosopher Francis Bacon was the first to expound on this crucial dichotomy in his important and aptly titled essay, "On Simulation and Dissimulation," where he applied it to interpersonal relations.

This two-part simulation-versus-dissimulation notion was first applied to military theory (introducing the two terms "positive" and "negative" camouflage) in a memo written in 1918 by Col. Richard Meinertzhagen.[1] In 1920, British artist-painter Solomon J. Solomon was the first to apply the two-part simulation-versus-dissimulation notion to military theory. He did this in his pioneering book, *Strategic Camouflage*.

Modern mathematicians Von Neumann and Morgenstern flirted with this two-fold concept for all deception, writing in their revolutionary book that introduced the world to the *Theory of Games*.

1. Timothy Ferris, *The Mind's Sky: Human Intelligence in a Cosmic Context*, New York: Bantam, 1992, pp. 185, 193.

Of the two possible motives for bluffing, the first is the desire to give a (false) impression of strength in (real) weakness; the second is the desire to give a (false) impression of weakness in (real) strength. Both are instances of inverted signaling — i.e. of misleading the opponent.

Finally, American psychologist Paul Ekman wrote that "There are two primary ways to lie: to *conceal* and to *falsify*." He noted, correctly, that specialists in military deception also make this same distinction.[2]

Dissimulation is hiding the real. It is covert. Half of a deception is concealed from the target. Its task is to conceal—or at least obscure—the truth. Operationally, dissimulation is done by hiding one or more of the characteristics that make up the distinctive pattern of the real thing.

Simulation is showing the false. It is overt. Half of a deception is presented to the target. Its task is to pretend, portray, and profess an intended lie. Simulation is done by showing one or more characteristics that comprise the distinctive pattern of a false thing.

Both simulation and dissimulation are always present together in any single act of deception. Nothing is ever "just" hidden; something is always shown in its stead, even if only implicitly. Thus, the homeowner who hides money in the cookie jar is pretending (showing) there is no money at home.[3] It is the two in combination that misdirect the attention and interest of the target, inducing it to form misperceptions (false hypotheses) about the real nature of what is impending.

Dissimulation and simulation are also widely referred to as denial and deception (D&D). However, we find that model less precise than defining deception as having two elements: dissimulation and simulation.

2. John von Neumann, Oskar Morgenstern, *Theory of Games and Economic Behavior,* NJ: Princeton University Press, 1944, p. 189.
3. A 1987 British Gallup poll found that 16% of Britons kept substantial sums of money hidden in their homes.

CHAPTER 3

On Simulation and Dissimulation[1]

Sir Francis Bacon

This short essay by the remarkable early English philosopher of science, lawyer, and intelligence officer states for the first time the taxonomy of hiding and showing. The essay distinguishes three degrees of hiding oneself—secrecy, where a man makes himself invisible; dissimulation, a negative position where he gives evidence that he is something other than himself; and simulation, giving affirmative evidence of what that other self is. He then lists three advantages of simulation and dissimulation and their three disadvantages.

This idea of hiding-and-showing as a linked pair (dichotomy) was already old in 1597 when Francis Bacon raised this everyman's notion to the level of philosophy in this essay. The reader will appreciate that this seemingly simple concept of simultaneously hiding one thing and showing something else in its place is the basis for both the theory of deception and all attempts to analyze deceptions.

It is no accident that we owe the refinement of this concept to Bacon rather than some other Elizabethan humanist. He was, after all, a great admirer of Machiavelli. Moreover, like his Italian model, Bacon was an historian, writer, diplomat, and an ambitious politician whose great ups and greater downs came at the whims of tyrants. Additionally, Bacon had early experience as a secret agent, since age 16, in the cutthroat politics of the English court.

Dissimulation is but a faint kind of Policy or Wisdome; for it asketh a strong wit and a strong heart to know when to tell truth, & to doe it. Therfore it is the weaker sort of politics that are the great Dissemblers.

Tacitus saith: Livia sorted well with the arts of her husband, & dissimulation of her sonne: attributing arts or policy to Augustus, & dissimulation to Tiberius. And againe, when Mucianus encourageth Vespasian to take arms against Vitellius, he saith: We rise not against the piercing judgment of Augustus; nor the extreme caution or closenesse of Tiberius. These properties, of Arts or Policy, and Dissimulation or Closenesse, are indeed habits and faculties severall, and to be distinguished. For if a man have that penetration of judgment as he can discerne what things are to be laid open, and what to be secretted, and what to be shewed at halfe lights, and to whom, and when, (which indeed are arts of State, and arts of Life, as Tacitus well calleth them); to him, a habit of Dissimulation is a hindrance and a poorenesse. But if a man cannot obtaine to that judgment, then it is left to him, generally, to be close, and a Dissembler. For where a man cannot choose or vary in particulars, there it is good to take the safest and wariest way in generall; like the going softly by one that cannot well see. Certainly the ablest men that ever were have had all an opennesse and francknesse of dealing; and a name of certainty and veracity; but then they were like horses well managed; for they could tell

1. Sir Francis Bacon, "On Simulation and Dissimulation" in *Essays of Francis Bacon*, 1627.

passing well when to stop or turne: and at such times when they thought the case indeed required Dissimulation, if then they used it, it came to passe that the former opinion spred abroad of their good faith and clearnesse of dealing made them almost invisible.

There be three degrees of this hiding and vailing of a man's selfe. The first Closenesse, Reservation, and Secrecy; when a man leaveth himself without observation, or without hold to be taken, what he is. The second, Dissimulation, in the negative; when a man lets fall signes and arguments, that he is not that he is. And the third Simulation, in the affirmative; when a man industriously and expressly faigns, and pretends to be that he is not.

For the first of these, Secrecy; it is indeed, the vertue of a Confessour. And assuredly, the secret man, heareth many Confessions. For who will open himselfe to a blab or a babler? But if a man be thought secret, it inviteth discoverie; as the more close aire sucketh in the more open: and as in confession the revealing is not for worldly use, but for the ease of a man's heart, so secret men come to the knowledge of many things in that kinde; while men rather discharge their mindes then impart their mindes. In few words, mysteries are due to secrecy. Besides (to say truth) nakednesse is uncomely, as well in minde as body; and it addeth no small reverence to men's manners, and actions, if they be not altogether open. As for talkers and futile persons, they are commonly vaine, and credulous withall. For he that talketh what he knoweth, will also talke what he knoweth not. Therefore set it downe; *that an habit of Secrecy is both politick and morall.* And in this part it is good that a man's face give his tongue leave to speake. For the discovery of a man's selfe by the tracts of his countenance is a great weaknesse and betraying; by how much it is many times more marked and beleeved then a man's words.

For the second, which is Dissimulation. It followeth many times upon Secrecy by a necessity: so that he that will be secret must be a Dissembler in some degree. For men are too cunning to suffer a man to keepe an indifferent carriage betweene both, and to be secret, without swaying the balance on either side. They will so beset a man with questions, and draw him on, and picke it out of him, that without an absurd silence, he must shew an inclination one way; or if he doe not, they will gather as much by his silence as by his speech. As for equivocations, or oraculous speeches, they cannot hold out long. So that no man can be secret, except he give himselfe a little scope of Dissimulation; which is, as it were, but the skirts or traine of Secrecy.

But for the third degree, which is Simulation and false profession; that I hold more culpable, and lesse politicke; except it be in great and rare matters. And therefore a generall custome of Simulation (which is this last degree) is a vice, rising either of a naturall falsenesse or fearefulnesse, or of a minde that hath some maine faults; which because a man must needs disguise, it maketh him practise Simulation in other things lest his hand should be out of use.

The great advantages of Simulation and Dissimulation are three. First, to lay asleepe opposition, and to surprize. For where a man's intentions are published, it is an alarum to call up all that are against them. The second is, to reserve to a man's selfe, a faire retreat. For if a man engage himselfe by a manifest declaration, he must goe through or take a fall. The third is, the better to discover the minde of another. For to him that opens himselfe men will hardly shew themselves adverse; but will (faire) let him goe on, and turne their freedome of speech to freedome of thought. And therefore it is a good shrewd proverb of the Spaniard; *Tell a lye, and finde a troth.*

CHAPTER 4

The 8 Principles of Security and the 8 Principles of Deception[1]

R.V. Jones

This excerpt by R.V. Jones is a concise, clear specification of the ways to hide something real (dissimulation) and the corresponding ways to show something false (simulation) in its place.

The Principles of Security

I hope that, without being historically comprehensive, the examples that I have given above will illustrate the principles of deception as far as I know them. As a first step, what has to be done is to prevent the enemy from deducing at least one of the following:

1. Where you are, and/or where he is.
2. What weapons and forces you have at your disposal.
3. What you intend to do.
4. Where you intend to do it.
5. When you intend to do it.
6. How you intend to do it.
7. Your knowledge of the enemy's intentions and techniques.
8. How successful the enemy's operations are.

All these objectives are in a sense negative, but they may be sufficient to achieve surprise. Basically they involve security rather than deception. Wolfe, for example, surprised Montcalm at Quebec because Montcalm had not thought that the Heights of Abraham could be scaled;[2] and the introduction of centimetric radar helped to change the balance in the battle of the Atlantic because the Germans were unaware that we had developed this new technique.

The Principles of Deception

These negative objectives are the function of the security organisation: each objective has a positive counterpart, which properly constitutes deception. The corresponding objectives of deception therefore are that the enemy must be persuaded to deduce:

1. Reginald V. Jones, "Intelligence and Deception," in Robert L. Pfaltzgraff, Jr., Uri Ra'anan, and Warren H. Milberg (editors), *Intelligence Policy and National Security,* Hamden, CT.: Archon Books, 1981, pp. 17-19.
2. Similarly, the Japanese surprised the British at Singapore, but there was an element also of self-deception in the British command. See N. F. Dixon, *On the Psychology of Military Incompetence,* London: Jonathan Cape, 1976.

1. You are somewhere else, and/or he is somewhere else.
2. Your weapons and forces are different from what they are.
3. You intend to do something else.
4. You intend to do it elsewhere.
5. You intend to do it at a different time.
6. You intend to do it in a different manner.
7. Your knowledge of the enemy is either greater or less than actually is.
8. His operations are either more or less successful than they actually are.

In short, you are seeking to provide your adversary with clues in his several channels of observation which are consistent with one another, but from which he will build up a false picture of reality. You hope then that, basing his own actions on this picture, he will at some crucial time either act incorrectly and so make himself vulnerable, or he will fail to take advantage of a situation which is in fact favourable to him but which he assesses as unfavourable.

There is scope for artistry in the devising and presentation of clues. While these should be consistent in every channel by which the victim can derive evidence, they should not be too obvious. For example, if bogus espionage reports are to be fabricated, they should include the kinds of error and misinterpretation a genuine spy is likely to make, so that the analytical officer on the other side will be led to feel that he is getting at the truth by eliminating the errors introduced by faulty observation or interpretation on the part of the spy. The picture that the analyst builds up will then be all the more convincing. As a member of the *Double Cross* operation said: "An item of information that the analyst has worked out for himself is worth ten that he has been told". Or the supposed spy may report an item faithfully and correctly, but, add that he himself does not believe it. I call this the *Herodotus touch* because of the statement in his History that he could scarcely credit what the Phoenicians had reported when they sailed round Southern Africa, "The sun rose on the other side," which subsequent scholars have taken as twin testimony to the bona fides of Herodotus and to the feat of seamanship on the part of the Phoenicians.

A further touch of artistry in deception is to provide an alternative to your true intentions so valid that if your adversary detects it as a hoax, you can then switch to it as your major plan and exploit the fact that he has discounted it as a serious operation.

CHAPTER 5

Ruses and Stratagems of War[1]

General Sir A.P. Wavell

Always mystify and mislead the enemy.
— Stonewall Jackson

General Archibald Wavell had originally issued this brief paper as Commander-in-Chief India, in July 1942 "to stimulate commanders of all grades to consider methods of deceiving the enemy, by outlining means which have proved successful in the past."

This is the classic appreciation of deception by the one officer who directly and single-handedly transmitted Britain's institutional memory of military deception between the two world wars. He brought its theory and practice from WWI to WWII and from the North African front to the China-Burma-India Theatre and to London GHQ where it resonated with Churchill's own views and soon with the United States, Britain's ally. What Wavell said more than sixty years ago holds true today.

Ruses and Stratagems of War

Possibly because the British character is normally simple and straightforward, more probably because our military training is stereotyped and unimaginative, deception of an enemy does not seem to come naturally to us. Hence we are apt to suffer in the field through lack of guile and to fall too easily into the enemy's traps and to miss opportunities of setting traps of our own.

Some years ago in a public lecture I referred to the definition by a distinguished soldier of his ideal infantryman as "athlete, stalker, marksman", and said that my ideal infantryman was "cat-burglar, poacher, gunman". The point of my definition, as against the other, was that the characters I named risk their life and liberty in the exercise of their profession and have to defend them with their wits, as does the soldier on service, while the athlete, stalker and marksman do not. In fact one is a peace definition of a soldier, the other for war.

The object of this note is to stimulate commanders of all grades to consider methods of deceiving the enemy, by outlining means which have proved successful in the past.

Practically all ruses and stratagems in war are variations or developments of a few simple tricks that have been practised by man on man since man has hunted man, i.e. since the existence of the human race. They can be roughly divided under the following heads (with the modern equivalent suggested in brackets):

1. Archibald (General Sir) Wavell, Brief Paper as Commander in Chief in India, July, 1942. See also Archibald (General Sir) Wavell, *Speaking Generally: Broadcasts, Orders and Addresses in Time of War* (1939–1943), London: Macmillan, 1946, pp. 80-83.

- False information or disguise ("Camouflage")
- Feigned retreat ("booby traps")
- Encouragement of treachery ("Fifth Column")
- Weakening of morale ("war of nerves")

To convey false information to the enemy by some means or other is the commonest trick of all. It has many variations. To hoist false colours was a frequent ruse in older naval warfare; it has its counterpart today in Q-ships or disguised raiders.

Deception may be achieved by word of mouth. The Greek Sinon, posing as a deserter, persuaded the Trojans to pull the Wooden Horse inside the walls of Troy. The Wooden Horse itself has its modern equivalent in the German capture of Bergen and Narvik by soldiers concealed in apparently harmless merchant ships. Similarly in medieval times a castle was once captured by soldiers hidden under brushwood in a cart.

Two of Napoleon's Marshals secured an important bridge over the Danube simply by walking across and assuring the enemy guard at the other end that an armistice had been declared; meanwhile a party crept up behind them and finally rushed the enemy end of the bridge. There have been instances in this war of parties being bluffed into surrender by persuading them that they were surrounded, etc.

Doing the same thing many times till the enemy is accustomed to it and then suddenly doing something quite different at the same time of day is sometimes effective in securing surprise.

Camouflage is the modern term for methods of concealment which have been practised by savages and others for many hundreds of years. In Shakespeare's *Macbeth* Malcolm made good use of the wood of Birnam; his instructions to his troops would, suitably paraphrased, be equally appropriate today:

Let every soldier hew him down a bough
And bear't before him: thereby shall we shadow
The numbers of our host, and make discovery
Err in report of us.

The effect of this ruse on Macbeth's morale was decisive.

The trick of putting one's cap or helmet on a stick and thrusting it out to draw the enemy's fire probably dates from the first day that warriors wore head-dresses.

The stratagem of feigned retreat to induce the enemy to leave a strong position and become disorganized in pursuit is a very old one. In the classic fight between the three Horatii and the three Curiatii, the surviving Roman by purposeful flight got his opponents strung out and was then able to kill them one by one. The feigned withdrawal of the Greeks put the Trojans off their guard; and a pretended retreat of the Norman horsemen broke up the Saxon formation and was the deciding factor at Hastings (1066). In 1918 General Gouraud disorganized a German attack at Rheims by a temporary withdrawal.

Drawing the enemy on to a minefield and the use of "booby traps" are modern developments of this ruse.

So-called Fifth Column methods are very old. Rahab of Jericho, Delilah of Gaza and Jael the Kenite were early Fifth Columnists; and military history is full of instances of troops or commanders being bought or suborned by the enemy.

Propaganda by wireless or by dropping leaflets is only a modern means of applying a very old technique for defeating an enemy.

Means of playing on the enemy's fears by terrifying appearance or by unusual noise are amongst the very oldest devices of war. The savage painted himself and shouted war-cries to terrify his enemies. Gideon in his famous night attack, described in the seventh chapter of Judges, made his small force wave torches and shout battle-cries to alarm their foes. Bahram of Persia ("that great hunter" of FitzGerald's Rubaiyat) once stampeded his opponent's cavalry during a night attack by tying nose-bags full of rattling stones round the necks of his own horses. The whistling bomb of the Nazi, the crackers and shouts of the small Japanese party which has penetrated to the rear and wishes to simulate a large force, are modern developments.

These few notes show that there is nothing new about the general principles of deceiving the enemy. Every commander of any grade should constantly be considering methods of misleading his opponent, of playing on his fears and of disturbing his mental balance.

Belisarius, Justinian's famous general (c. AD 555), was possibly the greatest master of stratagem who has held the highest commands. He won two almost bloodless victories, at Carchemish and outside Constantinople, by a mixture of bluff and daring which persuaded his opponents he was much stronger than he actually was. He also made frequent use of feigned withdrawals and ambushes.

Perhaps the most elementary principle of all deception is to attract the enemy's attention to what you wish him to see and to distract his attention from what you do not wish him to see. It is by these methods that the skillful conjurer obtains his results.

NOTE: Many of Wavell's techniques may also be classified as psychological warfare. Deception and psychological warfare often overlap.

CHAPTER 6

The Problem of Deception[1]

Cynthia Grabo

Grabo was a professional American intelligence analyst with the U.S. Army from 1942–49, then with the interagency National Watch Committee from 1950 to 1975, and finally with its successor organization, the Strategic Warning Staff until her retirement in 1980.

Unlike Wohlstetter, Handel, Vertzberger and the many other pessimists, Grabo leans toward the optimistic view that intelligence analysis can filter the noise (and redundancies) from the relevant signals to detect and give timely warning of hostile events—even when those events are cloaked by deception. This excerpt is from the declassified article originally written by Grabo in the early 1970s and condensed in 2002 by Jan Goldman.

The Problem of Deception

Confidence that a study of history and of techniques and principles of indications analysis will enable us to come to the right judgment of the adversary's intentions fades as one contemplates the chilling prospect of deception. There is no single facet of the warning problem so unpredictable, and yet so potentially damaging in its effect, as deception. Nor is confidence in our ability to penetrate the sophisticated deception effort in any way restored by a diligent study of examples. On the contrary, such a study will only reinforce a conclusion that the most brilliant analysis may founder in the face of deception and that the most expert and experienced among us on occasion may be as vulnerable as the novice.

The Infrequency and Neglect of Deception

There can be no question that, at least until quite recently, deception has been one of the least understood, least researched and least studied aspects of both history and intelligence. Military historians often have not even perceived the role which deception has played in the outcome of some major military operations. Indeed, the revelation in recent years of the part that deception played in World War II has led to a wholly new understanding of the history of that conflict. What accounts for the neglect of such an important subject?

One reason for the scant attention to deception almost certainly is its rarity. If true warning problems are seldom encountered, useful examples of deception are rarer still, and indeed a number of major crises of recent years seemingly have involved relatively little if any deception. A second, and related, factor is that the deception effort is likely to be the most secret and tightly held aspect of any operation and that countries often

1. Cynthia M. Grabo, *Anticipating Surprise: Analysis for Strategic Warning*, Washington, D.C.: Joint Military Intelligence College, Center for Strategic Intelligence Research, 2002, pp. 119–128.

have been reluctant, even after the fact, to relax security on the deception plan, even when other aspects of the operation are fairly well known. The exceptions, in which the deception operation has been recorded for our benefit and study, usually have been the result of the publication of articles or memoirs by participants in the plan, or the declassification of war records, usually well after the event. Deception tends to be forgotten and neglected between wars because it is usually not an instrument of peace. Few countries have made a practice of extensive or elaborate deception in time of peace. There are some exceptions to this, particularly in the field of counterintelligence and espionage in which deception is routinely practiced.

One reason why active deception is reserved for the exceptional situation involving national security interests is that success in deception is heavily dependent on its rarity and on the prior establishment of credibility. Any country that constantly or even frequently disseminates falsehoods would rapidly lose credibility and acceptance with other nations, and with its own populace. It is one thing to be highly security conscious and not to reveal much, and quite another to engage in an active deception effort to mislead. The most effective deceptions are by those whom we have come to trust, or at least who have been relatively truthful in their dealings with us over a period of years. Thus the true deception operation, at least a major and sophisticated one, usually is reserved only for that critical situation in the life of the nation when it is most essential to conceal one's intent. This will usually be in preparation for or in time of war.

Principles, Techniques and Effectiveness of Deception

The principle of deception, most simply stated, is to induce the adversary to make the wrong choice; or, as General Sherman put it, the trick is to place the victim on the horns of a dilemma and then to impale him on the one of your choosing. If this is left entirely to chance, the probability of the enemy's making the right or wrong choice will be in direct ratio to the number of alternatives which he perceives as equally viable. Although surprise can result from sheer misunderstanding, "the possibility of surprise through misunderstanding diminishes nearly to the vanishing point as one considers the more elaborate strategic operations."[2] Therefore, the planner must develop one or more plausible alternatives as bait for his victim and then employ a range of stratagems to mislead him. "The ultimate goal of stratagem is to make the enemy quite certain, very decisive and wrong."[3] If this ideal cannot be achieved (and this writer believes that it would be a rare situation in which such total deception could be achieved), the mere presenting of alternative solutions nonetheless will serve to confuse the adversary and lead him to disperse his effort or to make at least a partially wrong response.

In other words the best stratagem is the one that generates a set of warning signals susceptible to alternative, or better yet, optional interpretations, where the intended solution is implausible in terms of the victim's prior experience and knowledge while the false solution (or solutions) is plausible. If the victim does not suspect the possibility that deception may be operating he will inevitably be gulled. If he suspects deception, he has only four courses open to him: These are, in summary:

2. Bart Whaley, *Stratagem: Deception and Surprise in War,* Norwood, MA: Artech House, 2007, p. 133. The most thorough, published exposition on military deception is in *Strategic Military Deception,* Donald C. Daniel and Katherine L. Herbig, editors, New York: Pergamon Press, 1982.
3. Whaley, 135.

1. To act as if no deception is being used.
2. To give equal weight to all perceived solutions (in violation of the principle of economy of force).
3. To engage in random behavior, risking success or failure on blind guesswork.
4. To panic, which paradoxically may offer as good a chance of success as the "rational" course in 3.[4]

Thus, even a primitive deception effort will, by threatening various alternatives, create enough uncertainty to distract the most wily opponent and force him either to disperse his effort or gamble on being right. Further, Whaley concludes, in a judgment of greatest importance for warning, that even the most masterful deceivers have proved to be easy dupes for more primitive efforts. "Indeed, this is a general finding of my study—that is, the deceiver is almost always successful regardless of the sophistication of his victim in the same art. On the face of it, this seems an intolerable conclusion, one offending common sense. Yet it is the irrefutable conclusion of the historical evidence."[5]

A related, and also unexpected, finding of Whaley's study is that only a small repertoire of stratagems is necessary "to insure surprise after surprise." The fact that the victim may be familiar with specific ruses "does not necessarily reduce much less destroy their efficacy. This can be predicted from the theory, which postulates that it is the misdirection supplied by selective planting of false signals that yields surprise and not the specific communications channels (that is, ruses) used."[6] In other words, the same tricks can be used over and over again, and stratagem can be effective with only a small number of basic ruses or scenarios.

Whaley goes on to note that, as between security and deception, deception is by far the most effective in achieving surprise since the only important security in this case will be the protection of the deception plan itself, which usually needs to be revealed only to a very small number of individuals. If the security on the deception plan is tight enough, security on the rest of the operation can be outright slovenly, and "the most efficient stratagems calculatedly utilize known inefficiencies in general operational security." Whaley cites some examples of the extreme security maintained on deception plans, which the warning analyst should well heed, since it will upset the accepted theory that enemy plans may be learned from full confessions of high-ranking prisoners or defectors, or from interception of valid communications, authentic war plans, and the like. Thus, in preparation for the Pearl Harbor attack, the Japanese Navy issued a war plan on 5 November which gave full and accurate details of the planned attacks on the Philippines and Southeast Asia but which omitted any reference to the Pearl Harbor missions of the Navy, this portion of the order having been communicated only verbally. In the Suez attack in 1956, the entire British military staff from the Allied CinC on down were not informed on the collusion of the UK and France with Israel, so tightly was this held. In the Korean war, the U.S. planned an amphibious feint (the so-called Kojo feint) which only the most senior commanders knew to be a bluff; even the planners and commanders of the naval bombardment and carrier strike forces thought the operation was real and behaved accordingly. Thus, the misleading of one's own people has been an important feature in many deceptions, with the unwitting participants in the plan convincingly carrying out their roles in good faith, thus contributing materially

4. Whaley, 142–143.
5. Whaley, 146.
6. Whaley, 228.

to the success of the operation. So effective has security been on deception operations, that Whaley concludes that there have been almost no cases in which the deception plan itself was prematurely disclosed to the victim.

Types of Deception

This subject may be approached in a number of ways. Whaley identifies five specific varieties of military deception: intention (whether an attack or operation will occur at all), time, place, strength, style (the form the operation takes, weapons used, and so forth).

For strategic warning, the subject of this book, it will be obvious that the first of these (intention) is the most important. Indeed, some might say that this is the only variety of deception which should properly be defined as strategic, the other types above being essentially tactical problems. In fact, however, strategic warning or the perception of the adversary's intention often does fall victim to one or more of the other foregoing varieties of deception as well. Thus, in the Tet offensive of 1968, we were less the victims of misperception of the enemy's intention as such (it was obvious that attacks of some type and scope were in preparation) than of the other factors. We greatly underestimated the strength of the attacks; we were astounded at some of the places (particularly cities) in which the attacks occurred; we misperceived the style of the offensive in some degree (that is, the extent of covert infiltration of saboteurs and troop units, again particularly into the major cities); and there was something of a misestimate of the timing of the attacks in that it was generally assumed that they would be launched before or after the holidays rather than during them (a factor which accounted for so many South Vietnamese troops being on leave and for the lax security). Thus, it was all these misperceptions of the enemy's planning and intentions which contributed to the surprise—and initial success—of the Tet offensive. We were the victims of a combination of effective security, enemy deception and self-deception.

The history of warfare is filled with examples of the achievement of surprise in time, place or strength, or a combination of them. Whaley finds that, of the examples which he studied in which surprise was achieved, the most common mode was place (72 percent), followed by time (66 percent), and strength (57 percent). The least frequent type of surprise which he found was style, which prevailed in 25 percent of the cases he analyzed. There are nonetheless some very famous examples, including the dropping of the first atomic weapon on Hiroshima, and the introduction of Soviet strategic missiles into Cuba.

We may close this very inadequate discussion of this approach to types of surprise and deception by observing that one of the greatest and most successful military surprises in history, the Pearl Harbor attack, involved at least four of these modes. The United States had not correctly perceived the Japanese intention to attack U.S. territory at all and thus to bring the U.S. into the war—a step which logically appeared to be a gross strategic miscalculation, as indeed it was. The place of attack was not perceived, since the great bulk of the evidence pointed to Japanese attacks in Southeast Asia (which were in fact initiated almost simultaneously). The time of the attack contributed greatly to its success, Sunday morning having been deliberately chosen because the bulk of the U.S. warships would then normally be in port. The strength of the attack of course was not anticipated (since it was not expected at all where it occurred), security and deception having effectively screened the movements of the Japanese task force.

A second approach to types of surprise and deception, which is somewhat broader and perhaps more pertinent to strategic warning, is to examine the various methods or measures which may be used to achieve one or more of the foregoing types of surprise. We may identify roughly five of these: *security, political deception, cover, active military deception, confusion and disinformation.*

1. *Security* in itself is not strictly speaking a type of deception, in that it involves no active measures to mislead the adversary to a false conclusion, but is designed only to conceal preparations for attack. Thus the sophisticated analyst should take care to distinguish normal or routine security measures from true deception. Nonetheless, the line between deception and security is narrow and the two are very often confused. Moreover, an effective security program often can do much to mislead or deceive the intended victim of attack even if no more sophisticated measures are undertaken. Although security alone will not normally lead the adversary to undertake the wrong preparations or to deploy forces incorrectly, it may lead him to undertake very inadequate countermeasures or even to fail to alert his forces at all if security is totally effective.

In general, the greater the number of military measures which must be undertaken for the operation, and the larger the mobilization and deployment of forces required, the less likely it is that security alone can mislead. Whaley cites the views of Clausewitz that the high visibility of large-scale operations makes their concealment unlikely, and that true surprise is therefore more likely to be achieved in the realm of tactics than in strategy. This in fact has been borne out in recent examples. Although it was possible in large measure to conceal the military deployments required for the closure of the Berlin sector borders, it was not possible to conceal those for the invasion of Czechoslovakia, and in fact the USSR made no particularly great effort to do so. Some writers have argued that modern collection systems and communications will make security measures even less effective in the future—and this, would appear likely to be the case. Thus, the prospects are that various forms of active or deliberate deception will assume even more importance if surprise is to be achieved.

2. Preeminent among such methods is *political deception*—probably the easiest of all deception measures and possibly the most common. While political means may be used to promote tactical surprise, this method is of particular value as a strategic measure to conceal intent. Moreover it is one of the most economical means of deception and one in which the likelihood of disclosure is remote, since so few people need be involved in the plan. There are a variety of political deception tactics, of which we will note a few:

The direct or indirect falsehood may be put forth through diplomatic channels, official statements, the press or other media. In its simplest and most crude form, the state simply denies that it has any intent whatever of doing what it is preparing to do and asserts that all such charges are false—a method sometimes used, particularly if the stakes are very high. The more subtle method of the indirect falsehood is often preferred, however, and permits the leadership to maintain some degree of credibility after the event, or at least to deny charges of outright prevarication. This tactic was used by the USSR in a number of its public statements prior to the Cuban missile crisis, for example in the celebrated TASS statement of 11 September 1962 in which the USSR stated that all weapons being sent to Cuba were "designed exclusively for defensive purposes," and that there was "no need" for the USSR to deploy its missiles to any other country.

Another method of political deception which has often been used, particularly to lull suspicions in the relatively short term as final preparations for the attack are being made, is to offer to enter into "negotiations" to discuss the matter at issue when in fact there is no intention of reaching any sort of agreement. This tactic was used by the USSR on the eve of the counterattack to suppress the Hungarian revolt in November 1956, when Soviet officers opened negotiations with the Hungarians on Soviet "troop withdrawal." A form of this ruse was also used by the North Koreans for about two weeks before the attack on South Korea in 1950 when they issued "peace proposals" calling for a single national election.

Whaley identifies a slightly different form of this deception tactic, which is to lead the adversary to believe that the firm decision to attack is actually bluff. "This is a fairly common type of ruse, one intended to restore the initiative and insure surprise by implying that options other than war are still open, thereby concealing the full urgency of a crisis and encouraging the intended victim in the belief that he has more time and more options than is, in fact, the case."[7] He notes that this ruse was used at Port Arthur, at Pearl Harbor, in the German attack on the USSR in 1941, by the British in the attack at Alamein in 1942, and in the Israeli attack on Egypt in 1967.

A somewhat similar and relatively subtle form of political deception is to downplay the seriousness of the situation in diplomacy and in public statements in an effort to create the impression that the nation does not consider its vital interests at stake, or that its relations with the intended victim are pretty good or even improving. This may result in a quite sudden shift in propaganda to a more conciliatory tone, and friendly gestures to the adversary, after the decision or at least contingency decision to attack has already been reached. This is a quite common tactic, and one in which dictatorships are usually masters, particularly since their complete control of the press makes a shift in the propaganda line so easy. The USSR employed this tactic for weeks and even months prior to its attack on Japanese forces in Manchuria in August 1945, when it undertook an ostensible easing of tensions with Japan and began to be "almost cordial" to the Japanese Ambassador in Moscow, while the buildup of forces for the attack was under way in the Far East. The effort to deceive by political means will often entail not only the deception of many of one's own people, but may extend on occasion even to the leadership of allied nations, if the issue is of sufficient importance. And true practitioners of the art of deception even have been known to deceive their superiors (by failing to inform them of their plans)—although clearly this is a risky business undertaken only in the interests of tactical surprise for a specific military operation when war already is in progress.

3. *Cover* (here meaning the "cover plan" or "cover story") is a form of military deception which should be distinguished from active military deception, although it may often be used in conjunction with it. Cover will be used when it may be presumed that the military buildup itself cannot be concealed from the adversary, and its purpose therefore is to offer some seemingly plausible explanation (other than planned aggression) for the observable military activity. It may involve simply the putting out of false statements about the scale or purpose of the military buildup in order to conceal the real intention by attributing the military preparations to something else. Throughout history, the most usual explanation offered has been that the troops are "on maneuvers," although it is possible to think of other pretexts which might sometimes be used to explain troop movements, such as an alleged civil disturbance or disaster in a border area. The likelihood that the pretext of maneuvers would be used by the USSR to mask preparations for aggression was long recognized by Western intelligence, and the USSR

7. Whaley A548.

and its Warsaw Pact allies also professed to believe that NATO exercises could serve as a cover for attack.

Despite our presumed understanding of this tactic, the USSR achieved at least partial success with its several announcements during July and August of 1968 that its troops were engaged in various "exercises" in the western USSR and Eastern Europe. In fact, there were no bona fide exercises and the sole activity under way was the mobilization and deployment of Soviet and Warsaw Pact forces for the invasion of Czechoslovakia.

4. *Active military deception* is at once the most difficult form of deception to carry out, at least on any large scale, and also one of the most effective and successful. If security and political deception measures are most effective in lulling suspicions as to intent, active military deception is the primary means whereby the adversary is led to misdeploy his forces and to prepare for an attack at the wrong place and the wrong time. Even when strategic deception has failed, or was never possible in the first place, positive military deception has proved enormously effective in achieving tactical surprise, and hence in gaining victory and/or greatly reducing the attacker's casualties in the operation. Whaley in his treatise has compiled some impressive statistics on the effectiveness and rewards of positive deception operations, some of which have been so valuable and successful as literally to affect the course of history (as in the Normandy invasion).

The successful military deception operation may range from a relatively simple hoax or feint to a highly complex series of interrelated and mutually consistent measures all designed to create the wrong impression in the mind of the adversary (or to support his original but false conceptions) as to timing, nature, strength and place of the attack. Among the recognized techniques of active military deception are:

- Camouflage of military movements and of new military installations
- Maintenance of dummy equipment at vacated installations or in areas of the front where the attack is not to occur
- The simulation of a great deal of activity using only a few pieces of military equipment moving about
- The use of noisemakers or recordings to simulate a lot of activity
- The planting of seemingly valid, but actually false, military orders in the hands of the adversary
- The sending out of "defectors" with seemingly plausible but false stories
- The use of doubled agents for the same purpose
- The sending of invalid military messages by radio in the clear or in ciphers which the adversary is known to be reading
- The maintenance of normal garrison communications while the units themselves deploy under radio silence
- The establishment of entirely spurious radio nets to simulate the presence of forces which do not exist at all or to convey an impression of a buildup of forces in some area other than the planned attack
- A concentration of reconnaissance, bombing or artillery fire in an area other than the area of attack, or at least the equalization of such activity over a wide area so that the actual area of attack is not discernible from such preparatory measures

- False announcements or other deception as to the whereabouts of leading commanders
- Obvious training exercises for a type of attack (such as amphibious) which is not planned
- False designations for military units
- Actual deployments or feints by ground or naval units to simulate attack in the wrong area
- The use of enemy uniforms and other insignia
- Announcements that leaves are being granted on the eve of attack, or even the actual issuance of numerous passes for a day or so just prior to attack

The above list does not exhaust the tricks and ruses which have been devised and successfully used in military operations. Such active deception measures of course are often supplemented by political and propaganda deception measures, cover stories and extremely tight security on the real military operation. Thus the effect of the measures collectively can be the total misleading of the adversary as to the coming attack, even sometimes when he has accepted its likelihood and indeed may be well prepared for it in other respects.

It is obvious that a number of ruses cited above would be of limited use, and indeed could be counterproductive, in a strategic deception designed to conceal that an attack is planned at all, or in any area. In such cases, one does not wish to stir up a lot of military activity, or plant false documents about impending attacks, which will only arouse suspicions and stir the enemy's intelligence services into greater collection efforts. Some measures, such as bombing and artillery fire or even highly obvious and unusual reconnaissance, cannot be undertaken at all before hostilities have begun. For these reasons, some of the time-honored devices of military deception would not be used prior to an initial surprise attack which opens a war, the attack with which strategic warning is particularly concerned. At the same time, the reader can easily see that a substantial number of the tactics cited above could be most effectively applied to deceive us in a period prior to the initial attack. Among the ruses which should particularly concern us are: communications deception, especially the maintenance of normal communications accompanied by radio silence on deployments; planted military orders and other documents; the use of false defectors and doubled agents; and any of the other measures which might be used effectively to distract us from concentrating on the preparations for the real attack. For we may be reasonably certain that the greater and more important the operation, the greater and more sophisticated will be the positive deception effort. The fact that we have encountered relatively few cases of active military deception since World War II should not reassure us—in fact, it only increases our vulnerability.

5. *Confusion and disinformation* probably rank second only to political deception in the ease with which they can be used to mislead and distract the opposition. Confusion and disinformation tactics do not have to be highly sophisticated to be successful, although of course they may be. Even an elementary program to flood the market with a mass of conflicting stories and reports can be highly effective in distracting the time and attention of analysts and their superiors from the reliable intelligence on which they should be concentrating their efforts. Particularly if a crisis atmosphere already exists, as is highly likely, and some of the reports are sensational but have some degree

of plausibility, they can prove to be a tremendous distraction. If the volume of such planted information is large enough, the analytical system can literally be overwhelmed to a degree that some important and valid facts become lost in the mill, and others are not accorded their proper weight.

Moreover, such a mass of material compounds immeasurably the problem of analyst fatigue, always a factor in crisis situations, and may tend to generate a series of "cry wolf" alarms which will reduce the credibility of the authentic warning when or if it is received.

A conspicuous example of the damage that can be done by a large volume of false or unevaluated information was in the Chinese intervention in Korea in October-November 1950. This is not to say that the Chinese themselves necessarily had devised a sophisticated or extensive disinformation program. It is probable that a high percentage of the mass of spurious and contradictory reports which so confused the situation that summer and fall was never planted by the communists at all but was rather the product of the several highly productive paper mills in the Far East.

Most of those who have examined the intelligence failure that year have given altogether too little, if indeed any, attention to the adverse effects of the volume of this spurious material on the analytical process. Regardless of the origins of the material in this case, something of the same problem could surely arise again in another crisis should our adversaries choose to exercise their full capabilities to employ such tactics.

CHAPTER 7

Deception Maxims: Fact and Folklore[1]

L. Daniel Maxim

Dr. L. Daniel Maxim earned a Ph.D. from New York University in engineering and science. After a brief but important role contributing to deception studies, he subsequently went on to serve as President of Everest Consulting Associates. An active member of the Coast Guard Auxiliary, his continued research has specialized in toxicology and epidemiology.

We only list Maxim's ten concise maxims, thereby condensing his 50 pages to a more acceptable two.

1. *Magruder's Principle:* It is generally easier to induce a target to maintain a pre-existing belief than to deceive him for the purpose of changing his belief. The Germans did this to the Americans in their winter offensive in the Ardennes in 1944, Operation WACHT AM RHINE [Watch on the Rhine]. Even that code name suggested a defensive operation, which is what the Americans assumed would occur.
2. There are limitations to human information processing that are deceptively exploitable.
   - The "law of small number" means that you should not make conclusions based on a small set of data; there's no statistical certainty in doing so.
   - There is a frequent inability of targets to detect small changes in friendly indicators, even if the cumulative change over time is large. This is the basis for using conditioning (cry-wolf) as a deceptive technique.
3. *Multiple Forms of Surprise:* You can achieve surprise in the following categories: size, activity, location, unit, time, equipment (SALUTE), intent, and style (the manner/intensity with which you execute missions).
4. *Jones' Lemma:* Deception generally becomes more difficult as the number of sources available to the target to confirm the "real" increases. However, the greater the number of sources that can be deceptively manipulated, the greater the chance that you can provide the target "confirming, all-source intelligence."
5. *Choice of Types of Deception:* Ambiguity-reducing deceptions can be employed to make the enemy quite certain, very decisive, and wrong. Ambiguity-enhancing deceptions can be employed to mask your efforts when the target already has in his possession some elements of "friendly indicator truth," thus creating "noise."
6. *Husbanding of Deception Assets:* It may be wise to withhold the employment of deception capabilities until the stakes are high. The Soviets know we're revitalizing our deception capabilities, so let their intelligence-collection and decision-

1. Central Intelligence Agency, Office of Research and Development, Deception Research Program, *Deception Maxims: Fact and Folklore*, Washington: Office of Research and Development, Central Intelligence Agency, 1980, pp. 4-40.

cycle folks continually contend with "our threat," while friendly commanders employ it at the time and place of their choosing.

7. *Sequencing Rule:* Deception activities should be sequenced to maximize the portrayal of the deception story for as long as possible. Unit activities indicating the true mission should be masked (operational security, or OPSEC) to the last possible instant.
8. *Importance of Feedback:* An intelligence collection scheme should be employed to determine if the deception is being adopted, rejected, or deceptively countered. Deception-related Primary Intelligence Requirements/Intelligence Requirements (PIR/IRs) should be nominated and Named Areas of Interest/Tactical Areas of Interest (NAI/TAIs) established to facilitate feedback on and exploitation of the deception.
9. *The Monkey's Paw:* Deception efforts may produce subtle, unwanted reactions from the target and friendly units. Effect proper coordination in order to ensure that deceptions don't result in unit fratricide. Also, make sure that the deception objective is framed in terms of what you want the target to do, rather than think. The 23rd Headquarters-Special Troops was a Top Secret Organization attached to the U.S. 12th Army Group Headquarters in World War II. This 1100-man unit conducted 21 deception operations from 1944–1945. In Operation BREST, it portrayed an armor attack build-up, which was apparently believed by the Germans, because—due to lack of coordination—an actual armored unit tried to attack in that area. In another similar operation, the weakened German division opposite the phony armor build-up believed the story. However, the German commander, feeling that he was about to be overrun by U.S. armor, launched a spoiling attack—definitely not what the American commander had wanted him to do.
10. *Care in the Design of Planned Placement of Deceptive Material:* Generally, if the target's intelligence collection system has to "work for" the deceptive indicators you want to portray, the greater the likelihood he'll accept them as "truth." You can't baldly "announce" what you're doing, or he'll be suspicious.

CHAPTER 8

The Principle of Naturalness

Barton Whaley

I think anything out of the ordinary routine of life well worth reporting.
—Sherlock Holmes in The Hound of the Baskervilles (1901)

Each time an animal shifts from its natural rhythm, there is a reason, and thus an opportunity for us to ask why.
—Mark Elbroch, Mammal Tracks and Sign (2003)

To protect against the nemesis of inconsistency the magician deploys a cloak of *naturalness*. For magicians, *naturalness* is the result of making each sleight or other conjuring move simulate some normal gesture, action, or posture. Similarly, they build all their tricked props (gimmicks) to simulate common objects. They do this precisely in order to avoid having the spectators notice the secret actions or suspect the gimmick.

Note that the word naturalness, like many others in conjuror's jargon, has a meaning quite oblique to that in standard English; in this case, naturalness means *feigned naturalness*. Most top-class performers will go to great effort to develop this quality in both the design and practice of their tricks. Success in this is, perhaps, the highest test of their art. Indeed, as the great Swedish-American sleight-of-hand artist, Nate Leipzig, confessed, "There are a lot of good tricks which I never touch for some moves in them are not natural."

The practice of "naturalness" is itself a deceptive technique to lull suspicion. As magician Jean Foley said in 1924, "Naturalness is the mother of misdirection." Audiences look for tricky moves and overlook those that seem natural even when they aren't. Naturalness can also be contrived by conditioning an audience to accept certain moves as being innocent. For example, the magician may need one unfamiliar and hence unnatural move to work a particularly effective trick. She will overcome this telltale by introducing it several times during a previous trick where it will be seen as truly innocent. Therefore it is either accepted without notice or, at worst, noticed but quickly dismissed as merely some idiosyncratic gesture of the performer, something to do with general performing style and not with some specific secret method. By repetition, the unnatural has become natural, a new deception asset to be used when needed.

The concept of naturalness was kicking around in French magic circles since 1853 when Ponsin added it to his friend Decremps' list of the "Immortal Principles," urging magicians to use only natural moves in both sleights and feints. And French master illusionist Robert Houdin perpetuated this rule in 1868 in advising that "The more simple and natural the movements of the performer, the less likely is the spectator to detect the trick." However the word and concept have come to be associated with the late Dai Vernon because this modern Canadian master sleight-of-hand artist made it the keystone of his teaching and performance. Vernon began to work up this notion around 1903 when at age 8 or 9 he got his first book on how to cheat at cards, *Artifice Ruse and Subterfuge at the Card Table: A Treatise on the Science and Art of Manipulating Cards.* Of all the many books on the subject, this one by the mysterious S. W. Erdnase was (and still is) the best by far; Vernon quickly memorized this superb textbook. Erdnase points out that all deceptive actions "must be in perfect harmony with the usual procedure of the game." In addition, the cheat must observe the "inviolable rule" of "uniformity of action," namely, "Any departure from his customary manner of holding, shuffling, cutting or dealing the cards may be noticed, and is consequently avoided." Thus "Whether the procedure is true or 'blind' the same apparent action is maintained throughout."[1]

The problem of simulated naturalness in magic has been analyzed in depth by the late Spanish master, Arturo de Ascanio, in *The Psychology of Palming.*[2] There, he argues that a playing card is never suspected of being palmed, even for a long time, if the performer maintains naturalness in all gestures, posture, and actions with the "sinning hand." He explains that "The attentive eye of the spectator will not see anything, it will *glide* over the palming hand, like the look of a warlike observer over a well hidden gun: nothing unusual is seen ...."

To achieve this apparent but false naturalness requires analyzing each trick in minute detail. Each weak point must be found and every discrepancy in it identified. Then each discrepancy either must be hidden within some natural gesture, posture, or action or must be shown to have some plausible but false reason. For example, relaxed handling can keep a palmed coin hidden indefinitely. Similarly, a two-headed coin may be openly shown or a gaffed card shown as an implicit part of an otherwise honest packet of cards.

Magicians who use unnatural moves are engaging in "conjuring for conjurors," which entertains only when the spectators are other magicians ever-eager for a new effect or a new method at whatever cost in naturalness.

Some military deceivers use this technique of simulated naturalness. For example, mobilization is conventionally interpreted by intelligence analysts as a preliminary to war. To hide this common telltale, the Egyptian Army mobilized and partially mobilized several times during the months before their attack on Israel in 1973. The Israeli analysts, partly lulled, discounted the final genuine mobilization that came at Yom Kippur.

Three spectacular tactical military examples were the *three successive* British and Anzac withdrawals from their beachheads on Gallipoli during the 31 days between December 8, 1915, and January 9, 1916. They did this in the face of a disciplined and aggressive force of 120,000 Turkish and German troops that had the beaches in full view from the cliffs. London expected to lose between 25,000 and 40,000 men out of their total force of 120,000 plus all material. However, Lt.-Gen. Sir Charles Monro ordered that all visual and audible signs of activity outside the tents simulate normalcy while actual withdrawals took place under cover of night. The three withdrawals had gone undetected and unopposed—his ruse had worked. With only 10 casualties, Monro

1. S. W. Erdnase, *Artifice Ruse and Subterfuge at the Card Table: A Treatise on the Science and Art of Manipulating Cards,* Chicago, 1902, pp. 11, 22. All subsequent reprints give the main title as *The Expert at the Card Table.* See also Gerritt M. Evans, *How Gamblers Win: or, the Advantage Player's Manual,* New York: Gerritt M. Evans & Co., 1865, pp. 82–83.
2. Arturo de Ascanio, *The Psychology of Palming,* Madrid, 1982, pp. 5–6.

got his entire Allied force of 120,000 safely evacuated—plus nearly 95% of its artillery and horses.[3]

Conclusion

Examples of systematic and comprehensive application of the Principle of Naturalness are rare in military practice and almost entirely overlooked by military theory and military doctrine. I suggest that these three gaps be promptly filled. The most obvious lesson: Keep your deception plans simple, if you expect to keep them so seemingly "natural" that the enemy won't attract suspicion.

3. The only case study of these extraordinary events is Whaley, *Stratagem,* 1969, pp. A56–A65.

CHAPTER 9

Tactical Deception in Air-Land Warfare

Charles Fowler and Robert F. Nesbit

Charles Fowler was a member and served as chairman of the Defense Science Board (DSB). This article is based on a 1982–83 report done by the DSB. Fowler went on to serve as president of C.A. Fowler Associates.

Robert F. Nesbit earned his Master's degree in electrical engineering from Northeastern University and his was Vice President of the Center for Integrated Intelligence Systems at the Mitre Corporation at the time this article was written. In that position, he led the center serving the Department of Defense for the integration of intelligence systems.

Deception in a multiplicity of forms has been a part of nearly all military operations. The literature is replete with examples throughout the history of warfare. In his 2,500-year-old classic, *The Art of War*, Sun Tzu states that "all warfare is based upon deception" and also notes, "Rapidity is the essence of war; take advantage of the enemy's unreadiness, make your way by unexpected routes, and attack unguarded spots."[1] The story of the Trojan Horse is familiar to all.

Even George Washington practiced deception in the Revolutionary War by making his troop movements toward Yorktown appear to Sir Henry Clinton to be a ruse concealing a real attack on New York via Staten Island. This complex set of operations (also involving Lafayette, Cornwallis, Nathaniel Greene and the French and British navies) is described in MGen Edmund Thompson's article "Intelligence at Yorktown."[2] From today's perspective, the most astounding part of the story concerns General Green's capture of some encrypted correspondence from Cornwallis to his subordinates. Greene sent the captured messages to the Continental Congress! Upon receipt, Congressman James Lovell solved the code in four days and continued to decrypt subsequent messages for the duration of the campaign. This story caused some wag to make two observations: (1) That may well have been the last useful thing Congress did, and (2) if this happened today, the next morning's edition of the *Washington Post* would print the message.

HISTORICAL PERSPECTIVE

Deception operations can range in size from small unit operations, such as the MiG traps employed in the Vietnam War, to massive operations involving essentially an entire theatre, such as Operation Fortitude practiced in preparation for the Normandy invasion. Several of the more successful military deception operations in the 20th century are summarized below:

• "The Haversack Trick" was employed in the 1917 war between Great Britain and Turkey in Palestine. A British major named Meinhertzhagen deceived the Turks into thinking the British attack would be by an amphibious landing at Gaza on November 4, when the real attack was to take place at Beersheba on October 31. Major Meinhertzhagen distributed this false data by "coded" message after he had allowed the Turks to solve one of the radio codes, and then by intentionally losing his haversack that contained his lunch wrapped in official documents "confirming" the November 4 attack on Gaza. He also conducted a large effort designed to "find" the lost haversack. The Turks were caught completely off guard with their troops in the wrong place by the October 3 attack. The battleline that had been stalled for months was quickly crossed by the British and by November 9, they were in Jerusalem.[4]

• The German Army achieved complete strategic and tactical surprise in its 1941 invasion of Russia, largely as a result of a carefully crafted and explicitly detailed deception plan. The preparations for Operation Barbarosa were meticulously covered by surrounding them with a plethora of false rumors, poorly "encoded" misinformation, counterespionage operations and a believable cover story that Barbarosa itself was a diversion for the invasion of England. The German Army moved over 400 mi in only four weeks against the poorly prepared Russians and destroyed 1,400 Russian aircraft on the ground in just the first day.[5] As Roberta Wohlstetter noted in her study of Pearl Harbor, "To understand the fact of surprise it is necessary to examine the characteristics of the noise as well as the signals that after the event are clearly seen to herald the attack."[6]

• "Operation Fortitude" was designed to achieve surprise in the Allied landing at Normandy in World War II. The plan was to keep the German Reserve Army at Calais by making them think that would be the location of the Allied invasion of Europe. To do this, the Allies assembled a large number of decoys across from the Pas de Calais, including decoy gliders, landing craft, tanks, trucks, guns, etc. In addition, dummy communication nets were set up to look like a command headquarters, and troops were marched in during the day, back out at night and back in the next day. "Special means" (agents) were also used to leak the deception story. Due to this deception operation, the Allies were able to land the largest invasion force in history, with the Germans thinking (for the next six weeks) that it was only a diversion.[7]

• One of the more famous deception schemes occurred in 1943 — "The Man Who Never Was."[8] The British dressed a body as a Royal Marine Major and dropped the corpse into the sea off the coast of Spain. The corpse carried a briefcase containing documents indicating that the British planned to invade Greece. The intent of the deception was to cover the actual planned invasion of Sicily. The scheme caused the Germans to divert an entire armored division to Greece.

• The Soviets conducted a massive operation code named "Bagration" against the Germans in 1944. The offensive involved 175 Soviet divisions over a 900-km front in Belorussia. So-

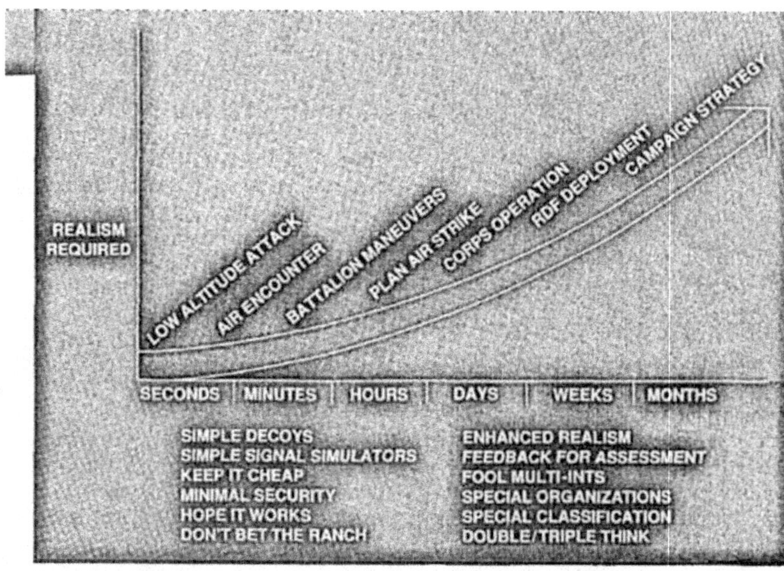

Fig. 1 The relationship between analysis time and deception fidelity.

---

*Authors' Note: This paper derives in significant part from a 1982-83 Defense Science Board (DSB) study for the chairman of the Joint Chiefs of Staff, Gen David Jones.[3] Norman Augustine, the chairman of the DSB, considered the findings and recommendations to be of such significance that he wrote in his memo forwarding the report to the secretary of defense, "It is my belief that this may be the most important report I submitted to you during my three-year chairmanship of the board." The succeeding chairman, Gen John Vessey, directed that the study be widely briefed and distributed throughout the services and commands. As a result, it had a major impact on US actions in this area.

The authors were participants in that DSB effort and have followed this field with interest since. Besides the authors of this article, participants in the study included Dr. Davis B. Bobrow, the late GEN William E. DePuy, USA (ret.), Gen Robert J. Dixon, USAF (ret.), Dr. Eugene G. Fubini, Dr. Robert Hermann, ADM Isaac C. Kidd, Jr., USN (ret.), Dr. Joshua Lederberg, LG Phillip D. Shutler, USMC (ret.) and the late Dr. Clarence H. Stewart. CAPT Joseph A. Muka, Jr., USN, served as executive secretary; COL Wayne B. Davis, USA, was the DSB Secretariat point of contact; Walter Jajko was the DUSDP point of contact; and John M. Porter was the USDRE point of contact. We would like to acknowldege the contributions of these other members; of DSB Chairman Augustine, who helped initiate the effort and also made a number of valuable suggestions; of of Gen Paul Gorman, who as a special assistant to the CJCS was an active participant in the early days of the effort and made several useful inputs.

While fully acknowledging these contributions, we assume total responsibility for the extraction, update and expansion of the DSB work. We further note that this paper represents our views and does not purport to represent those of the DOD, the DSB or any other person or organization.

viet plans paid considerable attention to cover, deception, security and surprise. In order to move 50 rifle divisions from other areas of the front, they established a 25-km deep security zone from Leningrad to the Black Sea to prevent German reconnaissance. To mislead the Germans concerning the offensive in the central Belorussia region, they "demonstrated" a large-scale concentration of forces on the southern Ukrainian front and undertook an offensive on the northern Leningrad portion of the front. These efforts succeeded in masking the real region of attack and resulted in a Soviet westward advance of 550-600 km.

Fig. 2 Decoys can be effective deception tools. Here, the plastic jeep on the right looks very much like the real jeep on the left. (US Army photo)

Some of the best examples of electronic deception are included in R.V. Jones accounts of British operations in World War II.

• German U-boats were being fixed generally by highly sensitive COMINT, and then localized by radar. But the Germans installed a radar detector called "Metox" on their subs which "saw" the radar before the radar could "see" the U-boat. Thus warned, the subs could submerge and escape. Jones not only changed the frequency of the search radar so that "Metox" could not pick up the signal, but he also provided British-controlled, German agents with two bits of spurious "intelligence." One was that the British Navy had abandoned radar for an infrared detector, and the other was that "Metox" itself radiated a signal on which British aircraft could home. As a result:

> Not only were the Germans very slow in realizing that we were using [changed frequency] radar... but they also developed a most ingenious paint for their U-boats to camouflage them against infrared as well as against visible light....When the Germans investigated, they found there was indeed radiation coming out of the Metox receiver, and they went to some trouble to suppress it....When the Germans finally realized that we were using radar after all, they fortunately blamed it for all their U-boat sinkings..., even going so far to say that this one invention changed the balance in the Battle of the Atlantic, and our [COMINT] feat remained secure.[9]

• In 1941, Jones provided cover and deception for the deployment of a new radio navigation aid for RAF bombers, a synchronous pulse system known as "Gee." He manipulated nomenclature, type numbers and equipment markings to foreshadow installation of a new piece of telegraph or voice communications gear in RAF bombers, so as to cover from German scrutiny of downed bombers or interrogation of crews the impending installation of the navigation aid. Further, Gee transmitter stations were camouflaged as radar, and irregularities were introduced into the Gee pulses to make the German electronic intelligence (ELINT) task more difficult. But these might not have worked had Jones not also authored a deception scheme to persuade the Germans that the British were about to install Lorenz beam, radio-direction-finding equipment in the bombers, for which he invented the name "Jay beams" — a real system intended for limited homing *vice* target designating use. Through British intelligence as well as German agents under its control, Jones provided Berlin with a report of an eavesdropped conversation between two RAF officers in which "Jay beams" were to pathfind to targets, and reports of special lectures in RAF bomber squadrons by England's well-renowned expert on Lorenz beams. Thus led off the scent, the Germans failed to detect Gee when it was put in use (March 1942), and the British enjoyed five months of successful operations before the Germans began jamming Gee. Jones went on to say, "A final twist of the story was that when the Germans realized that they had been hoaxed they ceased to pay any attention to the Jay beams, which continued to work throughout the war, and thus provided a useful homing service for our bombers when more sophisticated aids were jammed."[9]

• The Egyptians used a "conditioning" form of deception to achieve surprise in crossing the Suez Canal in 1973. Over several years they conducted repeated training exercises which the Israelis responded to at first, but eventually determined were routine. The buildup in September 1973 was designed to appear as a defensive training exercise to make the Israelis think the Egyptians feared reprisals for the Munich killings and some more recent kidnappings. Through this deception, the Egyptians gained surprise and easily crossed the canal.[10]

• Operation Bolo in Vietnam was very successful in luring North Vietnam MiGs into a trap. US Air Force F-4s were disguised as the less maneuverable, air-ground attack F-105s by using unique F-105 call signs and communication patterns and by carrying ECM pods previously used only on the F105. The North Vietnamese challenged the "F-105s"; in a short 12-minute air battle, the F-4s achieved complete surprise and shot down seven MiGs.[11]

DEFINITIONS AND SCOPE

US Joint Chiefs of Staff Publication 1 defines the intent of military deception and breaks the subject into three categories.[12] "Military Deception" is defined as DOD actions executed to mislead foreign decision makers, causing them to derive and accept desired appreciation of military capabilities, intentions, operations or other activities that evoke foreign actions that contribute to the originator's objectives. The three categories of military deception include:

a) Strategic Military Deception: Military deception planned and executed to result in foreign national policies and actions that support the originator's national objectives, policies and strategic military plans.

b) Tactical Military Deception: Military deception planned and executed by and in support of operational commanders against the pertinent threat, to result in opposing operational actions favorable to the originator's plans and operations.

the central region; this was the necessary feedback required for any major operation. Meanwhile, under cover of night and with complete radio silence, the Aggressor Force — using trains, narrow mountain roads at night and telephone communications — came down in two salients at each side of the exercise area and had the Defensive Force cut off and essentially surrounded while the defenders were awaiting the expected frontal assault.

There are several interesting points to be made from this example.

• It shows that good commanders do include deception in their activities. They are innovative and the forces do it well.

• It points out that denial of the real activity (by use of darkness and emission control) is as important as the deception activities.

• The activity suggested by the deception was believable. In fact, it was so logical it was not challenged or subjected to skeptical scrutiny.

• The defensive force did not use its SOTAS for several reasons. Had it done so, it would have been readily apparent that the real activities were on the flanks and not in the central region. An important point here is that a very successful, believable deception activity against the set of sensors used by the Defensive Force would not have worked against a more complete sensor set *unless* special provisions had been made to deceive, jam or remove the additional sensors.

As is often the case, while the exercise results indicate the tremendous benefits of deception, they also raise a number of questions. The Aggressor Force commander and his small planning staff had a number of months to work out the details of this operations plan, including the major deception activity. It is not at all clear that they would have the wherewithal to carry out this level of planning on other than an exercise basis. Nor is it clear that those who carried out the planning were aware of its vulnerability to SOTAS observation.

The conclusion to be drawn from this and other examples of tactical deception activities is that deception planning can be quite involved and complex; its execution can be quite risky, but its benefits can be overwhelming.

It should be noted for the record that the commander of the Aggressor Force who was personally involved in the innovation and guided the detailed planning of this very successful operation was MG Glen Otis, subsequently promoted to commander of TRADOC and commander, US Army, Europe at the rank of four-star general.

FUNDAMENTALS OF DECEPTION

There seem to be several fundamentals of successful tactical deception. We have dignified them here by calling them "rules."

First Rule: *To be effective, a deception operation must be one that causes the enemy to believe what he expects.* To put it another way, whereas the real operation can be, in fact should be, unusual, different, even downright unbelievable (see rule number six), the deception plan must be straightforward, sensible and essentially obvious to one's opponent.

This point is made frequently in descriptions of successful deception activities. For example, in his description of Yorktown, General Thompson notes:

> Washington fully understood the essence of deception — to make an action appear to coincide with the enemy's preconception.
>
> In this case, as Washington's agents informed him, Clinton fervently believed he was about to be attacked in New York; so the ruse

Decoys are not limited to Western forces. The Iraqis used plastic scud decoys like the one pictured here to fool Coalition pilots.

devised by Washington was to make the shift of the Continental Army toward Yorktown appear to be a ruse concealing a real attack on New York via Staten Island. Only too late would Clinton realize that his conceptions of ruse and reality were actually reversed.[2]

It has been asserted that a successful deception must contain 95% truth. This in turn suggests a corollary to this first rule. "Perhaps the bodyguard of lies would always be surrounded by a bodyguard of truth." There is another aspect of this, of course. What a person or group expects to happen is strongly dependent on his background and culture and his view of his opponent's background and culture. One student of deception put it this way: study and find what the opponent is a sucker for, what beliefs and biases are built into his society and culture. An important additional benefit of successful deception is that it can be a means to reduce an enemy's confidence in his own battle performance as well as enemy troop confidence in the capacity of their commanders and fellow units.

An Israeli officer who participated in military planning, including deception activities, stressed the need to understand thoroughly the opposing force right down to the level of individual commander characteristics so that one knew what they would consider "logical." The Israelis continue to do this for each country with which they might come into conflict.

Second Rule: *Timely feedback is an essential element of all major deception operations.* Feedback on the enemy's reaction to any deception activity is critically important if one's own position would be vulnerable should the opponent not be fooled.

Traditional means of observing reaction to deception activities have ranged from HUMINT reports by scouts or agents to intercepted communications. With current sensor capabilities, direct observation in real- or near-real-time from stand-off distances can permit observation of an opponent's reaction to deception activities and his observation of one's true activities.

Third Rule: *Deception must be integrated with operations.* The deception plan should never be created independently from the operations plan. The practice of having a separate deception planning staff is clearly wrong. To give further emphasis to this point, there have been many cases of impromptu or "ad-libbed" deception operations that, at best, never worked and, more often than not, were counterproductive. Some Vietnam observers estimate that for every deception operation that worked there were 10 or more that didn't because they weren't tied into operations and there was incomplete knowledge of the enemy's capabilities to observe what was happening.

To put it another way, operation and deception plans must complement and supplement each other. The overall activity must not only provide believable indicators of the false operation, but must deny believable indicators of the real operations.

Fourth Rule: *Denial of information on the true activities is also essential; it will depend, in significant part, on stealth and C^3 countermeasures (C^3CM) activities.* As noted earlier, it is one thing to provide inputs to a sensor to indicate activity where none exists, and it's a different thing, in most cases, to hide the real activity.

For example, carefully constructed false radio transmissions fed to an intercept system can frequently deceive an opponent — but only if there are no contradictory emissions from the real activity. This in turn requires carrying out the real operation in a manner such that radio, radar and any other electromagnetic radiations cannot be intercepted by the opponent; a practice known as emission control (EMCON). The use of covert or low-probability-of-intercept emitters or sometimes complete radio/radar silence is required.

Other forms of stealth include use of night, cloud cover or camouflage to hide from visual observations. Some sensors are particularly difficult to deceive or deny — a Joint STARS-like stand-off moving target indicator (MTI) radar, for example. In such cases, C^3CM such as jamming or attack (to destroy the sensor or drive it out of range) may be required. There is an intimate relation between tactical deception operations and stealth and C^3CM activities.

It is important to note that current sensor systems, unless countered, would have permitted the detection and foiling of essentially every successful tactical deception operation ever attempted. This has major implications for both tactical deception activities and for the sensor capabilities that a country and its allies actually deploy.

Fifth Rule: *The realism required for any deception activity is a function of the sensor and analysis capabilities available to the opponent* and *the time available to analyze the situation, disseminate the data to the appropriate points and take appropriate actions.* In a low altitude attack, for example, the

[Continued on page 76]

pilot has very little time — a matter of seconds — to tell the difference between a decoy tank and a real tank. Thus, the quality of the decoy can be very crude and the deception will still be successful. In contrast, if one wished to indicate a build-up of tanks in a port area where there was a matter of days or longer to examine photographs of the tanks, of track prints in the earth, etc., then a high degree of realism is required. **Figure 1** indicates this effect graphically.

For operations with little exposure time, minimal realism is required. Simple, cheap decoys and signal simulators can be used. There are no strict security requirements and really no need for complex feedback. One should expect that the deception will work, but should not be largely dependent upon its success.

As one moves to the right and upward on the curve in **Figure 1,** the need for realism increases dramatically, as does the need for feedback for assessment and for deceiving intelligence sources. This would seem to call for special organization and classification to reduce the risks of compromise and, finally, the kind of double-think, triple-think analysis that depends in substantial part on the detailed understanding of the society and culture of one's opponent.

Sixth Rule: *The most effective deception will be imaginative and creative; it cannot be "legislated" or "ordered"; and it* must *not become stereotyped or "bureaucratized."* Deception is not a subject that lends itself to strict rules, detailed regulations or regimented procedures. In fact, the worst situation would be one where "standard deception practices" were developed and institutionalized. In describing his plans for renewed emphasis on tactical deception throughout the Army, then Chief of Staff General Edward C. "Shy" Meyer noted that it would require special attention to prevent the handbook from containing something like "place one decoy tank 10 m right and 50 m below the road intersection." Military deception is best accomplished by bright, inventive (and tricky) commanders who are given the necessary tools and the latitude to use them as they see fit.

A key point to remember when reviewing the application of these rules in the modern era of improved sensors is that while the requirements for successful deception have changed, and will continue to change as the adversary acquires new sensor/intelligence capabilities, so have the opportunities. Each new sensor capability and each new communication channel to convey information and commands adds to the difficulty in carrying out unobserved operations and creating believable deception operations. On the other hand, most of these new capabilities have vulnerabilities and thus provide opportunities for deception.

The final point that stands out after looking at the overall subject of tactical deception is that commanders and their troops must be alert to and greatly concerned about being deceived by an opponent. This in turn leads to a fundamental observation: *The military group that is not devoting appropriate efforts to include tactics, R&D and plotting and scheming in general for deception is almost certain to be vulnerable to being deceived itself.*

This point was illustrated several times when deception practices were re-introduced in US exercises in 1983-84. In the first air exercises, simple disinformation over clear radio channels was assumed to be accurate and acted upon, leaving the deceived side very vulnerable. Two-dimensional tank decoys were completely successful the first few times they were used.

Fig. 3 Used effectively in Operation Desert Storm, this two-dimensional M1A1 Tank decoy is intended to deplete enemy ammunition in close combat operations. (US Army photo)

But in subsequent exercises, the deceived side was much more skeptical and regularly used cross checks for all suspicious information and situations. Any group skillful in and equipped for deception operations is less likely to be deceived. However, this knowledgeable skepticism has to be kept sharp, which presents a real problem with the frequent turnover in personnel. A prime objective of deception/counter-deception activities therefore should be a continuing effort to elevate the awareness of one's forces to the means, likelihood and impact of deception operations, both friendly and hostile.

At this point, a note about the two-dimensional tank decoys just mentioned is in order. Decoys are a particularly simple and attractive tool to be used in tactical deception operations. Their value will only increase as commercial satellite imaging systems become readily available on a worldwide basis.

Realistic decoys can be constructed in a variety of forms, including:

- inflatable decoys, which can be assembled by a few people with a blower kit in an hour or so
- skeletal decoys, a fairly low-fidelity approach consisting of a frame of wood or aluminum with canvas or nylon stretched over it
- compressed foam decoys, made from a polyurethane, rather expensive, material that can be stored in a box more than 10 times smaller than its deployed size.

• plastic decoys, a lightweight and potentially very high fidelity approach
• multispectral decoys, with which the addition of IR and radar "signatures" can be added to most types of decoys.

As examples, shown in **Figure 2** are a real Jeep (on the left) and a high fidelity rigid plastic decoy Jeep (on the right). The only obvious discriminators are the headlights and the lack of dirt on the decoy's tires. An Army critic said there was another discriminator: the Jeep on the right cost more than the one on the left.

In contrast, a much lower fidelity, two-dimensional M1A1 tank decoy is shown in **Figure 3.** Used effectively in Desert Storm, this decoy is intended to deplete enemy ammunition in close-combat operations. It weighs only 21 lb — including the hammer to drive in the stakes.

Figure 4 shows an inflatable, full-size F-15 decoy. Notice to the left of the F-15 is an inflatable R-9 refueling truck parked on a false operating surface.

Finally, in **Figure 5** is a decoy pontoon bridge comprised of floats carrying corner reflectors designed to fool a standard resolution imaging radar.

SENSORS, EQUIPMENT AND SCENARIOS

We have noted that as the capability of an enemy's sensor set increases, the difficulty of deceiving these sensors also increases. Instead of "difficult" we should perhaps say the "technical challenge" of deceiving the advanced sensors increases. It is quite possible (some say probable) that, unless proper precautions are taken, overall vulnerability to deception could increase as one's sensors become more sophisticated. The converse of this is certainly true in the extreme case — an army with no sensors is impossible to deceive. The likelihood of future armies having no sensors is growing much smaller, particularly given the appearance of systems such as the French satellite imaging system SPOT and our own Landsat, whose data are normally available to anyone. Further, several other countries will deploy similar systems in the relatively near future.

With the growing array of sensors now in military inventories around the world, here are some important requirements for creating and executing good deception/denial plans:
• knowledge of an adversary's tactics, sensors, C^3 system, decision cycle time and deception approaches

Fig. 4 Inflatable decoys can be particularly effective in deceiving high-speed low altitude attacks by enemy fighter aircraft. (US Air Force photo)

• equipment for concealment, deception and C^3CM
• planning aids and database.

Compiling a Database

It is necessary to have ready access to intelligence on enemy capabilities so one can tailor his deception and real operations to take those capabilities into account. This is becoming very complex, given the necessity for operations almost anywhere in the world. To obtain the information necessary to compile the required database, the following steps should be taken:
• create a model of the expected observables
• design a collection plan
• record the visible, IR and electronic observables of the operation (e.g., a division mobilization; an air squadron deployment) on film and tape (including the radar observables)
• analyze the tapes and compare to and revise the model
• repeat if and as necessary
• determine which signatures must be generated and which suppressed to make a successful deception
• obtain equipment and create the procedures and operations required to implement the deception.

The deception database would include a handbook of "enemy" and internationally available sensors, listing the capabilities and limitations of each (e.g., the ephemeris for relevant satellites, frequency coverage and location accuracy of SIGINT aircraft; characteristics of C^3 elements for getting the information to users; etc.). In addition, the database should include a good description in the appropriate form of friendly force elements and their characteristics.

Automated planning aids would allow: ready access to the database; interplay between friendly and hostile force characteristics; and the opposition's capabilities and cycle time. They also might provide a means for indicating necessary feedback quality and time criticality of the total operation.

Cover, Concealment, Camouflage and Stealth

Cover, concealment, camouflage and deception are generally subsumed under tactical deception because, as noted earlier, denial of the real activity is as important as creating the false picture.

There is a direct parallel in reducing electromagnetic radiation signatures and reducing optical/IR signatures. One might use the term "electromagnetic camouflage." The analogy holds in covertness as well as generating a confusing environment. It is probably useful to think in terms of stealth, cover, concealment and camouflage

Fig. 5 This decoy pontoon bridge is composed of floats carrying corner reflectors, and is designed to fool a standard-resolution imaging radar. (US Army photo)

across the board. This could include reducing electromagnetic backscatter cross sections from radar through optical frequencies and introducing confusion of the signature in the electromagnetic environment from radio through optical frequencies by means of covertness, noisy environments, difficult to distinguish patterns, etc.

During WWII, British interest and concern was so great they enlisted the services of a famous magician, Jasper Maskelyne, to help with concealment and decoy design. His many contributions are described in Reference 14.

It is a complex operation to either conceal or simulate the movement of an air squadron or an army/marine brigade. There are many signatures and there could be a large array of stand-off sensors "looking at" and "listening to" the area of interest. Before the development of stand-off sensors, the job could be accomplished by denying overflight of reconnaissance aircraft; by use of darkness and cloud cover and by EMCON combined with misleading radio communications. But this situation has changed somewhat and will change greatly with the deployment of more and better sensors — especially stand-off sensors. A new emphasis on "cover" techniques and operations is required.

Multispectral camouflage covers may introduce elements of "stealth" to combat vehicles, especially during transit on roads (that is, if the material is not suitable for use in combat due either to its fragility or being incompatible with combat operations for other reasons, such as visibility, restriction of motion, etc.). Significant reduction in radar cross section could have a major impact on target signatures performance (e.g., wheels versus tracks) and even on the basic performance of such systems as Joint STARS, ASARS and combat surveillance radars in general.

Equipment Design with Cover and Deception in Mind

It's very important in introducing or designing new systems to think of the cover and deception requirements. That is, how can operations be effectively conducted without using the system during the critical deception phase, and how do you efficiently simulate its signature in the non-real force?

Fortunately, the critical need for most systems occurs during the battles, i.e., *after* the forces are largely deployed, and not during the lengthy deployment phase. Further, those systems which have vulnerabilities for exploitation almost always offer an opportunity for deception. A few units simulating a large operation — combined with careful EMCON of the real force — will probably continue to be an important aspect of deception operations.

Decoy aircraft and emitters simulating operations could be an important factor in increasing the survivability of forward-based VSTOL aircraft. A small number of Harrier aircraft (six or so) with a minimal amount of support equipment (communications, simple navigation aids, etc.) are sometimes forward based. Such a small deployment is well suited to employing cover and deception. Simple Harrier decoys coupled with a few signal generators could realistically simulate a false forward location; the real deployment, using camouflage and EMCON, could be hidden from many nations' sensors.

DECEPTION IN DESERT STORM

The Persian Gulf War in 1991 was replete with the effective use of cover and deception — on both sides. The Iraqi forces emphasized tactical cover, and the United States focused on operational deception.

Iraq apparently learned a valuable lesson from the 1981 Israeli raid that destroyed the Osirik nuclear reactor. Ten years later in Desert Storm, Iraq's nuclear program escaped virtually intact. Through an elaborate combination of cover, deception, mobility, hardening and dispersal, the Iraqis were able to shield their key equipment and facilities from the intense coalition air strikes.[15] Their efforts in this regard continued after the war in a series of widely reported attempts to circumvent the International Atomic Energy Agency inspections. They employed sophisticated measures such as operating laboratories underneath what appeared to be bombed out buildings and crude measures such as moving key equipment into a truck which drove around a facility while the inspectors were inside.

Iraqi Scud operations were also conducted under elaborate cover schemes involving multiple hide positions, operating in darkness, use of decoys and emphasizing high mobility and short exposure times. Secretary of Defense William Perry commented recently that during the war with Iraq "we spent a lot of time and energy blowing up dummy Scuds."[16]

Iraq also made a great effort throughout its forces to minimize RF transmissions — so much so that some have concluded the Iraqi army committed "EMCON suicide."

On the other side of the battleline, deception took a different form. Shortly after Iraq invaded Kuwait, the US intelligence community prepared a handbook of Iraqi sensors, which was essential to deception planning.

Prior to the start of the air war, the US-led coalition made a concerted effort to accustom Iraqi air surveillance operators to seeing nighttime air operations patterns similar to what would be used for the initial attack. During Desert Shield large training operations were regularly conducted (usually on Wednesday nights) involving AWACS, Rivet Joint, Tankers, ABCCC and interceptors.[15] The intent was to achieve an element of surprise at the start of the air war — which began on a Wednesday night. The first night's raid was an enormous success, particularly devastating to the Iraqi air defenses.

Of course, the most elaborate and successful use of operational deception came in the cleverly crafted plan for the ground war. The famous "left hook" executed by the VII Corps devastated the Republican Guard divisions. By false radio messages, misleading troop deployments and deceptive rehearsals, Iraq had been carefully lead to expect several logical attacks: (a) an amphibious assault from the east; (b) a move north toward Baghdad through the Euphrates Valley by the XVIII Airborne Corps; and/or (c) a direct assault from Marine divisions from the south on Kuwait City.

The real main operation, the movement of a very large force 200 km to the west, was hidden from the Iraqis by "taking out their eyes" (as General Norman Schwarzkopf put it), by obtaining Russian and French cooperation (whose satellite photo systems could readily see a movement of this size) and by disciplined emission control and night operations. Aided by the speed with which the VII Corps proceeded from the west through "unnavigable desert," coalition ground forces achieved near total surprise and a stunning victory over the Republican Guard Divisions.

Just as the Reforger commander used his SOTAS to verify that his deception plan was working, General Schwarzkopf used his Joint STARS to ascertain the Iraqis did not reposition their forces to counter the left hook. Twenty-four hours before the ground attack was launched, Joint STARS spotted one unit starting to move from its position to the west; air strikes were called in, causing the unit to move off the road into fixed positions.

It is especially noted that the "left hook" deception operation was characterized by false indications that were logical and believable; completely integrated real and deception operations; accurate and timely feedback on enemy reactions; denial of the real operation accomplished by a combination of stealth and elimination of relevant sensors and associated C^3 systems; sufficient realism in the deception activity to convince the remaining Iraqis intelligence systems to keep forces largely in place until it was too late to react to the real operation; and, finally, boldness and creativity.

In short, the spectacularly successful "left hook" followed all the "rules" for an effective deception operation.

CONCLUSIONS AND RECOMMENDATIONS

Deception played a major role in achieving the rapid and overwhelming victory in the Gulf War. But a large part of the five-month Desert Shield period was required for preparation. If military leaders are to be prepared to plan such operations in a shorter period of time, we must do the following:

1)The intelligence community should provide and keep current a handbook for each threat country listing its various sensors and their capabilities to observe friendly "signatures"; a handbook describing all the relevant international sensor systems, their capabilities and availability; and a handbook describing each threat country's deception activities and equipment, both to help the operating forces and as a calibration for the deceivers.

2)The "signatures" (physical, optical/IR and electronic) of significant air and ground operations should be measured and analyzed to provide the information needed for the creation of deception and the denial of real operations.

3)Appropriate physical decoys (e.g., tanks, aircraft, etc.) and electromagnetic simulators (communications, radars, etc.) should be procured.

4)Relevant stealth and C^3CM systems and techniques necessary for denial of the real operations should be developed.

5)Means for providing feedback on enemy reactions must be established and used.

6)Deception (including the use of decoys and simulators for deception, stealth and C^3CM for denial and a system for feedback) should be made an integral part of all exercises.

7)Appropriate databases and planning aids should be developed and used in these exercises to acquire the training and experience to generate meaningful ops/deception plans rapidly.

8)Appropriate training, academic and R&D programs should be established.

If we do these things, our forces should have the knowledge and preparation needed to carry out successful deception planning and execution in a timely manner, even with the worldwide growth in sensor capabilities. An additional benefit will be a heightened sensitivity of our forces to deception activities by an opponent. ■

REFERENCES

1. Sun-Tzu, *The Art of War*, 6th C. BC, various translations, including those published by Oxford University Press, Random House and Delacorte Press.
2. Thompson, Maj Gen Edmund R., *Intelligence at Yorktown*, Defense 81
3. Defense Science Board, *Task Force on Command Support — Tactical Deception in Airland Warfare*, June 1983
4. Gardner, Brian, *Allenby of Arabia, Lawrence's General*, Coward-McCann, New York, NY, 1966
5. Whaley, Barton, *Codeword BARBAROSSA*, MIT Press, Cambridge, 1973
6. Wohlstetter, Roberta, *Pearl Harbor: Warning and Decision*, Stanford University Press, 1962
7. Norman, Albert, *Operation Overlord Design and Reality: the Allied Invasion of Western Europe*, Military Service Publishing Company, Harrisburg, PA, 1952
8. Montagu, Hon. Ewen, *The Man Who Never Was*, Lippincott, Philadelphia, PA, 1954
9. Jones, R.V., *Most Secret War*, H. Hamilton, London, UK, 1978
10. Allen, Peter, *The Yom Kippur War*, Scribner, New York, NY, 1982
11. Momyer, William W., *Airpower in Three Wars*, US Government Printing Office, 1983
12. Joint Chiefs of Staff, *JCS Publication 1-02, Department of Defense Dictionary of Military and Associated Terms*, US Government Printing Office, 1994
13. Daniel, Donald C. and Katherine L. Herbig, editors, *Strategic Military Deception*, Pergamon Press, New York, NY, 1982
14. Fisher, David, *The War Magician*, Cowan-McCann, 1983.
15. Cohen, Eliot A., director, *Gulf War Air Power Survey*, US Government Printing Office, 1993
16. *Jane's Defence Weekly*, February 18, 1995

Charles A. Fowler *is president of C.A. Fowler Associates, a member and past chairman of the Defense Science Board and a past member of the JED Advisory Board.*

Robert F. Nesbit *is vice president of the Center for Integrated Intelligence Systems at the Mitre Corp.*

Section III: Myths of Deception

Introduction

First, let's dispel a few of the myths that inhibit our ability to understand deception. The following quotations illustrate "The Tangled Web vs. the KISS Principle."

O, what a tangled web we weave,
When first we practise to deceive!
—Sir Walter Scott, Marmion (1808), Canto 6, Stanza 17

"Keep it simple, stupid" — KISS — is our constant reminder.
—Kelly Johnson, lead engineer of Lockheed's top secret "Skunk Works,"
More Than My Share of It All (1985), 161

"O what a tangled web we weave
When first we practise to deceive!
But when we've practised quite a while
How vastly we improve our style!"
—J. R. Pope (died 1941), "A Word of Encouragement" in *Punch*

This section exposes one general and three specific myths about deception. The first, "The Tangled Web vs. the KISS Principle," is particularly important because it tends to inhibit commanders and their staffs from adopting deceptive strategies and tactics. The next three readings are exposés of specific myths and one outright fabrication that have distorted the historical record.

CHAPTER 10

Meinertzhagen's False Claim to the Haversack Ruse (1917)

Barton Whaley

> *If you do not like the past, change it. ... Changing the past, especially through forgery and myth-making, may well be the most serious of all the abuses of history.*
> —William L. Burton[1]

The Meinertzhagen myth began soon after the end of World War One. Richard Meinertzhagen was a major in the British Army in the early 1920s. However, he was well into his 40s and his long but undistinguished military career was failing. He did not like his past. So he began to change it.

Even in a nation that celebrates its eccentrics, Richard Meinertzhagen was an outstanding British example, and in the narrow world of military deception and intelligence he was a giant—until 1993. There had been occasional derogatory rumors, tightly held among colleagues and acquaintances, during his life but none in print until he'd been safely dead for four years. In 1971, journalist and former MI-6 station chief Malcolm Muggeridge publicly labeled his first cousin "a fantasist full of tall stories which varied in the telling and were often, to me, totally incredible, though all calculated to make him appear a man of stupendous courage, resourcefulness and ruthlessness." However, being undocumented, all of these otherwise yellow-flag allegations had been dismissed out of hand and, it can be argued, rightly so. Then came 1993, the vintage year when a few researchers began to back bare assertions of Meinertzhagen's deceptions with some solid scraps of checkable evidence. Oliver and Boyd pioneered the hard job of winnowing the facts of his life from the fictions. However, Meiner, as Lawrence of Arabia called him, was the antithesis of most legendary figures — those quiet doers on whom we *impose* heroic fictions. He actively *invented and promoted* his own legend. Indeed, far more so than such other tellers of tall tales, as his friend Lawrence or the likes of Ernest Hemingway or Orson Welles. Where their kind would exaggerate and embellish an already good story, Meinertzhagen preferred creative invention, and to such a degree that he became more myth than legend, more fake than fact, more lie than truth. The prime example:

> "Spent today in deceiving the enemy. I have been busy lately compiling a dummy Staff Officer's notebook containing all sorts of nonsense about our plans and difficulties. Today I took it out to the country north-west of Beersheba with a view to passing it on to

1. William L. Burton, "The Use and Abuse of History," *The American Historical Association Newsletter*, Vol. 20, No. 2, February 1982.

the enemy without exciting suspicion. ... I was well mounted and near Girheir I found a Turkish patrol who at once gave chase. I galloped away for a mile or so and then they pulled up, so I stopped, dismounted and had a shot at them at about 600 yards. That was too much for them and they at once resumed the chase, blazing away harmlessly all the time. Now was my chance, and in my effort to mount I loosened my haversack, field-glasses, water-bottle, dropped my rifle, previously stained with some fresh blood from my horse, and in fact did everything to make them believe I was hit and that my flight was disorderly. They had now approached close enough and I made off, dropping the haversack which contained the notebook and various maps, my lunch, etc. I saw one of them stop and pick up the haversack and rifle, so I now went like the wind for home and soon gave them the slip, well satisfied with what I had done and that my deception had been successful. If only they act on the contents of the notebook, we shall do great things."
— *Meinertzhagen's Diary entry for 10 Oct 1917 at Rafa, Palestine.*[2]

This is an excerpt from the mendacious diary-memoir of Britain's most imaginative deception planning-operating impostor of the Great War, serving in the East African and Palestine campaigns. His lies inspired the unwitting General Wavell, who at the outset of WWII became the mentor of Brigadier Dudley Clarke and Peter Fleming and thence to Col. John Henry Bevan and Sir John Masterman in London and from them to the American Army.

Although Meinertzhagen falsely claims credit for all (or virtually all) of the extensive material on deception and intelligence operations, it does serve to show that these activities were in use not just at the time of his later interpolations, but also where independently verified (probably contemporarily) with the historical events described. Consequently, his book retains some marginal value as the basis for hypotheses about these matters and as leads to where to start research.

2. Richard Meinertzhagen, *Army Diary: 1899–1926*, London: Oliver and Boyd, 1960, pp. 222–223, 283–286.

CHAPTER 11

The Ultra Secret: Enigma Unwrapped A Review of F.W. Winterbotham's *The Ultra Secret*[1]

David Kahn

David Kahn received his doctorate degree from Oxford University in 1974 and worked as an op-ed editor for Newsday until 1998.

Wing-Commander "Freddie" Winterbotham's The Ultra Secret[2] *became a stunning best-seller when it first appeared in 1974. Although wildly applauded at the time, Kahn, America's world-class amateur historian of cryptology, is one of the few who recognized some of the very serious weak parts of Winterbotham's memoir.*

This book reveals the greatest secret of World War II and intelligence buffs. But it has to be read with caution.

"The Ultra Secret" tells how the British and the Americans exploited the information they obtained from cracking German messages enciphered with a cipher machine named the "Enigma." So valuable was this intelligence that it was given a special security classification, "Ultra," which the intelligence itself came to be called. The author, an R.A.F. officer, was put in charge of distributing Ultra under tight security to Churchill and to commands around the world. Winterbotham therefore saw much of the output and in this book has correlated it with the events of the war.

The stories he tells are revelations.

During the Battle of Britain, Ultra told the R.A.F. Fighter Command well in advance of radar detection how many bombers would be thrown against England and when. This enabled the British to parcel out their few fighters so that some would always be available to attack an oncoming wave. These tactics denied the Germans command of the air over England and consequently any possibility of invasion.

During the campaigns in North Africa, Ultra kept Gen. B.L. Montgomery informed fairly exactly of Gen. Erwin Rommel's order of battle and, in some cases, of his plans. It also enabled the British to know when supply ships would sail from Italy—and to sink them, thus eventually starving Rommel of

1. David Kahn, "The Ultra Secret: Enigma Unwrapped," *New York Times*, December 29, 1974. A review of F.W. Winterbotham, *The Ultra Secret*, UK: Littlehampton Book Services, Ltd., 1974. From *The New York Times*, December 29
2. F.W. Winterbotham, *The Ultra Secret*, New York: Dell, 1975.

vital fuel. Another intercept led to the Battle of Cape Matapan, which turned the Mediterranean from an Italian to a British lake.

The tide of the Battle of the Atlantic turned when Ultra dug deep into the naval Enigma in 1943 and revealed where the U-boats met their milch-cow supply submarines. Throughout the tough fighting in Normandy, Ultra delivered masses of intercepts from Hitler's messages on down, often within hours of their dispatch. This, "probably Ultra's greatest triumph," Winterbotham says, led to the "destruction of a large part of the German Army in the west."

Dozens of such stories crowd "The Ultra Secret" which is filled as well with sketches of the famous as Winterbotham, this bringer of good and bad tidings, saw them (Churchill was always polite). This makes exciting reading, and it constantly provides fresh insights into some of the best-known episodes of the war, for even the official historians did not have access to Ultra intercepts. The new material makes the book essential to the historiography of World War II.

But all is not exactly as Winterbotham tells it. He exaggerates the importance of Ultra, calling it "decisive" and writing as though it alone won the war.

Everyone now agrees that Ultra was of supreme importance, and that without it the war would have lasted longer. Even Gen. Mark Clark, criticized here for not exploiting Ultra properly during the Italian campaign, acknowledges that the reading of some Hitler signals saved his neck during the Anzio landings. But neither Marshal of the Royal Air Force Sir John Slessor, who wrote the foreward, nor Vice Admiral Sir Norman Denning, who was in charge of the Admiralty's U-boat tracking room, would say, in answer to my questions, that without Ultra, Britain would have lost the battles of Britain and the Atlantic.

Winterbotham, however, seems often to suggest that merely cracking the Engima sufficed to win the war. Of course it did not: otherwise things would have been a lot easier. But though Winterbotham himself sometimes gives cases where knowledge of German signals could not affect a battle, usually for lack of men or guns, cases where no messages were intercepted, and also cases where a change of plan falsified Ultra information, his attitude of Ultra-won-all negates them.

This tone is the basic flaw of the book, the reason the general reader needs to salt its information with knowledge of how wars are won. It is why the book is not history but merely a contribution to it. One that has to be checked, at that.

Winterbotham has written from memory 30 years after the events, using messages in the German archives to refresh his recollection of intercepts where he could, so it is not surprising that errors stipple the text. The American solutions of the Japanese diplomatic and naval cryptosystems had nothing to do with the breaking of Enigma. Winterbotham's attributing the original Enigma solution to information from a Polish employee of the cipher machine factory cheats the Poles of credit for one of the great cipher solutions of history. The facts are these:

On July 15, 1928, Polish cryptanalysts noticed a decided change in the letter frequencies of German army cryptograms, which they were intercepting. The Poles quickly concluded that the Germans had begun using the Enigma, which was invented and publicly sold early in the 1920's. Purchase of one of the commercial models showed that the Reichswehr had altered it for secrecy.

In 1932, the Polish Biuro Szyfrow (cipher bureau) got additional manpower in the form of three young mathematicians, Henryk Zygalski, Marian Rejewski and Jerzy Rozycki. They had achieved a partial solution in their office, hidden in the forest of Pyry outside Warsaw, when Poland's French allies furnished some key Enigma documents. Major Gustave Bertrand of French cryptographic espionage had obtained them from a Reichswehr cipher unit employee, Hans-thilo Schmidt, who wanted money. (Bertrand has told this story in his book, "Enigma.") With this help, the Poles completed their solution, and on July 26, 1939, presented two reconstructions of the machine to the French and two to the English.

These enabled the British codebreaking unit at Bletchley, a small town 50 miles northwest of London, to solve the later variations of the machine and other machines used for different branches of the German armed forces. Security forbade Winterbotham from recounting these details, but he properly and generously credits the achievement. To generate up-to-the-minute solutions for these other machines, incidentally, the Bletchley geniuses evolved perhaps the first modern electronic computer, which they nicknamed the "Colossus."

Why has the story remained under tight wraps so long? It seems that after World War II, Britain gathered up as many of the tens of thousands of Enigmas as she could find and later sold them to some of the emerging nations. Presumably if she could read Enigma messages in 1940, she could do so in 1950. Only recently have these countries replaced their Enigmas with new cryptosystems.

CHAPTER 12

The Ultra Variations: A Review of Anthony Cave Brown's *Bodyguard of Lies*[1]

Michael Howard

This was the earliest overview of Allied deception operations against Germany in WWII. It involved seven years of research that covered a wealth of obscure documents and choice interviews. At the time it was published, it was a generally useful first breakthrough into large parts of the previously classified literature. However, it was seriously marred by many gaps and several important errors of fact or interpretation. The most outrageous example was Brown's faulty interpretation that, in 1940, Churchill had deliberately sacrificed the city of Coventry to German bombers rather than reveal the fact that the British were reading the German codes, a false story that has managed to stay ahead of the evidence, having been retold as recently as 2004 by American columnist George Will. However, the then U.S. President Carter's Press Secretary, Jody Powell, was raving about this book to anyone who'd listen—strong evidence of how little understood military deception had then been at the highest level of U.S. government became clear. Conversely, all the experts (Trevor-Roper, Michael Howard, Pforzheimer, Wheatley, Constantinides, and J. Ransom Clark) were properly unhappy with the book's high rate of errors of both fact and interpretation. Brown consistently proved in this and later books that his understanding of intelligence and military deception were considerably less than he and the too-credulous main-line media reviewers believed. Fortunately, Brown's book has been recently superceded by Holt (2004).

Mr. Brown (often incorrectly double-barrel surnamed "Cave Brown"), an Englishman, served in the RAF as a photographer, then as a newspaper reporter and foreign correspondent, most notably for the Daily Mail (London) in the 1950s and early 1960s. He freelanced after 1962. Brown was famous for his outrageous expense accounts fueled by a sybaretic lifestyle. He settled permanently in the United States in 1969 where he lived until his death in 2006.

The "Ultra" secret is secret no longer. It seeped out gradually: there was never a single moment at which it was revealed, although publishers and journalist greet each new book of "ultra" memoirs as if it were a blinding flash of illumination. Unlike the "Double Cross system" – the total British control of German agents in the United Kingdom about which only a handful of people knew anything until Sir John Masterman broke ranks and told all – British interception and reading of German radio communications was known about by many thousands of people at the time and assumed to be taking place by anyone with any understanding of intelligence operations and signals systems.

1. Michael Howard, "The Ultra Variations," *Times Literary Supplement*, May 28, 1976. A review of Anthony Cave Brown, *Bodyguard of Lies*, New York: Dell, 1975.

The activities of the civil-military teams of cryptanalysts at Bletchley Park were a very open secret by the end of the war, and referred to with increasing frankness as the war years became more distant.

What was not, and perhaps is still not, generally appreciated was the success with which the British had penetrated German ciphers, and the way in which they were able to turn their knowledge to operational effect. Central to this success was the capture or reconstruction of the "Enigma" machines on which the German High Command enciphered its most sensitive traffic and which it believed to remain secure until the very end of the war. The most sensational revelation of this was made by Group Captain Winterbotham, an RAF liaison officer who dealt with this material at Bletchley, in his book *The Ultra Secret* (1974). But David Kahn had already dealt with "Enigma" in *The Code Breakers* (1967; and Gustave Bertrand's *Enigma* (1973) has given us what surely must be a definitive account of the Allies' acquisition and operation of the machine.

Few of these works give any idea of the continuous baffling complexity of cipher breaking even once the "Enigma" machines had been reconstructed. Different branches of the German services used different keys: the settings were constantly changed: and though an occasional operational coup put the cryptanalysts in possession of a set of keys, it was exceptional for them to be able to read any set of traffic currently for any prolonged period. Nor was "Enigma" the only system used by the enemy. There were others no less ingenious which cryptanalysts had to solve without any adventitious aids. So the generalization that the Allies were "reading enemy signals traffic throughout the war" is very misleading. We were reading some of it most of the time and most of it some of the time. The operational contribution of the knowledge so gleaned still cannot be assessed from open sources.

Something has been revealed, however, about the importance of "Ultra" for deception operations. To put it very briefly, our reading of German traffic gave us some idea of what they feared we might do, which enabled our deception staff to play on those fears. Even more important, it enabled them to see how the enemy was swallowing our deception measures. A signal from a double agent could be traced from his Abwehr controlling officer upward through German channels of communication till it reached the intelligence authorities at OKW, and down again through the distribution of information and directives to the various Commands. No less important, we knew what information was being derived from German agents in neutral capitals who were not under British control. Without "Ultra" and the "Double Cross system", Allied deception could never have played the central part which it did in ensuring that the overwhelming majority of allied offensive operations against Germany (Salerno being the striking exception) secured almost complete tactical surprise. Some of the work of the deception staffs has become public knowledge. The story of Operation Mincemeat in spring 1943, when bogus documents indicating an imminent assault on Greece rather than Sicily were planted on the Germans by means of an allegedly drowned liaison officer washed up on the beaches of Spain, has already been told by one of those chiefly responsible, Ewen Montagu, in his book *The Man Who Never Was*. The massive deception operations mounted in 1944 to persuade the Germans that the main allied assault was being prepared against the Pas de Calais rather than Normandy were partially described by Chester Wilmot in *The Struggle for Europe*, and the role played by the double agents was filled in by Sir John Masterman in *The Double Cross System*. But these were only the most spectacular successes of a continuous activity. From 1942 onwards not a single military operation of any magnitude was mounted by the Western allies in the European theatre without a deception element being included as an intrinsic part of the plan. Bogus radio traffic, visual deception by dummies to foil aerial reconnaissance, deliberately planted rumours, reports by double agents: all were as carefully organized and orchestrated as the movements of the forces themselves.

These are the activities which Anthony Cave Brown sets out to chronicle in *Bodyguard of Lies,* or at least the activities which he originally set out to chronicle when he began work, by his own account, over twelve years ago. In the writing, however, the book has swollen to such gargantuan proportions, and contains so many lengthy digressions on such matters as the German Resistance, the French Resistance, and the personalities involved in them, that the original object is lost for scores of pages at a time.

This is a pity, because the history of deception in itself is difficult enough to disentangle, without access to information which still remains classified. It has been made a great deal easier by the release in Washington of a great quantity of material which is still withheld in this country; but for purely British activities, particularly those relating to "Ultra" and "Double Cross", the researcher is still dependent very largely on personal revelations of a very uneven kind. Sir John Masterman's book counts virtually as a primary source, since it was compiled from the original documents while the author was still working with MI 5. Group Captain Winterbotham, on the other hand, was writing from memory thirty years after the event; while such autobiographies as that of Dusan Popov, one of the greatest double agents of the war, contain an entertaining quantity of straightforward fiction. There are in addition the personal memories of the surviving participants, for what they are worth after a third of a century, but these are not worth much unless they can be checked against more contemporary evidence. From the German side fortunately there is very complete documentation; and for those with a nose for such things unexpected treasures can be turned up in the records of the Navy and the RAF.

In spite of these difficulties Mr. Cave Brown has worked like a Trojan and come across an enormous quantity of information. For his sheer pertinacity he deserves very great credit indeed. What he has not done however is even to try to evaluate how much of this information is relevant, and how much reliable. It is hard to see why he includes information about, for example, German agents in Ireland, the assassination of Heydrich (in great detail), or the raid on German heavy water supplies in Norway – unless he hopes that this will help to sell the book.

But more important than his irrelevances is his total lack of critical acumen. Any story gets bunged down, whatever its source. No distinction is drawn between the very accurate accounts of deception operations for Overlord in the SHAEF files, the highly questionable assertions of self-justifying or self-inflating autobiographers, and the *obiter dicta* of elderly gentlemen in their anecdotage. Down they all go, facts, gossip, rumour, irrelevances, speculation, to be smothered in a rich sauce of romantic prose ("In its long history of doom and intrigue many fanciful figures had passed through the Portcullis Gate of Edinburgh Castle – Claverhouse and the marquess of Argyll, Queen Margaret and the Duke of Gordon...") and peppered liberally with solecisms, inaccuracies, and straightforward howlers.

Some of the slips are just silly and venial. Bradfield College is nowhere near Rugby, the Royal Patriotic Schools are not in Battersea, General Mason-Macfarlane did not negotiate the Italian surrender, Liddell Hart was not an adviser to Eisenhower nor Asher Lee to Churchill, Sir James Marshall Cornwall was not Vice-Chief of the Imperial General Staff in 1940, and so on. None of these matter much, save as indicators of the sloppiness of Mr. Cave Brown's work. Other errors, though more understandable, are dangerously misleading. "Special Means" did not include, as Mr. Cave Brown would have us believe, "a wide variety of surreptitious, sometimes murderous, always intricate operations of covert warfare". It was a term which British intelligence applied very specifically to the transmission of information to German intelligence, mainly through double agents: surreptitious and intricate certainly, but hardly murderous. Nor did the Twenty Committee "control" any double agents – much less operate an interrogation centre. The Twenty Committee was simply a liaison body between the armed forces, MI 5 and MI 6. Double agents, as Hugh Trevor-Roper has already patiently explained in the *New York Review of Books* for February 19, were controlled by B-1a, a branch of MI 5, and the credit for their successful manipulation is due primarily to the head of that section, Colonel "Tar" Robertson. To write the story of deception without even a mention of this organization is really to produce Hamlet without the prince.

As for the solecisms, they are hilarious. The Beaufort Hunt, we are informed, "was one of the most influential political groups in England ... as much a political conspiracy as a sport". Mr. Ewen Montagu of Mincemeat fame, who went on to become a highly controversial figure as Chairman of Quarter Sessions and who was never even elevated to the High court Bench, became, we are told, "one of England's great jurists". The movements of the Guards Armoured Division were falsely reported to the Germans because, according to Mr. Cave Brown, Rundstedt was expected to believe that "no general as blue-blooded as the commanding officer of the Guards Armoured, General Sir Allan Henry Shafto Adair (the sixth baronet, Harrow, Grenadier Guards, late governor of Harrow and Ensign of the King's Bodyguard of the Yeomen of the Guard), would ever allow himself to be very far from the main battle at H Hour". Connoisseurs will regret the omission from the British edition of the splendid description in the American edition of this book of the gentle Colonel Oliver Stanley as being "a man of towering anger. Had he not blackballed the Aga Khan for the Turf club?" There are times when the astral body of the late Daisy Ashford seems to take over from Mr. Cave Brown and insert into his massive narrative a kind of military supplement to *The Young Visiters*.

Mr. Cave Brown has in fact a very substantial chip on his shoulder about the British aristocracy. His book is not so much popular history as populist history. No opportunity is missed of taking a crack at that poor old Aunt Sally, the British upper classes. The remarkable military record of Eton College is chronicled (with typical irrelevance) with the comment "it was said afterwards that Eton had produced lions with brains, but its critics claimed that it also produced leaders who were unscrupulous, opportunistic, and concerned only with the preservation of their class and Empire". Well, fair enough. No Harrovian would disagree. But on it goes. The skill and dedication of the British deception staff is ascribed to "a malevolence perhaps born of the realization that, if they failed, their class would not survive". In their attempt to keep their existence secret, we are told "they failed, just as they would fail to survive as a class". Finally their object is described as:

> not only the defeat of Hitler; it was also the preservation of the Empire and of Britain as a world leader. Who could foresee that in winning a great victory over the most proficient military machine in the world ... the secret bureaus of England would be unable to preserve the very entity they were sworn to maintain – the power of London?

All this tells us at least as much about Mr. Cave Brown's social prejudices as it does about the group he is trying to describe. The historian can only regret that he should have allowed his vision to be so distorted, whether by a personal hang-up or by a desire to play to the gallery in the Middle West. The lengths to which he goes to resurrect all the old, discredited Anglophobe myths of the 1950s—that British Mediterranean strategy was motivated by a desire to retain a strangehold on Arabian oil, that Brooke deliberately plotted to place Eisenhower in a position of elevated impotence, leaving all the significant jobs to the British-rather suggests the latter.

None of it contributes notably to the history of deception which Mr. Cave Brown set out to tell. Neither do his long digressions on the group which he terms "the Schwarze Kapelle", the loose association of opponents to the regime within the German establishment, of which Canaris, chief of the Abwehr, was supposed to be the leader. The only serious reason for dragging this in would be to explain the truly astonishing inefficiency of the Abwehr's operations against Britain. But Mr. Cave Brown goes far beyond this in trying to prove a tortuous theory that, had it not been for "Ultra", which gave them all the information they needed, the Allies would have responded more readily to the Schwarze Kappelle's overtures; in which case their plots against Hitler might have succeeded, von Stauffenberg's bomb would (presumably) have exploded to better effect, and the war been brought to an earlier and more satisfactory end.

This at least appears to be the theory that Mr. Cave Brown is trying to prove; but so lush is his prose, so voluminous are his anecdotes, so labyrinthine are his digressions that it is never entirely easy to see what he is getting at. The same applies to his forays into the history of the French Resistance. The breaking of the Prosper network may have been due to its premature alerting as part of an Allied deception operation, Starkey, in the summer of 1943, but then again it may not. Mr. Cave Brown starts by asserting that it was.

> When the leaders of the resistance were stirred into activity in support of "Starkey", the Germans reacted with savagery and cunning to destroy certain key clandestine organizations in France...A deception operation that might otherwise have been written off as a rather muddied rehearsal for a performance that was still many months in the future had been transformed into a tragedy.

A few pages later this opinion is revised. The Resistance leaders, we are told, were captured "as a result of the lack of security which attended their activities—*some* of which they had been instructed to take in support of Starkey [my italics]—and the diligence of the S[icherheits] D[ienst]". That may well be nearer the mark. But Mr. Cave Brown is unwise to step into the Serbonian bog of Resistance history without preparing himself rather more carefully, and one awaits with interest what the experts will have to say about this part of his work.

It is rare however for Mr. Cave Brown to come out with a judgment even so forthright and unequivocal as this. More frequently he leaves himself open a line of retreat. Some reported statement we are told "may have been close to the truth", or "may not have been wrong". "There may", we are told, for instance, "have been some truth in these charges"; or, alternatively, they "cannot be discounted". Such phrases as "It is conceivable that", or "there are many who maintain that", such rhetorical questions as "was there a connection?", such forensic artifices as "by a remarkable coincidence—if it was a coincidence" occur throughout the work. As a result Mr. Cave Brown further inflates his narrative with a wealth of sensational gossip for which he ultimately accepts no responsibility, instead of going through the no doubt tedious process of scrutinizing, evaluating and comparing his sources, discarding the evident rubbish, and producing a serious study based on evidence which he had satisfied himself to be correct.

Had Mr. Cave Brown done this, one might be inclined to take rather more seriously what he has to say about such controversial issues as the Battle of Britain, the German air raid on Coventry, the Dieppe raid in August 1942 and the British air raid on Nuremberg in March 1944: matters on which the British press have recently quoted his views as if they were gospel truth.

Let us take first the Battle of Britain. Thanks to "Ultra", claims Mr. Cave Brown, "Churchill knew that Hitler would not invade without aerial superiority, but he deliberately led the world to believe that Britain was in mortal peril in order to rally his own people to meet the threat and enlist the support and sympathy of the United States". The only source quoted for this remarkable statement is the recollections of Group Captain Winterbotham. Winterbotham is an insecure foundation for this, as for much else. In the summer of 1940 we were no doubt reading the Luftwaffe traffic, which provided an enormous tactical advantage in fighting the Battle of Britain. But were we reading the traffic of OKW which alone might have indicated Hitler's strategic intentions? And if we were, would it have revealed a clear intention on the part of Hitler not to invade before he had secured aerial supremacy? Did he indeed have any clear intentions at that time? Did one need to have "Ultra" to appreciate that the Germans could not invade unless they had aerial supremacy? Above all, how could "Ultra" or any other agency reveal that the Luftwaffe would not secure aerial supremacy, as indeed it very nearly did?

The ubiquitous Winterbotham is also the sole source for Mr. Cave Brown's claims about the Coventry raid on November 14, 1940, which have already received a great deal of publicity over here. No other authority is quoted for the categorical statement "Ultra gave Churchill at least forty-eight, possibly sixty hours' warning of the devastating raid that was planned for Coventry". And presumably Winterbotham is also the source for the following paragraph:

> But if no extraordinary defensive measures could be taken to protect Coventry, might not a confidential warning that their city was about to be attacked on a large scale be given to civic authorities and to the fire-fighting, ambulance and hospital services? Should not the population of the inner city, together with those in hospitals who could be moved, be evacuated? To all these propositions, Churchill said no; there must be no evacuations and no warnings...it would alert the German intelligence service to the fact that the British had foreknowledge of the raid.

Apart from his cheerful assumptions about the administrative feasibility of evacuating a hundred thousand people at forty-eight hours' notice, one finds in Mr. Cave Brown's narrative no attempt to check on whether Winterbotham's memory after thirty years was entirely reliable. Most of the records are still inaccessible, but a lot of the people involved are still around, and if Mr. Cave Brown had devoted a tithe of the time to questioning them that he did to exploring the ramifications of the Prosper network he might have come up with rather a different story.

Intelligence sources, no doubt "Ultra" foremost among them, did indeed alert the British as early as November 11 to German intentions of launching a major raid on the United Kingdom in the near future, but they gave no indication of what the target would be. It was in response to this original general warning that the RAF launched its ill-fated spoiling attack, Operation cold Water. It was not in fact until the afternoon of November 14 that firm intelligence came through that the raid was due that nights, and that Coventry was the target. This indeed is what Winterbotham states in his own book, whatever private communications he may have made to Mr. Cave Brown later. The effective warning which we had of the raid on Coventry was thus not forty-eight hours, let alone sixty. It was about four.

As for the Dieppe raid, Mr. Cave Brown alleges that Churchill deliberately allowed this to go ahead in full awareness of its sacrificial nature in order to persuade the Americans of the impracticability of opening a "Second Front" in north-west Europe in 1942. More important for our purposes, he states that the central British deception staff, the London Controlling Section, headed by Colonel Oliver Stanley, was allowed to mount what they believed to be a deception operation which actually focused German attention on Dieppe. When Colonel Stanley discovered how he had been used, we are told, he "resigned abruptly...He could no longer be a member of an entourage that would sacrifice or risk 5,000 loyal Empire troops to keep Russia in the war."

For this very remarkable allegation Mr. Cave Brown quotes no source at all, and a very little gentle digging would have revealed to him the absurdity of the whole story. We know from the German documents that von Rundstedt's headquarters was not particularly alerted to Dieppe as a possible target, although they were certainly kept on the qui vive along the whole of the coast of north-west Europe. In any case the contribution of the London Controlling Section to this, if indeed it made any at all, must have been minimal. Set up the previous autumn in response to a suggestion from Wavell, it was at this period under-staffed, inexperienced and entirely uninfluential. Stanley himself was very sceptical as to what it might achieve and he gave it no impetus. Several times he attempted to resign, and in fact did so in June, 1942, two months before the Dieppe raid, when the section was entirely reorganized under Colonel John Bevan. So much for that canard.

Let us in conclusion consider Mr. Cave Brown's most contentious allegations, those bearing on the British air raid on Nuremberg on March 30, 1944. This, it will be recalled, was a disaster. Allied bombers were directed in a single stream, under clear skies, against a target deep inside Germany, and suffered very heavy losses as a result. The official figure was 108 aircraft lost on operations, which Mr. Cave Brown inflates by reference to unnamed confidential and semi-official sources, to 178.

As always after a military catastrophe, there was an immediate crop of bitter rumours to the effect that the operation had been betrayed in advance-rumours which Mr. Cave Brown treats as hard evidence. The only corroboration he can produce is a reported statement by General de Guingand some thirty years after the event, to the effect that "on at least one occasion the deception people were authorized to reveal the target of a major air attack on a German city to the Germans beforehand, in order to reinforce the credibility of an agent who was to be used to mislead the German high command". This city he thought was Stuttgart but, says Mr. Cave Brown, "he could not be sure that it was not Nuremberg". General de Guingand has since categorically denied, in private correspondence, that the name of Nuremberg was mentioned at all. In any case as Chief of Staff of 21 Army Group he would have had no responsibility for and little connection with any such operation. The second pillar on which Mr. Cave Brown builds his case is also decidedly rickety.

What about the course of the raid itself? Does the evidence indicate that the Germans were expecting it? Fortunately we have a most careful analysis of all the evidence on both sides by an independent and highly reputable historian, Martin Middlebrook, who in his book *The Nuremberg Raid* reaches the unequivocal conclusion that they were not. Anyone who wishes to form a conclusion as to Mr. Cave Brown's qualities as a historian need only compare his chapter on the raid with Mr. Middlebrook's book and see how, while using such evidence which seems to prove his case, Mr. Cave Brown ignores everything that does not. So unready were the German night fighters that, so far from their laying an effective ambush, a third of the bomber force was past the fighter assembly beacon at Aachen before they intervened. Indeed the war diary of the German 1st Fighter Corps which Mr. Cave Brown claims to have used (though he calls it throughout 1st Air Corpos), and to which Mr. Middlebrook has kindly called my attention, states specifically: "The British decoy manoeuvres prevented an early recognition of the target....the 1st Fighter Corpos presumed that the real attack would begin on another target." In fact, as Mr. Middlebrook points out, hardly any German single-seater fighters reached Nuremberg, and the unit nearest to Nuremberg never even took off.

Mr. Cave Brown further states that "at least one prisoner claimed that he saw someone in the Luftwaffe had marked the route to be taken (to Nuremberg) on a wall map". The source of this statement is a work by David Irving. Mr. Middlebrook showed conclusively in his book, by reference to RAF records, that this prisoner never existed. Despite much publicity and intensive investigations, no RAF man has ever been found who claims to have been told this or a similar story by the Germans before the Nuremberg raid. The German officers concerned at the Oberursel interrogation centre have denied all knowledge of this and similar incidents to which Mr. Cave Brown refers; but the chapter in which Mr. Middlebrook exposes the falsity of these rumours is blandly ignored.

It might further be asked why, if an attack on this scale did have to be mounted to establish the credibility of a double agent (itself an implausible hypothesis), the RAF should have chosen Nuremberg, an objective at almost extreme range, rather than one closer at hand and easier to reach. Stuttgart was a far better target for this purpose, and two raids which were mounted against it at about the same time as the Nuremberg raid would have fitted de Guingand's description equally well—if indeed his memory is to be taken seriously at all. Mr. Cave Brown discounts these by claiming that the Nuremberg raid was mounted with the deliberate intention of bringing on a prolonged fighter battle, in order to increase the attrition of the Luftwaffe fighter strength before D-Day. For this purpose, says Mr. Cave Brown, the bomber force was accompanied by 70–80 Mosquito night fighters. But Mr. Middlebrook by careful analysis of RAF records has shown that altogether only fifty-five Mosquitos were in action that night. Thirty-six of them were on "intruder" operations against German airfields, sometimes many miles from the scene of the bomber operation. This left a total of nineteen fighters to escort the 700-strong bomber force. What kind of attrition does Mr. Cave Brown imagine that these were supposed to inflict?

Why, it may be asked, devote so much time to breaking this butterfly on the wheel? Mr. Cave Brown, it may be alleged (though not by his publishers, who liken him to Arthur Bryant, Winston Churchill, and, rather more plausibly, Frederick Forsyth), is not a serious historian: he is only a journalist writing popular history. Why take him to pieces in the *TLS*? Do any of his mistakes really matter to anyone except those bloodless pedants, the academic historians?

To this it must be replied that they matter a great deal to the friends and relations of those who died in Coventry, at Dieppe, and on the Nuremberg raid, allegedly to further the Machiavellian plots of Churchill and a gang of unscrupulous Old Etonians who were concerned only with the preservation of their own class and Empire; to say nothing of the courageous men and women who were killed in the French and German Resistance. Any historian, however popular or populist, owes it to the memory of these people to get things right, and to spare no labour in doing so. Besides, Mr. Cave Brown comes before us with the enthusiastic recommendations of American official historians who ought to know better; and his allegations have been put forward as if they were established and incontrovertible facts in British newspapers, including *The Sunday Times*, which certainly ought to know better. In spite of all that can be done to set the record straight, millions of people are now likely to believe, and hundreds of historians to repeat, that the Coventry raid was known about three days in advance but kept quiet; and that the Nuremberg raid was deliberately betrayed to the enemy. Perhaps the damage done by such books as this and by the journalists who allow themselves to be taken in by them can never be completely undone. Mud is easy enough to throw, and some of it always sticks.

The sad thing is that it is all so unnecessary. There is in this book all the material for a first-class study of Allied operational deception, if only the author had taken the trouble to winnow out the grain from the chaff. As it is, he has protected his truths, not by a bodyguard of lies exactly, but by so confusing a fog of half-truths, insinuations, romantic inventions and irrelevant digressions that the deception staffs would have recognized him as a genius at their profession and recruited him on the spot. But specialists in Second World War literature will know how to classify this work. "Farrago" is the *mot juste*.

Section IV:
Theories: How Deception Works

Introduction

How Deception Works

Developing and implementing successful stratagems necessitate understanding human psychological processes.[1] Field commanders make many mental calculations prior to major military campaigns and individual battles. Both history and logic reinforce the fact that the commander whose mental calculations prove more accurate than his opponent's tends to triumph in war, often regardless of numerical or material advantage. In other words, war is fought in the minds of men long before it takes place on the ground.[2]

Having established that stratagem is a psychological phenomenon, to create surprise inducing sufficient dysfunction into an opponent's analytical process to cause him to act contrary to his real interests becomes the goal. This necessitates the deliberate misrepresentation of reality in order to gain a competitive advantage.[3] Misrepresenting reality calls for understanding the reality in which the opponent exists.

B. H. Liddell Hart's strategy of the *indirect approach,* though not intended to be a tome on deception, offers insights into its psychology. Liddell Hart tells us that a direct approach to an objective along "lines of natural expectation" is costly and unlikely to succeed. The reason has less to do with numbers or material advantage than with the control and "stability or equilibrium" associated with preparation for an expected action. Accordingly, accommodating an opponent's expectations assures the *equilibrium* in his favor and stiffens his resistance and resolve. Alternatively, an indirect approach that takes advantage of expectations creates disequilibrium and surprise.[4]

Achieving surprise results from an opponent choosing the wrong course of action among alternatives.[5] In other words, the "lines of natural expectation" can be blurred by increasing the number of viable alternatives. To illustrate this point mathematically, pursuing a single, viable course of action, one out of one, provides the enemy with certainty regarding one's intentions. The enemy is unlikely to make the wrong decision or to be surprised. By simply

1. Bart Whaley clearly identifies deception as "a psychological notion" in his classic book, *Stratagem—Deception and Surprise in War.*
2. Sun Tzu discusses the many mental calculations that first take place within the "temple" for the successful commander. He reminds us that the commander who fails to subject his plan to mental calculation also fails to win wars.
3. D. C. Daniel and K. L. Herbig, "Propositions on Military Deception," in *Strategic Military Deception,* Daniel and Herbig (eds), New York: Pergamon Press 1982, p. 3.
4. B. H. Liddell Hart, *Letter to The Army Quarterly,* January 1928, p. 400.
5. It is entirely possible that the enemy will undermine his own efforts through faulty analysis or as a result of an overbearing, delusional decision maker, such as Joseph Stalin. However, deception is about inducing behavior, not hoping the enemy shoots himself in the foot.

increasing the viable courses of action to two, the enemy has a 50% chance of picking the wrong alternative—even without any use of deception. The geometric leap from certainty to even odds is nontrivial. Further increasing the number of viable alternatives increases the probability of the enemy choosing the wrong alternative. However, increasing the number of alternatives requires increasing the investment in deception, often at the expense of the actual operation. A balance must be attained. Still, the forfeiture of surprise is certain without the presence of more than one viable course of action.[6]

The goal of stratagem is to ensure that the victim is surprised. Accordingly, first it is necessary to present the enemy with multiple, viable alternatives. Only then can the strategist bias the probabilities of the alternatives through a process of hiding the real information and showing false information to induce the enemy to choose the wrong alternative. Though simple in conception, the devil lies in the details.[7]

6. Bart Whaley, *Stratagem—Deception and Surprise in War*, Norwood: MA, Artech House, 2007, p. 69.
7. Ibid, 73.

CHAPTER 13

"Window" as Written in *Most Secret War*[1]

R. V. Jones

This chapter is Dr. R.V. Jones's statement of his theory of "Spoof." This became the basis for the war of electronic measures, electronic countermeasures (ECM), electronic counter-countermeasures (ECCM), and to infinity or, if sooner, final victory.

NOTE: "D.T." was an early German term for radar.

OUR STUDIES of German radar were described in a series of reports of which the two most substantial were 'D.T.' (this was one of the original German names for radar, as ours was 'R.D.F.') dated 10th January 1942 and 'German Nightfighter Control' of 29th December 1942. As with the X-Gerät Report they summarized the knowledge we had gained during the preceding year, and ran to some twenty thousand words each. In making such comprehensive surveys we sometimes recognized weaknesses that might otherwise have gone unnoticed: for example, it was only when we put all the evidence together that we could come to the surprisingly negative conclusion that the Germans had no satisfactory method of identifying their own fighters, as we had in I.F.F. This, combined with their ponderous method of telling plots from the Giant Würzburgs to the Seeburg Table, in turn suggested that they might have difficulty in following a bomber that was not flying a straight and level course. This in fact sometimes happened: as the fighter approached the bomber, it was inevitably picked up by the 'red' Giant following the bomber; if the bomber suddenly turned away the 'red' Giant might continue to follow the fighter and plot it in mistake. This promptly suggested that when flying through the Kammhuber Line our bombers should take 'evasive action' as a matter of routine, and various manœuvres such as the 'corkscrew' (a sudden twisting dive of a thousand feet or so, and subsequent recovery) were adopted. So besides the intelligence itself, our Reports inevitably considered possible counters to the German defences.

In particular, the 'D.T.' Report had a section discussing weak points in the German system, and pointing to the use of spurious reflectors, among other possibilities but first pressing for direct attack on the main radar stations:

The Report then went on to develop the theory of Spoof:

'10.3.0 *Spoof*

10.3.1 R.D.F. [Radar] is a system which reports to the enemy the

1. From *Most Secret War: British Scientific Intelligence 1939–1950*, London: Hamish Hamilton, 1978, pp. 287–299. Reproduced by permission of Penguin Books, LTD.

position of your forces by means of the radio echoes to which they give rise. To neutralize its value to him, it must be made to confound him. This may be done in two ways: by persuading him that you are either (a) where you are not, or (b) not where you are. Owing to the nature and good cover of German R.D.F. it may be taken, at least for this argument, that any aircraft within striking distance of the enemy coast will be observed as a positive signal on the cathode ray tube of an early warning station. It is therefore almost impossible to avoid giving to the enemy the necessary evidence from which he can deduce your true position. The art of Spoof lies in so colouring his appreciation of this evidence that he comes to a false conclusion. The only method philosophically possible, when you are bound to give him a positive indication of your position, is to provide him with a requisite number of imitation positive indications. No imitation can be perfect without being the real thing, but it is surprising what can be done by dexterous suggestion.

10.3.2 To any one R.D.F. equipment, an aircraft appears as a radio echo on a specific frequency. The fundamental elements of R.D.F. Spoof are echoes from spurious objects, and phenomena which simulate echoes or obliterate them. Obliteration, either by deliberate jamming or by sowing a dense field of spurious reflectors, corresponds to a smoke screen, and although in itself it is not a Spoof, it can be so handled that spoofing may result. For example, if a smoke screen generally preceded an attack, then the enemy will expect an attack whenever he sees a smoke screen. Thus, it is not of great use to obliterate his R.D.F. wherever and whenever a heavy attack is intended: the obliteration will itself provide the necessary early warning for the enemy to prime his defences. If obliteration be physically possible, then it must also be used at times other than those preceding attacks, so that the enemy can either establish no correlation or, better, can be induced to make false correlation.

10.3.4 Apart from obliteration, there remains the possibility of producing swarms of pulses which react as echoes off a large formation of aircraft, either from a ground station, or more easily from a single aircraft. This in itself is true spoof, which can be handled in mass so as to produce a greater spoof.

10.3.5 The enemy knows that an aircraft will show an echo which moves in the same characteristic way whatever radio frequency he projects, and therefore if he has a sufficient number of frequencies in reserve, he can discard those which give suspect indications. It is

> therefore necessary to be prepared to cover, say, 80% of his available frequencies, although in the heat of operations if even only one half his stations indicated a large raid, it would take a very cool controller to disregard them in favour of the others who truthfully reported only a small raid. It is easy to fall into the trap of thinking that the enemy is omniscient and panoptic and hence of believing that no spoof could fool him. German R.D.F. personnel are only human, and even relatively modest spoof might succeed.

The final paragraph discussed the timing of countermeasures.

> 10.3.7 It is unwise to be squeamish about taking countermeasures against any enemy development because of the danger of reciprocation. The enemy is not altogether lacking in ingenuity, and has probably thought of most of the counters. The true reason against undertaking countermeasures is that ultimately the enemy will learn to overcome them, and that it is only during the period of his education that we shall reap the advantage. This period must be made to occur only when we want it, no matter what action the enemy may take against our own system in the meantime. When the correct time arrives there should be no hesitation. Spoofing may require a very considerable degree of effort to be successful, but its cost must be weighed only against the relevant D.T. effort and against the additional damage which we could do with its aid.

The background to the foregoing paragraph had been a remarkable 'squeamishness' on our part regarding technical countermeasures against German radar, as far as the Air side was concerned. This contrasted with the readiness with which tactical countermeasures such as concentration and evasive action had been adopted. It also contrasted with what the Royal Navy had been doing in jamming the German coast-watching radar ever since we had discovered it in October 1940. I was amazed to sit through meetings of the Radio Counter Measures Board right up to the end of 1942 discussing whether it was advisable for us to start an 'R.C.M. War' and so invite German retaliation by the jamming of our own radar, when our Navy was already jamming German radar, and when the Germans had already jammed our radar very successfully in the *Scharnhorst–Gneisnau* episode. Although I repeatedly pointed to these facts, I encountered a nebulous but strong reluctance against taking technical measures against Luftwaffe radar.

The reluctance went right back to the early days of Bawdsey, and at the heart of it was Watson-Watt. No trials of spurious reflectors had been made after I had suggested their use in 1937; and A. P. Rowe, who succeeded Watson-Watt as Superintendent at Bawdsey, told me after the war that he did not hear of the suggestion until 1941. Writing to me in 1962 Rowe said 'When I took over from W. W. at Bawdsey, I found it was "not done" to suggest that the whole affair would not work'.[1] In defence of Watson-Watt it could be argued that anything that might have thrown doubts on the value of radar, such as the early development of a successful jamming system, might have caused the Air Staff to lack confidence, and thus not to support radar enthusiastically. We might then not have had such an effective system ready in time for the Battle of Britain. But it could equally well be argued that, had we been up against a more thorough opponent who had developed a proper system of jamming, our 1940 radar would have been rendered powerless, and we should have been culpably unprepared to deal with such a situation.

Such arguments apart, the fact was that in 1941 and for almost the whole of 1942 we shrank from technical countermeasures, and the most generous interpretation of Watson-Watt's attitude was that he had developed a 'bridge on the River Kwai' attitude towards radar, and it hurt him emotionally to think of radar being neutralized, even German radar. Perhaps, remembering our old enmity, I was in turn committed to the idea of wrecking radar with clouds of spurious reflectors; but of course I would never have wanted to see our own radar wrecked. With some justice events were to lead to an ultimate confrontation between Watson-Watt and me on this very issue, and I will now trace their contorted evolution.

Among the various technical countermeasures to German radar, one possibility was to fit a warning device to every bomber, telling the crew whenever a Würzburg or a Lichtenstein transmission was directed at them, so that they could take evasive action. This came into use as 'Boozer' in the spring of 1943. A much more elaborate device was that of sending back amplified and extended echoes, to suggest a large formation of aircraft; this was used on a small scale, following my visit to No. 11 Group after the Bruneval raid (p. 243) from August 1942 onwards. On the first occasion when it was tried (6th August) eight Defiant aircraft off Portland produced such a response that thirty

[1] In the same letter Rowe wrote 'Are you being modest in not claiming to be the originator of Window?' It was such an obvious invention as to be hardly worth claiming, and in any event I had the earliest and strongest incentive for thinking of it because of the argument with Watson-Watt.

German fighters were sent up around Cherbourg to meet the imaginary threat. And on 17th August our 'Moonshine' Defiants with supporting aircraft managed to divert nearly 150 German fighters from an American Flying Fortress attack on Rouen. But the scheme was elaborate, and it would only work if a very limited number of Freyas had to be dealt with, so it was dropped after a few months, to be later revived for D-Day. We were able to use 'Moonshine' earlier than the simpler countermeasures, perhaps because it was too elaborate for the Germans to use on a large scale against us, and because our Moonshine aircraft could stay so far outside the German defences that there was little chance of their being captured. But the main reason probably was that the whole episode had occurred at the lower level initiated by my visit to No. 11 Group, and that formal approval was never sought at the highest level.

With the two major possibilities, jamming and spurious reflectors, the situation was different for their use would be on such a scale that formal approval would have to be obtained. Jamming appealed to me less, because it was cruder and moreover would jeopardize any aircraft carrying a jammer because it could be homed on by fighters carrying suitable receivers. Spurious reflectors would be simpler, and contained an element of hoaxing. The phenomenon on which they depended was that of resonance. If a reflector is made of a simple wire or strip of metal of length equal to half the wavelength used by the radar station, it resonates to the incoming radio waves and re-radiates them to such effect that it is roughly equivalent to a whole sheet of metal whose dimensions are a square which has sides equal in length to half a wavelength. Thus a few hundred such strips or wires would reflect as much energy as a whole Lancaster bomber. Originally I suggested that wires should be suspended from balloons, because the long wavelengths that were usual in 1937 would require lengths of at least 10 feet; but we found that the predominant wavelengths in the German radar that we had to counter were about 50 centimetres, so each wire or strip need only be 25 centimetres long, and could be made light enough to fall through the air at a slow rate, and thus remain active for many minutes.

As our knowledge of German radar built up during 1941 I pressed the idea again, especially with Lindemann, and it finally resulted in trials being done late in 1941 and early 1942. They were undertaken under Robert Cockburn's direction at Swanage by Mrs. Joan Curran, now Lady Curran. Her results were all that we expected, and she tried various forms of reflector ranging from wires to leaflets, each roughly the size of a page in a notebook, on which, as a refinement, propaganda could be printed. The form that we finally favoured was a strip about 25 centimetres long and between 1 and 2 centimetres wide. The material was produced and made up into packets each weighing about a pound, and the idea was that the leading aircraft in a bomber stream would throw them out at the rate of one every minute or so, to produce the radar equivalent of a smoke-screen through which succeeding aircraft could fly. So much progress was made, after the years of delay, that by April 1942 enough material had been produced for it to be used by Bomber Command. It was given the code name 'Window' by A. P. Rowe, the Superintendent of T.R.E., and the scheme went up to the Chiefs of Staff on 27th April 1942 for their sanction.

The Official History *The Strategic Air Offensive Against Germany 1939–45* suggests (Volume 1, pp. 400–401) that Window was thought of at this stage as primarily a weapon against radar-controlled A-A guns, but this was certainly not so in my mind. If the Chiefs of Staff took this view, it must have been because of the Bomber Command obsession with anti-aircraft guns rather than nightfighters. For the first three years of the war the Command seemed to be far more concerned about guns, and it needed much convincing that the fighters were the greater threat, my own conclusion being that 70% of our losses were due to fighters at a time when the Command would not agree to a higher figure than 50%. So wary were they of the guns that the Command Intelligence Officer told me they never routed a raid over the island of Overflakkee because of the possible effect on the morale of the crews.

With the Chiefs of Staff's approval, the Window packets were, I believe, actually loaded on to bombers for a raid in May 1942, and were then off-loaded again because there was a last-minute prohibition on their use. What had happened in the meantime was that my old Oxford colleague Derek Jackson had heard of them for the first time: he had given up research for Active Service at the beginning of the war and had become one of the most skilled nightfighter radar operators in Fighter Command, and he was now the Command's Airborne Radar Officer. Naturally, he was concerned about the reduction in our own nightfighting efficiency if the Germans were to use Window against us. He was always able to put his arguments with remarkable force, and on this occasion he called on Lindemann and swung the latter round from supporting Window to opposing its use. Lindemann went to Sinclair, convinced that Window should be withheld, and succeeded in getting its introduction deferred until the effect on our own nightfighters had been evaluated. Lindemann now had Jackson bombarding him on one side and me on the other. For the time being Jackson prevailed; and he was, of course, strongly supported by Watson-Watt.

All through the summer of 1942 the battle went on. I was quite sure that the arguments against using Window were ill-founded, because there was no prospect of the Germans attacking us on anything like the scale that we were now mounting against them. What was left of the German bomber force was mainly tied up in Russia. The only good argument for withholding Window was the one that had been given in my main report on German radar, which was that we should pick our time when its introduction would have greatest effect; we could not expect this effect to last indefinitely, as the Germans gradually found ways of getting round the difficulties that it presented for them, and it could be debated whether or not the optimum time had yet come. But I was sure that we were delaying too long.

I thought that my chance had finally come when in October 1942 I received a report which came from a Danish agent who had heard two Luftwaffe women personnel talking in a railway train. The report ran:

> A. A considerable number of women have lately been taken on for work in aircraft control stations. Formerly this work was done by men, but since this summer women are being used and their number is constantly increasing.
>
> B. Operations of night fighters are controlled entirely from such stations, the pilot being almost a passenger. The direction of flight, altitude and even the moment to open fire are given (in latter case by a man) from the ground stations which are equipped with special detectors (or predictors) for this purpose. The principle on which such detectors (or predictors) work has something to do with the metal in an aircraft.
>
> C. Informant has heard of an instance when a British machine in the Rhineland deceived a German control station by throwing out aluminium dust and then changing its altitude. As the dust cloud was nearer the control station than the plane, the detectors (or predictors) guided the German night fighters in this direction and they are said even to have opened fire on the dust cloud.

It was an extraordinary Intelligence situation, and unique in my experience, because the value of the report was independent of whether it was genuine or 'a plant'. The whole of the opposition to our use of Window had been based on the argument that although the idea was obvious, it had not occurred to the Germans. And yet, now we had the report, either it was genuine, in which case Luftwaffe personnel were discussing the effects of Window or it was 'a plant', in which case the German Security Service had shown that they were well aware of the possibility of Window. In either case, it proved that someone in Germany knew of the Window principle, and therefore the argument based on the premiss that the Germans did not know of it was now substantially demolished.

I at once wrote a report about it, and circulated it to those concerned on 24th October 1942. I telephoned Bomber Command and told B. G. Dickins, the Head of the Bomber Command Operational Research Section, that if the Command would now press for using Window I had ample evidence to demolish the opposition. Portal called a meeting for 4th November, and I made a point of seeing Lindemann to try to talk him round before the meeting and to give him the chance of changing his ground. This he refused to do, dismissing my report as merely being something two W.A.A.F.'s had said to one another in a train. When I told him that I would have to oppose him at the meeting he said 'If you do that, you will find Tizard and me united against you!' I could not help replying 'If I've achieved that, by God, I've achieved something!' and we parted in laughter. Actually, he must have been quite misinformed about Tizard's attitude, for the latter had already minuted that my report had completed the argument for the immediate introduction of Window.

The day of the meeting came, and I was looking forward to the fight. I knew that I could expect opposition from Lindemann and also from Sholto Douglas, the C.-in-C. of Fighter Command and, above all, from Watson-Watt. But with Bomber Command and the evidence, I thought that we must win. And had Tizard been there it would have been a very interesting battle. But for some reason he was absent, and Bomber Command, represented by Sir Robert Saundby, my train-playing host of two months before, collapsed on me. Saundby said that his C.-in-C. did not wish to press for Window at the present time since he considered that other countermeasures, the jamming of the Freyas with a device known as 'Mandrel' and the interference with the German radio telephonic controls, would be enough for the time being.

The battle was therefore left to me alone, and I heard Lindemann witheringly talk of the irresponsibility of basing major policy on what someone had happened to overhear two W.A.A.F.'s saying in a train. I therefore lost hands down, and in a last defiance I asked Portal whether he would now advise our Security Services to call off the national effort that was being expended in propaganda to prevent careless talk in our own country. For this was the one example in the whole of my experience where careless talk had really been important, and all that Lord Cherwell could do was to dismiss it as talk between two women in a train. Portal saw that I was indignant, and so he added a recommendation that we should now work out in detail just what the effect of Window both on the Germans and on our own defences might be. We should work out the amount of Window that we should need to drop on a typical raid, and estimate the effect that it would have in reducing our casualties. We should also work out how much Window would be needed by the Germans to produce something like the same effect on our defences. To get all sides of the case presented, he asked that Fighter Command, Bomber Command, and I should make independent estimates, after which we should have a further meeting.

In the meantime, jamming of the Freyas was started in December 1942, using the Mandrel jammers. Some of these were carried in Defiant aircraft in the form of an airborne 'screen' intended to blind the Freyas to our approach. This was at first partly successful, in that, with the Freyas blinded it was much more difficult for the Giant Würzburgs to be directed on to individual bombers; but the Germans overcame the problem in various ways, one of which was to change the wavelengths of the Freyas to others which the Mandrels could not jam. The other countermeasures brought into use at the same time was the jamming of the German ground to nightfighter communications by transmitters in our bombers, each using a microphone to broadcast the noise of its engines so as to superimpose these on the German conversations. I was rather worried when I heard that the code name given to this device was 'Tinsel' because it might suggest to any German Intelligence officer the metal foil strips that we were still withholding.

As for computing of the amount of Window required, we were to assume a raid of a particular magnitude on a typical target, and work out how much Window material would be needed to render the German radar system ineffective. Derek Jackson at Fighter Command was the first to come up with an answer: his figure was 84 tons. He also calculated that one ton of Window used by the Germans would wreck our own radar system. Charles Frank and I made our own calculation, pointing out that Jackson had calculated the worst case, and that if one adopted assumptions that were reasonable in the light of our special knowledge of the German radar system, the figure was much more probably 12 tons instead of Jackson's 84. Dickins at Bomber Command cautiously waited until Jackson and I had both made our bids, and then produced a calculation showing that 48 tons would be needed.

Our estimate had been completed by 4th January 1943, but Portal did not reconvene the Window meeting until 2nd April 1943. I knew from Lindemann's secretary that he was still adamant, and that he was going into the meeting to repeat his opposition to the use of Window. However, he was not called on to speak until all three calculations had been presented. On behalf of Fighter Command, now headed by Trafford Leigh Mallory, Derek Jackson explained how he came to a figure of 84 tons, whilst one ton would suffice for the Germans if they wanted to knock our radar out. Portal then called on me, and I showed where I thought the Fighter Command estimates were too stringent. He then called on Harris, who deflected the question to Dickins, who gave a longish account of how Bomber Command had arrived at its figure. When he had finished, Portal simply asked 'What has Bomber Command done beyond adding the Fighter Command and A.D.I. (Science) figures together and dividing by two?'

At this point, Watson-Watt intervened, saying that he would like to give a warning based on his experience, and that he considered Jackson's figure of 84 tons much too low, although he agreed that one ton would be sufficient to knock our own system out. His argument was that no radio or radar device in his experience had been more than 20–30% efficient in its initial stages of use and therefore Jackson's figure of 84 tons should be multiplied by at least three if we were to be sure that the German radar would be knocked out, because of the inevitable mistakes that we would make in using Window properly.

As he spoke, I could hardly believe the extent to which he had left himself open, for up to that time I had thought of him as a coldly calculating opponent who would keep himself completely covered. As soon as he had finished I intervened and said to Portal 'Are we to accept this, Sir? S.A.T. [i.e. Watson-Watt] has said that no electronic device in his experience is more then 20–30% efficient in the early stages of its use. If so, why does he apply this to our use of Window against the Germans, but not to the Germans' use of Window against us?' Portal hesitated for a moment as he grappled with my point, and turned to Lindemann for a clarification. Lindemann must have seen at the same moment that Watson-Watt's position was indefensible, and he explained to Portal the import of what I had said. Portal looked sharply round, pointed his finger at Watson-Watt and said 'S.A.T., you're clean-bowled!' To Watson-Watt's credit, he replied 'Not bowled, sir, but caught at the wicket, perhaps!' and that was the end, or almost the end, of the argument that had started between us nearly six years before, for Portal would have no further debate. He said that he would go straight to the Prime Minister to ask permission to use Window as soon as possible, which was understood to be 1st May.

But there was further delay, probably brought about by continued opposition by Fighter Command and Watson-Watt, for the Chiefs of Staff decided that Window should not be used until Sicily had been invaded. I have never been able to follow the logic of this argument. Portal now pressed for a further meeting of the Chiefs of Staff, with the Prime Minister in the chair, and this took place on 23rd June 1943. It was only the second time I had seen Churchill since the beams meeting of three years before. The intervening occasion had also been concerned, I think, with a discussion of the night defences. I remember that occasion because most of us had assembled in the ante-room to the Cabinet Room, waiting for Winston to appear, when I noticed out of the corner of my eye an individual in a boiler suit come padding into the room; I imagined him to be a Ministry of Works maintenance engineer who had inadvertently strayed into the room full of Generals while looking for some domestic installation such as a hot water radiator. It seemed kindest to pretend not to notice the poor chap, in the hope that he would realize his error and quietly disappear; but to my surprise, he turned out to be the Prime Minister in his famous siren suit and, picking me out from the group, he came straight up to me and said 'Mr. Jones, very glad to have you here!'

At this final Window meeting, Watson-Watt and I were once again in opposition, for Portal let me state most of the Air Staff case for using Window. Watson-Watt again emphasized the damage to our own night defences but Churchill then turned to Leigh Mallory, head of Fighter Command, pointing out that he was the man who would have to 'carry the can' if our defences failed, and what did he think? Leigh Mallory very decently gave the opinion that even though his defences might be neutralized he was now convinced that the advantage lay with saving the casualties in Bomber Command, and that he would take the responsibility. That concluded the argument, and Churchill said 'Very well, let us open the Window!' And just as Churchill could say that my appearance at his meeting in June 1940 reminded him of the passage in the *Ingoldsby Legend* that starts 'But now one Mr. Jones comes forth and depones', so could I now counter on Watson-Watt's behalf from Jane Austen's *Mr. Woodhouse*: 'Open the windows! But, surely Mr. Churchill, nobody would think of opening the windows at Randalls. Nobody could be so imprudent!'

We were still to wait until successful landings were made in Sicily, but this was now a matter of only a few weeks. At the end of the meeting Churchill called me over to him, and asked me, 'Has Mr. Sandys seen you yet?' I well knew the import of his question, which must wait till a later chapter, and I had to tell him, 'No, Sir!' He frowned deeply and said, 'Very well. I shall call a Staff Conference next week! Hold yourself in readiness!'

No doubt there would be many who would have some claim to credit if Window was the success we expected it to be, but it was equally clear that if it failed there was one above all others who would get the blame. This was I, for not only had I started the idea but, with fateful justice, I had fought the case right through to the catching out of Watson-Watt and now had scored again with Churchill. Night after night I thought about ways in which Window could go wrong. Might we, for example, be dropping so many metal strips that these would get into the air intake of subsequent bombers and clog their engines? Would the bomber crews drop the packets at the prescribed rates, or would they use too much at one time and too little at another?

I decided that the best hope for good Window discipline was to have the bomber crews fully understand what they had to do, and the best way of informing them was to have some of my own officers lecture on the German night defences to every bomber station. I had at least two very good lecturers in Hugh Smith and Edward Wright (an old friend, and Professor of Mathematics at Aberdeen, who had recently joined me), but the problem was to get them into the Command. In this, I was rather less than honest. I knew the Commander-in-Chief strongly resisted any direct contact between the Air Staff and his bomber crews although the crews were expected to send in reports of any encounters they had with German nightfighters, or any other experiences that might throw light on the German defences, and most of these reports ultimately came to me. I therefore telephoned not the Commander-in-Chief but the amiable Saundby and told him that I thought that I could get much more out of the crews' reports if two of my chaps could go round the Command and talk to them. At the same time, I said, it would be an opportunity for the crews to hear from us about the latest state of the night defences. 'That would be difficult,' Saundby said, 'because the C.-in-C. doesn't like anyone from the Air Staff going out to his stations!' To this I replied, 'These are good sensible chaps that I'd be sending. One of them is a magnificent model railway maker. In the office here I have a model that he has made of a Great Western County tank engine, in No. 1 gauge. It is coal-fired and super-heated, and you can come along any time yourself and we'll get it going for you. You can then see whether he'd be alright in going round the Command.' Saundby immediately said 'Oh, if he's a chap like that—that's different. When do you want them to go?' And so Smith and Wright spent a hectic fortnight each of them lecturing in at least two stations, and sometimes three, every day.

There was one further attempt to stop the introduction of Window. I do not know whether it was engineered by Watson-Watt, but somehow Herbert Morrison heard about it and was characteristically alarmed by the possibility that we should sustain civilian casualties if the Germans in turn used Window against us. Churchill had therefore to call yet another meeting, on 15th July, when Morrison threatened to raise the question of Window in the War Cabinet. Churchill replied that the matter was too technical for the Cabinet, and that he was personally prepared to accept the responsibility for the decision to introduce Window. It was to be used from 23rd July onwards.

So we were at last to employ Window. Although the only evidence that the Germans must be aware of the principle was that given in the report from Denmark, it was enough to make me wonder whether the Germans were keeping Window in abeyance because they feared that if they were to use it and we were to retaliate, their defences would suffer much more than ours. Following my example with the imaginary German beam conference in 1940, my colleague and ally (later Air Chief Marshal Sir) John Whitworth Jones wrote the minutes of an imaginary German Air Staff conference on the topic 'Operation Fenster' (window, in German) in which he portrayed the Luftwaffe's dilemma. After the war we found this to be very near the truth: from Bruneval onwards they had expected us to use Window, and were mystified by our not employing it or any other technical countermeasure against the Würzburgs, even after we had started to jam the Freyas. They themselves made trials first on the Düppel Estate near Berlin and later over the Baltic in 1942. The effects were so dramatically disastrous that Goering ordered all the relevant reports to be destroyed, and he forbade any further mention of the matter in case the British might acquire the idea. So for more than a year both we and the Germans had hesitated to use Window against one another in the fear of losing on the exchange.

CHAPTER 14

The Psychology of Deception[1]

Ray Hyman

This article is possibly the most concise account of the psychology of deception, and is a valuable introductory survey of the history and theory of the entire field from con artists to gambling cheats and magicians. Professor Hyman concludes that a "strong" psychology of deception is possible, although he suggests a cross-disciplinary theory is apt to be most effective.

Ray Hyman, born in 1928, was inspired to take up magic at age 7 by a birthday box of tricks from his father. He worked his way through college as a professional mentalist, doing mind reading, palmistry, and hypnotism. In 1953, he earned a Ph.D. from Johns Hopkins University. In 1961, Hyman moved to the University of Oregon where he was a professor of psychology from 1964 until emerited. There, among other things, he taught a seminar on the "Psychology of Deception." He is a semi-professional platform magician and mentalist.

Hyman is a noted skeptical investigator of psychic phenomena and a founding member (since 1976) of the Committee for the Scientific Investigation of the Paranormal (CSICOP) and served on the editorial board of its journal, "The Skeptical Enquirer."

INTRODUCTION

Deception implies that an agent acts or speaks so as to induce a false belief in a target or victim. Even if we restrict the definition to cases of deliberate or intentional deception by the agent, the domain of activities and situations covered by deception is enormous and heterogeneous. Deception includes practical jokes, forgery, imposture, conjuring, confidence games, consumer and health fraud, military and strategic deception, white lies, feints and ploys in games and sports, gambling scams, psychic hoaxes, and much more.

These varied examples can be categorized as deception because they share *formal* properties. But when we speak about a psychology of deception, we

1. Ray Hyman, "The Psychology of Deception," *Annual Review of Psychology*, Vol. 40, 1989, pp. 133–154. Used, with permission, from the *Annual Reviews of Psychology*, Volume 40 © 1989 by Annual Reviews www.annualreviews.org.

raise the possibility that these examples also might share coherent psychological properties. When we refer to the psychology of perception or the psychology of memory we have in mind the possibility of a discipline that promises to hold together conceptually. Thus in one sense—what I will call the strong sense—"the psychology of deception" would be a field of inquiry that deals with coherent psychological phenomena.

No psychology of deception exists in this strong sense. Psychologists and other scholars do study, and speculate about, aspects of deception. Some have even offered preliminary theories about deception. But no one has managed so far to demonstrate that a single, coherent framework can meaningfully account for the psychological issues involved in the various types of deception.

Although we do not know if a psychology of deception in the strong sense is feasible, it is a likely possibility in a weak sense—i.e. a psychology of deception analogous to a psychology of work or psychology of law. In the latter cases, "psychology of" does not imply that either work or law can be characterized in terms of a coherent set of interrelated psychological propositions. Instead, the implication is that various aspects of work, for example, can be illuminated or understood in terms of various subfields of psychology. Employee selection and placement can be dealt with in terms of individual differences and psychometrics; training can be related to the psychology of learning; and other aspects can be related to principles and theories from such subareas as motivation, decision-making, problem-solving, and social psychology. Many aspects of deception can also be treated in terms of psychology's subareas. Indeed, such work is already taking place, especially in lie detection.

In this chapter I do not attempt to provide either systematic or comprehensive coverage of the domain of deception. Right up to the deadline for submission of this chapter, I kept discovering new caches of literature on deception. I have managed to read only a fraction of the articles I gathered in preparation for this task. Thus my objectives are to provide a preliminary overview of the domain and to suggest some of the questions and issues that might occupy a future psychology of deception.

HISTORICAL APPROACHES TO A PSYCHOLOGY OF DECEPTION

The Early Promise for a Psychology of Deception

The early years of psychology's existence as an independent science offered the strong possibility for a psychology of deception. During the late 1800s four major psychologists published articles on the topic. All four—Jastrow, Dessoir, Binet, and Triplett—used the conjurer as their model. Although those early accounts relied heavily on the prevailing associationistic psychology, the principles they illustrate have aged surprisingly well and harmonize with the contemporary view of cognitive psychology.

After 1900, however, the psychology of deception disappeared. A good guess is that it disappeared for the same reasons that attention, imagery, and other mentalistically oriented topics did. A psychology of deception is, of necessity, a mentalistic psychology. Human deception deals with the correspondences between internal representations and external reality. The behavioristic psychology that dominated American psychology from the early 1900s until the cognitive revolution in the late 1950s had no room for mentalism of any kind, including the intentionalism inherent in a psychology of deception.

Jastrow's Psychology of Deception

Joseph Jastrow (1900) included a chapter on "The Psychology of Deception" in his book *Fact and Fable in Psychology*. Jastrow pointed to the obvious illusory nature of knowledge based on sensory information. Helmholtz's controversial doctrine of unconscious inference formed the basis for much of Jastrow's argument. Our perceptions, according to this doctrine, are not based directly on sensory inputs, but rather on inferences that we unconsciously make from these inputs. Such a doctrine can easily accommodate illusory perceptions. Most contemporary cognitive psychologists and cognitive scientists accept, as almost axiomatic, that perception is highly inferential. However, the later Gestalt psychologists vigorously rejected the idea of unconscious inference. Today, those psychologists who call themselves neo-Gibsonians also strongly reject the idea that perception is indirect and inferential.

Those who reject the doctrine of unconscious inference do so because, among other things, they believe it fails to account for why perception, under most everyday conditions, is typically veridical. Whether or not the doctrine fails to account for veridical cognitions, it does provide a good account of nonveridical perception. In contemporary terms, Jastrow's position assumes that every cognition involves a contribution of the observer. The sensory input provides some constraints, but the perceiver automatically corrects for any deficiencies in the data by interpreting the inputs in terms of strong assumptions and expectations. Under ordinary and familiar circumstances these unconscious inferences serve as veridical guides to what is actually the case. But under special conditions they can badly lead us astray. This, in a nutshell, is Jastrow's argument. It accords well with contemporary cognitive accounts.

Jastrow explained illusions within this framework. He wrote that "we are creatures of the average; we are adjusted for the most probable event; our organism has acquired the habits impressed upon it by the most frequent experiences; and this has induced an inherent logical necessity to interpret a new experience by the old, an unfamiliar by the familiar." Jastrow here anticipated the ideas of both adaptation-level and schema theory.

Although many illusions result from the inferential nature of perception, Jastrow realized that unconscious inferences, unlike conscious ones, are inevitable. Even when we are aware that what we are experiencing is illusory, we are still subject to the illusion. However, our knowledge and experience enable us consciously to correct for such errors. But the ability to correct for such errors comes from having previously acquired the correct knowledge. This is why, according to Jastrow, children cannot adequately separate appearance from reality. He pointed out that when children view a spoon half immersed in water they believe it is really bent. It is only fairly recently that psychologists (such as Flavell) have begun studying the development of the distinction between appearance and reality.

Jastrow devoted the bulk of his chapter to the analysis of conjuring tricks. To some degree their success depends upon the audience's ignorance of the tricks' technical details, but it depends much more upon such psychological factors as misdirected attention, controlled expectations, and suggestiveness.

The conjurer's victims know, of course, that they are being deceived by skill and illusion. They also have nothing to lose. Jastrow contrasted this situation with the fraudulent spiritualistic practices of his day. "Spiritualistic phenomena present a perfect mine for illustrations of the psychology of deception." Unlike the conjurer's audience, the sitters at a seance believe that what they are witnessing may truly be paranormal. Jastrow believed it is this attitude that enables the medium to produce "miracles".

After presenting careful analyses of how the conjurer and spiritualistic medium succeed with their deceptions, Jastrow admitted that these deceptions may differ in important ways from con games and other deceptions. He suggested that the latter may be too varied and complex to be comprehended easily within the same framework. Jastrow ended on a note of optimism. He predicted that the spread of education, the increase in technical and other knowledge, and the understanding of the psychology of deception will greatly diminish the possibilities for deception.

Dessoir and Binet

Dessoir (1893) and Binet (1896) provided accounts similar to Jastrow's. All three psychologists carefully studied conjurers and their art. Both Binet and Jastrow enlisted the active cooperation of the most famous magicians of the period. Dessoir pointed out that anyone could purchase books and instructions from magic dealers. But knowing the mechanics and details of the trick neither protects you from being fooled by it nor enables you successfully to perform it. The real secret, according to Dessoir, lies in what he calls the "psychological kernel." Among other things this involves the ability to convince spectators that you really hold an orange in your left hand when it actually remains in your right. Dessoir claimed that this ability is at least partly inherited. The successful conjurer guides the thoughts of the onlookers to the desired conclusions. Today, we might call this the invited inference.

Dessoir, Jastrow, Binet, and Triplett acknowledged implicitly what many contemporary magicians make explicit. A conjurer should lead spectators into the desired inference rather than assert directly what he or she wants them to believe. For example, the amateur magician's statement, "I have here an ordinary deck of cards," violates the principle of the invited inference. Experienced magicians avoid calling attention to the "ordinariness" of the prepared deck. Instead, they simply produce it and handle it as if it were ordinary. Their onlookers spontaneously infer that it is what it seems to be.

Dessoir made one claim that Triplett (1900) questioned. He suggested that uneducated spectators were more difficult to deceive than the educated. Dessoir reasoned that the uneducated person, on guard against being presumed unintelligent, actively resists deception. The educated person, on the other hand, has no need to be defensive and therefore surrenders to the illusion. Triplett countered that uneducated persons might be more readily tricked because they lacked the technical knowledge to understand how the illusion was accomplished.

Regardless of who is right in this regard, the issue of whether some categories of people are more easily fooled than others has occupied both deceivers and scholars of deception. Conjurers, for example, believe that children below the age of five cannot be deceived with the same effects that succeed with adults. This belief is consistent with the findings and theories of contemporary developmental psychology.

Following Jastrow, Dessoir applied the principles of successful conjuring to an analysis of the successes achieved by fraudulent spiritualistic mediums.

Binet (1896) also based his analysis upon the laws of association and the doctrine that the mind interprets the information it gets from the senses. He argued that the study of prestidigitation was an important way to learn how the mind perceives. Using Sully's distinction between active and passive illusions, Binet classified conjuring illusions as passive. By this he meant that they were experienced in the same way by all normal minds.

In another sense, however, conjuring illusions require the cooperation of the public. The public attend the conjuring performance expecting to be deceived. Like Jastrow and Dessoir, Binet recognized that sleight-of-hand succeeds not by quickness but by misdirection. Binet pioneered in the use of photography to analyze sleight-of-hand.

Triplett

Triplett (1900) did his doctoral dissertation on the psychology of conjuring deceptions. A few years ago, Triplett's claim that conjuring originated in a universal instinct of deception would have appeared quaint. However, with the advent of sociobiology and the renewed interest in nonhuman deception, Triplett's discussion of deception in the context of natural selection no longer seems out of place.

Triplett believed that conjuring developed to help primitive man impress and gain control over his contemporaries. He characterized the instinct of deception as "grafted" upon the religious instinct. Human deception, according to Triplett, traces its roots back to such things as a protective mimicry in nature. He points out that there can be no mimicry without a creature to be deceived.

Triplett collected more than 300 observations of spontaneous fooling or deceiving by children, most by children under three years of age. Unfortunately Triplett supplied no details, so we do not know how sophisticated these deceptions were or if they would be classified as true deceptions by today's standards. Triplett also referred to "animistic tendencies of savage people" and to the power, wealth, and self-importance that deception provided the priest-conjurer.

Having shed its original priestly and supernatural context, modern conjuring is, according to Triplett, motivated by the same elements found in other types of play. His analysis of why conjuring succeeds relies on the same arguments used by Dessoir, Binet, and Jastrow. Triplett, however, sets out his analysis in a more systematic manner. He discusses the principles under the rubrics of Attention, Perception, and Suggestions and Association.

Under Attention, Triplett spells out the classic rules of conjuring: Never reveal in advance the nature of the effect, and never repeat the same trick twice on the same occasion (at least not using the same method). Both these rules recognize the fact that if spectators do not know in advance how to direct their attention then their chances of detecting how the trick was done are greatly lessened. Although the basis for these two rules seems obvious, we will see how they consistently play an important role in enabling fraudulent psychics to fool otherwise competent scientists.

Triplett concluded his discussion of the role of attention with a comment that surprisingly suits contemporary approaches to the psychology of deception. "We may summarize this section," he wrote, "with the observation that the fixed mental habits, evolved for useful purposes, to avoid being surprised and deceived, are the very agents employed by the conjurer to this end."

With slight changes in terminology several of the points made by Jastrow, Binet, Dessoir, and Triplett are consistent with the ideas of contemporary cognitive psychology. These include the role of attention and ways to control that attention, the role of expectations and suggestions, the difference between passive and active attention (today's top-down and bottom-up processing), schema theory, priming, and the idea that children before the age of five lack the cognitive structures and knowledge to understand the difference between appearance and reality.

Triplett and others, by the way, tested some of their ideas experimentally. Triplett performed an experiment in which he tossed a silver dollar, a tennis ball, or an apple into the air. This was done before pupils in grades four, five, and seven. The experimenter sat behind the teacher's desk, threw the ball into the air, caught it, and let his/her hands sink behind the desk. On the first trial the ball was tossed three feet in the air. On the second trial, it was tossed four to five feet. On the return of the ball after the second toss, it was dropped between the experimenter's legs and a third toss was mimed. The students were asked to write down what they saw on the third trial. Of the 165 subjects, 78 (47%) wrote that they had seen the ball go up on the third toss and then disappear. Triplett obtained even greater deception when he repeated the experiment under less brilliant lighting conditions.

Although these early forerunners of a psychology of deception focused on conjuring, in many ways their analyses anticipated other aspects of the contemporary scene, attempting to place deception in both biological and developmental contexts. Like some current researchers, too, they applied their findings to account for the success of fraudulent psychics.

THE CONTEMPORARY SCENE

The Psychological Data Base

My computer search of the APA data base under the topic of "deception" gleaned 722 references covering the period 1966–1986. Because this data base excludes books and articles in chapters, and because many journals are not covered, this outcome underestimates the number of relevant publications.

Using abstracts, I categorized the articles. Pieces on lying and lie-detection constituted 39% of the total. Articles dealing with the ethical and methodological issues involved in using deception in psychological experiments comprised 16% of the total. Other categories with relatively large representation involved developmental approaches to deception, evaluation of lie scales on personality inventories, and self-deception (mostly philosophical treatments). The remaining articles dealt with animal deception, legal matters, therapy, the Machiavellian Scale, cheating on tests, and the pathology of lying. The number of articles in the ethics-of-deception-in-experiments category hit a peak of approximately 15 per year in the middle 1970s and has subsequently declined to approximately 3–5 per year. The number of articles on lying and lie detection has tended to increase each year during this period.

Informal Survey of the Remaining Literature on Deception

The psychological data base contains only a small fraction of the literature on deception. Many important psychological articles on deception have appeared in sources not covered by the data base. Many major categories of deception (e.g. military and strategic deception) are not represented or are poorly represented (e.g. nonhuman deception). Several publications on swindles and confidence games, medical and psychiatric aspects of deception, fraud in psychical research, sociology of deviance, and self-deception escaped inclusion in the psychology data base. Indeed, I keep finding new sources for references on deception.

CATEGORIES AND EXAMPLES OF DECEPTION

In the absence of a systematic framework for, or a theory of, deception I cannot formally justify including the examples below and the exclusion of others. Nevertheless, I hope my choices fairly sample the varieties of deception and highlight a representative set of questions and issues.

Nonhuman Deception

Some authors might argue that nonhuman deception is a contradiction in terms because deception is uniquely human. Such a position depends partly on definition and partly on assumptions about animal cognition. A minimal definition of deception (cf Mitchell 1986) might include the following: an organism *S* emits a signal (behavior or communication) that is registered by an organism *R*. *R* believes that the signal indicates that *X* is the case and behaves appropriately, when in fact *X* is not the case. Mitchell would add that *S* should desire or benefit from *R's* behaving as if *X* were the case. In this form, Mitchell's definition easily accommodates many kinds of nonhuman activity. However, many writers insist that *S* should *intend* that *R* falsely believe that *X* is the case.

For many, the requirement that the false belief be intentionally created would rule out nonhuman forms of "deception"—e.g. mimicry, deceptive signaling among fireflies, and the like. Most of us would not concede to infrahuman creatures the required level of cognitive complexity. But what about pongoid apes such as chimpanzees? Here the situation is not so clear. Chevalier-Skolnikoff (1986) among others has argued that adult chimpanzees develop the cognitive capacity for at least some forms of intentional deception. Investigators have reported many examples of apparent deception among nonhumans (cf Mitchell & Thompson 1986). So far, however, determined skeptics have managed to account for the observed behavior without resort to intention (Lewin 1987). On the other hand, philosophers such as Dennett, working closely with naturalists, believe they can devise criteria that will detect intentional deception if it occurs among nonhumans (Dennett 1987).

Regardless of whether nonhumans can intentionally deceive, they do engage in intricate forms of behavior that function as deceptions. Do these deceit-like behaviors have any relation to human deception? If not, what makes human deception unique? What is the biological basis, if any, of human deception?

In this chapter I can only supply a few examples to suggest the nature and scope of the problem.

ASHLEY'S DOG The philosopher Dennett (1978) wanted to know if animals other than humans were second-order intentional systems. An intentional system in the philosophical sense is one that has mental states (such as desires and beliefs) about things other than themselves. Dennett has little doubt that many nonhumans are first-order intentional systems in this sense. A second-order intentional system is one that can have beliefs, desires, and other intentions about other beliefs, desires, and intentions. To *intentionally* deceive someone, for example, the deceiver must be a second-order intentional system (I use "intentionally," here, in both the philosophical and everyday senses.) The deceiver must have beliefs about the intended victim's beliefs.

Dennett requested friends to supply him with examples of nonhuman second-order intentional systems. Peter Ashley provided this compelling instance:

> One evening I was sitting in a chair at my home, the *only* chair my dog is allowed to sleep in. The dog was lying in front of me, whimpering. She was getting nowhere in her tyring to "convince" me to give up the chair to her. Her next move is the most interesting, nay, the *only* interesting part of the story. She stood up, and went to the front door where I could still easily see her. She scratched the door, giving me the impression that she had given up trying to get the chair and had decided to go out. However, as soon as I reached the door to let her out, she ran back across the room and climbed into her chair, the chair she had "forced" me to leave (Dennett 1978, pp. 274–75).

If we attribute to the dog the *intention* that Ashley *believe* she *wants* to go out, then we are dealing not merely with a second-order, but with at least a third-order intentional system. Dennett concedes, however, that we can provide a lower-level account of the dog's behavior. The dog might simply be a behaviorist, associating her trip to the door with Ashley's leaving the seat. The dog might have this first-order belief without harboring intentions or beliefs about Ashley's mental states.

A simpler account might be as follows. The dog has a hierarchy of desires. Her currently strongest desire is to get into her chair, but this is blocked because her master is sitting in it. Consequently, her next highest desire is activated: she tries to get outside. When Ashley vacates the chair to let her out, however, she suddenly becomes aware that her first desire can now be fulfilled.

THE BROKEN-WING DISPLAY Dennett compares the preceding example with the case of low-nesting shore birds such as the plover, which appear to feign a broken wing to lure a predator away from the nest. If anything, such deceptive behavior appears more sophisticated than that exhibited by Ashley's dog. Even more intricate and devious is the "deception" that occurs both within and between species of fireflies (Lloyd 1986).

No one suggests that plovers or fireflies intentionally conceive and carry out their "deceptions," using cognitive operations similar to those of humans. Indeed, we tend to dismiss such seemingly complicated social interactions as instinctive. But as Dennett (1987) suggests, the behavior of these nonhumans is in some sense no less intelligent. The sophisticated deceptions work and contribute to the survival of the species. In these instances the "intelligence" is not located in the individual cognitive system of each plover and each firefly. Rather, it originates in the fine-tuning of inherited capacities by natural selection over thousands or millions of generations.

The preceding paragraph implies that we do not need intention—at least not higher-order intention—for successful deception. If this is so, can human deception be explained without recourse to intention? Or is human deception qualitatively different from the deceptions that pervade the nonhuman animal world?

DECEPTION AMONG CHIMPANZEES One place to look for continuity between nonhuman and human deception is among the pongoid apes, our closest animal relatives. Those who study chimpanzees have looked for examples of intentional deception. As De Waal (1986) acknowledges, however, it is almost impossible to prove unequivocally that a given act was intentional. Investigators have observed apparently deceptive behaviors that seem intentional. De Waal (1986) classifies such behaviors as: (*a*) camouflage, (*b*) feigning interest, (*c*) feigning a mood, (*d*) signal correction, and (*e*) falsification.

De Waal (1986) supplies a typical example of chimpanzee deception observed at the Arnhem zoo. Dandy, a young male, was seen walking over a spot where some grapefruits had been buried. Because Dandy did not stop or look around as he passed the spot, the observers thought he had not noticed the grapefruits. However, some three hours later, when the other chimpanzees were asleep, Dandy ran directly to the spot, dug the grapefruits up, and ate them.

De Waal acknowledges that alternative scenarios can account for such behavior without attributing deceptive intent to the chimpanzee. He makes the

out, however, she suddenly becomes aware that her first desire can now be fulfilled.

THE BROKEN-WING DISPLAY Dennett compares the preceding example with the case of low-nesting shore birds such as the plover, which appear to feign a broken wing to lure a predator away from the nest. If anything, such deceptive behavior appears more sophisticated than that exhibited by Ashley's dog. Even more intricate and devious is the "deception" that occurs both within and between species of fireflies (Lloyd 1986).

No one suggests that plovers or fireflies intentionally conceive and carry out their "deceptions," using cognitive operations similar to those of humans. Indeed, we tend to dismiss such seemingly complicated social interactions as instinctive. But as Dennett (1987) suggests, the behavior of these nonhumans is in some sense no less intelligent. The sophisticated deceptions work and contribute to the survival of the species. In these instances the "intelligence" is not located in the individual cognitive system of each plover and each firefly. Rather, it originates in the fine-tuning of inherited capacities by natural selection over thousands or millions of generations.

The preceding paragraph implies that we do not need intention—at least not higher-order intention—for successful deception. If this is so, can human deception be explained without recourse to intention? Or is human deception qualitatively different from the deceptions that pervade the nonhuman animal world?

DECEPTION AMONG CHIMPANZEES One place to look for continuity between nonhuman and human deception is among the pongoid apes, our closest animal relatives. Those who study chimpanzees have looked for examples of intentional deception. As De Waal (1986) acknowledges, however, it is almost impossible to prove unequivocally that a given act was intentional. Investigators have observed apparently deceptive behaviors that seem intentional. De Waal (1986) classifies such behaviors as: (*a*) camouflage, (*b*) feigning interest, (*c*) feigning a mood, (*d*) signal correction, and (*e*) falsification.

De Waal (1986) supplies a typical example of chimpanzee deception observed at the Arnhem zoo. Dandy, a young male, was seen walking over a spot where some grapefruits had been buried. Because Dandy did not stop or look around as he passed the spot, the observers thought he had not noticed the grapefruits. However, some three hours later, when the other chimpanzees were asleep, Dandy ran directly to the spot, dug the grapefruits up, and ate them.

De Waal acknowledges that alternative scenarios can account for such behavior without attributing deceptive intent to the chimpanzee. He makes the cognitive competence around age five, but several years of experience and practice are needed before they can apply their deceptive competence effectively.

Jastrow (1900) and Triplett (1900), in agreement with conjurers, wrote that magicians could not fool young children in the same way that they could fool adults. They suggested this was because children had not yet developed notions of causality. When the conjurer appears to remove and then restore an assistant's head, older children and adults are baffled. Younger children, however, lacking both causal structures and knowledge of physiology, do not realize that a genuine decapitation would be fatal. The conjurer's problem is not that young children cannot be fooled, but that they cannot recognize and appreciate being fooled.

Young children, as we have seen, do attempt deceptions, but these are transparent and do not succeed. Presumably this is because they cannot fully appreciate how the situation appears to their intended victim. To deceive successfully, the deceiver must be able to take the perspective of the victim.

Both the appearance-reality distinction and the capacity for perspective-taking have been investigated by developmental psychologists. The evidence indicates that children of three years and younger cannot distinguish appearance from reality or take the perspective of another (Flavell et al 1983; Taylor & Flavell 1984). The same evidence suggests that children have acquired both of these cognitive skills by the time they reach the first grade. Indeed, Flavell et al (1983) show that these two abilities are intercorrelated and develop together.

However, the total set of cognitive tools needed for deception may be much larger and not in place until early adolescence. Newman (1986) argues that investigators may have failed to distinguish between cooperative communication, based on shared knowledge, and perspective-taking. To communicate successfully on the basis of assumed mutual knowledge, he suggests, does not require the child actively to represent the situation as perceived by the other. However, to deceive intentionally, the child must actively represent the victim's situation and keep it separate from the true situation.

Chandler (1988) believes that first-graders can appreciate that others may interpret a *situation* differently from the way they themselves do, but this need not imply a full recognition that two individuals can construe the *same information* differently. The first-grader might assume that another individual lacks the information needed to interpret the situation properly, and that if the missing information became available the other would, of necessity, find the single true interpretation. According to Chandler, only in early adolescence does the child fully appreciate the possibility that, given the same information, the other person can still construe the situation differently. Chandler's analysis seems to imply that it is not until early adolescence that children achieve the capacity for intentional deception.

The theory that children achieve the cognitive structures necessary for intentional deception in early adolescence may help account for the many and striking instances of psychic fraud by young teenagers. Nicol (1979) reviews many examples of such adolescent fraud, ranging from the fraudulent pretense to being manipulated by witches to the creation of poltergeist phenomena. I describe once such case in the next section.

Psychic Fraud

From its inception, psychical research has been plagued by problems of fraud (Hyman 1981a, 1985). When we read about how a fraudulent medium or psychic fooled an investigator (often a trained scientist), the success of the trick seems incredible. We react similarly to tales of confidence swindles and scientific hoaxes. But our assumption that we, unlike the victimized investigator, would have seen through the trick is based on hindsight and is probably unwarranted. Reading about the swindle differs essentially from being involved at the time of its occurrence.

Zimbardo et al (1977) refer to this attitude as "the illusion of personal invulnerability" or the "not me syndrome." We express it when we disparage the victims of deception and hoaxes as gullible.

> Such an orientation is dangerous because it alienates us from the human condition. By setting ourselves apart from others, we do not learn the important message from their experience—namely, that the source of their suffering, loss of face, or fall from grace may lie not in their personal weaknesses nor in the stars, but in *the power of the situation* (p. 3).

Rather than dismiss cases of successful deception as due to naiveté, we should look upon them as opportunities for discovering those situational and other factors that enabled the deception to succeed. Delanoy (1987) provides some valuable insights into factors that make the psychic pretender's task easier than it might otherwise be. She supplies us with an unusually frank account of her experiences in testing an alleged psychic who participated in over 20 testing sessions at the University of Edinburgh.

Tim was 17 years old at the time of the tests. He claimed that he had begun bending metal with the power of his mind at age four. Delanoy describes him as "an almost ideal subject." He was cooperative, seemed willing to submit to any controls, and even suggested additional controls. During the first few sessions, which were deliberately informal so as to allow the subject to acclimate to the laboratory conditions, Tim managed to bend several objects, but never when he was being carefully observed. For the next four sessions, the investigators employed controls suggested to them by the magician James Randi. The results were similar. Tim produced bent objects but never when the strict controls were in force.

During sessions 9 and 10 a visiting parapsychologist tested Tim on a special PK (psychokinesis) machine alleged to be fraud-proof. Tim succeeded on some trials. Although the parapsychologist had failed to verify that all the machine's controls had been operative during Tim's successes, he deemed the results promising. Once Tim brought a sealed plastic cube to the laboratory. He claimed to have psychically altered a metal band inside the cube. The investigators mailed the cube to Randi. Randi returned the cube with a description of how it had obviously been tampered with.

Further sessions with Tim produced similar inconclusive results.

After having experienced several sessions with Tim, Delanoy listed several reasons for being suspicious of his abilities. He had never bent metal under controlled conditions. Not only did the evidence suggest that he had tampered with the plastic cube, but, in relation to the same incident, he had apparently lied about his witnesses. In addition, he had shown no interest in trying his abilities with a similar, but fraud-proofed cube. He also had shown suspicious behavior in connection with some other tests.

Despite these reasons, Delanoy was reluctant to give up on the possibility that Tim had real powers.

> However, there were also several factors which would point towards the subject's genuineness. He had succeeded in producing positive results on JI's piezo-electric crystal organ machine, at a time when the artifact controls were activated (although they had not been adequately tested for proper functioning). Indeed, JI did find the subject's results to be most promising. Other factors included: (1) he had suggested a method of metal-bending (placing a cupped hand over a small object placed on a flat surface) which was approved by Mr. Randi as being "fraud-proof"; (2) he had suggested filming his metal-bending in the ganzfeld using an infrared camera; (3) he was always very willing to work with any controls which we suggested; (4) he had invested a great deal of time in working with us and seemed quite willing to continue doing so; and (5) his general demeanor was consistently cooperative and openly friendly. This last factor requires some elaboration. During the time we had worked together I felt I had come to know Tim fairly well. We had established what appeared to be an honest, friendly and trusting rapport. If he was fraudulent I thought he either deserved an Academy Award for a most convincing performance or that he was prone to severe self-delusions, at the least (p. 252).

Delanoy refers to the generally accepted ideal in psychical research of having "a friendly and open rapport" with the subjects. She points out that parapsychologists "need" their subjects. Without them, of course, they have no basis for collecting evidence. Finally, she acknowledges she was probably biased towards believing Tim.

Although the Edinburgh parapsychologists devoted 20 sessions and approximately 60 hr to investigating Tim, they were spared the embarrassment of endorsing his "powers." They had sought and followed the advice of magician and psychic investigator Randi about how to control for possible trickery. Tim never managed to produce results under the strictly controlled conditions.

During the final sessions, the Edinburgh investigators decided to try to resolve their doubts about Tim. Up to that point they had never hidden anything in their procedures from Tim. But for these sessions they set up a hidden camera. The hidden camera produced evidence of blatant trickery on Tim's part. When confronted with this evidence, Tim confessed. He explained how he had practiced conjuring since childhood and had deliberately offered his services to the parapsychologists to see if he could trick them.

Delanoy believes her experiences with Tim are typical. Many believe they can judge the honesty of their subjects. The evidence, however, indicates that despite such confidence, humans cannot accurately distinguish honest from dishonest communications (DePaulo et al 1985; Zuckerman et al 1986). Although she had several objective reasons for suspecting Tim's claims, she managed to find even more compelling reasons to believe them. She could not believe, for example, that a 17-year-old student could dissimulate so convincingly, and she'd been misled by Tim's suggesting fraud-proof tests, even though he never succeeded at them.

Tim's metal-bending "talents" had been stimulated by the highly publicized feats of the alleged psychic Uri Geller in the 1970s. Geller was investigated and endorsed by many prominent scientists (for useful accounts see Marks & Kammann 1980; Panati 1976; Randi 1982). Yet, not one of these scientists ever observed Geller to bend metal under controlled conditions (Hyman 1976). Why were otherwise competent scientists willing to endorse the paranormal powers of Geller when he, like Tim, could not perform under properly controlled conditions?

Insights into how this could happen come from an article published in *Nature* by a team of prestigious British physicists who had observed Geller and were obviously impressed (Hasted et al 1975). They admitted that they had not observed Geller's feats under rigorous, fraud-proof conditions, but pointed out that in investigating paranormal phenomena it was counterproductive to insist upon rigor at all costs. Unlike the typical physical phenomena of the scientific laboratory, the alleged psychokinetic powers of Geller emanated from a human mind. This implied that the relationship between the experimenter and the "psychic" was very important. Indeed, this was why they rejected having a magician present during the experiments: Magicians tended to be skeptical and generated a negative attitude that blocked the occurrence of psychic phenomena.

These same investigators also found that it was important for all present to be relaxed and to seriously want the "psychic" to succeed. Just as important, they added, was open-mindedness. They found phenomena more likely to happen when no one was looking attentively for any specific outcome. "In addition, matters seem to be greatly facilitated when the experimental arrangement is aesthetically or imaginatively appealing to the person with apparent psychokinetic powers."

Hasted et al conceded that such conditions make it difficult to prevent trickery. But they optimistically believed this could nevertheless be done.

> How, then, are we to avoid the possibility of being tricked? It should be possible to design experimental arrangements which are beyond any reasonable possibility of trickery, and which magicians will generally acknowledge to be so. In the first stages of our work we did in fact present Mr. Geller with several such arrangements, but these proved aesthetically unappealing to him (Hasted et al 1976, p. 194).

Cynics may note how convenient it was for Geller that fraud-proofing turned out to be "aesthetically unappealing," and that the conditions Hasted et al listed as conducive to Geller's success are just those conditions that, according to Binet, Dessoir, Jastrow, and Triplett, enable the conjurer to fool onlookers.

To comprehend how such prominent scientists can paint themselves into such a corner we must view the situation from their perspective. When they wrote their article, they had already become convinced of Geller's paranormal powers. They realized that no such powers had ever yet survived scrutiny by scientific methods. From their perspective, then, the major task was to find a way to keep the powers they credited from fading under investigation. If they could find conditions that enabled the "psychic" to produce his phenomena reliably in the laboratory, then they could later bring in the skeptics and use more traditional scientific methods.

The flaw in this strategy, of course, is that if the alleged psychic is a fraud, the investigators have played into his or her hand.

Other Cases

I briefly mention a few other types of deceptions in this section.

DECEPTION IN SCIENCE Deception in science has recently become a topic of interest (Bechtel & Pearson 1985; Broad & Wade 1982; Kohn 1986). The focus tends be on the possible causes of data manufacturing, plagiarism, and other deceptions by seemingly talented scientists. But an equally interesting question is how deceptions—originating within science or without—succeed in fooling competent scientists. The preceding section discussed how scientists have been fooled by fraudulent psychics.

An interesting example of another sort is that of Piltdown Man (Blinderman 1986; Gould 1980; Millar 1974; Weiner 1980). Who carried out this hoax? The obvious suspect is Charles Dawson, the man who originally "discovered" the fossils and, along with Sir Arthur Smith Woodward, announced the discovery of an ancient fossil man with a very human braincase and a decidedly ape-like jaw. From 1912 until 1953, Piltdown Man was accepted as a legitimate fossil and, at least in the earlier period, was considered the likely ancestor of modern man. Many writers, unable to believe that a country solicitor and amateur scientist could so successfully have fooled the greatest archaeologists, continue to look for a suspect with the knowledge and skills to carry out the hoax. This search for a more worthy deceiver seems motivated by the same elitism that refuses to accept Shakespeare as the author of great literature.

In fact, though, the Piltdown hoax was not skillfully executed. When Weiner first raised the question that fraud might have been involved, he asked himself on what evidence everyone had accepted the association of the ape-like jaw with the Piltdown skull. As an anatomist, he decided that the only basis for accepting this match was the fact that the molars were worn flat—a feature never previously witnessed in ape's teeth. Weiner and Le Gros Clark made a list of the features that would be exhibited by molars that had been worn flat naturally. When they went to London to examine the Piltdown jaw, they found that the molars deviated from the criteria on every count (Le Gros Clark 1956). For example, under natural stresses the first molar is normally more severely worn than the second. The first and second molars of the Piltdown jaw were worn to the same degree. The molar teeth of the lower jaw normally wear down more severely on the outer than the inner side. The Piltdown molars were worn down in exactly the opposite direction. Because enamel and dentine wear at different rates, the worn molars normally exhibit shallow concavities. The surfaces of the Piltdown molars were worn flat without any differences between the enamel and the dentine. With normal wear, the edges of the molars are rounded and beveled. The Piltdown molars had edges that were sharp and unbeveled. In addition, microscopic inspection revealed that the surfaces of the molars were scored with telltale crisscross scratches.

These and other obviously artifactual signs were glaringly obvious to Weiner and Le Gros Clark.

> Indeed, so obvious did they seem that it may well be asked how was it that they had escaped notice before? The answer is really quite simple—they had never been looked for. The history of scientific discovery is replete with examples of the obvious being missed because it had not been looked for, and the present instance is just one more example: nobody previously ever examined the Piltdown jaw with the idea of a possible forgery in mind (Le Gros Clark 1956, p. 145).

The Piltdown forgery succeeded not so much because of the skill of the forger, but because scientists, like other humans, are heuristic systems. They focus on what matters and do not waste time on what can be taken for granted. Such a heuristic works well most of time. But when scientists deal with a nature that is intelligent and responsive to their activities, then their normal mode of operation makes them especially vulnerable to deception.

SWINDLES AND CONFIDENCE GAMES The literature on swindling and confidence games is enormous and rich in instructive examples. The classic work on confidence games is Maurer's *The American Confidence Man* (1974), a slightly updated version of his 1944 book *The Big Con*. Additional sources of information include Henderson (1985), Nash (1976), and Ortiz (1984). Hankiss (1980) and Leff (1976) attempt to provide theoretical frameworks for understanding confidence swindles. Leff's little-known book is particularly successful in using dramaturgical, social psychological, and microeconomic theory to provide a coherent account of con games, swindling, and deceptive selling techniques.

Accounts of confidence games, because of their often bizarre characteristics, frequently elicit variants of the "not me syndrome." Listeners or readers cannot believe they can be victimized by such scams. Yet the victims are often bankers, successful business people, and even con artists themselves. In 1925 the famous confidence man Victor Lustig managed to sell the Eiffel Tower to successful Parisians not once, but twice. How he did this is particularly instructive (Mason 1978).

LYING AND LIE-DETECTION I cannot attempt here to review the literature on lying and lie-detection. The research and the literature on this topic contribute by far the largest number of articles to the psychological data base. The issues are important and relevant to many ethical, legal, and other societal concerns. For this reason the topic is characterized by vigorous controversies. The issues are complex, and this topic deserves a chapter of its own in some future volume. The following references provide an introduction to this literature: Ben-Shakhar et al 1982; DePaulo et al 1985; Ekman 1985; Lykken 1981, 1985, 1987a,b; Raskin 1988; Raskin & Kricher 1987; Waid & Orne 1982; Zuckerman et al 1986.

ADDITIONAL TOPICS Practical jokes (Smith 1956), self-deception (Krebs et al 1988; Sackeim & Gur 1985; Quattrone & Tversky 1984), the Munchausen Syndrome (Pankratz 1981), and strategic and military deception (Daniel & Herbig 1982; Gooch & Perlmutter 1982) are among a number of additional topics that provide instructive examples of deception.

BORDERLINE AND RELATED CASES

Without a theoretical framework, it is difficult to decide what elucidates deception. Many types of deception are interesting because they occur both as deliberate deceptions and as innocent mistakes or self-deceptions. Most character readers, for example, are as much victims of subjective validation as are their clients (Hyman 1977, 1981). But a subset of these practitioners deliberately employ deception to gain information and convince their clients of their powers. The metal bending of Uri Geller and his many imitators is the result of deliberate trickery. But the metal bending that occurs at spoon-bending parties (Houck 1984) typically results from self-delusion. The owner of the famous horse Clever Hans (Pfungst 1911) apparently was unconsciously cuing his horse, but most owners of allegedly intelligent or mind-reading animals deliberately use deceptive cuing. In addition to these ambiguous cases, there is the enormous class of human errors, mistakes, delusions, and self-deceptions.

SUMMARY AND CONCLUSIONS

When I undertook to write this chapter, I thought hopes for a psychology of deception in the strong sense were unrealistic. A psychology of deception in the weak sense, however, seemed well within reach. One value of trying to apply contemporary psychological theories to the understanding of deception could be the testing and sharpening of the theories.

I tried not to prejudge the issue. I avoided providing a systematic framework for organizing the material in this chapter, instead, simply discussing salient topics in deception. My purpose was to display the richness of the domain and the heterogeneity of issues that a potential theory of deception might have to deal with.

Although I have managed to read only a small fraction of the literature, I have become optimistic about the possibility of a coherent psychology of deception in the strong sense. (This optimism may itself be a form of self-deception.) I have discovered articles that go beyond their specific area of deception and attempt to generalize to other areas—e.g. DePaulo et al (1985), Krebs et al (1988), Flavell et al (1983), Chandler (1988), Heuer (1982), Whaley (1982), Leff (1976), and the collection edited by Mitchell & Thompson (1986). The works of philosophers such as Dennett (1987) and Freyd's (1983) study of the consequences for cognitive structures of shared knowledge are examples of theoretical efforts highly relevant to a future psychology of deception.

A more theoretical chapter than mine might have discussed the fact that most of what we know about deception belongs to "folk psychology." Some contemporary philosophers have expressed doubts about constructing a psychology on the basis of such information (Churchland 1984). Other philosophers believe that using folk psychology as our point of departure may not be such a bad idea (Flanagan 1984; Goldman 1986). Such an issue matters for a psychology of deception, because the folk psychologies of deceivers in various domains might well be what a psychology of deception might be trying to explain.

What would a psychology of deception in the strong sense look like? Would it be biologically based? Or would it be functional and close to existing folk psychologies? Deception is obviously interpersonal in nature. At one level the units would be individuals interacting in a social context. Does this mean that a theoretical account of deception has to use concepts and units from social sciences that focus on levels of function broader than those psychologists deal with?

Perhaps it is premature to try to answer such questions. But it does seem that some aspects of successful deception can be explained in terms of concepts that are basically biological. Other aspects seem to require cognitively oriented accounts. Many of the more interesting deceptions (e.g. confidence games) seem to require accounts at a social level. Whether a single coherent theory can provide an adequate account of this range of phenomena remains to be seen. One possibility is that such a theory will develop concepts and structures of its own that will cut across levels currently kept separate by existing disciplines.

Literature Cited

Bechtel, H. K., Pearson, W. 1985. Deviant scientists and scientific deviance. *Deviant Behav.* 6:237–52

Ben-Shakhar, G., Lieblich, I., Bar-Hillel, M. 1982. An evaluation of polygraphers' judgments: a review from a decision theoretic perspective. *J. Appl. Psychol.* 67:701–13

Blinderman, C. 1986. *The Piltdown Inquest.* Buffalo, NY: Prometheus Books

Binet, A. 1896. Psychology of prestidigitation. *Annual Report of the Board of Regents of the Smithsonian Institution, 1894,* pp. 555–71. Washington, DC: GPO

Broad, W., Wade, N. 1982. *Betrayers of the Truth.* New York: Simon & Schuster

Chandler, M. 1988. Doubt and developing theories of mind. In *Developing Theories of Mind,* ed. J. Astington, P. Harris, D. Olson. New York: Cambridge Univ. Press. In press

Chevalier-Skolnikoff, S. 1986. An exploration of the ontogeny of deception in human beings and nonhuman primates. See Mitchell & Thompson 1986, pp. 205–20

Churchland, P. 1984. *Matter and Consciousness.* Cambridge, MA: MIT Press

Daniel, D. C., Herbig, K. L., eds. 1982. *Strategic Military Deception.* New York: Pergamon

De Waal, F. 1986. Deception in the natural communication of chimpanzees. See Mitchell & Thompson 1986, pp. 221–44

Delanoy, D. L. 1987. Work with a fraudulent PK metal-bending subject. *J. Soc. Psychical Res.* 54:247–56

Dennett, D. C. 1978. *Brainstorms: Philosophical Essays on Mind and Psychology.* Montgomery, VT: Bradford Books

Dennett, D. C. 1987. *The Intentional Stance.* Cambridge, MA: MIT Press

DePaulo, B. M., Stone, J. I., Lassiter, G. D. 1985. Deceiving and detecting deceit. In *The Self in Social Life,* ed. B.R. Schlenker, pp. 323–70. New York: McGraw-Hill

Dessoir, M. 1893. The psychology of legerdemain. *The Open Court,* 7:3599–3602, 3608–11, 3616–19, 3626–27, 3633–34

Ekman, P. 1985. *Telling Lies: Clues to Deceit in the Marketplace, Politics, and Marriage.* New York: W. W. Norton & Co.

Flanagan, O. J. 1984. *The Science of The Mind.* Cambridge, MA: MIT Press

Flavell, J. H., Green, F. L., Flavell, E. R. 1983. Development of knowledge about the appearance-reality distinction. *Monogr. Soc. Res. Child Dev.,* Ser. 212, p. 51 (1)

Freyd, J. J. 1983. Shareability: the social psychology of epistemology. *Cognit. Sci.* 7: 191–210

Goldman, A. I. 1986. *Epistemology and Cognition.* Cambridge, MA: Harvard Univ. Press

Gooch, J., Perlmutter, A., eds. 1982. Military deception and strategic surprise. *The J. Strategic Stud.* 5 (1):(Spec. iss.)

Gould, S. J. 1980. The Piltdown conspiracy. *Nat. Hist.* 89:8, 10–11, 14, 16, 18, 20, 22, 25–26, 28

Hankiss, A. 1980. Games con men play: the semiosis of deceptive interaction. *J. Commun.* 30:104–12

Hasted, J. B., Bohm, D. J., Bastin, E. W., O'Regan, B., Taylor, J. G. 1975. Scientists confronting the paranormal. *Nature* 254:470–73. Reprinted 1976 in *The Geller Papers,* ed. C. Panati, pp. 183–96. Boston, MA: Houghton-Mifflin

Henderson, M. A. 1985. *How Con Games Work.* Secaucus, NY: Citadel

Heuer, R. J. 1982. Cognitive factors in deception and counterdeception. See Daniel & Herbig 1982, pp. 31–69

Houck, J. 1984. PK party history. In *Proceedings of the Symposium on Applications of Anomalous Phenomena,* ed. C. B. S. Jones, pp. 501–14. Alexandria, VA: Kaman Tempo

Hyman, R. 1976. [Review of *The Geller papers.*] *The Zetetic (The Skeptical Inquirer)* 1:73–80

Hyman, R. 1977. Cold reading: how to convince strangers that you know all about them. *The Zetetic (The Skeptical Inquirer)* 1:18–37

Hyman, R. 1981a. Scientists and psychics. In *Science and the Paranormal: Probing the Essence of the Supernatural,* ed. G. O. Abell, B. Singer, pp. 119–41. New York: Charles Scribner's Sons

Hyman, R. 1981b. The psychic reading. In *The Clever Hans Phenomenon: Communication with Horses, Whales, Apes and People,* ed. T. A. Sebeok, R. Rosenthal, pp. 169–81. New York: The NY Acad. Sci.

Hyman, R. 1985. A critical historical overview of parapsychology. In *A Skeptic's Handbook of Parapsychology,* ed. P. Kurtz, pp. 3–96. Buffalo, NY: Prometheus Books

Jastrow, J. 1900. *Fact and Fable in Psychology.* Cambridge, MA: Riverside Press

Kohn, A. 1986. *False Prophets: Fraud and Error in Science and Medicine.* New York: Blackwell

Krebs, D., Denton, K., Higgins, N. C. 1988. On the evolution of self-knowledge and self-deception. In *Sociobiological Perspectives on Human Development,* ed. K. McDonald, pp. 105–29. New York: Springer-Verlag. In press

Leff, A. A. 1976. *Swindling and Selling.* New York: The Free Press

Le Gros Clark, W. 1956. The exposure of the Piltdown forgery. *Proc. R. Inst. GB* 36:138–51

Lewin, R. 1987. Do animals read minds, tell lies? *Science* 238:1350–51

Lloyd, J. E. 1986. Firefly communication and deception: "Oh, what a tangled Web." See Mitchell & Thompson 1986, pp. 113–28

Lykken, D. T. 1981. *A Tremor in the Blood: Uses and Abuses of the Lie Detector.* New York: McGraw-Hill

Lykken, D. T. 1985. The probity of the polygraph. In *The Psychology of Evidence and Trial Procedure,* ed. S. M. Kassin, L. S. Wrightsman, pp. 95–123: Beverly Hills, CA: Sage

Lykken, D. T. 1987a. The validity of tests: caveat emptor. *Jurimetrics* 27:263–70

Lykken, D. T. 1987b. Reply to Raskin & Kircher. *Jurimetrics* 27:278–82

Triplett, N. 1900. The psychology of conjuring deceptions. *Am. J. Psychol.* 11:439–510

Vasek, M. E. 1986. Lying as a skill: the development of deception in children. See Mitchell & Thompson 1986, pp. 271–92

Waid, W. M., Orne, M. T. 1982. The physiological detection of deception. *Am. Sci.* 70:402–9

Weiner, J. S. 1980. *The Piltdown Forgery.* New York: Dover

Whaley, B. 1982. Toward a general theory of deception. *J. Strategic Stud.* 5:178–92

Zimbardo, P. G., Ebbesen, E. B., Maslach, C. 1977. *Influencing Attitudes and Changing Behavior.* Reading, MA: Addison-Wesley. (2nd ed.)

Zuckerman, M., DePaulo, B. M., Rosenthal, R. 1986. Humans as deceivers and lie detectors. In *Nonverbal Communication in the Clinical Context,* ed. P. D. Blanck, R. Buck, R. Rosenthal, pp. 13–35. University Park, PA: Penn. State Univ. Press

CHAPTER 15

"Killing No Murder" The Strategy of Indirect Approach[1]

B.H. Liddell Hart

This letter by Liddell Hart, dated "12th November, 1927," ends with the first appearance in print of his theory and term, "The Strategy of Indirect Approach." There he previews the key elements of this theory in phrases and even whole sentences that will echo through all his later writings:

> *More and more clearly has the fact emerged that a direct approach to the object or objective along the 'line of natural expectation' has ever tended to negative results. The reason being that the strength of an enemy country or force lies far less in its numbers or resources than in its stability or equilibrium — of control, moral, and supply. ... To move along the 'line of natural expectation' is to consolidate the enemy's equilibrium, and by stiffening it to augment its resisting power.*

By stressing "expectation," Liddell Hart surely implies our enemy's perception—estimate or preconception—of our most likely course of action.

> *In contrast, the decisive victories in military history have come from the strategy of indirect approach, wherein the dislocation of the enemy's moral, mental or material balance is the vital prelude to an attempt at his overthrow.*

Here, Liddell Hart consolidates the previous point by leaning toward the psychological element.

> *This theory is inclusive of but wider than the [Camon-Napoleon] manœuvre sur les derrières.... Where the one is concerned primarily with the logistical moves, the other seeks to probe deeper to the psychological foundations.*

Here, Liddell Hart explicitly separates the physical and psychological parts of his theory and elevates the latter. In other words, he is clear that although "indirect" can include the specific direction of movement of military units in a physical environment (geospatial battle space), its general meaning is the psychological one of perceived expectations. Critics of his Indirect Approach theory consistently go off-target by focusing narrowly on the geometrical vector implied by the word "indirect" and overlooking its overriding stress upon the psychological element of expectedness.

> *The strategy of indirect approach is, indeed, the highest and widest fulfilment of the principle of surprise.*

1. B.H. Liddell Hart, "Killing No Murder," *The Army Quarterly*, Vol. 15, January 1928, Correspondence with the Editors, pp. 396-401.

Where surprise is precisely the enemy's revised perception that his initial expectations have been overturned.

"KILLING NO MURDER"*

CORRESPONDENCE WITH THE EDITORS

(1)

GENTLEMEN,

Your editorial, Professor Spenser Wilkinson's article, and your review constitute a triple cross-examination of certain aspects of my book, " The Remaking of Modern Armies," which it would be discourteous to ignore. Indeed, they compel an answer—all the more because in them abuse is not made a substitute for argument. Indeed, if I may be permitted, I would congratulate you on their good taste in discriminating between the language traditionally permissible in criticizing, on the one hand, a collective policy or body, and, on the other, a named individual. This discrimination is too often unobserved in military controversy—in recent months we have had examples in certain published diaries and, again, in a book wherein faulty history and logic were camouflaged by such " personal " vulgarities as calling Mr. Churchill a " damp squib."

In contrast, I recognize that your editorial—like the other articles—is not only " fair comment " but, on the whole, a fair commentary. It does, however, convey to the reader that my theme is a one-sided refutation of battle as a means of victory, whereas it is rather an argument to remedy the lopsidedness which has arisen through over-emphasis on battle as the *all-important* means to victory. My considered summing up was that, even with the new potentialities of the air weapon, " It may prove impossible to reach the moral objective without battle, but the positive and negative examples from history serve to show that, even so, the best way to bring the enemy's armed forces to battle under conditions favourable to oneself is by striking at the moral objective. . . ."

A further unintentional misinterpretation occurs when you say

* *See* " Killing no Murder: An Examination of Some New Theories of War," by Spenser Wilkinson, late Chichele Professor of Military History in the University f Oxford, *Army Quarterly*, October, 1927.

that " it is somewhat disquieting to find that an English writer should countenance such a wholesale system of murder " as air attacks on unfortified towns. A careful analysis of my argument would show that it is divided into separate and distinct compartments. The first is a statement of the effects of air attack, with an illustration of what would happen to us, not an advocacy that we should do it to a foe. The second is a survey of the military, economic and ethical objections to such a method. *Advocacy* is reserved for the third, and it is then an advocacy that in our application of such a method we should use non-lethal gases instead of high explosive, following an argument that the infliction of death and destruction recoils on one's own head. It appears to my poor sense of logic that this is an argument not for, but against " the indiscriminate slaughter of civilians."

I turn to Professor Spencer Wilkinson's article, the title of which, I confess, gives me the impression that it was perhaps chosen for its effect rather than for its applicability. There is much in the article with which I agree—it could not be otherwise where the author is perhaps the profoundest of all contemporary students of military history. But I must challenge him on certain points.

He suggests that the excessive losses and duration of the war might be attributed, " with equal reason," to our failure to concentrate sufficiently on the Western Front. I suggest that this reasoning arises from an undue mental concentration on the British effort. It overlooks the fact that the Germans at Ypres in Oct.–Nov. 1914, and the French in their 1915 offensives brought to bear a greater superiority of strength over the defenders than we could perhaps ever have hoped for, and yet failed signally to break the resisting power of modern defence. I feel that we did, and do, inexcusably disregard these lessons of 1915—which attested the impregnability of modern defence in sectors where there is a definite expectation of attack. In the light of French and German experience it is difficult to find any warrant for the belief that a few or even a considerable number of extra infantry divisions would have made any material difference to our results.

Secondly, I prefer the verdict in the " Official History of the War," Vol. III, and reinforced by Falkenhayn's admissions, on the potentialities of the Dardanelles *conception*—as distinct from the half-hearted method—to that implied by Professor Spenser Wilkinson. To this question I revert, *indirectly*, later.

Professor Wilkinson does not interpret me quite fairly when he says that I ascribe the military doctrine of 1914-1918 to Clausewitz.

I endeavoured to make it clear that I ascribed it to his disciples, but held him responsible for giving a dangerous bias to his wide survey of the objects of war—by excessive stress on the destruction of the enemy's armed forces as the primary and paramount step towards victory.

That this destruction came to be visualized in the World War as a destruction of the enemy's troops, rather than the dislocation of his military organism, can be proved by reference to the memoirs and letters of those who directed our strategy—with their constant emphasis upon "killing Germans." Compared with this purpose there was small concentration of thought upon the far deeper effects of paralysing the enemy's brain, physically or psychologically, nerve system and arteries.

Professor Wilkinson then disputes my statement that the German Army's objective in 1870 was Paris, not the French Army. If he will refer to the war diaries of Frederick III he will find clear evidence that Paris was the objective, and that the encirclement of the French at Sedan came about as an almost casually decided incident *en route*.

Next, he argues that any attempt of a tank force to strike at the command and communication centres in the enemy's rear "will hardly be practicable in the next war, because" it "will immediately be met by the enemy's tanks. . . ." I would reply that the more mobile an instrument the greater is its power of evasion, as has been proved by cavalry in the remoter past and by aircraft in the recent past.

Professor Wilkinson, basing his faith on the last war, doubts the decisive effect of an air blow, "jumping" the army, against the moral objective. It seems to me perilous reasoning to base such a judgment on the puny attempts made in a war where the mind and energy of the warring powers were concentrated on killing the enemy's soldiers, and where the bulk of the adolescent air forces were utilized as mere auxiliaries to this task. Even as it was, Professor Wilkinson might feel less confidence if he had examined the statistics of the drop in the output of munition factories on the nights of air raid warnings, or had seen the population of a great town such as Hull, for example, stream out nightly into the fields.

But I maintain that the effect of a general strike—with unrestrained sabotage—or an earthquake is a better guide to the dislocation and paralysis which would be produced by air blows against the docks, railways and munition centres of a country. All these might at the outset be proclaimed " out of bounds " to civilians in the enemy country, as justifiable military objectives. At the same time Professor Wilkinson's claim for the immunity of civilians strikes me as discordant with his emphasis on the fact that modern war is carried on by the whole energy and resources of a nation.

I endorse, and have endorsed in my book, his care for the maintenance of ethical standards, but I would remind him that faulty logic has caused as much suffering as deliberate brutality. If one could discard national bias and acquire an historical point of view, there might appear little difference between our blockade and the German submarine campaign, except that between a lingering and a quick death, or that in the one case non-combatants could avoid going on the menaced ships, whereas in the other they could not leave the blockaded country. I should be happy if I could feel that the sum of suffering to women and children caused by the blockade was less than that caused by the German submarine and air attacks. I agree with Professor Wilkinson that a sense of righteous action is vital to the moral of the soldier and sailor, but both the sense and the moral are inspired more by faith than by logic, and most of the atrocities in history have been carried out by men whose moral was fortified by their conviction of acting righteously—witness the religious wars.

From the article I pass to the review. Your reviewer asks when I will produce a comprehensive and systematic statement of my proposals for the mechanized British Army of the future. In the present fluctuating state of technical design it is more rational to deal with the needs of the moment and to give direction signs for the future than to draft a rigidly detailed plan—until we seriously begin the conversion of our present machine-gun fodder into armoured units. Then, such a book will be justified. Whether it can be undertaken is another matter, for a still more pertinent answer to your reviewer's question is—" When the British officer acquires the book-buying habit." For his reputation is that of being one of the worst book-buying classes even in a country which has a notoriously poor demand for serious books. At present hardly any publisher can afford to bring out a book appealing only to the technical military public. And if Clausewitz had been a living Englishman his work " On War " would either have remained unwritten or his publisher would have filed his petition in bankruptcy. The editor of a military review should have a keen appreciation of this reasoning. It is true that in consequence the military reviews become the channel of publishing compressed theses which would otherwise be expounded within the wide spacious dimensions of a book, but the editor may well feel that quality of matter is an inadequate consolation for paucity of circulation.

Here ends my reply to the questions raised in your editorial, article and review, but I would fain close on a more constructive note. For destruction is only of value as a preparation of the ground for fresh construction. Professor Wilkinson's challenge is a spur to hasten the elaboration and fuller explanations of certain aspects of my theory of war. From the fruits of my humble efforts to explore the field of military history certain deductions have gradually ripened, and some at least have now reached a stage where they may be crystallized into thesis form. Among these is the theory of what may be termed " The Strategy of Indirect Approach." More and more clearly has the fact emerged that a direct approach to the object or objective along the " line of natural expectation " has ever tended to negative results. The reason being that the strength of an enemy country or force lies far less in its numbers or resources than in its stability or equilibrium—of control, moral and supply. The former are but the flesh covering the framework of bones and ligaments. To move along the " line of natural expectation " is to consolidate the enemy's equilibrium, and by stiffening it to augment its resisting power.

In war as in wrestling, the attempt to throw the opponent without loosening his foothold and balance can only result in self-exhaustion, increasing in disproportionate ratio to the effective strain put on him. Victory by such a method can only be possible through the advantage of an enormous margin of superior strength, and even so lacks decisiveness. It is a fundamental violation of the law of economy of force.

In contrast, the decisive victories in military history have come from the strategy of indirect approach, wherein the dislocation of the enemy's moral, mental or material balance is the vital prelude to an attempt at his overthrow. The strategy of indirect approach is, indeed, the highest and widest fulfilment of the principle of surprise.

Among a host of examples the strategy of Scipio's campaign in Africa, Sulla's strategy in Italy, Jenghiz Khan's against the Kharismian Empire, Sabutai's against Hungary, Wallenstein's in 1632, Marlborough's on the Danube, Bonaparte's which led up to Lodi, Grant's Vicksburg campaign, and Sherman's unduly neglected but decisive " march to the sea " and thence to Virginia, may be quoted as illustrative of the varied forms of the " indirect approach."

For this theory is inclusive of but wider than the *manœuvre sur les derrières* which that great student General Camon established as the key-method of Napoleon in his conduct of operations. Where the one is concerned primarily with the logistical moves, the other seeks to probe deeper to the psychological foundations and in so doing finds, even in the sphere of operations, an underlying connection with many strategical plans which have no outward resemblance to a *manœuvre sur les derrières*—but are none the less definitely vital examples of the " strategy of indirect approach." This brief explanation must for the moment suffice to indicate the tenor of the thesis, which is now " on the stocks " and will, I hope, be launched in a few months.

Yours faithfully,

B. H. Liddell-Hart.

12th November, 1927.

CHAPTER 16

Cognitive Factors in Deception and Counterdeception[1]

Richards J. Heuer, Jr.

Richards J. Heuer, Jr. served with the CIA for forty-five years. He was recruited while a graduate student at UC Berkeley by former CIA Director Richard Helms. Heuer is most known for his book, "Psychology of Intelligence Analysis," where he discussed the human mind's weaknesses in dealing with uncertainty. Heuer's advice to the intelligence community has been in support of creating an environment that promotes and rewards critical thinking.

Introduction

To be successful, deception must achieve a desired impact upon the thinking of the deception target, either a national or military decision maker or the intelligence analyst working for the decision-maker. The chances of success are enhanced the more a deceiver understands about the thought processes of the target leaders or analysts. Conversely, the chances of avoiding deception increase the more one understands one's own information processing capabilities and limitations. In examining these judgmental processes, one can either determine the propensities and predilections of individual persons or examine those factors that most men and women seem to have in common. This paper takes the latter approach.

That human beings often make erroneous judgments is self-evident from our daily experience, and it has been demonstrated by many psychological experiments. Military, political and economic issues involving interaction with other nations are among the most complex analytical problems. This complexity is normally exacerbated by lack of information concerning some critical elements of a problem and a large volume of fragmentary, ambiguous and even erroneous information concerning other elements. Judgments must be made in the face of great uncertainty.

Over 20 years ago, Herbert Simon advanced the concept of "bounded" or limited rationality.[1] Because of limits in our mental capacity, he argued, the human mind cannot cope directly with the complexity of the world. Rather we construct in our mind a simplified model of reality and then work with this mental model. We behave rationally within the confines of our mental model, but this model is generally not very well adapted to the requirements of the real world.

Simon's theory of bounded rationality was stimulated by earlier psychological research on perception, memory, attention span, and reasoning capacity that documents

1. Richards J. Heuer, Jr., "Cognitive Factors in Deception and Counterdeception," in *Multidisciplinary Perspectives on Military Deception*, Donald C. Caniels, et al., editors, United States Naval Postgraduate School, May, 1980, pp. 45–101.

limitations in our "mental machinery." A principal thesis underlying more recent research has been that these limitations cause us to employ various simplifying strategies when processing information to make judgments and decisions. Psychologists have conducted many experiments to identify these strategies and to show how—at least in laboratory situations—they affect our judgment and decisions. Students of international relations, particularly Robert Jervis,[2] have conducted historical research to document instances in which political and military decisions appear to have been significantly influenced by these psychological variables.

This research provides substantial experimental and historical evidence to support Jervis' conclusion that "perceptions of the world and of other actors diverge from reality in patterns that we can detect and for reasons that we can understand."[3] These patterns of erroneous perception and judgment are frequently called "biases." A bias, as the term is used here, is an error in judgment that is consistent and predictable. It is not predictable in the sense that all persons under the same circumstances will make the same error all the time. Rather, it is predictable in a statistical sense, in that given a large number of cases most people will be influenced by this tendency most of the time.

One can identify several types of biases. Motivational biases result from the influence on judgment of our ambitions and fears, and the need to perceive our past behavior as commendable and consistent. The functional roles and circumstances of organizations generate patterns of biased organizational judgment.[4] Cultural biases are rooted in predispositions inherent in one's cultural values and heritage.

This paper deals only with perceptual and cognitive biases. Perceptual biases arise from the nature of the process by which we perceive the world about us, and they limit the accuracy of our perceptions. Cognitive biases result not from any intellectual or emotional predisposition toward a certain judgment, but simply from the way the mind tends to work. They influence how we estimate probabilities, evaluate evidence, and attribute causality.

Of the diverse forms of bias, we have opted to discuss perceptual and cognitive baises for two reasons. They are the most general forms of bias, presumably affecting all persons regardless of cultural background or organizational affiliation. Cognitive biases are also the least well known, for most research on these biases is of recent origin.

The paper is divided into three sections, one dealing with perceptual biases, one with cognitive biases, and a concluding section that includes discussion of the broad problem of countering deception. Not all perceptual and cognitive biases are discussed here, for we have selected just those that seem most relevant to the problem of deception. The first two sections present the biases sequentially with a concluding discussion relating them to the deception problem. These discussions are principally from the point of view of the deception planner, for countering deception involves basically different problems discussed primarily in the conclusion. An appendix contains a brief summary of all the biases and their implications.

A word of caution before proceeding further. How humans perceive and process information to arrive at analytical judgments is not fully understood. Although the evidence presented here is persuasive, it is not conclusive. The intention of this paper is not to deliver the last word in psychological theory, for psychologists differ among themselves just as much as historians, intelligence analysts or priests. The purpose is to

2. For example, Richard Betts contends that military intelligence has an organizational bias toward "worst-case" analysis in analyses that support procurement planning and "best-case" analysis in evaluating the results of military operations.

describe the current state of knowledge from a perspective that views human cognitive processes as a critical variable in explaining fallible human judgment. The aim is to learn something about how people make judgments, and to provide guidelines concerning how to exploit this knowledge to deceive others or prevent being deceived ourselves. The guidance is limited to principles that will generally be helpful, not firm rules that guarantee an optimal result, for judgment is not guided by algorithms that ensure a correct answer.

Biases in Perception

The process of perception links the individual to his or her environment and is critical to accurate understanding of the world about us. We tend to think of perception as a passive process: we see, hear, smell, taste or feel stimuli that impinge upon our senses. If we are at all objective, we record what is actually there.

Yet perception is demonstrably an active rather than a passive process; it constructs rather than records "reality." Perception implies understanding as well as awareness. It is a process of inference in which the individual constructs his or her own version of "reality" on the basis of information provided by the senses. This sensory input is mediated by complex and poorly understood mental processes that determine which information we attend to, how we organize it, and the meaning we attribute to it. Thus what we perceive and how readily we perceive it is strongly influenced by our past experience, education, cultural values and role requirements, as well as by the stimuli recorded by our receptor organs. This should shake some traditional assumptions about "objectivity" in analysis.

Figure 1 Perception is a process of constructing reality rather than recording it.

Expectations Condition Perception

Many experiments have been conducted to demonstrate the extraordinary extent to which the information obtained by an observer depends upon the observer's own expectations, assumptions and preconception. For example, when you looked at Figure 1, above, what did you see? Did you note that the article is written twice in each of the three phrases.

This is commonly overlooked, because perception is influenced by our expectations about how these familiar phrases are normally written. If you perceived Figure 1 correctly, you have exceptional powers of observation, were lucky, or have seen the figure before. This simple experiment demonstrates one of the most fundamental principles concerning perception: we tend to perceive what we expect to perceive. A corollary of this principle is that it takes more information, and more unambiguous information, to recognize an unexpected phenomenon than an expected one.

Another classical experiment to demonstrate the influence of expectations on perception used playing cards, some of which were gimmicked so the spades were red and the hearts black. Pictures of the cards were flashed briefly on a screen and, needless to say, the test subjects identified the normal cards more quickly and accurately than the anomalous ones. After test subjects became aware of the existence of red spades and black hearts, their performance with the gimmicked cards improved but still did not approach the speed or accuracy with which normal cards could be identified.[5] This shows that patterns of expectation become so deeply embedded that they continue to influence perceptions even when we are alerted to and try to take account of the existence of data that do not fit our preconceptions. Trying to be objective does not guarantee accurate perception.

The position of the test subject identifying playing cards is analogous to that of the intelligence analyst or government leader trying to make sense of the paper flow that crosses his desk. What is actually perceived in that paper flow, as well as how it is interpreted, depends in part, at least, on the analyst's patterns of expectation. We do not have expectations just about the color of hearts and spades. We have a set of assumptions and expectations about the motivations of people and the processes of government in foreign countries. Events consistent with these expectations are perceived and processed easily; those which contradict prevailing expectations tend to be ignored or distorted in perception. Of course, this distortion is a subconscious or pre-conscious process, as illustrated by how you presumably ignored the extra words in the triangles in Figure 1.

This tendency to perceive what we <u>expect</u> is far more important than any tendency to perceive what we <u>want</u>. In fact, there may be no real tendency toward wishful thinking. The commonly cited evidence supporting the claim that people tend to perceive what they want to perceive can generally be explained equally well by the expectancy thesis or the availability bias (to be discussed later).[6]

Expectations have many diverse sources, including past experience, professional training, and cultural and organizational norms. All these influences predispose us to pay particular attention to certain kinds of information and to organize and interpret this information in certain ways. Perception is also influenced by the context in which it occurs. Different circumstances evoke different sets of expectations. We are more attuned to hearing footsteps behind us when walking in an alley at night than along a city street in daytime, and the meaning we attribute to the sound of footsteps will vary under these differing circumstances. A military intelligence analyst is similarly tuned to perceive indicators of potential conflict. When the evidence is ambiguous, as is commonly the case in intelligence analysis, this predisposition increases the likelihood the indicators will be perceived accurately when they in fact exist and escape the attention of other observers, but it also increases the chances they will be perceived erroneously when they are not really there.

Patterns of expectation, rooted in past experience and training, tell us, subconsciously, what to look for, what is important, and how to interpret what we see. These patterns form a "mind set" that predisposes us to think in certain ways. A mind set is akin to a screen or lens through which we perceive the world. For example, the following truisms have been part of the mind set of intelligence analysts:

- A totalitarian government enjoying the support of effective military and security organizations cannot be overthrown by popular opinion.
- When the position of a dictatorial ruler is threatened, he will defend his position with force if necessary.
- The principal threat to friendly governments comes from the left, not the right.

These premises were part of the lens through which U.S. policy makers and intelligence analysts alike perceived developments in Iran in 1978 prior to the fall of the Shah. They had a significant impact on where analysts focused their attention, what they expected to happen, and how they interpreted the unfolding events. That all these "truisms" were proven wrong in that instance is perhaps no small part of the explanation why the Shah's demise took the United States government by surprise.

There is a tendency to think of a mind set as something bad, to be avoided. One should have an open mind and be influenced by the facts rather than by preconceived notions! But there is no such thing as "the facts of the case." There is only a very selective subset of the overall mass of data to which we have been subjected that we take as facts and judge to be relevant to the question at issue. Actually, mind sets are neither good nor bad; they are unavoidable. There is no conceivable way of coping with the volume of stimuli that impinge upon our senses, or with the volume and complexity of the data we have to analyze, without some kind of simplifying preconceptions about what to expect, what is important, and what is related to what. "There is a grain of truth in the otherwise pernicious maxim that an open mind is an empty mind."[7] Objective analysis is not achieved by avoiding preconceptions (that would be ignorance or self-delusion), but by making our basic assumptions and reasoning as explicit as possible so they can be challenged by others and we can ourselves examine their validity.

Perceptions Resist Change

One of the most important characteristics of perceptions is that they are quick to form, but resistant to change. Once we have perceived an object, event or situation and formed some judgment about its essential characteristics, we are biased toward continuing to perceive it in the same manner even though the object of our perception may change.

Figure 2 illustrates this principle by showing part of a longer series of progressively modified drawings that change almost imperceptibly from a man into a woman.[8] The right hand drawing in the top row, when viewed alone, has equal probability of being perceived as a man or a woman. When test subjects are shown the entire series of drawings, one by one, their perception of this intermediate drawing is biased according to which end of the series they started from. Test subjects who start by viewing a picture that is clearly a man are biased in favor of continuing to see a man long after an "objective observer" (i.e., an observer looking at a single picture) recognizes that the man is now a woman. Similarly, test subjects who start at the woman end of the series are biased in favor of continuing to see a woman. Once an observer has formed an image, that is, once he or she has developed a mind set or expectation concerning the phenomenon being observed, this conditions future perception of that phenomenon. This is the basis for yet another general principle of perception: new information is assimilated to existing images.

Figure 2 Impressions resist change.

This principle explains why gradual, evolutionary change often goes unnoticed. It also explains the phenomenon that an intelligence analyst assigned to work on a topic or country for the first time may generate accurate insights that have been overlooked by experienced analysts who have worked on the same problem for ten years. A fresh perspective is sometimes useful, for past experience can handicap as well as aid analysis. This tendency to assimilate new information to pre-existing images is greater "the more ambiguous the information, the more confident the actor is of validity of his image, and the greater his commitment to the established view."[9]

Figure 3 provides the reader an opportunity to test for him or herself the persistence of established Images. Look at the picture. What do you see—an old woman or a young woman? Now look again to see if you can visually and mentally reorganize the data to form a different image—that of a young woman if your original perception was of an old woman, or of the old woman if you first perceived the young one.[10] Do not look at the footnote unless you need clues to help you identify the other image.[3] Again, this illustrates the principle that perceptions are quick to form but resistant to change.

When you have seen Figure 3 from both perspectives, try shifting back and forth from one perspective to the other. Do you notice some initial difficulty in making this switch? One of the most difficult mental feats is to take a familiar body of data and reorganize it visually or mentally to perceive it from a different perspective. Yet this is what intelligence analysts are constantly required to do. In order to understand international interactions, we must understand the situation as it appears to each of the opposing forces, and constantly shift back and forth from one perspective to the other as we try to fathom how each side interprets an ongoing series of interactions. Trying to perceive Soviet as well as U.S. interpretations of international events is comparable to seeing both the old and young women in Figure 3; once we have perceived events one way, we tend to resist alternate perspectives.

3. The old woman's nose, mouth and eye are, respectively, the young woman's chin, necklace and ear. The old woman is seen in profile looking left. The young woman is also looking left, but we see her mainly from behind so most facial features are not visible; her eyelash, nose and curve of her cheek may be seen just above the old woman's nose.

Figure 3 It is difficult to look at the same data from different perspectives.

Impact of Ambiguity

Initial exposure to ambiguous or blurred stimuli interferes with accurate perception even after more and better information becomes available. This effect has been demonstrated experimentally by projecting onto a screen pictures of common, everyday things such as a dog standing on grass, a fire hydrant, and an aerial view of a highway cloverleaf intersection.[11] The initial projection was blurred in varying degrees, and the pictures were then brought into focus slowly to determine at what point test subjects could identify them correctly. This experiment showed two things. First, those who started viewing the pictures when they were most out of focus had more difficulty identifying them when they became clearer than those who started viewing at a less blurred stage. In other words, the greater the initial blur, the clearer the picture had to be before people could recognize it. Second, the longer time that people were exposed to a blurred picture, the clearer the picture had to be made before they could recognize it.

What happened in this experiment, and what presumably happens in real life with cognitive as well as visual perceptions, is that despite ambiguous stimuli we form some sort of tentative hypothesis about what it is we are seeing. The longer we are exposed to the ambiguous data, the greater confidence we develop in this initial and perhaps erroneous impression, so the greater the impact this initial impression has on our subsequent perceptions. For a time as the picture becomes clearer, there is no obvious contradiction; the new data is assimilated to our previous image, and the initial interpretation is maintained until the contradiction becomes so obvious that it forces itself upon our consciousness. The early but incorrect impression tends to persist because the amount of information necessary to invalidate a perception is considerably greater than the amount of information required to form an initial impression. The problem is not that there is any inherent difficulty in grasping new perceptions or new ideas, but that established perceptions are so difficult to lose. Thus inaccurate perceptions generated by ambiguous data may persist even after additional information has been received to clarify the initial ambiguity. One might seek to limit the adverse impact of this tendency by suspending judgment for as long as possible as new information is being received.

Implications for Intelligence Analysis

Understanding the ways in which perception is commonly distorted has significant implications for comprehending the nature and limitations of intelligence analysis. If we consider the circumstances under which accurate perception is most difficult, we find these are exactly the circumstances under which intelligence analysis is generally conducted—dealing with highly ambiguous situations on the basis of information that is processed incrementally under pressure for early judgment. This is a recipe for inaccurate perception. That intelligence analysts perform as well as they do is testimony to their generally sound judgment, training and dedication in performing an extremely difficult task.

Intelligence seeks to illuminate the unknown. Almost by definition, intelligence analysis deals with highly ambiguous situations. Yet we have seen that the greater the ambiguity of the stimuli, the greater the impact of expectations and pre-existing images on the perception of that stimuli. Thus despite maximum striving for objectivity, the intelligence analyst's own preconceptions are likely to exert a greater impact on the analytical product than in other fields where the analyst is working with less ambiguous and less discordant information.

Moreover, the intelligence analyst is among the first to look at new problems at an early stage when the evidence is very fuzzy indeed. The analyst then follows a problem as additional increments of evidence are received and the picture gradually clarifies—much as the test subjects in the experiment demonstrating that initial exposure to blurred stimuli interferes with accurate perception even after more and better information becomes available. If the results of this experiment can be generalized to apply to intelligence analysts, it suggests that because the analyst starts observing a potential problem situation at its early and most unclear stage, he or she is at a disadvantage as compared with others—for example, policy makers—whose first exposure may come at a later stage when more and better information is available.

The receipt of information in small increments over time also facilitates assimilation of this information to the analyst's existing views. No one item of information may be sufficient to prompt the analyst to change his view. The cumulative message inherent in many pieces of information is not examined as a whole. The Intelligence Community review of community performance before the 1973 Arab-Israeli War noted:

> The problem of incremental analysis—especially as it applies to the current intelligence process—was also at work in the period preceding hostilities. Analysts, according to their own accounts, were often preceding on the basis of the day's take, hastily comparing it with material received the previous day. They then produced in 'assembly line fashion' items which may have reflected perceptive intuition but which [did not] accrue from a systematic consideration of an accumulated body of integrated evidence.

And finally, the intelligence analyst operates in an environment that exerts strong pressures for premature judgment. Policy makers' needs for interpretive analysis are greatest within at most two or three days after a new event occurs. The system requires the intelligence analyst to make an almost instant diagnosis before sufficient hard information becomes available to make a well-grounded judgment. This diagnosis can only be based upon the analyst's preconceptions concerning how and why events normally transpire in a given society.

As time passes and more information is received, a fresh look at all the evidence might suggest a different explanation. Yet we have seen from the various perception experiments that an early judgment adversely affects the formation of future perceptions. Once an observer thinks he or she knows what is happening, this perception tends to resist change. The new information received incrementally fits easily into the analyst's previous image. This perceptual bias is reinforced by organizational pressures favoring consistent interpretation, for once the analyst has committed him or herself in writing, both the analyst and the organization has a vested interest in maintaining the original diagnosis.

Implications for Deception

One overwhelming conclusion stands out with respect to deception: it is far easier to lead a target astray by reinforcing the target's existing beliefs, thus causing the target to ignore the contrary evidence of one's true intent, than it is to persuade a target to change his or her mind.

Military operations possess a certain logic. Terrain, weather, supplies and the relative balance of forces often suggest optimal tactics or strategy. Yet, if the preferred alternative is equally obvious to the enemy, these advantages can be offset by the enemy's counter-preparations. Thus planners of military operations may use deception to conceal their true intent, and in doing so they are faced with two basic alternatives. They can plan to attack in a place, time and manner most expected by the enemy, while seeking through deception to achieve surprise by changing the enemy's expectations. Or they can reinforce the enemy's expectations while planning a surprise attack in a different place, time or manner.

The tendencies to perceive what we expect to perceive and to assimilate new information to existing images make it far easier to reinforce a target's existing beliefs than to change them. Deceptions that follow this principle seldom fail, for the odds are then strongly in favor of the deceiver. The human capacity to rationalize contradictory evidence is easily sufficient to outweigh the pernicious effects of security leaks and uncontrolled channels of information that deception planners might otherwise fear are certain to compromise their efforts.

Deceptions that require persuading a target of something he is not already predisposed to believe should be avoided if at all possible. If nonetheless required by the operational situation, the chances for success may be enhanced by following a simple sequencing rule. Because of the tendency to integrate new information into existing beliefs, the first goal in any effort to change beliefs must be to ensure that the target is at least considering seriously the desired alternative hypothesis. This may require initiating the deception with strong and obvious evidence that forces the desired conclusion to be at least considered seriously by intelligence analysts and policy makers. This is then followed in quick succession by additional supporting evidence that leads the target to a reasoned conclusion in favor of the desired alternative.

The opposite tactic, which seems incorrect from a psychological point of view, would be to save the more dramatic evidence until after the stage has been set by transmitting a number of supporting messages. The expectation is that the target initially attributes little importance to the supporting messages, but once the key is received the other pieces are perceived to fall into place forming a coherent and persuasive picture. The weakness of this tactic is that the target may have failed to notice, forgotten or misinterpreted the earlier evidence, for information that does not fit neatly into an existing hypothesis tends to be ignored or misperceived. Intelligence analysts and policy makers are commonly confronted with a large amount of discordant information. They have only a limited capacity to sort and store discordant or seemingly irrelevant information in memory in a manner that makes it possible to recall it for the evaluation of hypotheses that are not now under consideration.

Planning and implementing a deception typically involves a major investment of time, energy and ego. When people make such an investment in preparing a message, they tend to overestimate how clear this message will be to the receiver. This results from the importance of context in perceiving and interpreting a signal; when a message is placed in a different context it assumes a different meaning. The message developed by the deception planners is understood by them in the context of the endless meetings in which alternatives were weighed and details worked out. They are so familiar with their own thinking that they risk overlooking the degree to which the message is clear to them only because they know what to look for.

The target of a deception is likely to have a different agenda of concerns, different predispositions and a different information base than the deception planners. Normally this will lead to a different interpretation of messages. If the deception planners have sufficient understanding of the target's situation and thinking, messages may be planned to take advantage of the particular context in which they will be received, but in practice the target may miss many clues the deceiver sets out for him and may assign considerable weight to factors the deceiver regards as trivial or to information of which the deceiver is wholly unaware. To the extent that the deception signals reinforce the target's expectations, there is a large margin for error and these miscalculations have little impact. If the goal is to change the target's mind, however, they may be critical.

It is not by accident that discussion to this point has concerned implications for the perpetration of deception rather than its detection. The counterdeception problem is extremely difficult. When should we disbelieve our eyes and ears and the seemingly logical conclusions of our mind? When should we second guess, and say to ourselves that since most of the evidence points to X, then Y must be true rather than X? Even in the absence of deliberate deception, the evidence at the most critical junctures is typically so ambiguous that the proper conclusion is far from obvious. To the extent that we cannot believe the evidence, the analytical problem becomes even more complicated. The problem of detecting deception is not simply a matter of accurate perception. It is embedded in the much larger problem of effective intelligence collection and analysis, and we reserve discussion of these points until the end of this paper.

Cognitive Biases

The cognitive biases discussed here are grouped according to whether they affect the estimation of probabilities, the evaluation of evidence, or the attribution of causality.

Biases in Estimating Probabilities

Estimating probabilities is important because we live in a probabilistic world. Social, political, military and economic developments are not rigidly determined but occur or fail to occur with some degree of probability. Decision makers cannot be certain of the outcome of their actions, so they must weigh the probabilities of alternative outcomes. The information on which these decisions are based also involves many uncertainties expressed in probabilistic terms. The intelligence analyst, for example, is constantly assessing probabilities with respect to the intentions of foreign leaders, the capability of military forces, the future consequences of current events, or the credibility of sources.

Typically, these probability judgments are expressed in imprecise terms such as possibly, probably or very likely terms that unfortunately have different meanings to different people. But the issue here is not whether communication and decision making can or should be improved by replacing those verbal qualifiers with numerical ranges of probability. It is whether the estimates themselves are influenced by systematic biases that affect their accuracy. Research by experimental psychologists suggests that this is in fact the case. Knowledge of these biases may be useful to those planning deception or seeking to avoid it.

Availability Bias

One of the simplified rules of thumb we use in making probability estimates is known as the availability rule. In this sense availability refers to imaginability or retrievability. Psychologists have shown that two of the cues we use in judging the probability of an event are 1) the ease with which we can imagine relevant instances of the event, and 2) the number or frequency of such events that we can easily remember.[12] In other words, we are using the availability rule of thumb whenever we estimate frequency or probability on the basis of how easily we can recall or imagine instances of whatever it is we are trying to estimate.

Normally this works quite well. If one thing actually occurs more frequently and therefore is more probable than another, we probably will be able to recall more instances of it. Events that are likely to occur generally are easier to imagine than unlikely events. We are constantly making inferences based on these assumptions. We estimate the probability of successful deception by recalling historical examples of deception under similar circumstances. We estimate the probability that a politician will lose an election by imagining ways in which he may lose popular support. Although this generally works well, we are often led astray because the ease with which things come to mind is influenced by many factors, such as emotional saliency, vividness and how recently we have been exposed to them, all of which may be unrelated to the correct probability. When this happens, our judgment is biased in favor of the probability of those events that are most available. For example, the Soviet assessment of the likelihood that Germany may once again become a military threat to Soviet interests seems clearly biased by the ready availability of vivid memories of the Second World War.

Intelligence analysts often have difficulty estimating the likelihood of low probability events, especially when those events have potentially very serious consequences. For example, what is the likelihood of civil war in Canada, perhaps even including Soviet or Cuban assistance to Quebec, during the next ten years? Or the likelihood of an aggressively anti-American, Castro-like government coming to power in Mexico? It is difficult for us to imagine such developments, so we assign them a very low probability, but imaginability is most likely irrelevant to an accurate assessment of the probability that either of these developments will actually occur. To the extent that our estimate is influenced by ready imaginability rather than by a full analysis of the causal factors at work, we are likely to underestimate the true probability.

Sino-Soviet reconciliation is another low probability, high significance event, but here the availability bias is likely to cause people to overestimate the probability of it actually happening. This is because it is so easy to imagine such a development and what impact it would have on U.S. policy. In fact, our memory of having been taken by surprise by the Sino-Soviet split causes many people to be preoccupied by the possibility of reconciliation. Analysts working full time on this question are considering the operative causal factors, not making quick and easy inferences on the basis of imaginability. But the policymaker or generalist who does not have the time or the information to go into details must unconsciously take shortcuts, and the obvious shortcut is to use the availability rule of thumb for making inferences about probability.

Anchoring Bias

Another strategy that people seem to use intuitively and unconsciously to simplify the task of mentally processing complex information is called "anchoring." Some natural starting point is used as a first approximation to the desired judgment. This starting point is then adjusted, based on the results of additional information or analysis. Typically, however, the starting point serves as an anchor or drag that reduces the amount of adjustment, so that the final estimate remains closer to the starting point than it ought to be.

Anchoring has been demonstrated by asking a group of test subjects to estimate one or more known quantities, for example, the percentage of people in the United States who are age 55 or older. In an experiment that used this question, the test subjects were given starting percentages that were selected randomly—they were drawn out of a bowl—and were then asked to adjust these arbitrary starting points until they reached their best estimate in response to the question. Because of insufficient adjustment, those who started out with an estimate that was too high ended with higher estimates than those who started with an estimate that was too low, and vice versa. Even the totally arbitrary starting points acted as an anchor, causing drag or inertia that inhibited full adjustment of estimates to the point that the test subjects would otherwise have considered desirable.[13]

Policy makers and intelligence analysts deal with dynamic situations. They must continually review their estimates in response to changes in the situation or the receipt of previously unavailable information. Ideally, there should be a direct correlation between changes in the situation and/or new information and changes in the estimate, but such is frequently not the case. There is much evidence to suggest that people do not change their judgments enough. Once an estimate is made, thinking becomes anchored and moves only within a narrow range around that spot.

Overconfidence Bias

Problems of perception and bias might be less serious if people had a better appreciation for the limits of their own knowledge. Many tests have been conducted that show people have difficulty expressing accurately the degree of uncertainty in what they know or believe. People tend to be overconfident about how much they know.

The Subjective Probabilities Assessment Test (SPAT) uses 75 general knowledge questions with known answers taken from an almanac. Test subjects are asked not just to answer the questions, but for each question to also assign a probability that shows how confident they are that their answer is the correct one. For questions with two possible answers, a 50% probability indicates complete uncertainty about which of the two answers is correct, while 100% indicates absolute certainty that the chosen answer is the proper one.

Performance on the SPAT test is not measured by the number of correct answers, but by the "calibration" between assessed probability that the answer is correct and the actual number of correct answers at each probability level. If a subject indicates 70% certainty on ten questions, then a perfect score would be seven correct and three incorrect answers. Under such circumstances, we would say that the person's subjective or intuitive feeling of certainty is perfectly calibrated with the reality of his performance. On the other hand, only five or six correct answers would indicate overconfidence and eight or nine correct answers under-confidence in assessing the accuracy of one's knowledge.

Figure 4 shows the results from testing almost 1,000 DIA and CIA intelligence analysts and managers.[14] For questions on which analysts expressed 100% confidence, the median analyst was correct only 76% of the time.

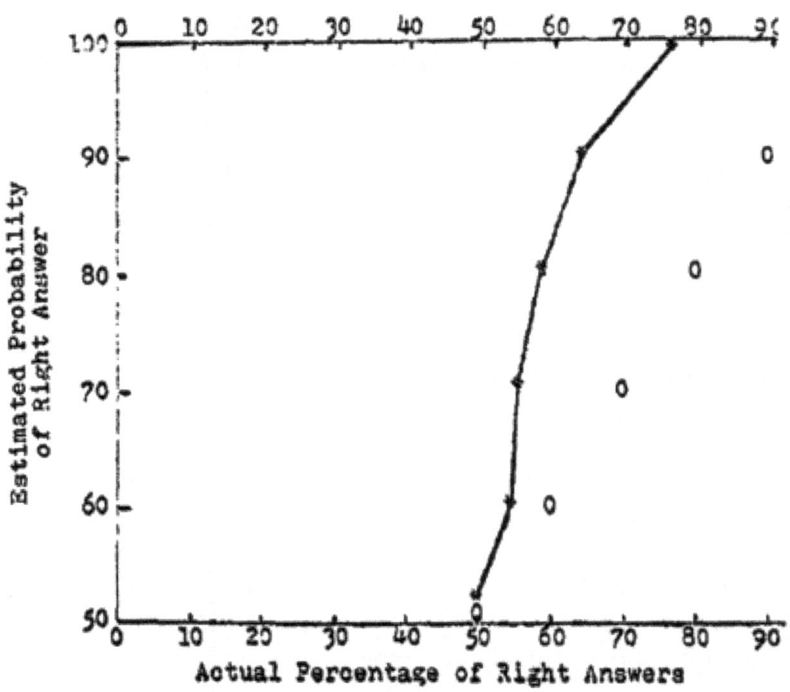

Figure 4

When expressed confidence was 90%, the judgment was correct on only 632 of the questions, and so on as indicated in the graph. In brief, the test demonstrated that the intuitive feelings of certainty of intelligence analysts do not conform with reality. The analysts are markedly overconfident of how much they know.

Very similar results have been found with test groups not affiliated with the intelligence community.[15] Experiments have shown that the overconfidence bias is greater for difficult questions than for easy ones. Training involving feedback on one's performance improves the calibration between expressed confidence and actual performance; weather forecasters perform exceptionally well on the SPAT test, doubtless because they commonly receive rapid feedback on the accuracy of their probability judgments. On the other hand, neither superior intellectual ability nor expertise in the subject matter of the questions has been found to improve performance in assessing the certainty of one's knowledge.

The intelligence analyst's subjective overconfidence may not be translated into overconfident intelligence judgments. Organizational and motivational incentives also influence how the analyst expresses his level of confidence when writing an intelligence report. Hedging to avoid embarrassment in the event of error and overwriting to mask ignorance are not uncommon. The bias toward overconfidence applies to the private feelings of analysts, not necessarily to the way they present these feeling in formal intelligence products.

Implications for Deception

Availability bias may make analysts believe that strategic deception is more common than it really is, and thus cause them to be more disposed to perceive it. Successful cases of deception are far more salient, and consequently more available for recall in memory, than cases in which deception was not employed under comparable circumstances. Deception attracts both the popular imagination and the attention of historians, while the absence of deception in strategic operations does not. When an analyst is faced with a situation in which deception may or may not be employed, his or her estimate of the probability of deception is influenced by this easy retrievability of past instances of deception.

The availability bias also suggests that employees of watch offices will tend to overestimate the probability of whatever it is they are watching for. Having been briefed and trained to recognize certain indicators, and having imagined and rehearsed scenarios that include the watched-for developments, it is not surprising that the watched-for developments are at the forefront of their minds as they try to forecast the future course of events. To the extent that the watched-for development is judged more probable, the perceptual bias of seeing what we expect to see also plays a greater role.

If the goal of a deception is to induce ambiguity or to persuade the watch officers that what they are watching for is <u>not</u> happening, e.g., that there is no intent to attack when an attack is in fact planned, a watch office is an extremely difficult deception target. On the other hand, it may be possible to exploit the watch officers' preconceptions, for example, as part of a plan to exploit the cry wolf syndrome. The watch office might be provoked to issue an alert of impending attack several times when no attack is in fact planned, so that future alerts will be received more skeptically. In this procedure, the availability of the attack scenario is countered by building up in the watch officers another availability—the memory of recent false alarms.

Policy makers and others seeking to avoid deception would do well to keep the availability bias in mind and to check the extent to which it influences their thinking on critical issues. If their judgments of the likelihood of future events are determined by imaginability or retrievability rather than by causal analysis, they should have little confidence in these judgments and should seek an independent assessment based on more systematic analysis.

The significance of the anchoring bias to the deception planner depends upon the type of deception being planned. If the goal is to change a target's previous beliefs, anchoring will facilitate achieving this objective.

Because the anchoring bias tends to prevent analysts from revising their estimates as much as they should when new information is received, analysts seeking to avoid surprise will generally wish to try to overcome this bias. The prognosis is not favorable. In one experiment, the bias persisted even after test subjects had been given feedback to show the bias and after they had been urged to try to overcome this tendency in answering a new set of estimation questions.[16] This is a common finding in experiments dealing with cognitive biases; the biases persist even after test subjects are informed of them and instructed to try to avoid them or compensate for them.

One possible technique for avoiding the anchoring bias, to weigh anchor so to speak, may be to ignore one's own or others' earlier judgments and rethink a problem from scratch. In other words, consciously avoid using any prior judgment as a starting point. There is no experimental evidence to show that this is possible or that it will work, but it certainly seems worth trying. Alternatively, it is sometimes possible to avoid human error by employing formal statistical procedures. Bayesian statistical analysis, for example, can be used to revise prior judgments on the basis of new information in a way that is designed to avoid any anchoring bias.[17]

Overconfidence exacerbates the impact of all the biases. Although a written estimate may have been hedged, if the analyst's subjective feeling is one of overconfidence and satisfaction with his or her estimative performance, there will be few efforts to improve. For the deception planner, the implications of a target's overconfidence are the same as the implications of the anchoring bias. It is one more obstacle to overcome in inducing a target to change his or her mind, but it can be readily exploited if the objective is to reinforce the target's existing convictions. In ambiguity-inducing deceptions, the intent is to reduce the target's confidence.

Biases in Evaluation of Evidence

Collection and evaluation of evidence are crucial steps in analysis. Are there systematic biases in the way we handle evidence? We have seen in our discussion of perception that new information tends to be assimilated to existing images. Thus the order in which we receive information affects our judgment. Evidence received early in an investigation has a greater impact on our thinking than evidence received after our impressions have already formed. At present, however, we are concerned with several other problems as discussed below.

Oversensitivity to Consistency

Consistency is normally an appropriate guideline for evaluating evidence. We formulate alternative explanations or predictions and select the one which encompasses the greatest amount of evidence within a logically consistent scenario. When very little evidence is available, however, we tend to be oversensitive to consistency. We have more confidence in conclusions drawn from a very small body of consistent information than from a larger body of less consistent data. This is incorrect, because conclusions drawn from very small samples are highly unreliable.

Test subjects were asked to predict students' class standing on the basis of grades obtained in the freshman year, and to indicate the amount of confidence they had in their predictions. The predictions were almost identical when based on a single B in one course as when based on an A in one course and a C in another, but there was a significant difference in level of confidence. Subjects expressed far more confidence when predicting from a single grade than from an inconsistent pair of grades. This is not justifiable statistically.[18]

Similarly, a government leader is likely to have more confidence in a recommendation reached unanimously by a group of three advisors than in a recommendation concurred in by 10 members of a 12-man panel. This, too, is inconsistent with the laws of statistical probability. When an intelligence analyst has little data on a certain subject, but all the data are consistent, the analyst is likely to overestimate the degree of confidence he or she should have in the judgment drawn from that data.

Absence of Evidence

One of the significant differences between intelligence analysis and most academic research concerns degree of control, in determining the data that are used. The academic researcher generally tries to define his or her research problem as one for which the data are known to be available. The intelligence or policy analyst is generally drawing inferences from very incomplete data. He or she must work with the evidence at hand and somehow take into account the fact that much relevant information is known to be missing.

Ideally, intelligence analysts should recognize that relevant evidence is lacking and be able to factor this into their calculations, estimating the potential impact of the missing data and adjusting confidence in their judgment downward in recognition that key information is unavailable. Unfortunately, this ideal may not be the norm. "Out of sight, out of mind" may be a better description of the impact of gaps in the evidence.

This problem can be demonstrated using a fault tree, which is a schematic drawing showing all the things that might go wrong with any endeavor. Fault trees are often used to study the fallibility of complex systems such as a nuclear reactor or space capsule. Figure 5 is a fault tree showing all the reasons why an automobile might not start.

The "car won't start" fault tree in Figure 5 was shown to several groups of experienced mechanics.[19] One group was shown the full tree and asked to imagine 100 cases in which a car won't start. Members of this group were then asked to estimate how many of the 100 cases were attributable to each of the seven major branches of the tree, that is, to battery failure, ignition system failure, etc. A second group of mechanics was shown only an incomplete version of the tree; three major branches were omitted in order to test how sensitive the test subjects were to what was left out.

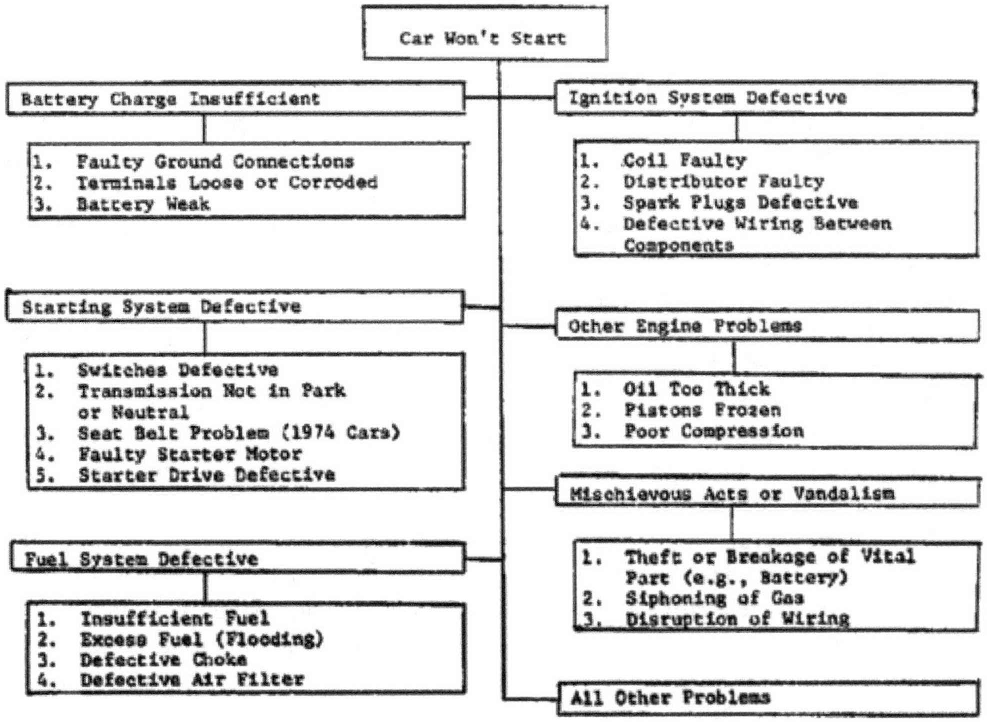

Figure 5

If the mechanics' judgment had been fully sensitive to the missing information, then the number of cases of failure that would normally be attributed to the omitted branches should have been added to the "Other Problems" category. In practice, however, the "Other Problems" category was increased by only half as much as it should have been, indicating that the mechanics shown the incomplete tree were unable to fully recognize and incorporate into their judgments the fact that some of the causes for a car not starting were missing from the fault tree. When the same experiment was run with nonmechanics, the effect of the missing branches was much greater.

As compared with most questions of intelligence analysis, the "car won't start" experiment involved rather simple analytical judgments. That the presentation of relevant variables in the abbreviated fault tree was incomplete could and should have been recognized easily by the experienced mechanics selected as test subjects. That the mechanics performed so poorly on this experiment suggests that intelligence analysts may have similar problems. Missing data is a normal characteristic of intelligence problems, and it is probably more difficult to recognize and incorporate the missing data in judgment in abstract intelligence problems than in the more concrete "car won't start" experiment.

Persistence of Impressions Based on Discredited Evidence

Impressions tend to persist even after the evidence that has created those impressions is fully discredited. Psychologists have become interested in this phenomenon because many of their experiments require that the test subjects be deceived, for example, that they be made to believe they were successful or unsuccessful in performing some task or that they possess certain abilities or personality traits when this is not in fact the case.

Professional ethics require that test subjects be disabused of these false impressions at the end of the experiment, but this has proven surprisingly difficult to achieve.

Students' erroneous impressions concerning their logical problem-solving abilities persevered even after the students were informed that manipulation of good or poor teaching performance had virtually guaranteed their success or failure.[20] Similarly, test subjects asked to distinguish true from fictitious suicide notes were given feedback that had no relationship to actual performance; the test subjects had been randomly divided into two groups, with members of one group being given the impression of above average success and the other of relative failure at this task. The subjects' erroneous impressions of the difficulty of the task and of their own performance persisted even after they were informed of the deception, that is, informed that their alleged performance had been preordained by their assignment to one or the other test group. Moreover, the same phenomenon was found among observers of the experiment as well as the immediate participants.[21] The impressions persisted even after the evidence on which they were based was fully discredited.

There are several cognitive processes that might account for this phenomenon. The previously mentioned tendency to interpret new information in the context of pre-existing impressions is relevant here but probably not sufficient to explain why the pre-existing impression cannot be eradicated even when the new information authoritatively discredits the evidence on which it is based. An interesting but speculative explanation draws on the strong human tendency to seek causal explanations.

When evidence is first received, it is perceived within a context that implies causal connections between the evidence and some antecedents that explain the evidence. The stronger the perceived causal linkage between the evidence and its antecedents, the stronger the impression created by the evidence. Thus in the experiment with suicide notes, one test subject attributed her apparent success in distinguishing real from fictitious notes to her empathetic personality and the insights she gained from the writings of a novelist who committed suicide. Another ascribed her apparent failure to lack of familiarity with people who might contemplate suicide. Even after learning that the feedback concerning their performance was invalid, these subjects retained this plausible basis for inferring that they were either well or poorly qualified for the task. Thus their initial impressions of task difficulty and of their own ability remained unchanged.[22]

In more general terms, when evidence is received, we postulate a set of causal connections that explains this evidence. Even though the evidence may subsequently be discredited, the causal linkages remain plausible and may be seen as sufficient to imply the existence of an event even in the absence of the now-discredited evidence. The previously preceived causal linkage comes easily to mind. It is a readily "available" (note previous discussion of availability bias) explanation that makes the event seem more likely than it would have appeared prior to receipt of the discredited evidence.

Implications for Deception

The bias favoring a small amount of consistent information over a large body of less consistent data supports the common maxim in deception operations that the deceiver should control as many information channels as possible in order to reduce the amount of discrepant information available to the target. Deception can be effective even with a small amount of information as long as the target does not receive contradictory data. Not only should the notional picture be consistent, but the deceiver should actively discredit the real picture as well. To achieve maximum consistency, it is necessary to discredit the true as well as build up the false.

To deception planners, the "car won't start" experiment suggests that deception is unlikely to fail because of information that is not provided. The absence of evidence is often overlooked, so errors of omission will be less serious than errors of commission. Conversely, the analyst attempting to detect deception would be well advised to consider carefully what information is missing. If the enemy were planning X, what would be the observable consequences of this plan, what is the likelihood this evidence could in fact be observed, and what inferences should be drawn from the fact that certain evidence is not observed?

Neither of the above conclusions is at all surprising. The persistence of impressions based on discredited evidence, however, does have counter-intuitive implications. The impressions created by information fed through a double agent may persist even after the opposition learns that its agent has come under control and that information from this source cannot be trusted. If we give credence to information and it affects our thinking and we subsequently learn that this information was deliberately leaked by an enemy, this subsequent knowledge does not necessarily reduce the impact of the initial report. Once information rings a bell, so to speak, the bell cannot be "unrung."

The ambiguity of most real world situations contributes to the operation of this perseverance phenomenon. Rarely in the real world is evidence so thoroughly discredited as is possible in the experimental laboratory. Assume, for example, that you receive a report that an intelligence agent you run has come under hostile control. Assume further that you have formed a number of impressions on the basis of reporting from this agent.

It is easy to rationalize the perseverance of these impressions by arguing that the information was true despite the agent being under hostile control, or by doubting the validity of the report claiming the agent is under control. In the latter case, the phenomenon of impression perseverance may itself affect evaluation of the evidence that supposedly discredits the impression; it is because we retain our initial impression that we disbelieve the new evidence.

It is a truism that security is an essential element of successful deception. If the deception is undertaken to protect the security of an operational plan, compromise of the deception might be worse than no deception at all, for it could attract attention to the true plan. While security is obviously desirable, it may not be quite as essential as past deception planners have believed, for there are cognitive factors that help reduce the adverse consequences of security leaks. The persistence of impressions based on discredited evidence is one of these. Others dealing with the human capacity to rationalise contradictory evidence have been discussed in the perceptual biases section. There is empirical evidence to support this conclusion. Of the 63 cases of strategic surprise or deception studied by Barton Whaley, none had perfect security. Some more or less specific warnings were present in every case, yet surprise or deception was successful nonetheless.[23]

Biases in the Perception of Causality

We cannot see causation in the same sense that we see a desk or a tree. Even when we observe one billiard ball strike another and then observe the previously stationary ball begin to move, we are not seeing causation. The most we can see is the juxtaposition of events in time and space. The perception of causation results only from a complex process of inference, not from direct observation. As other forms of inference, it is subject to systematic biases. The two biases discussed in this section increase the likelihood that analysts will perceive deception when it is not in fact present.

Bias Toward Causal Explanations

We have a deep psychological need to understand our environment. Understanding implies order, so we arrange our observations into regular patterns and relationships. Happenings that we cannot understand may be attributed to God's will or to fate, which is somehow preordained, for we resist the thought that outcomes may be determined by forces that interact in random, unpredictable ways. People generally do not accept the notion of chance or randomness. Even dice players behave as though they exert some control over the outcome of a throw of dice.[24]

Because of this need to impose order on our environment, we may seek and see patterns that actually are not there. Some recent research in paleobiology seems to illustrate this tendency. A group of paleobiologists has developed a computer program to simulate evolutionary changes in animal species over time. But the transitions from one time period to the next are not determined by natural selection or any other regular process; they are determined by computer-generated random numbers. The patterns that are produced by this program are very similar to the patterns in nature that paleobiologists have been trying to understand. Events that seem, intuitively, to have a very strong pattern, were in fact generated by random processes.[25] This suggests that there may, in fact, be no valid causal explanation of evolution.

B.F. Skinner noted a similar phenomenon in the course of experiments with the behavioral conditioning of pigeons. The normal pattern of these experiments was that the pigeons were given positive reinforcement, in the form of food, whenever they pecked on the proper lever at the proper time. To obtain the food regularly, they had to learn to peck in a certain sequence. Skinner demonstrated that the pigeons "learned" and followed a pattern even when the food was actually dispensed randomly.[26]

These examples suggest that in military and foreign affairs, where the patterns are at best very difficult to fathom, there may be many events for which there is no valid causal explanation. Our bias against randomness as an explanation may cause us to impose a pattern on these events so that we see causal relationships that are not in fact there. It clearly does not do much for our ego as analysts to admit that some of the things we are called upon to explain might be caused by random processes such as the random numbers used in the paleobiologists' computer program. And it is certainly unlikely that the customers for our intelligence would appreciate such an explanation, so there are motivational biases that reinforce the cognitive bias favoring order over randomness.

The need to perceive order and reason in the world around us causes us to overestimate the extent to which other countries or other people are pursuing a coherent, rational, goal-maximizing policy. We tend to see the actions of other governments as the intentional result of central direction and planning, and to overlook the fact that the same behavior might be more accurately explained by accident, blunder, the unintended consequence of well-intentioned policy, improperly executed orders, bargaining among semi-independent bureaucratic entities, or following standard operating procedures under inappropriate circumstances.

Internal vs. External Causes of Behavior

Attribution theory is a sub-field of psychology dealing with how we assess the causes of behavior. Most research in attribution theory employs a basic dichotomy between internal and external causes of behavior. Internal causes include a person's attitude, beliefs and personality. External causes include such factors as incentives and constraints, role requirements, or difficulty of a task. Attribution theory examines the circumstances under which we attribute behavior to either internal or external causes. Such differences in attribution may have significant consequences for behavior, as our response to friendly or unfriendly actions of other persons may be quite different if we attribute the behavior to the nature of the person than if we see the behavior as resulting from situational constraints over which that person had little control.

The fundamental attributional error is to overestimate the importance of personal traits and dispositions in determining behavior. When we observe another's behavior, we are too quick to infer broad personal qualities or dispositions from this behavior and to expect that these same dispositions will determine the actor's behavior in other contexts. The so-called "Peter Principle" is a simple illustration of this bias. An employee is promoted to his or her level of incompetence because the supervisor attributes the employee's promotion-meriting performance exclusively to personal ability and assumes that this personal characteristic will continue to determine performance despite changes in the situational context. Much research into personality traits, however, shows that personal traits are not consistent determinants of behavior; which trait predominates at any given time is heavily dependent upon the situational context in which the behavior takes place.

Most interesting for our purposes, however, is that susceptibility to this attributional error depends upon whether we are examining our own behavior or observing the behavior of others. We tend to attribute the behavior of others to the nature of the person, while we see our own behavior as conditioned by the nature of the situation in which we find ourselves.[27]

This bias is partially explained by differences in information available to actors and observers. In evaluating our own behavior, we compare our present behavior with our own past behavior in similar or different contexts. This past behavior is well known to us, so it is easy to compare the impact of different situations on our behavior over time. This causes us to focus on the nature of the situation as the principal variable explaining differences in our own behavior. The observer of another person, on the other hand, typically lacks this depth of knowledge of the other person's behavior in other circumstances. So the observer's orientation is to examine how the actor's behavior compares with the behavior of other <u>persons</u> under similar circumstances. This prompts a focus on the nature of the person rather than on the nature of the situation. Other differences in perspective between actor and observer may also contribute to this bias.

I know of no experimental evidence that this bias applies to our perception of the behavior of countries as well as the behavior of individuals, but such an extrapolation seems plausible and is supported by personal experience. Reportedly one of the persistent differences between intelligence analysts responsible for the Soviet Union and those responsible for China and working on Sino-Soviet relations is this: Soviet analysts tend to attribute Chinese behavior to the nature of the Chinese, while they see Soviet options as circumscribed by many situational constraints. Chinese analysts tend to take the opposite view, that is, that the Russians behave like Russians while Chinese actions are the product of the situation in which the Chinese find themselves.[28]

Thus familiarity, either with oneself or the country for which one is responsible, produces empathy and understanding, and attribution of behavior to external circumstances rather than to the nature of the actor. Lack of information concerning the past behavior and current circumstances of an actor, or lack of empathy for whatever reason, causes us to perceive that actor's behavior as stemming from the nature of the actor. As with all the cognitive biases, we are describing a tendency, not a black and white rule that applies to all people in all cases. In assessing the behavior of others, we normally do make some allowance for situational pressures and role requirements, but this allowance is often insufficient.

A principal implication for international relations is that this bias sows the seeds of mistrust and misunderstanding, as countries have quite different perceptions of the causes of each other's behavior. There are also several corollaries and related biases that are quite relevant to the analysis of international affairs.[29]

When we fall prey to the attributional bias of judging another country's behavior to be more heavily influenced by the nature of the people or the leaders than is in fact the case, we tend to perceive this state as more hostile than it really is. If actions that adversely affect our interests are attributed to the predispositions and attitudes of the other country, we perceive these actions as expressing hostility. If, however, the other nation's actions are actually responsive to situational constraints, it is unnecessary to assume hostile intent. Similarly, attribution of behavior to personal or national characteristics and the assumption that these characteristics are consistent over time leads to the perception of behavior as inflexible and unchanging. Conversely, to the extent that behavior is attributed to external circumstances, it is perceived as flexible and subject to influence by our own actions.

Implications for Deception

Deception planners need to avoid these biases relating to causality in order to evaluate accurately the situation in which they find themselves and to estimate how a target is likely to respond to whatever information is provided. But the most direct relevance of these biases to the question of deception is their impact on the analyst seeking to detect and avoid deception. Both biases tend to make analysts perceive deception when it is not really there.

Deception is an example par excellence of a policy that is centrally directed, well planned, and highly coherent and rational. As a causal explanation, deception is intrinsically satisfying precisely because it is so orderly and rational. When other persuasive explanations are not available (perhaps because the phenomena we are seeking to explain were actually caused by mistakes, failure to follow orders, or other factors unknown to us), deception offers a convenient and easy explanation. It is convenient because intelligence analysts are generally sensitive to the possibility of deception, and its detection is often taken as indicative of sophisticated, penetrating analysis. It is easy because almost any evidence can be rationalized to fit the deception hypothesis; in fact, one might argue that once deception has been raised as a serious possibility, this hypothesis is almost immune to disconfirmation. While deception is by no means an uncommon phenomenon, I suspect that our bias toward seeing events as part of an orderly pattern leads us to perceive deception more frequently than is warranted.

This tendency to perceive deception is reinforced by the bias toward perceiving the behavior of others as caused by the nature of the person rather than by situational constraints. When another person or government employs deception, we commonly (although perhaps erroneously) consider this decision to stem primarily from the nature of the person or government rather than the nature of the situation in which this person or government finds itself. It is satisfying to attribute deviousness and malevolence to our enemies. And if they are devious and malevolent, of course they will engage in deception. Deception is "them" acting, not just responding to events. When we observe activity that we do not otherwise understand, deception may be a more attractive explanation than to simply admit that we have insufficient information or understanding of the situation,

Our recollection of the many cases in which deception has proven successful makes us think we should be more rather than less sensitive to the possibility of deception. But the fact that deception is generally successful does not mean people are insufficiently disposed to perceive it, or that increased alertness will protect us from being taken in. Deception is generally successful despite our predisposition to perceive it, not because of any tendency to overlook this possibility. As we shall see shortly, greater alertness to deception increases our vulnerability to the most common form of deception. To determine whether governments tend to be oversensitive to deception or not sensitive enough, it is insufficient to look only at the many cases in which countries have been deceived. It is equally necessary to examine cases in which they have perceived deception erroneously. Such cases may be equally common but are seldom documented as they are intrinsically less interesting to historians.

Conclusion

We have examined a number of perceptual and cognitive biases and their implications for strategic deception and counterdeception. For quick reference, the biases and their implications are summarized in tabular form in the Appendix. Three primary conclusions emerge from this examination.

1. Perceptual and cognitive biases strongly favor the deceiver as long as the goal of deception is to reinforce a target's preconceptions or to simply create ambiguity and doubt about the deceiver's intentions. Under these circumstances, which are by far the most common forms of deception,[30] the deceiver clearly holds most of the cards. If the situation is such that the deceiver can achieve planned goals only by changing the target's preconceptions, however, the target is shielded by many of the same perceptual and cognitive biases that otherwise work to his or her disadvantage.
2. While security is obviously desirable for any deception plan, perfect security is rarely attained and deceptions succeed without it. When the deception is planned to reinforce preconceptions, the target's ability to rationalize discrepant information tends to offset security leaks and uncontrolled channels of information. Even after a source of information has been discredited, impressions created by the information from that source tend to persist.
3. Analysts are generally predisposed to perceive deception. Instances of successful deception are far easier to recall than cases in which deception was not employed

> under similar circumstances, and this sensitizes us to the possibility of deception. We are attracted to deception as an explanation for otherwise incongruous events because the deception explanation enables us to impose order and reason on a disorderly world, and because it enables us to attribute deviousness and malevolence to our enemies. These factors sometimes cause us to perceive deception when it is not really present.

One might think that the analysts' predisposition to perceive deception would offset the advantages we have attributed to the deceiver, but such is not the case. Deception is generally successful despite the target's alertness. According to Barton Whaley's analysis of 68 cases of strategic surprise or deception between 1914 and 1968, deception was successful in 91% of the cases in which it was attempted.[31] As in so many other fields, major advantages accrue to the actor who seizes the initiative, rather than to the reactor who seeks to parry the initiatives of others. Whaley's finding highlights the unenviable position of the intelligence analyst seeking to avoid being deceived, and the remainder of our conclusion is devoted to a closer look at the analyst's plight and what might be done to alleviate it.

The fundamental problem involved in avoiding deception, and avoiding intelligence surprise in general, is the problem of determining when to change our mind in response to new information that does not jibe with our current conception. If we are unreceptive to new information, we cannot learn from experience or keep abreast of changing circumstances and situations. If we are too receptive, we are unduly influenced by the most recent report or the latest short-term trend. There is no magic formula that tells us when to discount discrepant evidence and when to revise our thinking to take it into account. But as a general rule, we err more often on the side of being too wedded to our established views and thus too quick to reject information that does not fit these views, than on the side of being too quick to revise our beliefs. Thus, most of us would do well to be more open to evidence and ideas that are at variance with our preconceptions.

In his study of surprise attacks, Abraham Ben-Zvi identified two kinds of information on which estimates of impending conflict might be based—strategic assumptions and tactical indicators.[32] Examples of strategic assumptions include the U.S. belief in 1941 that Japan wished to avoid war at all costs because it recognized U.S. military superiority, and the Israeli belief in 1973 that the Arabs would not attack Israel as long as they lacked sufficient airpower to secure control of the skies. Such preconceptions are based on a large body of interrelated evidence and have usually been held for a long time. Tactical indicators are the specific reports concerning preparations or intent to initiate hostile action, or more generally, specific evidence from current events that indicates the direction in which events are moving. This distinction between strategic assumptions and tactical indicators is very similar to the distinction we have been making between pre-existing beliefs and new information.

Ben-Zvi studied five cases of intelligence failure to foresee a surprise attack: Pearl Harbor, German attack on the Soviet Union in 1941, Chinese intervention in the Korean war, Chinese attack on India in 1962, and the Arab attack on Israel in 1973. He found that in each case tactical indicators of impending attack were present but were discounted because they conflicted with analysts' and policy makers' preconceptions.[4]

4. CIA post mortems on cases of intelligence failure have also found that information that would have permitted an accurate assessment was generally available, but that this information assumed significance only with the benefit of hindsight.

The strategic assumptions were not revised in the light of the increasing flow of contrary tactical information. Ben-Zvi argues that whenever strategic assumptions of intention to attack and tactical indicators of impending attack converge, an immediate threat is perceived and appropriate preparations are made. But when there is a divergence between strategic assumptions and tactical indicators, the strategic assumptions always prevail. Thus despite the evidence of preparations for an attack, the actual attack comes as a "surprise," as in the five cases analyzed. Ben-Zvi concludes that tactical indicators should be given increased weight in the decision making process.

This may well be appropriate advice. It certainly accords with our conclusion that people err most often by being too quick to reject new information that does not conform to their preconceptions. But Ben-Zvi does not consider cases in which alarming tactical indicators have been properly discounted as maneuvers, bluff or deception rather than as indicators of impending attack. Ascribing more weight to tactical indicators in all cases will increase the frequency of false alarms, and this too entails costs. While we should in general be more open to changing our minds as a result of discrepant tactical or other information, in any single case it is impossible to know a priori whether we should revise our estimate or stick with a long-established view.

Alertness to the possibility of deception can influence the degree of one's openness to new information, but not necessarily in a desirable direction. The impetus for changing one's estimate of the situation can only come from the recognition of an incompatibility between a present estimate and some new evidence. If people can explain new evidence to their own satisfaction with little change in their existing beliefs, they will rarely feel the need for drastic revision of these beliefs. Deception provides a readily "available" explanation for discrepant evidence; if the evidence does not fit one's preconceptions, it may be dismissed as deception. Further, the more alert or suspicious one is of deception, the more readily available is this explanation. Alertness to deception presumably prompts a more careful and systematic review of the evidence. But anticipation of deception also leads the analyst to be more skeptical of all the evidence, and to the extent that evidence is deemed unreliable, the analyst's preconceptions must play a greater role in determining which evidence to believe. This leads to a paradox: the more alert we are to deception, the more likely we are to be deceived.

Actually, this paradox applies only to the type of deception in which the deceiver's goal is to exploit and reinforce our preconceptions. If the deceiver's goal is to sow confusion or make us change our mind, it will be to our advantage to ignore the evidence and stand by our preconceptions.

The problem of how to detect deception is not generically different from other common problems of intelligence analysis. It is, for example, very similar to the general problem of early warning. From a cognitive perspective, there are no prescriptions that apply uniquely to the deception problem. Consideration of ways to improve our ability to detect deception is a part of the much broader problem of improving intelligence analysis in general.

While cognitive psychology does not provide direct insights on how to detect deception, it can be of indirect assistance. By better understanding how our mind processes information, including the diverse perceptual and cognitive biases to which we are subject, we can hope to compensate for some of these basic problems in human information processing. We can identify situations in which our normal faith in our impressions should be suspended, and in which some more systematic means of handling the evidence may be appropriate. We can also identify guidelines concerning the types of analytical methods that may be most useful in supplementing intuitive judgment.

A common factor in cases of successful deception, and in most cases of intelligence surprise in general, is that analysts have become fixed in a mind set that does not respond effectively to discrepant information. Thus methods for breaking mind sets are particularly relevant to the problem of detecting deception. This includes such practices as competitive analysis, use of a devil's advocate to analyze alternative scenarios, interdisciplinary brainstorming and other techniques that facilitate the identification and analysis of alternative hypotheses.

Current research suggests that people perform poorly at generating a full set of hypotheses.[33] If the correct hypothesis is not even formulated for consideration, there is clearly little chance of making an accurate estimate. Formation of alternative hypotheses and identification of the indicators and observables associated with each hypothesis helps direct an economical search for information. The hypotheses also serve as an organizational structure for storage and recall of information in memory.

There is a strong tendency to view the significance of evidence in terms of the degree to which it supports, contradicts or seems irrelevant to what we already believe to be the case. We overlook the fact the evidence we think of as supporting our case may also be quite consistent with several alternative hypotheses, so we draw from the evidence false confirmation of our pre-existing beliefs. We can avoid this by evaluating the evidence in terms of its diagnosticity in helping revise our estimates of the relative likelihood of each hypothesis.

As we saw in the old woman-young lady experiment, it is difficult to look at the same data from several different perspectives. Yet this is exactly what is required for the simultaneous evaluation of multiple hypotheses. Some sort of methodological aid is useful to facilitate this task, as an aid to memory and to help integrate the many probabilistic judgments that are required. There are a variety of computer programs available for this purpose,[34] but significant benefits might also be obtained by simple paper-and-pencil techniques.

Our intention in these final paragraphs has been to be suggestive, not prescriptive. Traditional, intuitive methods of analysis have not been sufficiently effective in detecting deception, so it is necessary to explore other alternatives. We have tried to point out some useful directions for this exploratory effort, but a fuller discussion of analytical methodology goes far beyond the scope of this paper.

Appendix

Review of Biases and Their Implications for Deception

Bias	Implication
Perceptual	*Biases*
We tend to see what we expect to see. It takes more information to recognize an unexpected phenomenon than an expected one.	It is far easier to reinforce a target's existing preconceptions than to change those beliefs.

Perceptions are quick to form but resistant to change. Once we have formed an impression about an object, event or situation, we are biased toward continuing to perceive it in the same way.	It is far easier to reinforce a target's existing preconceptions than to change them. Ability to rationalize contradictory information may offset hazards of security leaks or uncontrolled channels.
Initial exposure to ambiguous or blurred stimuli interferes with accurate perception even after more and better information becomes available.	Impact of information can be affected by the sequence used in feeding it to a target.
Biases in Estimating Probabilities	
We estimate probability according to how easily we can imagine an event or recollect instances of the event. Known as availability bias.	Because deception is more salient than absence of deception, analysts will believe deception is more common than it really is, and thus be more disposed to perceive it. Employees of watch offices will generally overestimate the probability of whatever they are watching for. This leads to the cry wolf syndrome that can be exploited by deception planners. Analysts seeking to avoid surprise should not make judgments based on availability.
We estimate probability by seizing upon some natural starting point as a first approximation, then adjust the estimate incrementally in response to new information or further analysis. Normally we do not adjust enough. Known as anchoring bias.	Because people do not revise their judgments enough as new information is received, it is easier to conduct a deception aimed at reinforcing the target's existing preconceptions than to change these beliefs.
In translating our subjective feelings of certainty into a probability estimate, we are generally overconfident about how much we know.	Overconfidence exacerbates the impact of all the biases, for if we are satisfied with our judgments we will make fewer efforts to improve them. An overconfident target is easier to deceive when the deception involves exploiting existing preconceptions.
Biases in Evaluating Evidence	
We are biased toward seeing events as part of an orderly, causal pattern. Randomness, accident and error tend to be rejected as explanations for observed events. We overestimate the extent to which other people or countries are pursuing a coherent, rational goal-maximizing policy.	We are predisposed to perceive deception even when it is not really there. As a causal explanation, deception is intrinsically satisfying because it is so orderly and rational.
We tend to attribute the behavior of others to the nature of the person or country, while we see our own behavior as conditioned by the nature of the situation in which we find ourselves.	We are predisposed to believe our enemies will engage in deception. It is satisfying to attribute deviousness and malevolence to our enemies, and if they are devious and malevolent, of course they will engage in deception.

Endnotes

1. Herbert A. Simon, Models of Man: Social and National (New York: Wiley, 1957).
2. Robert Jervis, Perception and Misperception in International Politics (Princeton, N.J.: Princeton University Press, 1976).
3. Ibid., p. 3.
4. Richard Betts, "Analysis, War and Decision: Why Intelligence Failures Are Inevitable," World Politics, XXXI (October, 1978), 65.

5. Jerome S. Bruner and Leo Postman, "On the Perception of Incongruity: A Paradigm," in Perception and Personality: A Symposium, eds. Jerome S. Bruner and David Krech. New York: Greenwood Press, 1968.
6. Jervis, op. cit., Chapter 10.
7. Betts, op. cit., p.84.
8. Drawings devised by Gerald Fisher in 1967.
9. Jervis, op. cit., p. 195.
10. This picture was originally published in Puck magazine in 1915 as a cartoon entitled "My Wife and My Mother-in-law."
11. Jerome S. Bruner and Mary C. Potter, "Interference in Visual Recognition," Science. 144 (1964), 424-425.
12. Amos Tversky and Daniel Kahneman, "Availability: A Heuristic for Judging Frequency and Probability," Cognitive Psychology. 5 (1973), 207-232.
13. Amos Tversky and Daniel Kahneman, "Anchoring and Calibration in the Assessment of Uncertain Quantities," Oregon Research Institute Research Bulletin, 12 (1972).
14. R. M. Cambridge and R. C. Shreckengost, Are You Sure? The Subjective Probability Assessment Test (Information Science Center Monograph, Office of Training, Central Intelligence Agency, March 1978).
15. Sarah Lichtenstein, Baruch Fischhoff and L. D. Phillips, Calibration of Probabilities: The State of the Art. Technical Report DDI-3 (Eugene, Oregon: Oregon Research Institute, 1976).
16. M. Alpert and H. Raiffa, "A Progress Report on the Training of Probability Assessors," Unpublished Manuscript (Cambridge, Mass.: Harvard University, 1968).
17. Nicholas Schweitzer, "Bayesian Analysis: Estimating the Probability of Middle East Conflict," Quantitative Approaches to Political Intelligence: The CIA Experience, ed. Richards J. Heuer, Jr. (Boulder, Colo.: Westview Press, 1978); Stephen J. Andriole, "Computer-Based Bayesian Forecasting Methodologies" and Richards J. Heuer, Jr., "Applications of Bayesian Inference in Political Intelligence," both in Gerald W. Hopple and James A. Kuhlman, eds., Expert Generated Data: Applications in International Affairs (Boulder, Colo.: Westview Press, Forthcoming 1980).
18. Daniel Kahneraan and Amos Tversky, "Intuitive Prediction: Biases and Corrective Procedures," Management Science, special issue devoted to forecasting applications, in press.
19. Baruch Fischhoff, Paul Slovic and Sarah Lichtenstein, Fault Trees: Sensitivity of Estimated Failure Probabilities to Problem Presentation, Technical Report PTR-1042-77-8 (Eugene, Oregon: Decision Research, August, 1977).
20. R. R. Lau, M. R. Lepper and L. Ross, "Persistence of Inaccurate and Discredited Personal Impressions: A Field Demonstration of Attributional Perseverance," paper presented at 56th Annual Meeting of the Western Psychological Association (Los Angeles, April 1976).
21. Lee Ross, Mark R. Lepper and Michael Hubbard,"Perseverance in SelfPerception and Social Perception: Biased Attributional Processes in the Debriefing Paradigm," Journal of Personality and Social Psychology, 32, 5 (1975), 880-892.
22. Lee Ross, Mark R. Lepper, Frit Strack and Julia Steinmetz, "Social Explanation and Social Expectation: Effects of Real and Hypothetical Explanations on Subjective Likelihood," Journal of Personality and Social Psychology. 33, 11 (1977), 818.
23. Barton Whaley, Stratagem. Deception and Surprise in War (MIT, Unpublished Manuscript, 1969), 164.
24. Ellen J. Langer, "The Psychology of Chance," Journal for the Theory of Social Behaviour. 7 (1977), 185-208.
25. Gina Bari Kolata, "Paleobiology: Random Events over Geological Time," Science. 189 (1975), 625-626.
26. B. F. Skinner, "Superstition In the Pigeon," Journal of Experimental Psychology. 38 (1948), 168-172.
27. B. F. Skinner, "Superstition In the Pigeon," Journal of Experimental Psychology. 38 (1948), 168-172.
28. Based on author's discussion with CIA analysts.
29. For a more detailed discussion of these, see Jervis, op. cit., Chapters 8 and 9.
30. Based on secondary analysis of Whaley, op. cit., Appendix B. Of the cases of deception studied by Whaley, 792 exploited the target's preconceptions.
31. Based on secondary analysis of Whaley, op. cit., Appendix B.
32. Abraham Ben-Zvi, "Hindsight and Foresight: A Conceptual Framework for the Analysis of Surprise Attacks," World Politics, XXVIII, 3 (1976), 381-395.

33. Thomas Mehle and Charles Gettys, "Pre-Decision Behavior: Are People Any Good at Coming Up with Hypotheses?", paper presented at the Bayes Conference (Los Angeles, February 1979). Additional research on this topic is underway at the Decision Processes Laboratory, University of Oklahoma.
34. Programs of this type for use in early warning and other intelligence problems have been developed by several contractors, in particular Decisions & Designs, Inc., McLean, Va. For several references, see footnote 17.

CHAPTER 17

The Theory of Practical Joking—Its Relevance to Physics[1]

R. V. Jones

Dr. Jones was a leading military deception planner, experimental physicist, and practical joker—three nominally separate activities that he links by theory in this landmark paper. By doing so, Jones extends his special theory of military deception into other disciplines in order to create the first general and systematic theory of deception.

Specifically, Jones argued that practical jokes and hoaxes work for the same reasons that military deception plans work. Both involve "induced incongruities" where, by presenting false evidence, the deceiver lets the victim "build up an incorrect but self-consistent world-picture," thus causing him to take actions that are incongruent with reality. Dr. William Harris took Jones's concept and proposed that it be reverse engineered to detect deception—a process he labeled "incongruity analysis" and, later "incongruity testing." Whaley then applied Incongruity Analysis as the basis of all deception detection or counterdeception.

1. R. V. Jones, "The Theory of Practical Joking: Its Relevance to Physics," *Bulletin of the Institute of Physics*, Vol. 8, June 1957, pp. 193-201. © Institute of Physics. Reproduced by permission of IOP Publishing.

The Theory of Practical Joking — Its Relevance to Physics*

By Professor R. V. JONES
C.B., C.B.E., M.A., D.Phil., F.Inst.P.

(*University of Aberdeen*)

AT FIRST SIGHT there may seem little relation between physics and practical joking. Indeed, I might never have observed their connection but for an incidental study of the life of James Clerk Maxwell. Two things, among many others, struck me. The first was the growth of his sense of fun from the primitive joke of the boy of six tripping up the maid with the tea tray to the refined, almost theoretical, jokes of his later life. The second was his mastery of analogy in physical thinking : already, at the age of twenty-four he had written a part-playful, part-serious essay on the theory of analogy which showed two of the main features of his mind. On the lighter side, he pointed out the relation between an analogy and a pun : in the former one truth lies under two expressions, and in the latter two truths lie under one expression. Hence from the theory of analogy one can by reciprocation deduce the theory of puns. To the more serious side of Maxwell's understanding of analogy I shall return later, but all this set me thinking about the possible connection between the theory of practical joking and physics. One factor which encouraged me was the high incidence of mischievous humour among physicists. Even Newton, it is recorded, caused trouble in his Lincolnshire village as a boy by flying at night a kite carrying a small lantern ; and in this century the spritely skill of the late Professor R. W. Wood and Professor G. Gamow is already legendary. While I hope to illustrate this paper with examples, I propose first to analyse (if this is not altogether too brutal a process) the essentials of a joke.

INCONGRUITIES

The crux of the simplest form of joke seems to be the production of an incongruity in the normal order of events. The schoolboy laughs, for example, at the sight of his master slipping on a banana skin because this is distinctly abnormal behaviour on the part of one whom the boy regards as dignified. And when we contemplate such a scene we can see that there may indeed be something in the psychologist's theory that laughter had its origin in shouts of triumph over a fallen enemy. Now in most jokes the incongruity is neither as simple nor as violent as this one ; and it is not necessary that it should cause pain. Hence, if we follow the career of an individual joker who is also a man of feeling, we find that his jokes grow more refined and less painful, while the incongruities mount.

We hear the story, for example, of Maxwell showing Kelvin some optical experiment, and inviting Kelvin to look through the eyepiece. Kelvin was surprised to find that, while the phenomenon described by Maxwell was undoubtedly there, so was a little human figure, the incongruity, dancing about. Kelvin could not help asking "Maxwell—but what is the little man there for ?" "Have another look, Thomson," said Maxwell, "and you should see." Kelvin had another look, but was no wiser. "Tell me, Maxwell," he said impatiently, "What

* Based on a lecture given to the Midland Branch of the Institute on February 12, 1957.

is he there for ?" . . . "Just for *fun*, Thomson," replied Maxwell. When we consider a simple incongruity of this type, we can see why this form of humour is sometimes described as "nonsense"; for "sense" implies the normal order of things, and in this order an incongruity makes "nonsense." A simple incongruity in the literature of physics is R. W. Wood's recording of the fact that he cleaned out an optical instrument by pushing his cat through it.

Incongruities are always lurking round the corner, to be called forth even by an errant comma. I saw a report of an address by a Low Church minister in which, speaking of vestments, he had said : "I want to wear no clothes which would distinguish me from any other man." The printer, however, had inserted a comma after "clothes," with impressive effect.

Even a change of dimension is sufficient to cause an incongruity. Lord Cherwell has a story of a scientist at Farnborough in World War I, who was so dismayed by the delays in ordering commercial equipment that when he wanted a dark-room lamp he made a pencil sketch of one, to be made up by the workshop. It availed him little, however, because a proper engineer's drawing had by regulation to be made in triplicate before the workshop would start. Weeks elapsed, and finally after a knock on his door two workmen wheeled in the largest dark-room lamp ever constructed. In making the workshop drawing the draughtsman had left out one dash, with the result that intended inches became actual feet. One of the classic incongruities of this type is that due to Benjamin Franklin in a letter to the Editor of a London newspaper in 1765, chaffing the English on their ignorance of America : "The grand leap of the Whale up the Falls of Niagara is esteemed, by all who have seen it, as one of the finest spectacles in Nature !" It is not, however, necessary for humour that the change of dimension should result in exaggeration. Understatement is at least as effective. We shall long remember Sir Winston Churchill's comment on General Weygand's 1940 foreboding that in three weeks England would have her neck wrung like a chicken : "Some chicken, some neck !"

A variation on the simple incongruity in humour is to produce a congruity where incongruity is normally expected. One does not expect, for example, any congruity about the names of joint authors of scientific papers. It was therefore rather a surprise to find a genuine paper by Alpher, Bethe and Gamow, dated April 1 in *The Physical Review* for 1948.* And one would be equally surprised to find a reference to ecclesiastical doctrine in a modern scientific paper ; but Gamow contrived to make a relevant reference to an address by Pope Pius XII in a *Physical Review* paper in 1952. Many puns fall in this unexpected congruence class, and are specially notable in naval signals. The most recent volume of the naval history of the 1939-45 war reminds us of the commander of one of our submarines who signalled to an accompanying destroyer at the beginning of an action "I intend to remain on the surface as long as possible" and received from the destroyer commander the reply "And so do I !"

The capacity to perceive congruence among apparent incongruities is one of the essentials of wit, and as an example we may take the incident at a peace conference after Waterloo when the French Marshals ostentatiously turned their backs on the Duke of Wellington as he entered the room. "I have seen their backs before," observed the Duke. Indeed, this capacity for perceiving congruences is akin to that of the search of science for underlying unity.

A further variation of humour is produced when a false incongruity is expected by the victim, and an incongruity then genuinely occurs which he promptly discounts. The late Sir Francis Simon had this happen to him when he was head of a laboratory in Germany. One night his research students were working with liquid hydrogen, and there was an explosion which damaged the laboratory some time after midnight. One of the research students telephoned the

* Dr. Gamow says in his *Creation of the Universe* that there was afterwards a rumour that "when the α, β, γ theory went temporarily on the rocks, Dr. Bethe seriously considered changing his name to Zacharias," and that a fourth collaborator stubbornly refused to change his name to Delter.

professor to inform him of the damage. All he could get from Sir Francis was an amiable "All right, I know what day it is!" It was the morning of April 1.

HOAXES

Simple incongruities, direct or inverted, can be humorous enough, but the more advanced jokes usually involve a period of preparation and induction, sometimes elaborate, before the incongruity becomes apparent. They are then called hoaxes. Maxwell's jokes were often simple in their preparation; he is credited with having engineered the advertisement of his Inaugural Lecture at Cambridge (which is still very well worth reading) in such a manner that only his undergraduate students heard of it, and he gave it to them alone. The senior members of the University merely saw that the new professor would deliver his first lecture on a particular day, and they attended in force. This lecture, however, was the first of his undergraduate course, and his delighted students enjoyed the experience of seeing Maxwell gravely expounding, though with a betraying twinkle in his eye, the difference between the Fahrenheit and Centigrade scales to men like Adams, Cayley, and Stokes.

With some hoaxes the period of induction of the victim may be extended. In this type, which is probably the most interesting philosophically, the object is to build up in the victim's mind a false world-picture which is temporarily consistent by any tests that he can apply to it, so that he ultimately takes action on it with confidence. The falseness of the picture is then starkly revealed by the incongruity which his action precipitates. It has not proved difficult, for example, to persuade a Doctor of Philosophy to lower his telephone carefully into a bucket of water in the belief that he was co-operating with the engineer in the telephone exchange in finding a leak to earth. The prior induction consisted of building up in his mind a picture of something being wrong with his telephone by such tactics as repeatedly ringing the bell and then ringing off as he answered.

As a further example, we may recall one of the works of a German physicist, Dr. Carl Bosch, who about 1934 was working as a research student in a laboratory which overlooked a block of flats. His studies revealed that one of the flats was occupied by a newspaper correspondent, and so he telephoned this victim, pretending to be his own professor. The "professor" announced that he had just perfected a television device which could be attached to a telephone, and which would enable the user to see the speaker at the other end. The newspaper man was incredulous, but the "professor" offered to give a demonstration; all the pressman had to do was to strike some attitude, and the voice on the telephone would tell him what he was doing. The telephone was, of course, in direct view of the laboratory, and so all the antics of the pressman were faithfully described. The result was an effusive article in the next day's paper and, subsequently, a bewildered conversation between the true professor and the pressman.

The induction of the victim can take many forms. One of the favourite ways is an acclimatization by slow change. R. W. Wood is said to have spent some time in a flat in Paris where he discovered that the lady in the flat below kept a tortoise in a window pen. Wood fashioned a collecting device from a broom-handle, and bought a supply of tortoises of dispersed sizes. While the lady was out shopping, Wood replaced her tortoise by one slightly larger. He repeated this operation each day until the growth of the tortoise became so obvious to its owner that she consulted Wood who, having first played a subsidiary joke by sending her to consult a Professor at the Sorbonne whom he considered to be devoid of humour, advised her to write to the press. When the tortoise had grown to such a size that several pressmen were taking a daily interest, Wood then reversed the process, and in a week or so the tortoise mysteriously contracted to its original dimensions.

Induced incongruities mutual to two victims are possible. Each victim is led to believe in a false world-picture which is nevertheless consistent with and complementary to the false world-picture of the other victim. A simple example is the device of privately telling each of two people whom one is about to introduce that the other is a good fellow who has, however, been going through a severe nervous strain, and who is therefore apt to get both irritable and rude if contradicted. It is therefore advisable to humour him by agreeing with all that he says. The two victims then go to great lengths to agree with one another, and separate in the conviction that each has handled the other extremely well.

HOAXES IN WAR

Induced incongruities have a high place in warfare, where if the enemy can be induced to take incorrect action the war may be advantageously affected. A stratagem in which some of my wartime colleagues were involved is now well known as "The man who never was." These same colleagues also worked with me in some technical deceptions, of which one was the persuasion of the Germans in 1943 that our successes against the U-boats were due not to centimetric radar but to a fictitious infra-red detector. We gained some valuable months while the Germans invented a beautiful anti-infra-red paint and failed to find the true causes of their losses. The paint, incidentally, was a Christiansen filter of powdered glass in a transparent matrix over a black base. The filter "peaked" in the near infra-red, so that incident radiation in this region went through and was absorbed in the underlying black. Visible light was scattered back by the filter, which thus gave a light grey appearance to the eye, but was black to the near infra-red. This simulated admirably the reflecting power of water, and thus camouflaged the U-boat. It was afterwards reported that the inventor of the paint was Dr. Carl Bosch.

We must readily admit that not all the successful hoaxes in the war were British. The book *Operation Nordpol* tells of a hoax on the British sabotage organization by a German security officer, Colonel Giskes. Although he had captured several British agents, Giskes succeeded in making it appear that they were still at large, and persuaded us by counterfeit radio messages to keep them supplied with sabotage equipment. The peak of his achievement was when he persuaded us to drop them by parachute a supply of tennis balls instead of explosive, on the grounds that the agents were in touch with the King of the Belgians who was very fond of tennis, and that the balls would therefore make excellent douceurs in securing his aid. I hope that our admiration for such technique will always rise above the havoc of war. A sense of history ought to have warned us that tennis balls were a touchy international commodity ; an earlier consignment had led to Agincourt.

While on the international plane, I would tell you of another hoax in Intelligence, this time by a very gallant Frenchman whom I knew during the war. He was one of our agents, and on one of many occasions he parachuted into France with about half a dozen radio transmitters. He managed to hide them at his base, a farm, but he was faced with the problem of distributing them to his sub-agents. This was a dangerous operation, but he executed it in the following manner. He went into the nearby town, and made friends with the local German commander, to such effect that he was able to put to the German the proposition that he was in the black market and stood to make a considerable sum by selling a load of carrots. The only thing that was holding up the deal was his ability to deliver the carrots to the buyer ; he therefore offered the German half the profits if he would make available a German Army lorry with a driver. The German agreed, and the lorry called at the farm where the carrots lay—and where the radio sets were hidden. A large paper-covered package was carried on to the lorry, and covered by carrots. Naturally, the driver was suspicious ; our French friend, however, tore open one corner of the package before it was finally covered and drew out a packet of a hundred cigarettes. These he thrust into the driver's hands, hinting that the black-market carrots were a mere cover for even more black-market cigarettes, and promising the driver another hundred when everything was delivered. The driver was now content, both with the cigarettes and with the knowledge of the hoax that was being played on his superior, and duly delivered both carrots and package to the customer. The hero—and well he deserves the title—then went into the town a few days later and paid the German officer with money which had also been parachuted into France.

There still remained the problem of further delivery of the sets. If n had originally been dropped into France, one had been left at the first resting point, and $(n-1)$ were with the carrots. The hero therefore pointed out to the German commander that the whole transaction had been black-market, and that if only the commander could somehow discover it he was entitled to confiscate the carrots. What was more, our friend had a new customer, and could therefore repeat the whole operation to their mutual profit. The whole performance therefore took place again, the only difference being that there were $(n-2)$ sets concealed in the carrots this time. I do not know how many more times it was repeated before the carrots went rotten.

Before I turn to the more serious side of this lecture there is one further story from Physics in which the exact classification of the incongruity can be left as a problem to be worked out at leisure. It concerns Lord Kelvin's lectures at Glasgow, where he used to fire a bullet at a ballistic pendulum ; as an undergraduate at Oxford I had heard a story of how Kelvin missed on one occasion, with the result that the bullet went through a wall and smashed the blackboard of the lecturer next door. Kelvin rushed into the next room in some alarm to find the lecturer unscathed, and the class shouting "Missed him —try again, Bill." This experiment has now produced a further incident, and to avoid any doubt I wrote to Professor Dee for his own account of what happened. This is what he says :

" In the Quincentenary Celebrations here I had to lecture on the history of the Department. Of course Kelvin figured strongly in this. One of Kelvin's traditional experiments was to fire a rifle bullet at a very large ballistic pendulum. All his students regarded this as the highlight of the course. He was reputed to have the gun charged with a big dose of powder—the barrel is about half an inch internal diameter. I decided this experiment must be repeated but there was great alarm here that the barrel would burst and annihilate the front row (Principal and Senate). So I decided to use a modern rifle. I also decided to make it a double purpose experiment by using Kelvin's invention of the optical lever to display the pendulum swing to a large audience. On the night all went off well.

The next day I repeated the whole lecture to the ordinary class. Mr. Atkinson was the normal lecturer to this class and he had noticed that in referring to the dual purpose of the demonstration I used the phrase " . . . fitted a mirror to the pendulum so that I may kill two birds with one stone." After the explosion to my surprise a pigeon fell with a bloody splash on to a large white paper on the bench—our lecture room is very high. I tried to resolve the situation by saying " Well although Mr. Atkinson isn't lecturing to you to-day he appears to be behind the scenes somewhere. But he does seem to have failed to notice that I said *two* birds with one stone ! " Immediately a second pigeon splashed on the bench ! Whether this was due to a slip up in Atkinson's mechanical arrangements or to his brilliant anticipation of how I would react I don't really know but I always give him the credit of the second explanation.

Anyway the students loved it but I wonder how many would remember about the optical lever ? "

When the foregoing example has been evaluated, the following one may be considered. It was a well-known story a few years ago when we were inclined to laugh at Russian science. It concerns a Russian professor who trained a cricket to jump when commanded. He demonstrated this to his students and then said : " Gentlemen, we will now remove the cricket's hind legs." He did so, and again commanded the insect to jump. After repeated failure he said " Gentlemen, from this experiment we can surely conclude that when we remove a cricket's hind legs, we deprive it of its sense of hearing." At this point, the story normally ends, because we physicists think that the professor is a victim of a self-inflicted incongruity. However, a cricket really does have ears in its legs, and our laughter at the professor's folly is only just saved by the fact that they are in its fore legs. There is an obvious moral to be drawn.

TECHNICAL SPOOF IN WAR

I want to turn now to technical deception in war, as exemplified by our attempts to mislead the German night defences in their appreciation of our raiding intentions. The method here is that of the induced incongruity ; by a false presentation of evidence we wish the enemy controller to build up an incorrect but self-consistent world-picture, thus causing him to generate the incongruity of directing his nightfighters to some place where our bombers are not. I originally developed this "Theory of Spoof" in a wartime report ; the salient points, which have some interest in physical theory, are the following. As with all hoaxes the first thing is to put oneself in the victim's place (indeed, a good hoax requires a sympathetic nature), to see what evidence he has with which to construct and test his world-picture. In night aerial warfare in 1939-45, this evidence was mainly the presence of deflections in the trace of the cathode ray observing tube. Therefore any device which would give rise to such deflections could provide an element of Spoof. One such device was a jammer which would cause fluctuating deflections all the time, thus concealing the true deflection due to the echo from an aircraft. This, like a smoke screen, would render the enemy unaware that you are where you are. A more positive technique is to provide a false echo, and if possible to suppress the genuine one, thus giving him the impression that you are where you are not. The easiest way of providing a false echo is to drop packets of thin metal strips, cut to resonate to the enemy's radar transmissions. This is, of course, what we did in 1943. There is little time to tell now of the fortunes of this technique, but the packets were extremely successful, and they changed the character of air warfare at night. At first, the German controllers confused the individual packets with aircraft ; I can still remember the frustrated tones of one controller repeatedly ordering a packet to waggle its wings as a means of identification. Soon, however, the Germans gave up the attempt to make detailed interceptions, and tried to get a swarm of fighters into our bomber streams. We then used many tinfoil packets dropped by a few aircraft to provide the appearance of spoof raids, which lured the nightfighters off the track of our main raids.

At this stage we witnessed a few self-inflicted incongruities, where the German controllers generated their own delusions. I myself heard this happen with a Bomber Command raid on Bochum. Another example is recorded in General Galland's book *The First and the Last.* In those days when German radar was blotted out by the metal strips ("The fateful hour of the Luftwaffe"—Galland), Goering used sometimes to take personal control of the defences. With one American raid, above the clouds by daylight, Goering decided from the line of flight that the target would be the ball-bearing works at Schweinfurt. He therefore ordered his fighters to fly back towards this town. His ground observers then reported the noise of aircraft flying towards Schweinfurt ; actually these were his own fighters, which could not be identified above cloud, and Goering took the evidence as confirming his guess about the target. From then on the mistake became self-propagating. No bombs were dropped on Schweinfurt, and as some of the fighters overshot Goering looked for the next target further East. This was the Leuna works in Leipzig, and so he ordered his fighters there. Again the ground observers duly reported the stream of aircraft, and again there were no bombs on Leipzig. So Goering had another guess, this time the Skoda works at Pilsen in Czechoslovakia. On went the fighters at his command, until the cloud cover slowly broke up to show not an American aircraft in the sky. The American target had been Düren in the Rhineland, more than 300 miles away, and the Americans were safely on their way home long before the Reichsmarshall was disillusioned. It says much for Goering that he thereupon sent a signal to all units in which he congratulated himself and all participants on the " successful defeat of the air raid on the fortress of Köpenick "—the latter being a reference to one of the greatest hoaxes yet perpetrated, where before the first World War a private citizen disguised as an Army officer paralysed all the civil machinery of the garrison town of Köpenick.

As the war went on the Germans gradually found ways of distinguishing between echoes from metal foil packets and those from aircraft. The packets, for example, resonated to one particular frequency, and therefore they had a relatively poor response to another frequency. If two radar stations watched on widely separate frequencies, a genuine aircraft echo would be present on both, whereas the foil echo would appear only on one. The foil could, of course, be cut to different lengths, but as the number of frequencies was increased, the amount of foil needed was greater. Moreover there was a pronounced Doppler effect on the echo from an aircraft, with its high speed, but little effect on the echoes from the foil drifting with the wind. Thus, against an omniscient controller, we have to make the decoy echoes move with the speed of aircraft, and reflect different frequencies in the same way. This is easiest done by making a glider of the same size as the bomber. Then if we allow the enemy controller to use sound and infra-red detectors and other aids, we find that the only decoy which can mislead him into thinking that there is a British bomber flying through his defences is another British bomber flying through his defences.

Another example is one that I encountered earlier in what has been called " The Battle of the Beams " in 1940. Here the problem was to upset the navigation of the German night bombers, when they were flying along radio beams to their targets. The signals received by the pilots telling them to steer right or left were counterfeited in this country, and sometimes resulted in their flying on curvilinear courses. However, had the pilots had unlimited time of observation they could have detected that there was something wrong, even if we had exactly synchronized our transmitters with those of the Germans. The bombers were in general flying away from their own transmitters and towards ours, and so they would have received a Doppler beat from which they could have deduced that a second transmitter was active. If one allows the possibility of various fairly simple tests, which fortunately would take too long in actual warfare, one arrives at the conclusion that the only place for a second transmitter which will simulate the original exactly is coincident with the original and the counterfeit thus defeats its purpose.

In war, of course, neither controllers nor pilots are omniscient, and they have limited time and means of observation. They can therefore be misled ; but if we grant the possibility of unlimited observation, then we can see from these simple examples that no model can be perfect unless it is an exact replica of the original—and even then the perfection is spoilt by the fact that now two exist where there was one before.

The ease of detecting counterfeits is much greater when different channels of examination are used simultaneously. This is why telephonic hoaxes are so easy—there is no accompanying visual appearance to be counterfeited. Metal strips were most successful when only radar, and that of one frequency, was employed. Conversely, the most successful naval mines were those which would only detonate when several different kinds of signal, magnetic, hydrodynamic and acoustic, were received simultaneously. A decoy which simulates all these signals is getting very like a ship. From these considerations, incidentally, we can draw a rather important conclusion about the detection of targets in defence and attack : that as many different physical means of detection as possible should be used in parallel. It may therefore be better in some circumstances to develop two or three independent means of detection, instead of putting the same total effort into the development of one alone.

THE SUCCESSES AND LIMITATIONS OF PHYSICAL MODELS

Now the thesis that no model can be perfect is at the root of physical theory. In fact, when we realize that this is so, we ought to be rather impressed that models and analogies in physical theory have been so successful. When we say that we have "explained" some new physical fact, what we usually mean is that we have been able to interpret it in the light of our previous experience, i.e. to form an analogy between it and something we already know well. We "explain" the properties of gases on Kinetic Theory, by forming an analogy between gas molecules and ball bearings; and from the mechanical properties of ball bearings, which are a matter of everyday experience, we form a theory which predicts reasonably well the properties of gases. We began to feel we knew something about the atom when Rutherford formed his analogy between the electron structure and the solar system which we already knew, and when Niels Bohr formed his analogy between the stability of the nucleus and that of a charged liquid drop. More recently, the shell structure of the nucleus has had some analogy with the outer electronic structure, and Yukawa's prediction of the meson was from analogy between nuclear force fields and electromagnetism. There are many other examples which spring to mind, and the mechanistic view of the late nineteenth century was largely due to the misleadingly great success of analogy with mechanical models. The limitations of such models ought, however, to be realized, and indeed anyone who has tried practical model-making will realize that even a change of scale is enough to make a model break down. There is a limit, for example, which is reached in the making of model railway engines set by the atomic nature of matter. And anyone who tends to cling too enthusiastically to any one model in physics should heed the ninety-seventh Psalm: "Confounded are they that serve graven images." It is, of course, true that in physical observations we do not have unlimited time of observation, and it is therefore possible that a model may simulate completely all the properties of the original that we have been able to observe. Differences between the model and the original will, however, appear on further observation.

In physical thought we have in the main progressed by the process of inventing a model using components which are already well established in our experience to simulate the behaviour of entities in new fields of experience. In this connection it is interesting to note the theory of William James that ability to reason in new situations by analogy with previous experience is the true criterion of intelligence. Having regard to the limitations of models brought out by the preceding discussion, however, we must not be surprised when even the most successful models ultimately break down; by a breakdown we mean that some aspect of behaviour of the model leads us to a false prediction of a property of the entity. Thus the classical "paradox" about the wave and particle theories of radiation is no paradox at all. There are some properties of radiation which are best simulated by the model of a stream of corpuscles, miniaturized from a macroscopic concept of things the size of ball bearings. There are other properties which are best simulated by a model using wave motion miniaturized from concepts of mechanical waves at sea or on strings. Ought we to be surprised when either model breaks down? The truth is that light is not, and cannot in principle be, the same as either limited model, and we ought not to be shocked when we find that it can be simulated in part by each of two (and there might even be more) imperfect models. Each model is nevertheless extremely helpful in predicting some of the actual properties of light, and danger arises only when we forget its inherent limitations.

I have mentioned the breakdown of models caused by a change of scale, and we have, of course, had much experience in this century of the difficulties of making models of elementary particles using macroscopic concepts. The difficulty may become clearer if we imagine a thinking being whose everyday experience is with bodies of dimensions of the order of a million miles. Now, owing to the preponderance of gravitational forces, which would cause plastic flow inside large solid bodies, most of the regular bodies that he sees will be approximately spherical, or at most flat discs. It might be extremely difficult for him to visualize a body of cubical shape. Now if he discovered elementary elastic particles the size and shape of bricks, he might be hard put to it to "explain" the properties of bricks, such as the preferential ways in which they could be assembled together, or their rather peculiar collision properties, because there would be no analogy with bodies in his everyday experience.

It is fascinating to speculate conversely about phenomena involving large dimensions, because we may in fact have difficulty in visualizing the paramount forces. Just as on an atomic scale gravity is a negligible force, it is possible that on the galactic scale there is a force which is negligible in any laboratory experiment that we can perform. Such a force, for example, might be a small repulsion diminishing at some rate less than the inverse square. This would make it of comparable importance to gravity at great distances, and might help to explain some of the structural features of the galaxies and the universe. And it would be in general keeping with the present context if we revive Crabtree's eighteenth century concept of the Force of Levity.

We have seen in the last hundred years the rise of the mathematical model, where mathematical symbols and operations become the components. This is especially true of Heisenberg's matrix mechanics, and such models have been very profitable. With Maxwell we must admit that :

"There are, as I have said, some minds which can go on contemplating with satisfaction pure quantities presented to the eye by symbols, and to the mind in a form which none but mathematicians can conceive. There are others who feel more enjoyment in following geometrical forms, which they draw on paper, or build up in the empty space before them. Others, again, are not content unless they can project their whole physical energies into the scene which they conjure up. They learn at what a rate the planets rush through space, and they experience a delightful feeling of exhilaration. They calculate the forces with which the heavenly bodies pull at one another, and they feel their own muscles straining with the effort. To such men momentum, energy, mass are not mere abstract expressions of the results of scientific inquiry. They are words of power, which stir their souls like the memories of childhood.

For the sake of persons of these different types, scientific truth should be presented in different forms, and should be regarded as equally scientific, whether it appears in the robust form and the vivid colouring of a physical illustration, or in the tenuity and paleness of a symbolical expression."

We can therefore see physics proceeding by both mathematical and physical models. The new frontiers of to-day must be converted into the everyday experience of tomorrow, so that the younger physicists feel entirely at home in it, and can draw upon its phenomena to form in turn the basis of models for "explaining" the new phenomena which they will encounter. We might be tempted to ask whether there is any end to this process, but I for one am unable to answer. It will be time enough to think about that when we encounter a long period without advances.

With Bacon, I see the physicist in much the same position as an intelligence officer, trying to assemble and interpret the evidence of Nature. Because of his limited means of observation, he will often build up a false world-picture from biased clues ; but just as the intelligence officer can check his mental picture of the situation by making further observations, so can the physicist correct his model by further experiment. We often hoax and mislead ourselves when we attempt to understand the natural world, but we can usually recover the right track by enough effort in both theory and experiment. We should not be discouraged by the inevitable failure of models ; instead we should be grateful for their tremendous fruitfulness.

CHAPTER 18

Styles in Deception[1]

Amrom Katz

Amrom Katz was an early expert on space reconnaissance. Between 1941–1954 his research led to the improvement of photographic reconnaissance equipment. He went on to serve as the chief civilian physicist at an air reconnaissance lab at Wright-Patterson Air Force Base. In 1954, Katz joined the RAND Corporation as a senior scientist, where he focused on reconnaissance systems for airplanes and aircraft.

Styles in Deception

> *With arts voluptary I couple practices joculary; for the deceiving of the senses is one of the pleasures of the senses.*
> —Francis Bacon, The Advancement of Learning, 1605[2]

There are at least two types of deception. The first is illustrated by three-card Monte, a popular game at state fairs, carnivals, and elsewhere. The operator of the game (who is usually the winner in the short run and is always the winner in the long run) shuffles three cards on a table, with all three cards remaining in full view of the bettor. One of the cards is, perhaps, a queen of spades and is shown to the bettor. After turning the three cards face down, the operator moves them in and out of position. The trick is to deceive the bettor, in full view, about which card is the queen. Another version of this game is played with three walnut shells and a pea. This is what we can and will call surreptitious deception. The replacement of one type of ICBM in the silo by a larger, more advanced ICBM while we are presumably monitoring known missile sites is properly described as *surreptitious deployment.* However, the deployment of ICBMs by methods not known to us, in places we do not know about, leaving target signatures that are either undetectable, so nondescript as to catch no one's attention, unidentifiable, or misidentified should properly be called *covert deployment.* It is a good bet that the U.S. intelligence community would do much better at detecting surreptitious deployment of ICBMs than at detecting covert deployment. For exactly the same reason, the Soviet Union is more likely to deploy extra missiles covertly than to deploy them surreptitiously—if they choose to cheat at all.

We can arbitrarily divide camouflage-and-deception (C/D) methodology into two types: A and B. C/D Type A is very technical: encrypting telemetry, jamming receivers, putting out false signals, substituting space shots for MIRV testing. Type A is very modern, very technological, very elegant—a match for our skills, aptitudes, techniques, and style. We would expect the Soviets to do this if they do anything. Camouflage, elegant

1. Amrom Katz, "The Fabric of Verification: The Warp and the Woof," in *Verification and SALT: The Challenge of Strategic Deception*, William Potter, editor, Boulder, CO: Westview, 1980, pp. 211-212.
2. Sir Francis Bacon.

or otherwise, of standard methods of constructing missile sites would be included in this. It is what we expect the Soviets to do; it is a game we might win.

So why should *they* play this game? Because *we* prefer it, we're prepared for it, and we're good at it? These are precisely the reasons why they should choose an alternate course and go to C/D Type B. Type B is characterized by attempts to cause us to look in the wrong places, to give us a form of intelligence judo, to throw us a bone that we can chew on for a while (such as a poorly camouflaged site). In the meantime, the Soviets could be tearing off in another direction (without our following), so that we may miss an entire phenomenon. C/D Type B is very ingenious and characterized by a modus operandi that we are unprepared to believe in, expect, react to, detect, appreciate, understand, or be good at. In short, it's not our style. By temperament, style, preference, and mentality, our people are geared—and geared very well—to cope with and, perhaps, win at C/D Type A.

What about Type B? It requires another type of person who is not encountered often, who would expect the Soviets to be doing C/D Type B and would worry about it.

CHAPTER 19

The Process of Deception

Barton Whaley

The deception planning process is often implied in the literature on deception. Here, this process is made explicit.

The Process of Deception has, in the following sequence, 10 steps:

1. In planning a deception, the planner must know its *strategic goal.* For the magician, this is always simply to provide the audience with a pleasing and surprising effect or puzzle, for that is all they expect if he is to earn his keep. For the military commander, it may be to launch a surprise invasion of another country, effect a landing on an enemy beach with minimum initial opposition, get the jump on the enemy with a new attack or counter-attack, or simply get a reconnaissance or rescue party in and out of hostile territory with sufficiently low casualties to assure success. These goals pose the kinds of problems that the planner initially faces; they are the givens he must work within.
2. The planner must decide how he wants his target to *react* in such a given situation. The magician requires only that his audience concentrate their attention and interest on the portrayed effect to the exclusion of the secret method. For the military deception planner, however, the problem is more subtle, as best put by Britain's most experienced WWII deception planner, Brigadier Dudley Clarke, head of the so-called A Force deception planning team in Cairo:

   > In the first Deception Plan I ever tackled I learned a lesson of inestimable value. The scene was Abyssinia... Gen. Wavell wanted the Italians to think he was about to attack them from the south in order to draw off forces from those opposing him on the northern flank. The Deception went well enough—but the result was just the opposite of what Wavell wanted. The Italians drew back in the South, and sent what they could spare from there to reinforce the North, which was of course the true British objective. After that it became a creed in 'A' Force to ask a General 'What do you want the enemy to do?', and never 'What do you want him to think?'

3. Only then must the planner by himself—whether general or magician—decide what he wants the target to think about the facts or event, precisely what it is they should *perceive.*
4. He must decide specifically what is to be *hidden* about those facts or impending events and what is to be *shown* in their stead. In doing this he should remember the caveat that hiding and showing ideally take place simultaneously, as any deviation from simultaneity gives the target more time to discover the switch. In

military practice, hiding usually takes place prior to showing, only sometimes simultaneously, and fortunately rarely afterwards because of the extreme risk.

5. The planner now analyzes the *pattern* of the real thing to be hidden so as to identify its distinguishing characteristics (charcs), specifically which of these charcs must be deleted or added in order to give another pattern that suitably *masks, repackages,* or *dazzles.*
6. He does the same for the false thing to be shown to give another pattern that plausibly *mimicks, invents,* or *decoys.*
7. At this point the planner has designed a desired *effect* together with its hidden *method.* He must now explore the means available for presenting this effect to the target. The magician may be limited by the type of apparatus he has at hand, his ability to purchase or construct appropriate new apparatus, or his theatrical ability. Military commanders or practitioners of intelligence may be limited by their available deception assets and will often have too little time to acquire additional ones. They must make do or go back to Stage Four planning and try for an alternative design.
8. Having the effect and the means, the planning phase has ended and the operational phase begins. In magic, the planner is usually also the performer. In the military and intelligence fields, the deception planner usually hands over the plan of execution to operational units to present (sell) the *effect.*
9. The *channels* through which the false charcs (and patterns) are communicated must be ones open (directly or indirectly) to the target's sensors. A magician should not use a 'false count' of clicking coins before a deaf spectator. An intelligence officer should not plant disinformation in a newspaper unless he has reason to believe the enemy monitors that paper.
10. In order for the deception to succeed, the target must accept (buy) the *effect,* perceiving it as an *illusion.* Deception will fail at this point only if the target takes no notice of the presented effect, notices but judges it irrelevant, misconstrues its intended meaning, or detects its *method.* Conversely, the target will:
    - Take notice, if the effect is designed to attract his *attention*;
    - Find it relevant, if the effect can hold his *interest*;
    - Form the intended hypothesis about its meaning, if the projected pattern of charcs is *congruent* with patterns that are already part of his experience and memory; and
    - Fail to detect the deception, if none of the ever-present charcs that are *incongruent* are accessible to his sensors.

Effective deception planning must anticipate all four of these contingencies. And a wise deceiver will seek feedback, monitoring the target's responses, to assure that these four contingencies are being met.

Section V: Deception Traditions

Introduction

Deception Traditions

The use of guile in military and political affairs has been present in varying degrees, over time, in all cultures. In fact, the rules of the game have changed little since antiquity. Sun Tzu and Clausewitz now rank as the two quintessential philosophers of war and appear on every military officer's reading list. They seem to represent two important but distinct military traditions. The connection between politics and strategy is common to both, but Clausewitz's advocacy for battle and Sun Tzu's preference for deception in lieu of battle has inaccurately portrayed the Western way of war as preferring open battle to that of Oriental trickery. In actuality, there is little in Sun Tzu not found in Western military treatises.

At the same time that Sun Tzu declared that deception is the *Tao* of war, the Greek Xenophon was preaching that every general must be a contriver of stratagems and that stratagems produce the greatest effects in war. Devising stratagem has historically been linked to the definition of the good general. The difference between China and the West is that Sun Tzu offers a succinct handbook filled with handy maxims for memorization. In contrast, the obscure plodding style of a Thucydides is less accessible and much of Western military thought, though advocating many of the same principles as Sun Tzu, is tedious.[1]

Mao Zedong read Sun Tzu as did Che Guevara and other modern guerrilla warfare icons. The intensification of interest in Sun Tzu in the 1960's created a cult and the Western fascination with him continues with no end in sight.

The mostly accepted notion of Western war as heroic, face-to-face open battle, versus the Eastern ethos of trickery, deceit, misdirection and the avoidance of pitched battle, is deceptive in its own right. Both balance of power situations and asymmetrical power relationships promote stratagems. The Cold War nuclear balance fueled deception just as stratagems of weak actors against numerically, logistically, and technologically superior powers manifest themselves in the strategy and tactics of guerrilla warfare and terrorism.

The cult of Sun Tzu may exaggerate his unique merits. We should not ignore other cultural traditions of stratagem. Kautilya, the Hindu Machiavelli, composed his *Arthashastra*, allegedly *ca* 300 B.C. Nor should we ignore the Islamic tradition of stratagem

1. Everett L. Wheeler, "Stratagems in the Western Tradition: Reflections and Response," paper presented at the *Conference on Strategic Deception in Modern Democracies: Ethical, Legal, and Policy Challenges,* sponsored by the Triangle Institute for Security Studies, The United States Army War College, The U.S. Naval Academy, and Duke University's Kenan Institute for Ethics, Chapel Hill, NC, October 31–November 1, 2003.

captured in the book, *The Subtle Ruse: The Book of Arabic Wisdom and Guile*. An ancient Arab proverb asserts that a trick is worth more than a tribe.

The perception of Western honor and open war versus Oriental trickery paints an artificial picture. Chivalry and honor exist in the Chinese, Japanese, and Hindu traditions of war too. The Western belief in the "decisive battle" does not accurately represent what Western military treatises actually say. The "cult" often forgets British theorist G.F.R. Henderson, whose writings on Stonewall Jackson call for a return of maneuver, deception, and surprise. Henderson's students, Edmund Allenby in WWI and A.P. Wavell in WWII, would not forget their mentor's teachings. Likewise Winston Churchill became an advocate of stratagem. Nor should we overlook the writings of T.E. Lawrence, a theorist of guerrilla warfare. The inter-war years would see Liddell Hart, a reader of Sun Tzu, develop his strategy of the indirect approach that arguably opened the WWII stratagematic floodgates.[2]

We live in an age of stratagems and we also seem to be entering a new age of lawlessness in international behavior. Not accepting the universality of stratagem can result in self-deception.

2. Ibid.

A British Tradition of Stratagem: And Its Influence on American, Israeli, Indian and Pakistani Doctrine

CHAPTER 20

Conditions Making for Success and Failure of Denial and Deception: Democratic Regimes[1]

M. R. D. Foot

M. R. D. Foot was an army officer during the Second World War and was awarded the French Croix de Guerre for work with the SAS in Brittany. He taught politics and history at Oxford, and for six years was a professor of modern history at Manchester. Foot was the first editor of "Gladstone's Diaries" and has written—among other books—"SOE in France" (the official history), "Resistance," "MI9," (with J.M. Langley) and "Six Faces of Courage."

"...willing to overcome by fraud those whom he desired to subdue, because he was wont to say that it was the victory that brought the glory, not the method of achieving it."
—*Machiavelli*, on Castruccio Castracani, dictator of Lucca ca. 1325, *The Prince*

Introduction

This look at British deception begins with a glance back to Sun Tzu, Machiavelli, Marlborough, and Wellington, but deals mainly with the Second World War of 1939-45. In this conflict, General A. P. Wavell was the inspiration for a deception system whose central feature was to persuade the enemy that allied forces were much larger than they really were. In May 1944 the Germans believed that there were seventy-nine divisions waiting in England to invade Western Europe; in fact there were fifty-two.

Deception in the Mediterranean was run by Brigadier Dudley Clarke, and by Colonel J. H. Bevan in London. Their triumphs included *Mincemeat*, which persuaded the Germans that Sardinia, not Sicily, would be attacked next after the fall of Tunis; and *Fortitude*, which led them to believe that the Normandy landings were a diversion to draw troops away from a later main assault on the beaches south of Boulogne. Ultra secret intelligence provided an indispensable control over how well these plots were working.

A sub-plot of *Fortitude* was *Titanic* in which a handful of parachutists succeeded in decoying away the reserve regiment opposite

1. M. R. D. Foot, "Conditions Making for Success and Failure of Denial and Deception: Democratic Regimes," in *Strategic Denial and Deception*, Roy Godson and James J. Wirtz, editors, NJ: Transaction, 2002, pp. 95–114.

Omaha beach. The Germans wasted the whole morning of 6 June 1944 hunting down the decoys instead of counter-attacking the forces of General Bradley. An eyewitness account of the origins of this deception is provided here. *Fortitude North* was aimed directly at Hitler himself, and played on his liking for Andrew Thorne, the Grenadier Guards colonel who had been British military attaché in Berlin in 1932-35. By 1943, Thorne had been promoted to the rank of general and was commander-in-chief, Scotland. Hitler was persuaded, mainly through double agents, that Thorne intended to invade Norway. He kept 300,000 troops locked up there uselessly until the end of the war.

The principal lesson learned from these efforts was the need for absolute secrecy. This sits uneasily with parliamentary, congressional, or media government, but is essential for success in this type of war. Access to enemy ciphers was an additional asset that is unlikely to be available in the future. It also was a help to have the business run by officers from similar backgrounds who shared a common outlook, and it was indispensable for all the secret services involved to cooperate, thinking only of victory, and disregarding private service advantage.

Deception

Deception is far older than warfare. It is almost as old as combat, which goes back to before the Stone Age. The first primitive man or woman who thought how to feint when fighting a fellow human being originated a device now in common use by boxers, fencers, and footballers. It is strange that it should be regarded as something extraordinary by journalists and under-informed commentators on war.

Sun Tzu laid down, centuries before Christ was born, that "All warfare is based on deception," and that no general could accomplish anything worth doing without it.[1] Today, Sun Tzu's general principles of strategy seem to gain warm approval from staff colleges the world over. Developments in technology have made wars, especially between industrially developed powers, more devastating than they used to be; but the principles of how to fight them have not altered much in two and a half millennia.

"To such a name no praise can be high enough," says the epitaph on Nicolo Machiavelli's grave in Florence. In the early sixteenth century, Machiavelli put deceit high among the characteristics that would be necessary to a successful ruler; and even though he has become a byword for immorality, he remains favorite reading for those who seek greatness in politics—of which strategy is a part. Indeed, we have Churchill's word for it that at the summit, politics and strategy are one.

Marlborough and Churchill

The history of warfare is littered with examples of deception, most of them successful. There seems to be no limit to the credulity of bad generals. Yet, in a paper that concentrates on modern British deception efforts, one example from Queen Anne's reign is too striking to omit. In the summer of 1704, Marlborough finally secured the agreement of the Dutch to move his combined Anglo-Dutch army where he wanted, to defeat the French forces in Bavaria. He set off up the right bank of the Rhine, and so confused the French highest command about where he was really heading, that he was able to catch Marshal Tallard unreinforced at Blenheim and to inflict a severe defeat on him.[2] In a sub-continent where Churchillolatry is rife—I have affectionate recollections of the Other Other Club at Madison, Wisconsin—it is worth drawing attention to the huge stroke of luck for the free world that Churchill was out of office for ten years from 1929. He spent several years pondering the career of his great ancestor Marlborough, about whom he wrote a four-volume life. Working on this, he went through all the problems of coalition war, and its manifest advantages and disadvantages. He thus prepared himself, unconsciously, for the task that was to fall to him in May 1940. In what he regarded, in his childhood, as the Great War—the war against revolutionary and Napoleonic France—there had been ample scope for deception. For instance, in the mid-1790s Pitt flooded France with forged *assignats*, as a means of destabilizing the young republic's currency and thus bringing it down. However, this is perhaps an example of subversion, rather than deception.

Wellington

Wellington was certainly ready to deceive his enemies when he could. As the allied army formed up on the heights of Busaco on 27 September 1810, Sergeant Hamilton of the 43rd Light Infantry found his whole battalion required to change their coats with the Portuguese battalion beside them. Massena's scouts who came forward to reconnoiter allied movements encountered Wellington's attempts to confuse him about where the weak spots in the allied line were. In both Wellington's great sieges of Ciudad Rodrigo and of Badajoz in 1812, a false attack was mounted at the same time as the main assault to confuse the enemy garrison. This succeeded in both sieges.

The First World War

Churchill himself had had some practical experience of high command during the Great War of 1914-18. He was the politician in charge of the Royal Navy when it began. He was at least familiar with the Navy's arrangements for deceiving enemy fleets in battle; later he plunged himself deep in the controversies that followed the apparently inconclusive battle of Jutland (31 May-1 June, 1916) in the early 1920s, in which feinting had been among the many maneuvers used. The German High Seas Fleet, for instance, had left its W/T call sign operating as usual in harbor, to conceal the fact that it had actually gone to sea.

During World War I, deception was attempted in land warfare also, but the conditions of trench warfare were such, on the western front at least, that it was hardly possible to deceive the enemy about where any major blow was to fall. The concentrations of ammunition and men that were necessary to attempt any serious assault on the trench line that ran from the North Sea to the Swiss border were so vast that they were bound to be spotted. If air reconnaissance, which camouflage might baffle, did not pick it up, then spies, who could see beneath camouflage, could give ample warning to prepare countermeasures. Similarly, the attack on the Dardanelles, finally mounted on 25 April 1915, had been preceded by so many preliminary reconnaissances and bombardments that it forfeited surprise. Ludendorff, with the advantage of interior lines on the western front, tried all the same to bemuse Haig and Foch, and sometimes succeeded, but not for long.

The Interwar Years

The 1920s and 1930s were not the grandest years for the Foreign Office. Officials did their best to follow the principle laid down by Arthur Balfour, a former prime minister, when he was foreign secretary during Woodrow Wilson's presidency: "Open covenants, secretly arrived at." But in their secret negotiations, British diplomats were too often victims of German deception. They were profoundly mistaken, for example, about the real strength of the Luftwaffe in the middle 1930s. This was only sorted out when ultra secret intelligence about German capabilities in the air came on stream in late spring 1940. In the autumn of 1939, the entire British high command, right up to the war cabinet, was taken by an ingenious *Sicherheitsdienst* ploy. They believed they were in touch with a body of German generals disaffected from the Nazi high command, and successfully impersonated by Major Walter Schellenberg. It was he who kidnapped the British secret intermediaries at Venlo on the Dutch-German border on 9 November and accused them of having tried to blow up Hitler the day before. The sour taste left by the Venlo incident was probably the reason why, all through the war, the British resolutely rejected any approaches made to them by German diplomats or intelligence staff who were trying to bring the war to an end.

Two other major peacetime deceptions should be noted. They were in progress in 1934-38 while Churchill's *Marlborough* was coming out, and their results still resound today. One was orchestrated by that master of propaganda, Dr. Joseph Goebbels. It convinced most of the German people, and then a substantial slice of the populations of Europe and of North and South America, that the Nazi regime was doing wonders for Germany, particularly for its youth. In Goebbels's imagery, the Führer of Nazi Germany was the mildest, gentlest, and most benign of dictators. The horrors of the Holocaust persuaded most of us otherwise. Yet even today there are fanatics who deny that the Holocaust ever took place, and look back fondly on the antics of the Bund deutsches Mädel and on how kind Hitler was to children.

Lenin, too, was wonderful with children, and Lenin's disciple Stalin brought off the second, simultaneous feat of worldwide deception, rivaling that of Goebbels. Many of my own English contemporaries, and many a few years older than myself, were captivated by the prospect that Stalin was building a new society, the like of which had never been seen before, and which would establish the equality of mankind. The 1936 constitution he set up for the USSR looked marvelous; nobody outside Russia noticed that little clause which gave the Communist Party of the Soviet Union the right to override everything and everybody, if need arose. Of course, the methods by which the Bolsheviks had come to power had been coarse, but the rough edges, surely, would soon be smoothed out by time and care. The horrors of Beria's Gulag, which rivaled those of Himmler's concentration camps, were still more of a closed book to the outside world, and for much longer. (The Bolshevik ascent to power also included a nice stroke of deception. At the critical moment on the night Trotsky took over Petrograd, the cruiser Aurora fired only blanks. Live ammunition was withheld until the Winter Palace had fallen, by which time it was no longer needed.)

The perceptions of key players played a pivotal role during the Munich crisis of September 1938. The only moment at which there was any trace of faltering on the German side was when it was announced that the Royal Navy had mobilized. Chamberlain had no desire to go to war that autumn; but the mere impression that he might made his putative enemies ponder. It also is worth recalling that the personal impression made on Hitler by Chamberlain and Daladier at the Munich conference was of men so feeble that thereafter he felt confident that he could outface them and get his way. He strode ahead on his determined path; the next world war followed.

The Second World War

The immediate cause of World War II was a deceptive measure by Hitler known as the Gleiwitz incident. "I will provide a propagandistic *causus belli*. Its credibility does not matter," he told his generals on 22 August 1939. Several Abwehr and SS parties were told to raid various parts of the German-Polish frontier. One of the raids was carried out on the night of 25-26 August, when Hitler had originally wanted the war to begin. Six nights later, the radio station at Gleiwitz (now Gliwice), which was then well inside the German frontier, was raided by SS men in Polish uniforms. They pretended to be Poles taking it over. They broadcast a few inflammatory phrases, in Polish; and left some corpses behind (concentration camp prisoners, who could easily be spared) to impress the United States correspondents who were summoned next day. This provided a final excuse for the German invasion of Poland early that day, 1 September 1939.

Early in the war, the Royal Navy was able to bring off a deceptive stroke that was old in Admiral Nelson's day, but sufficed to deceive a Reichsmarine commander less versed in naval history. Captain Langsdorff took the pocket battleship *Admiral Graf Spee* into Montevideo after an indecisive action with three British cruisers on 13 December 1939. The least damaged of the cruisers hovered off the port, exchanging streams of flag signals with a supposedly more powerful friend over the horizon. There was no more powerful friend, but Langsdorff believed he would be outgunned if he came out, and sank his ship at the mouth of the River Plate, thus presenting the British with a badly needed victory.

Wavell and Clarke

As the war developed, British soldiers and policymakers settled down to organize deception seriously, inspired by the war's most intelligent, though not its luckiest, high commander, General (later Field-Marshal Earl) A. P. Wavell. Wavell had long pondered how best the British Empire, weakened as it had been by the scrapping of the Grand Fleet (at Churchill's insistence) in the 1920s, could stand up to the growing menaces of Bolshevism, Fascism and Nazism. He had concluded that trickery would have to be used as a substitute for strength, until real strength could be built up again.

Why was Wavell so devoted to deception? Because, as a member of General Edmund Allenby's intelligence staff, he had watched a now famous example of its use against the Turks in October 1917. Beatrice Webb's nephew Richard Meinertzhagen had succeeded in misleading the Turks around Gaza about which flank Allenby was about to attack, by dropping a bloodstained haversack full of false documents.

While he was commander in Palestine in 1936 (then a British mandate under the League of Nations), Wavell earmarked three young officers for special duties in the event of another major war. One was Orde Wingate, who later became famous on his way to his heroic, accidental death on the second Chindit expedition into Burma. The second was Tony Simonds, who died early in 1999. He organized wartime escapes from Eastern Europe to the Near East. The third, Dudley Clarke, wrote a book, *Seven Assignments*,[3] about the less interesting parts of his war, which included the formation of the commandos during the summer crisis of 1940. He never said anything in public about his role as Wavell's deception officer.

Late in 1940, Wavell secured Clarke at his own elbow in Cairo, where he had become commander-in-chief, Middle East. Clarke took over the running of deception, keeping Wavell informed by word of mouth about what he was up to. He called his own tiny staff "A Force," which meant nothing, as it was intended to do. Italy had by this time run to the help of the apparently invincible Germany. Clarke's main preoccupation was to convince these two enemies that Wavell's forces were substantially bigger than they were. Inflating the enemy's view of one's own order of battle was central to British successes on the deception front all through the rest of the war; it succeeded admirably. For the supreme instance, by May 1944, the Germans believed that there were seventy-nine divisions waiting in England to invade Western Europe, when in fact there were fifty-two.

Clarke was aided in tactical deceptions by a small team of camouflage experts of unusual skill, headed by Jasper Maskelyne. In the 1920s and 1930s, one of the main delights for a child in London had been to visit Maskelyne and Devant's conjuring theater. Jasper Maskelyne packed up his full skills and took them to the western desert. He could transform tanks into trucks, or trucks into tanks in a few minutes of wrestling with hoops and painted canvas. He even created inflatable warships, right up to battleship size. This caused no end of confusion to the Italian and German intelligence staffs trying to make sense of British strengths and dispositions. Coming from the world of entertainment, Maskelyne had no reservations about going public. He chronicled his achievements in an enjoyable book, *Magic-Top Secret,* published in 1949, long before academic inquiries into this sort of subject started.

Camouflage is not always impenetrable to air reconnaissance. Several times over, each side in the desert noticed that it was being deceived. Both the Royal Air Force and the Luftwaffe used to cherish the anecdote of the dummy aircraft that were bombed with wooden bombs.

Two examples of successful deception in the eastern Mediterranean are worth mentioning. While Montgomery was preparing for the second battle of El Alamein, in autumn 1942, his engineers built some fifteen miles of dummy water pipeline across the desert, running towards the southern end of his defensive line. The pipeline was clearly visible to German air reconnaissance. German engineers judged, as they were meant to, that it would be completed by mid-November. So Rommel went on leave, and was absent from his front when Montgomery's attack was launched against the other flank of the Axis army on 23 October. The second example was longer lasting. Through double agents and captured agents' sets played back, the British convinced the Germans that they had a sizeable army assembling in Syria, Lebanon, and Iraq that would invade the Balkan peninsula. They thus kept several more German divisions tied up on garrison duties in the Balkans than the realities of war would have justified.

Central Organization

A start had already been made in Whitehall with the organization of a central deception service, a group called the Inter-Service Security Board (ISSB), whose cover task was to allot code-names for impending major operations. Jo Holland, who from 1938 to 1940 worked in the war office on methods of irregular warfare, set up the ISSB, which formally was a sub-committee of the Joint Intelligence Committee (JIC). He also found the man to run it, Major E. E. Coombe. Coombe was a regular engineer, born in India in 1897, who had won the Distinguished Flying Cross on the Salonika front during the Great War, had been wounded in Dublin during the Troubles, and had been given his war office post to do research on any subject he chose because he had failed a medical examination for promotion to lieutenant colonel. In two years, he founded the escape service, the commandos, the Special Operations Executive (SOE), the subversion service, and the deception service, and organized the stay-behind parties who were to disrupt the German invasion of England that (thank goodness) never happened. His health was re-established and an ordinary battalion command in his regiment fell vacant. As a regular soldier he thought he ought to take it. So he bowed out of the secret war early, but he had made his mark.

Sir Michael Howard, in the decisive and indispensable book on British deception policy, *Strategic Deception* specifies that strategic deception was originated by one man: Wavell.[4] In the now quite large literature devoted to deception, much of it trivial or inaccurate, or both, Howard's book, the fifth volume of the series edited by F. H. Hinsley, shines like a beacon on a dark night. Howard's work, and it is almost the only one, bears witness to what was actually done, because it is founded on such archival material as the wartime deception service left behind, as well as the writer's unusually well-developed historical insight.[5]

Secrecy

Howard finished the draft of his book some years before it was allowed to appear in print. A few words in defense of secrecy may be allowed, in a country where the political public seems to believe nothing ought to remain secret for fifty minutes, let alone fifty years. Mrs. Margaret Thatcher (as she then was) refused to pass the book for publication, having learned as a young woman when she was a clerk in a solicitor's office, that it was always a mistake to reveal to the other side any weighty point that might be usable in court, unless the rules of court compelled one to do so. As Wavell had noticed, deception could become a powerful weapon in the armory of a weak power. Great Britain was no longer as strong as it had been in Mrs. Thatcher's girlhood; she saw no reason to explain to potential enemies how Britain might fight them. Once she had left office, the book could appear.

Wavell certainly insisted on the need for absolute secrecy where A Force's work was concerned. He carried this doctrine to India with him, when he became commander-in-chief there in July 1941. He set up an equivalent to A Force under Peter Fleming (elder brother of Ian, creator of James Bond), and let Fleming do what he could to deceive the Japanese. Like Clarke, Fleming told nobody anything he did not need to know. His main difficulties were caused by the incompetence of the Japanese intelligence staffs against whom he was working. They were often reluctant to take the bait he held out to them.

The London Controlling Section

Clarke visited London early in October 1941, and explained what he was doing to the chiefs of staff. They warmly approved, and at once placed Coombe's ISSB under a more senior personage: Colonel Oliver Stanley, MC, MP, a recent secretary of state for war, younger son of the seventeenth Earl of Derby who was nicknamed "King of Lancashire."

Stanley established that deception was a matter for the operations, rather than the intelligence, staffs to cope with, a vital point that needed to be settled early on. But his heart was not in it, and in May 1942, with Churchill's leave, he went back to full time politics. He was succeeded by Colonel J. H. Bevan, a stockbroker in peacetime, who remained the central figure in British deception planning for the rest of the war. Nominally, Bevan came under the future operations planning section of the joint planning staff. He had a small office directly beside the cabinet war rooms at the southeast corner of St. James's Park, but in fact he ran a virtually independent secret service, which came top in the inter-secret-service pecking order. (The war rooms are now, as part of the Imperial War Museum, open to public inspection.) Anything Bevan asked for, he got; he was too sensible and too honest a man to ask for what he knew would be extravagant or unobtainable.

Bevan's little group was called by another meaningless name, the London Controlling Section (LCS). He belonged, as Stanley did, to the old ruling class. He was the grandson of a former speaker of the House of Commons who had become a viscount, and he was related to many notables. For example, his cousin, the current viscount, had married a duke's daughter, and his own brother-in-law was Field-Marshal Alexander. Moreover, he had known Churchill when working in Paris as a junior staff officer in 1918, after surviving several years as an infantry subaltern in the Hertfordshire Regiment on the western front. The old ruling class had run the secret services for centuries (Walsingham's decipher expert, for instance, had unraveled Mary Queen of Scots); they had traditions of how the work ought to be done, which spread quite widely through the officer class, and formed a dense, and all but impenetrable, security screen. It was long regarded as the depth of bad form to put anything about such matters on paper, let alone discuss them in front of strangers.

Such a community was well placed for dealing with so secret a subject; and was not likely to blab out of turn about what it was doing. Secrecy is a recurring theme here; it has always been regarded, in England at least, as the indispensable condition for adequate deceptions. Peter Fleming perhaps carried the doctrine a shade far at

Jellore near Calcutta where the Royal Air Force's special duty aircraft were stationed. When he wanted something done by air on the deceptive front, he came down to the airfield himself, took the relevant flight commander, who himself would fly the operation, out for a walk, explained what he wanted in detail, and made sure no notes were taken.[6]

An awkwardness about coalition warfare needs mention here. If one is conducting deceptions in complete secrecy, what can be said to one's coalition allies? The British practice was to say nothing at all to the various governments in exile in London. That included de Gaulle's still formally unrecognized staffs in London and later in Algiers, and least of all de Gaulle himself, because his political ciphers were leaky. Nothing was said even to the Poles, whose ciphers even the cryptographers at Bletchley Park could not read; and not much to the USSR, on the ground that the Germans could probably read a good deal of Soviet cipher traffic. But the British were more forthcoming to the Americans, once the United States had entered the war. American officers sat on Bevan's small staff, and they, too, learned to keep their mouths shut outside the office.

Bevan was made privy to every secret he asked about. As Fleming once said, "It is impossible, or at any rate highly dangerous, to tell a lie until you know what the truth is going to be."[7] Bevan not only saw all the chiefs of staffs' minutes, but also was aware of the codebreaking achievements of Bletchley, and sat on the XX Committee that MI5 organized to look after the playing back of captured German agents.

Double Cross

All but one of the agents the Germans sent to Great Britain during the war were captured; most of them agreed to cooperate with their captors, while those who did not were tried and executed. The one exception was a wireless operator who committed suicide. His body was found with his unopened transmitter beside him.

The branch of MI5 that handled German agents, called B1a, came under Major T. A. Robertson, another man who never gave newspaper interviews.[8] He set up an inter-service committee to run these double agents, with representatives on it from all three of the armed forces, Combined Operations Headquarters (COHQ), and the intelligence and cipher services. I know from first-hand experience the secrecy surrounding Robertson's operation. I was posted to COHQ for eighteen months from August 1942 as an intelligence officer, and I shared a desk with a fellow-captain, G. G. Rice, who became a close friend (he was my best man when I married in 1945). Only after his death did I discover that he had been COHQ's man on the XX Committee. He never breathed a word to me of its very existence, a further instance of the watertight-compartmented security of English wartime life.

That the deception, decipher, and security services could all be of one mind, and work together amiably toward the same end, collaborating with a small sealed knot of officers close to the highest command, provided part of the secret of the British wartime successes in this field. Emphasis also must be laid on the decipher aspect. Thanks to the probably unique opportunities offered by the penetration of the Enigma and *Geheimschreiber* cipher systems, both of which the Germans regarded as wholly secure, the deception staff could dangle baits before the enemy, and find out exactly whether and when he nibbled at them.

"Mincemeat"

One of the masterpieces of the deception service was operation "Mincemeat" conducted in the summer of 1943. Tunis fell in May, with a larger haul of prisoners than even at Stalingrad. Obviously, the Anglo-American force in North Africa was preparing to invade Europe. Any bright child with a school atlas could see that Sicily was, equally obviously, the next step. Johnny Bevan decided all the same to try to make the Germans believe that preparations against Sicily were only a cover for assaults on Sardinia and on southern Greece.

The keystone to this deceptive arch was to float a body ashore in southwest Spain. Dressed as a major in the Royal Marines, he had a briefcase full of highly secret material, including personal letters to allied commanders in the Mediterranean from such personages in London as Mountbatten. Their content pointed unmistakably to Sardinia rather than Sicily. The briefcase fell into Abwehr hands with the connivance of the Spanish police. The letters were opened and re-sealed (a shade clumsily as the British were later able to establish the fact of opening). The chiefs of staff were able to telegraph to Churchill, who was in the United States at the time, "Mincemeat swallowed whole." The Germans dutifully reinforced Sardinia and the Peloponnese, leaving Sicily alone.

There was a literary as well as a strategic sequel. Duff Cooper, a minister at the time, became aware of this story, and five years after the war's end published a novel based on it called *Operation Heartbreak*.[9] The secret authorities were alarmed to hear rumors that word was spreading that the novel was in fact a true story. So the admiralty called on Ewen Montagu who had done all the detailed work preparing Mincemeat, and had then gone back to the profession of barrister. Dictating continuously for forty-eight hours, Montagu produced an account of his own, *The Man Who Never Was*[10] that deservedly became a world bestseller and took the wind out of Duff Cooper's sails.

Montagu put nothing in that book that government wanted to remain secret. After Bletchley's decipher operation became public knowledge in the early 1970s, Montagu was able to write a sequel, *Beyond Top Secret U*,[11] in which he explained how heavily the deception service had relied on deciphering Abwehr messages to gauge how well their ploys were really deceiving their enemies.

As for how time passed in the London Controlling Section, much is clear from J. C. Masterman's admirable *The Double Cross System*.[12] More came from the pen of a prolific writer of detective stories, Dennis Wheatley, whose *Deception Planners* gives a detailed account of its work (and includes a good photograph of Bevan).[13] Wheatley, when he joined, was a pilot officer, RAF, the lowest commissioned rank in all the armed forces. Yet, with a decent private income and a London club, he had entrée to the world of the old ruling class and could move among its members as one of them. By the end of the war he was a wing commander (equivalent to an army lieutenant-colonel); as Bevan's secretary, he had earned his promotions. He, too, emphasized that "as few people as possible, having regard to practical necessities, should be allowed into the secret of any deception plan."[14]

"Fortitude"

The main achievement of the LCS was operation "Bodyguard" renamed "Fortitude." The objective, of course, was to deceive the Germans about the time, place, and nature of the start of operation "Overlord," the allied re-invasion of Western Europe in 1944. It is now popularly misremembered as D-day; misremembered, because every operation had its D-day, not just the Normandy landings. Bevan went to Russia in good time, with an American colleague, Colonel William H. Baumer of the United States Joint Security Control, to discuss with the Russian staffs, who were great believers also in the power of deception, how Soviet and Anglo-American efforts were to be combined for the protection of Overlord. They met with the usual maddening delays, but in the end secured useful agreements.[15] Much of the detailed work for Fortitude was done by Roger Fleetwood Hesketh, a Lancashire notable, who wrote a full account of it at the end of the War. This was long kept secret, and was published only after his death with the simple title of the codename: *Fortitude*.[16]

Fortitude was a colossal success. It made the Normandy landings, tough as they were, succeed. Had Montgomery's left flank been counterattacked with vigor in mid-June 1944 by the formidable German forces east of the Seine, the invasion might well have failed. It is now taken for granted that those landings were bound to succeed because they did; in fact, the outcome was far from a certainty. Churchill, in particular, could never forget that at the landing on V beach at Gallipoli, for which he was responsible, the sea had run red with blood for thirty yards off shore on the evening of April 25, 1915. (This was related to me by a man who had been there, as an engineer corporal; he had risen to be quartermaster of my unit at the outbreak of the next world war.) Churchill, who knew this story, offered no guarantees that Neptune, the assault phase of Overlord, would not encounter a similar rate of casualty.

Fortitude prevented a disaster on the beaches by distracting and diverting the defenders. Through double agents and dummy radio traffic, the LCS succeeded in persuading the Germans that General Patton commanded a large, notional United States army in Kent and East Anglia, which was standing by to cross the Channel at the Straits of Dover. The German general staff had picked this as the most likely spot for an allied invasion, because air cover could there be densest. The landing further west, led by Montgomery under Eisenhower's command, was presented to the Germans as a diversion, intended to draw troops away from the threatened area round Calais.

"Titanic"

On the very night of the invasion, 5-6 June 1944, three tactical deceptions were mounted to confuse the enemy even further: "Glimmer," "Taxable," and "Titanic." Glimmer and Taxable were airborne spoofs. Two specially selected RAF squadrons, 218 and 617 (the Dambusters), each ran one of these operations. Their aircraft flew overlapping oblong courses, at a set speed, throwing out bundles of Window foil at precisely timed intervals. On the few German naval radar screens that R. V. Jones had left in action they appeared as large fleets; one heading for Cap d'Antifer, north of Le Havre, and the other for Cap Gris Nez, near Calais. They appeared to move at eight knots from midnight till first light when they were near the coast, and then vanished.[17] I chose the codename for operation *Titanic*, and may perhaps be allowed, for a couple of paragraphs, to appear as witness instead of historian. By May 1944, I was intelligence officer of the Special Air Service brigade, an international brigade made up of two British and two French parachute units, some Belgians, a few Yugoslavs, a few Austrians under Scots cover names, and a Pole who was my assistant. We were forbidden to land before D-day. Our task was to operate, in uniform, well behind the main battle line, disrupting German reinforcements and providing intelligence.

About mid-May, I was telephoned by a stranger: would I scramble? I did. He introduced himself by naming a fellow intelligence officer whom I knew well, on Eisenhower's intelligence staff, and with whom I had been dealing for months on various points. He gave his own extension number, but not his name (he may have been Colonel David Strangeways, a canon of the Church of England who died in 1998). I was to allot a codename to an operation that would take place on the night of D-1 to D-day for Neptune, and was to provide four (shortly reduced to two) parties for it, each consisting of an officer, a sergeant and three other parachutists. He specified that they were to come from the British component of the brigade. They were to report to Tempsford airfield, the base of the RAF's special duty squadrons, by 1800 hours on D-2, where they would be briefed. I checked back with my friend at Supreme Headquarters Allied Expeditionary Force (SHAEF), got the brigadier's approval, and had the parties duly told off. I was forbidden to go with them. Regimentally, Titanic is regarded as a failure, because of the ten men who went only a sergeant and a trooper came back. Tactically, it made a little more sense. The men were armed with Verylight pistols and Victrola gramophones playing recordings of small arms fire, interspersed with soldiers' chat. With each party went 500 dummy parachutists, supposed to explode on landing. The aim was to simulate two major airborne landings. One group landed near Jumièges on the lower Seine, and at least gave the commandant of Le Havre a horrible fright. He telegraphed to everyone he could think of that he had been cut off, and every senior German commander who visited his map room after breakfast on 6 June saw a flag in Le Havre marked *abgeschnittet*, a minor source of confusion.

The other party had a more direct impact. They landed southeast of Isigny, which, in turn, is at the southeast corner of the Cherbourg peninsula, and made such a stir that the local unit rang up divisional headquarters. The duty officer went to wake the general but he was away in Rennes at a senior officers' conference on how to repel airborne attacks. So he woke up the chief of staff instead. This was a regular army officer, brought up to counterattack automatically whenever attacked. He ordered out the reserve regiment of the division, which spent the morning of D-day beating the woods near Isigny for a major airborne landing that was not there, instead of being on hand to put General Bradley's landing at Omaha beach back into the sea.

For several weeks before the landings, as a security precaution, foreign diplomats in London had been forbidden to use their own ciphers. Only the Americans and the Russians (and, in secret, the London-based Poles) were exempt from this ban. On 7 June, foreign secretary Eden sent a message to Eisenhower that he proposed to end the ban at once. He was persuaded with difficulty to keep it going until the end of the month, in order to reinforce Fortitude.

An extra touch was provided, at this critical stage, by one of the star double agents, Garbo. In a report sent by radio on D+3, he drew the Germans' attention to the fact that no formation from the [notional] First United States Army Group had taken part in the landings, and that Patton's divisions were still poised for their assault. At the same time, SOE sent out a call for immediate mobilization to Belgian resistance, in a cipher the Germans read easily. Some German formations east of the Seine already ordered to western Normandy were therefore turned round, and directed towards the Calais neighborhood instead.[18]

"Fortitude North"

Besides the main Fortitude deception, there was a subplot called Fortitude North which deserves attention because it validated one of Bevan's golden rules: go for what you know for certain about your enemy, thus making him act (actions are far more important than beliefs) in the way you want. One of Bevan's friends was Andrew (Bulgy) Thorne of the Grenadier Guards who had been a colonel military attaché in Berlin from 1932 to 1935. As a matter of routine, he attended, in uniform, Hitler's first big diplomatic reception in 1933. Making the rounds, Hitler with his odd eye for detail noticed that Thorne was wearing—after his gallantry decorations—what the army called (after a once famous children's cartoon) Pip, Squeak, and Wilfred, the three main Great War service medals, with the 1914 star. Had he been at First Ypres? He had? Where was he on 30 October 1914? It turned out the two of them had been only about five hundred yards apart that day. Hitler, therefore, marked Thorne down as "ein alter Frontkämpfer," his highest praise, and was disappointed when Thorne was posted back to Aldershot to command the Guards Brigade two years later. One of the many reasons why he hesitated to invade England in the late summer of 1940 may have been because he discovered that Thorne was by then the corps commander against whom his armies would first have to fight upon landing. It was leaked to the Germans, through the double agents and just in advance of the newspaper reports, that by 1943 Thorne had become a full general, a knight, and commander in chief, Scottish command. To anyone in the know, this meant that his career as a field commander was over, but Bevan then took a hand. He leaked more and more bogus information through his controlled agents indicating that Thorne was preparing an invasion of Norway. Dummy landing craft on the east Scottish coast as well as dummy radio traffic, bore this out. Hitler therefore kept a garrison of over 300,000 men in Norway because he was afraid of what Thorne might do. None of these troops was moved to help avert imminent defeat either on the eastern or on the western front, throughout the last winter of the war.

Air Deceptions

Two other routine features of deception, both affecting the war in the air, were undertaken during most of the war against Hitler. To protect the secrecy of Bletchley's successes with the German and Italian codes, it was an inflexible rule that no target found through ultra-secret intercepts could be attacked, until it had been spotted by more usual means of reconnaissance. This often resulted in aircrew being instructed, for no stated reason, to search in a particular direction. Reconnaissance aircrews, though bright, were not unduly inquisitive, and there were no unfortunate repercussions.

Again, once Bomber Command had organized its system of marshaling aircraft over a selected target in a stream of carefully timed attacks, it was usual for a few Mosquito aircraft to go into Germany a few minutes before the main attack, and drop markers on some quite different target. This caused the German night fighter defenses to send their aircraft to the wrong parts of the country, reducing the main bomber stream's exposure to their attacks.

Gulf War

It is no wonder that General Norman Schwarzkopf made decisive use of deception during the Gulf War of 1991, threatening a seaborne attack on Saddam Hussein's left just before he delivered a devastating land attack on his right. Not the least of the general's successes was that he managed this without exasperating the American news media, which since their triumph against President Nixon over the Watergate affair in the early 1970s seemed to have advanced to a position of power.

Counterterror

The principles of deception are as usable against terrorists as they are against more conventional enemies. It is important for intelligence and operations sections to work completely hand in hand, without any reservations. What matters is to persuade the enemy to act to his own disadvantage, rather than to affect his thoughts. Double agents can be used to plant misinformation about proposed security precautions. Terror occasionally can be used against terrorists, so that weak links in their networks can be persuaded to change sides. Ideally, bomb-makers can be arrested as they are getting their bombs ready for use, rather than after the bombs have gone off.

Lessons Learned

The principal lessons learned from the British experience of deception are these:

- Deception is a matter for operational planning, but also needs incessant contact with reliable intelligence staffs and sources;
- Complete secrecy is indispensable, while the deception planners, in turn, must know everything they can about the intentions of their own side and of the enemy;
- It is helpful if the staff in charge all come from a single, cohesive background;
- Deception must persuade the enemy to act, rather than to think;
- All the staffs concerned, both directly and from other agencies, must be prepared to sink their interdepartmental differences and pursue the common aim of victory, without insisting that theirs is the only important contribution.

Notes

1. Sun Tzu, *The Art of War*, tr. Samuel B. Griffith (Oxford, UK: Clarendon Press, 1963 ed.), pp. 66-7.
2. See Winston Churchill, *Marlborough. His Life and Times* (London: George G. Harrap & Co. Ltd, 1934 ed.) 2.332-44.
3. Dudley Clarke, *Seven Assignments* (London: Jonathan Cape, 1948).
4. Sir Michael Howard, *Strategic Deception*, volume 5 of F. H. Hinsley, *British Intelligence in the Second World War* (London: HMSO, 1990).
5. The only other books fit to be mentioned in the same breath are R. V. Jones, *Most Secret War* (London: Hamilton, 1978) and *Reflections on Intelligence* (London: Heinemann, 1978, revised ed., 1989); and *Double Cross* J.C. Masterman, *The Double Cross System in the War of 1939 to 1945* (New Haven, CT: Yale University Press, 1972), a bowdlerized version of his official report to MI 5 at the end of the war. The rest of this paper relies heavily on Howard.
6. See Terence O'Brien, *The Moonlight War: The Story of Clandestine Operations in South-East Asia, 1944-5* (London: Colins (1987).
7. Ibid., p. 28.
8. Obituary, London *Times*, 16 May 1994, p. 19a.
9. Duff Cooper, *Operation Heartbreak, A Story* (London: Hart-Davis, 1950).
10. Ewen Montagu, *The Man Who Never Was* (London: Evans Bros., 1953).
11. Ewen Montagu, *Beyond Top Secret U* (London: P. Davies, 1977).
12. Masterman, *The Double Cross System*.
13. Dennis Wheatley, *The Deception Planners: My Secret War* (London: Hutchinson, 1980).
14. Ibid., p. 35.
15. Howard, *Strategic Deception*, pp. 111-2.
16. Roger Hesketh, *Fortitude: The D-Day Deception Campaign* (London: Little, Brown, 1999).
17. R. V. Jones, *Most Secret War*, 405-6; *Radar: A Wartime Miracle*, Recalled by Colin Latham & Anne Stobbs, and by Men & Women Who Played Their Part in It for the RAF (Phoenix Mill, Far Thrupp, Stroud, Gloucestershire, UK: Sutton Publishers, 1996) 137-40.
18. Masterman, *The Double Cross System*, pp. 156-8.

CHAPTER 21

Some Notes on the Organization of Deception (1944)[1]

Brigadier Dudley Clarke

Brigadier Dudley Clarke was an officer in the British Army, known as a pioneer of military deception operations during the Second World War. His ideas for combining fictional orders of battle, visual deception, and double agents helped define Allied deception strategy during the war. For example, in January 1941, Clarke began Operation Abeam, fabricating the existence of a British paratrooper regiment in the Mediterranean region. It would be two years before such troops reached the Mediterranean region, but Clarke played on Italian fears of an airborne assault.

This piece, long believed lost, was recently discovered in a U.S. archive.[2] *It is published here in its entirety for the first time.*

Some Notes on the Organization of Deception in the United States Forces

Deception of the enemy on a systematic, continuous and theater-wide basis was first started, so far as I am aware, by General Wavell in the Middle East at the end of 1940. The instrument that was devised to affect it was "A" Force; and if I personally were to start all over again with the experience of the four succeeding years before me I would still build up a machine on the same lines as the "A" Force of 1944. At the same time I am fully conscious that "A" Force was first shaped to meet the special needs of particular personalities and conditions and then developed to keep pace with a vastly changing strategy, so that, however well it may have fitted into the British Mediterranean War Machine, it provides by no means a set pattern for other nations and other theaters. We have in fact, both here and in N.W. Europe, seen failure in attempts to build up an American "A" Force. It seems clear, therefore, that the United States will need to design a special deception organization to fit their own characteristics and the peculiarities of their own theaters of operations. Nevertheless I am venturing to suggest that any such organization will only succeed if it is firmly based on certain principles, which I have become convinced from my own experience are sure foundations for this very specialized work. The aim of this paper is to present those principles in an objective and impersonal manner, so that the authorities concerned may avoid the very real danger of blunting this sensitive weapon by forcing it into place in a rigid military machine. Rather, I suggest, should the machine adapt itself to turn this new weapon to the best advantage against the enemy as it has successfully done with so many other innovations. I have stressed

1. Brigadier Dudley Clarke, *Some Notes on the Organization of Deception*, 1944.
2. As first reported in a 2005 revised printing of Thaddeus Holt, *The Deceivers*, NY: Scribner, 2004. The last four paragraphs had been previously published in Michael I. Handel (editor), *Strategic and Operational Deception in the Second World War*, London: Frank Cass, 1987, pp. 88–90.

this point at the beginning because I myself have had to meet a certain American intolerance of the apparently illogical situation of "A" Force in fitting neither into a square nor a round pigeon-hole. This comes, I know, from an understandable dislike of "private armies", and a neatness of mind which ranges each component of the war machine in its appropriate staff section. I am sure that a more sympathetic view would prevail if there were a clearer understanding of the conditions under which "A" Force came into the world and grew up, and I will therefore preface any thesis on principles with an attempt to describe in brief the background from which they have emerged.

"A" Force started under two handicaps of which any similar American organisation is happily free. The first handicap was a lack of precedent: Deception on a big scale had never been practiced before and it had to prove its worth at the same time as it was trying to find its feet. As a result the early "A" Force could only operate freely in the zones of operation of the more imaginative commanders who, amid the successive adversities of 1941, were prepared to give a trial to anything which offered them a prospect of help. This led to the organization extending in an unbalanced way, dependent upon "selling" itself, and inevitably shaped to suit the personalities concerned. Chief among the latter was of course General Wavell, who numbered amongst his characteristics a sense of security-mindedness far beyond anything we know at the present time. It may well be that we no longer need it to that degree; but in 1941 he was striving, with pitifully inadequate forces, simultaneously to recover a shattered army from Greece and Crete, to conquer the vast territory of Italian East Africa, to invade Vichy-French Syria, defeat the rebel forces of Iraq, and to launch a counter-offensive against the German-Italian Desert Armies on the Egyptian border. Small wonder, perhaps, that he placed security in the forefront of his policy, but to the budding "A" Force it represented the second major handicap and one which made the "selling" process all the more difficult. At that time we thought, rightly I believe; that deception would only succeed so long as the enemy was kept in ignorance of the fact that it was being practiced at all. From this it followed that only the smallest possible inner circle of our own people was initiated into it, and every effort was made to hide from all the rest the fact that any Deception Organisation even existed. Hence "A" Force started on the basis of a "Secret Service" — and in fact remained so until well on in 1943 — a circumstance which had the greatest influence on its growth and on the shape it finally assumed. In the early days the Commander of "A" Force received his instructions in every case direct from the Commander-in-Chief himself, either verbally or in the C-in-C's own handwriting, and dealt with nobody except his Chief of Staff and the two Directors of Operations and Intelligence. He held, as cover, an appointment on the C-in-C's personal staff, and no deception document was ever handled in GHQ outside the C-in-C's personal secretariat. When instructions had to be issued by GHQ to implement a deception scheme the C-in-C personally addressed them to the Army Commander concerned with the request that no one but his Chief of Staff should be informed of their real purpose. In consequence of this very elaborate protection "A" Force was able to preserve an almost complete incognito, and it was probably not until 1943 that the enemy began to detect the existence of a far-reaching deception machinery in the Allied ranks. On our side the incognito was dispelled to a certain extent during the operations against Sicily, when certain details of the cover plan for the landings had necessarily to be disseminated amongst a large force of British, American and Canadian troops to whom it was probably something of a novelty. Since then the Germans cannot fail to have realized a good deal of the deceptions that have been practiced upon them and, although the need for stringent secrecy regarding deception is still as great as ever, the organizations employed upon it no longer have to be hampered by the conditions which affected "A" Force so greatly at the start.

I have tried to picture this background in some detail in order to explain the fact that "A" force never started as a logical blueprint conception, but developed by trial and error to fit every kind of condition extending from defeat to victory. For all its rigidity the British Army is not a very logically minded institution, and often has succeeded by this process of trial and error in producing something which "works" no matter how illogical it may appear. The peculiar British system of brevet promotion is a good example which works extremely well, though it must seem strangely puzzling to an American. I will not, therefore, defend the logic of "A" force, but will claim that in shaping itself into something workable it has discovered a few basic facts which must be inseparable, from any other organization of the same type.

The first concerns the scope of the organization's activities and, in particular, the directions in which they should be focused. Until this is properly understood there will be a tendency to muddle deception with psychological warfare and even to suggest that the same instrument can serve both purposes. A moment's examination of the aims of the two will show this to be fundamentally unsound, and any attempt to mix both in practice will be highly dangerous. Nevertheless the danger is often present and is sometimes curiously difficult to dispel. The essential difference lies of course in the audience for whom the two organisations cater. Psychological warfare starts at the apex of a triangle and endeavors to spread its arms as wide as it can to embrace the broadest possibly base. It matters little if many of its audience can detect the origin of their messages nor if a privileged few can recognize distortion of the truth; its appeal is to the masses and it is unlikely to influence the thought or actions of the enlightened inner circles of the General Staff. Deception, on the other hand works in exactly the opposite way. It starts at the base of the triangle and concentrates its influence towards a single point at the apex; its essential aim is to conceal the origin of its messages by directing them upon this single point from as many different directions as possible. It cares little for the thoughts and actions of the masses, but it *must* penetrate directly into the innermost circles of all. Its audience is narrowed down to a small handful of individuals, as represented by the senior members of the enemy's Intelligence Staff, and sometimes even to a single individual in the person of the Head of that Intelligence Staff. If they can influence him to accept as true the evidence they have manufactured for his benefit, then they have accomplished their entire aim, since it is only through the Head of the Intelligence that any enemy commander receives the impression of his opponent upon which he has to base his plan of operation. It is necessary, therefore, that the single-purposeness of any deception machine should be recognized from the start and its shape dictated by the overriding need to concentrate every ounce of its diverse efforts upon that one ultimate target. As a corollary it follows that those who direct the deception machine must have an adequate knowledge of the small group of men on whom all their activities are focused, of their national characteristics, their language, thoughts and professional methods with all their strengths and their weaknesses.

It is this note on personalities which leads me to the next principle, which I firmly believe to be a foundation stone in the successful application of deception. Deception is essentially an art and not a science, and those who practice it must be recognized as falling into the category of artists and not of artisans. This is, I know, difficult to accept in professional military circles where it is widely believed that the art of war can be taught to the average educated man, even though he may have little aptitude for it. But, nevertheless, I am convinced it is true; and twice in "A" Force I have seen highly qualified and highly intelligence staff officers of the British Regular Army fail completely to cope with the work, although both did brilliantly afterwards on the Operations Staff. What they lacked was just the sheer ability to create, to make something out of nothing, to conceive their own original notion and then to clothe it with realities until eventually it would appear as a living fact. And, since that is precisely what the Deception Staff must do all the time, it follows that the art of creation is an essential attribute in all who are charged with such work. To expect those who have not this art to produce the required results will lead to risks beyond that of mere failure.

If this thesis is accepted it is easy to see why one brain — and why one alone — must be left unhampered to direct any one deception plan. It is after all little more than a drama played upon a vast stage, and the author and producer should be given as free a hand in the theater of war as in the other theater. (Also, of course, in both they must have the necessary qualifications to justify that confidence). It is not a bad parallel to compare a Commander in the field with the impresario who wants to mount a successful play at his theater. He decided on the type of play he wants—drama, comedy, musical, etc. and instructs an author to produce a script. Having accepted the script, he appoints a producer to mount the play. From that point onwards he may well leave everything else to those two, and look only to the results obtained. Provided these are satisfactory, the impressario who is not himself an author or producer, wisely leaves them to rule the cast, scenery, costumes and all else that goes to make the play. The wise Commander-in-Chief will follow the same example. In his case the matter is simplified by the fact that the head of his Deception Staff doubles the roles of author and producer. The Commander therefore tells him what sort of deception he needs, examines the plans produced for him with the required aim in view and, once the final version is approved, watches only the results and leaves all else to his specialist. In both peace and war, however, the Chief is the best judge of the results; in both cases he assesses them by the reactions of the audience (or the enemy), and should interfere in proportion to the degree in which they fail or succeed to achieve the object he himself has set.

And it is this mention of the "object" which brings me to the last of the principles I have tried to enunciate. For the theatrical impressario this presents no difficulty — all he wants is to see the audience moved to tears, laughter or rhythm in concert with the plan — but to the General it is a problem which merits most careful thought. His audience is the enemy and he alone must decide what he wants them to do — to advance? to withdraw? to thin out or to reinforce? Whatever he chooses, the main point is that his "object" [objective] must be to make the enemy DO something. It matters nothing what the enemy THINKS, it is only what he DOES that can affect the battle. It is therefore wrong, and always wrong, for any Commander to tell his Deception Staff to work out a plan "to make the enemy think we are going to do so-and-so." It may be that the plan will succeed but that the enemy will react to it in a totally unexpected way, upon which the Commander will probably blame the Deception Staff who have in fact produced exactly the results they set out for. It is this boomerang effect which has made many people apprehensive of using the deception weapon, and it cannot be stressed too strongly that, if used in the wrong way, it can prove a real danger. But there is one sure way to avoid any possible risk, and that is to get the OBJECT right. Given a correct "object", the deception plan may fail but it cannot in any way do harm. Give it a wrong "object" and it will invariably give wrong results. Our theatrical impressario after all will not attempt to dictate to the author the plot of the play, but that is precisely what the General does who tells his Deception Staff that he wants the enemy to be made to "think" something. It assumes a knowledge of the enemy's likely reactions which the Deception Staff should know from experience very much better than the General. It is for the latter to say what he wants them to do, and for the specialists to decide what the enemy must be made to think in order to induce them to act in the manner required. Perhaps an illustration will explain this best. In the early part of 1941 General Wavell wanted the Italian reserves drawn to the south in order to ease his entry into Northern Abyssinia. He considered this might be done by inducing them to reinforce the captured province of British Somaliland, and he gave instructions for a Deception Plan to be worked to persuade the Italians that we were about to invade Somaliland. Deception was new then and on the surface that appeared to all concerned to be a perfectly laudable object. The plan, innocently ignoring the real object of influencing the location of the enemy's reserves, was entirely successful; but the results were totally unexpected. In face of the threatened invasion, the Italians evacuated British Somaliland. Not only had General Wavell to draw upon his own meager forces to re-occupy the country, but the Italian garrison was freed to swell the forces in the north which were to block our advance at Keren. Had a different object been chosen, quite a different deception plan would have emerged and perhaps a quite different effect produced upon the actions of the enemy.

That concludes this brief review; and I will end by summarizing that to be successful any deception organisation needs:

a. To be so organized that it directs the whole of its efforts to influence the enemy's intelligence staff—and that alone.
b. To be composed of senior officers with a real knowledge of the intelligence staff that is to become their audience.
c. To be directed, as specialists in an art, by a Commander and Staff who tell them what results they require and who leave them unhampered to arrange the best means of obtaining those results.
d. To be given an object in terms of the manner in which the enemy is required to ACT in order to further the operational plans of their own Commander.

Provided these four principles are faithfully observed, it matters little how the organization is shaped and it can best take the form most suited to the nationality concerned and the theater of war affected.

Rear H.Q. "A" Force
/s/ D.W. Clarke
c/o G.H.Q. M.E.F.

D. W. CLARKE
30 October 1944

Brigadier Commander "A" Force

American Attempts at Deception: A Spotty Record

CHAPTER 22

The Actor's Tale[1]

Douglas Fairbanks, Jr.

Douglas Fairbanks, Jr. was commissioned as a reserve officer in the United States Navy at the onset of World War II and was assigned to a Commando staff in the United Kingdom. Having witnessed (and participated in) British training and cross-channel harassment operations emphasizing the military art of deception, Fairbanks attained a depth of understanding and appreciation of military deception then unheard of in the United States Navy. Upon returning from his military career, Fairbanks went into the family business of acting, and between 1954 and 1956 he made a number of half-hour programs as part of a syndicated series for television called "Douglas Fairbanks, Jr. Presents."

Douglas Fairbanks Junior was already a famous Hollywood star when America entered the war. It was somehow most appropriate that a man who had lived his life in the Hollywood Dream Factory should become part of the machinery of Allied Deception. After all, film technicians from Shepperton and Elstree Studios helped the British. The famous actor helped too. Before the Americans entered the war after the Japanese had bombed Pearl Harbor on December 7 1941, Fairbanks had worked tirelessly to persuade a reluctant American public that the fight in Europe was their fight too. Once America had joined the hostilities he joined up and spent the early part of the war as a Junior Grade Lieutenant on convoy duties. Later he was to become what he calls a 'two-striper', a full Lieutenant, escorting convoys to Malta and Murmansk. He sailed in destroyers, cruisers and finally in aircraft carriers. By the end of the war the list of the places in which he had fought sounds like a catalogue of the war itself.

'I was in the North Atlantic, Murmansk, Russia, North Cape and the Arctic Circle, across the Channel in France and then the Mediterranean, North Africa, Sicily, Greece, Salerno, Anzio, Naples, Corsica, back to France, you name it.

'My entree into deception came about as a result of my long standing family friendship with Lord Louis Mountbatten. He wanted some American representatives on his staff. There were already one or two, but he asked for more and he asked specifically for me as an old family friend. He knew I'd been at sea several times and thought that it might be interesting to get me involved in the dirty tricks department, in deception and tactical diversions. The relationship

1. Douglas Fairbanks Jr., "The Actor's Tale", in *Trojan Horses,* Martin Young and Robbie Stamp, editors, London: The Bodley Head, 1989, pp. 165–170. Reprinted by permission of The Random House Group Limited.

with Mountbatten went back as far as my childhood, because he and Lady Mountbatten spent their honeymoon with my father at his house in California. He was Godfather to one of my children, she was Godmother to another one. The families have always been close.

'I suppose it's like anywhere else. You go for the people you know and trust. If I'm making a film and I know somebody who's very talented in building ships down in Florida, I would still have to convert him from making ships, however good he was at that, into making movies. But at least I would already know that we spoke the same language. When I was in London, Mountbatten's influence and indeed my many connections meant that I had no problem being taken seriously but out in the field it was a little different. In fact it was a freakish, difficult position for me because being a name that was familiar to a lot of people, they would suspect that I was just there for show. Well, that was a good thing for them to think, that I was a decoration, a "do-nothing". But they were sometimes rather astonished to discover that I would go on actual operations. Mind you they didn't know that the operation was part of a deception plan. There was a sort of bewilderment on the part of others, "What the hell was I doing there?" In a way that reaction was useful because they'd never suspect that I was in something as subtle or as top secr as this. I know that on this score Mountbatten was worried at firs that I'd be a compromise to security because I'd be recognised. Tha was the case at first but eventually it worked the other way around

'In terms of sheer rank I was often out of my depth. I had suc a humble rank and yet I was moving in high circles. The Army wa able to make spot promotions so that people could fit into their jobs So people in the Army were promoted to Lieutenant, Captain up t Colonel and Brigadier. The Navy just wouldn't play that game. I jus stayed at my humble rank until promotion was due and then, little b little I moved up.

'I came to London to work on the planning side of deceptio and then travelled to Inverary in Scotland where they were workin on the tactical side, mostly for the Army. When I was up in Scotlan I was asked to consider how the army work on tactical deceptio could be adapted to naval warfare. I had been briefed to make particular study of sonic deception at sea and was also intrigue by the use of false radio broadcasts to confuse the enemy wireles interceptors. I came across people like Disney Barlow when I mad my way across to Laggan House in Ayrshire, where there were a sorts of experiments going on in sonic deception. Montgomery a El Alamein had some success with that at sea when he simulated a amphibious landing, which threw the German Commander, Gener Stumme, into a bit of a panic. But it was in Scotland, which was a lo colder, that I first learned all about it; how they made the recording and how they experimented behind trees to see what the reactio was. Then we had to adapt that to amphibious warfare and pu these devices on small boats. The amphibious side of tactical decep tion moved to a base in Southampton, where we built up a grou of Special Operations people who were involved in raiding, dirt tricks, deception, diversion, a little of everything that was off-bea I was a sort of liaison with the British for Special Operations. W had some small adventures across the Channel and finally we ha the North African invasion in November 1942 and establishe bases out there. On my trips out to the Middle East I met Dudle Clarke on a number of occasions. He was taciturn, pleasant, ver retiring, physically small, a deception in himself because he didn look very important. He looked to be an amiable little fellow, classic professional soldier, not very impressive to meet until yo started talking to him, then he became very impressive indeed. I fact it was Dudley Clarke who was sent out to Washington on mission to brief the American Chiefs of Staff on what he had bee doing out in the Middle East, in September 1942. Out in Cairo, I also knew Noel Wild, who was a very smart Hussar, Clarke's right hand man. I seem to remember that Clarke and Wild had their headquarters in what had been a brothel before the war. At one time or another I came across David Strangeways, David Hunt, and the conjuror Jasper Maskelyne, they were all involved, in one way or another, with Dudley Clarke, who really gave a lead to the whole thing.

'I was still doing a liaison job and in the early summer of 1943, after Rommel had been chased out of North Africa, the Americans established small sonic and tactical deception units at their bases in Algiers and Tunis. By this time I had gone back to the States for a while and was helping to train those units at a base in Virginia, before they were posted. Recruiting for the units was interesting. We went to the universities and asked for volunteers with backgrounds in electronics. They were not told what they were volunteering for except that it could be dangerous. There was no shortage of young men wanting to come along and see for themselves what it was all about. Once they had undergone their basic training, we had them shipped out to the Mediterranean and equipped with pre-recorded sonic deception devices.

'How did it work? Well, we would pre-record sounds of landing-craft, voices talking in the middle of the night, signals back and forth. At night, one or two boats would sound like twelve or fifteen. Of course, sound travels further over water than anywhere else. Or we might arrange for a single plane to fly overhead, giving out signals as though there were a whole flock of planes.

'And then we went into the visual end of it, in fact every kind of diversion on the tactical side. But that had to mix with the strategic, like the famous story of the "man who never was". Well, it all had to fit into a pattern. Except for the High Command on the American side, I was one of the few Americans who knew what was going on on the British side. First of all I liaised with the British and then tried to make our own people adapt to the deception plans.

'At the outset, the American High Command hated the entire idea of deception. The Chief of Naval Operations, Admiral Ernest King, thought it was "just boy scout stuff". His attitude was that if you were going to fight a war then just get ahead and fight it, don't do that kind of thing. They thought that War should be fought as War. Trojan Horse stuff was not for them. As far as they were concerned we were just a bunch of small boys wasting our time with tricks, instead of getting on with the business of fighting. Headquarters in Washington took a long time to be convinced of the value of deception but my Commander, Admiral Hewitt, Eisenhower and indeed Roosevelt himself simply saw the results the British were achieving and they were in close, friendly relations with the British, so there was some fairly high powered support, which won the day in the end.' Douglas Fairbanks Jnr was unusual in that despite being a member of the planning staff he was allowed to be both a planner and an operations man. Normally the information the planner carried in his head was considered too valuable to risk if he were taken prisoner.

'As I was the only one who had had experience with the British on these commando raids and on deception and diversionary operations, I would help to write the diversionary plan but there was no-one available to carry it out. They would assign a senior officer to do the job, but often he couldn't understand the kind of thing we were up to, so I had to change hats and go with them to advise the senior officer, which was very unconventional.'

But Douglas Fairbanks Jnr has no doubts about the theory and the practice of deception. This is how he summed it up:

'I suppose at first, in the very simplest way, the purpose is to confuse the enemy as to your target. For example, when we landed at Sicily in 1943 one of the earliest things we did was to make them think we were landing in Sardinia. Once it became obvious that we were going to land in Sicily then we planted the idea that it could be there, or there or there. Finally, we persuaded them we were going to land *there*, but actually we landed *here*. That was the trick. The second thing was to give an impression of much greater strength than we actually possessed, with dummy parachutists, an illusion of many more ships than we actually had and so on.

'At the beginning of 1944 I was appointed as representative on the Joint Security Council (JCS) which was the American organisation under the Joint Chiefs of Staff responsible for co-ordinating deception operations. It was the American equivalent of the LCS. In the run up to D-Day we co-ordinated with things happening in the Mediterranean on the strategic side, air traffic, radio traffic, movement of ships, false information and so forth. We also did some tactical deception with dummy craft and "beach-jumper" units. Beach-jumper was a codename for the American Tactical Diversionary Groups. They were small boat operations. It was the BJs who created the effects out of small boats of sounds of battles and bright lights, a sort of "son et lumière".

'There were two operations, Taxable and Glimmer, which although I was not involved personally, were for me a culmination of a lot of the amphibious deception, that I had been involved in throughout the war. The plan was to persuade the Germans early on D-Day itself that there were two very substantial convoys heading right for the French Coast. They were meant to draw attention away from the real landing zones. Taxable made its way towards the bit of coast between Cap d'Antifer and Fecamp and Glimmer steamed towards Boulogne. They used all sorts of things that had been developed during the war: something we called Chaff but the British called Window, thousands of little pieces of metal dropped out of the air which showed up on the German radar screens as a blizzard. There were balloons called Filberts, which had their own radar reflectors, and something called Moonshine, which just bounced the enemy's radar right back at him and magnified it too. Once the two fake fleets had almost reached the French coast, the next stage was sonic deception, and that involved loud speakers sending out all the noises of a huge invasion fleet. By the time the Germans came out to see what was going on the whole party had disappeared. I think that the Germans had really been led to believe that this was it, the big one, and in the early stages it did help to take some pressure off the Normandy beaches. I think anyone would have been fooled, because the deceptions were expertly worked out in detail. Sometimes the Germans might have suspected that one of these operations was not all it was got up to be, but they didn't dare take a chance. What made all this so impressive was the amount of resources that the Allies were prepared to throw at their deception plans. It's not enough if the enemy is disabused of what he thinks you're going to do. You must make him think positively that you are definitely going to do something else and that takes planning and manpower. Not only does he think I'm not going to hit with my right hand, but I make him think that I'm going to hit with my left hand. In actual fact, I'm going to kick him!'

CHAPTER 23

American Strategic Deception in the Pacific: 1942–44[1]

Katherine L. Herbig

Taking into consideration Japanese preoccupations and intentions, the strategic demands of being on the defensive, the shortcomings of the Japanese intelligence system, and the loss of means to act other than as the deception demanded, the U.S. victory in the Pacific War demonstrates the important role of U.S. strategic deception. Strategic deceptions like Wedlock were not perfect, and their failings can be as profitably studied for lessons as their successes. They were good enough, however, to reinforce concerns for areas of secondary importance, to generate additional reinforcements for these areas at the expense of reinforcing actual targets, and to waste flagging intelligence resources in sorting out superfluous scenarios from true ones. The U.S. was winning the Pacific War regardless of its deceptions, on the basis of its much larger pool of economic and human resources. Strategic deception helped to reduce the cost in lives by directing and holding Japanese forces away from the Marianas and, later on during Bluebird, from Okinawa, when these important strongholds were invaded. The history of the Pacific War is the more complete for having written into the familiar story the little-known chapter on U.S. strategic deception.

In standard American histories of the Second World War in the Pacific it has been the Japanese who are credited with the ability and the will to deceive their enemies, not the Americans. Japan's surprise attack on Pearl Harbor, which humiliated the US and catapulted the country into the war, became the popular archetype for Japanese treachery. When the investigations of the Pearl Harbor raid held during and after the war revealed that the attack itself had been supported by a clever deception campaign, this only reinforced the stereotype. The Japanese had simulated radio traffic from their invasion fleet which falsely placed it in the coastal waters south of Japan, not headed north-west toward Hawaii. They had planted false clues to their own unpreparedness, such as sending Japanese soldiers ashore on leave just hours before their ships sailed. They had reinforced garrisons in Manchuria to suggest a northern target for their obvious military preparations instead of the true southern one, and they had prolonged their show of diplomatic negotiations in Washington.[1]

1. Katherine L. Herbig, "American Strategic Deception in the Pacific: 1942–44," in *Strategic and Operational Deception in the Second World War*, Michael J. Handel, editor, Great Britain: Frank Cass & Co, 1987, pp. 260–264. Reprinted by permission of Frank Cass & Co. (Taylor& Francis Group).

Yet it was the United States, not Japan, that perfected the more sophisticated capacity for deception in the Pacific War. Eventually the US adopted deception on a scale which helped to shorten the war and to save American lives. While strategic deception came to characterize American strategy by the war's end, Japan's attempts to deceive diminished over time. Taking a lead from the British, whose elaborate deceptions in Europe contributed to the Allied victory there, the Americans tried to deceive the Japanese about their every strategic move during the last two years of the war. With the relatively recent release of some of the records dealing with deception, it is possible to begin the study of this hitherto obscure chapter in American military history. This history helps to clarify how in fact the United States won the war in the Pacific and what an important part deception played in the American, not the Japanese, war effort.

The United States built its capacity for strategic deception from nothing when the war began into a large program led by a trained cadre of specialists at its end. Thinking deceptively did not come easily or naturally to most American military officers, whose training stressed ideas of mass and position derived from nineteenth-century European theorists. Arranging to meet the enemy with overwhelming force, not weakening him in advance with deception, preoccupied mid-twentieth century American strategists. To some extent deception remained suspect in American military circles throughout the war, and those who understood its potential and wished to rely more on it found themselves repeatedly justifying their views.

None of the American military services had deception planning agencies when the war in the Pacific broke out. Deception needed an institutional home before its supporters could build the organization they needed. Since strategic deception demands access to the highest levels of leadership, and the authority to manipulate actual as well as bogus operations, the obvious place for a deception agency for the Pacific war was in the burgeoning bureaucracy of the Joint Chiefs of Staff (JCS) in Washington. In the fall of 1942 at the urging of the British, a small agency was designated two tasks: military security, its primary role, and only secondarily, the co-ordination of deception activities for the Allied operation Torch, the invasion of North Africa that would take place in November 1942. Called Joint Security Control, the deception agency suffered from a schizophrenic mission both to protect secrets and to create them, and it fought bureaucratic rivals throughout the war for more power and discretion for itself and more scope for its deception projects.[2]

During the first 18 months of the war, while a deception oversight agency gradually evolved and began to function in Washington, US Navy and Army commanders and their staffs handled deception for themselves. Deception by Americans in the Pacific in 1942 and for most of 1943 was at an operational level, linked to a specific campaign or impending clash with Japanese forces, and intended primarily to have local or short-term significance. From the beginning deception played a part in the American effort to regain the initiative in the Pacific and then to sustain the drive to Japan itself. The operational deceptions of the first months became the base of experience on which the later strategic deception programs were built.

After Pearl Harbor the US Navy stretched its scarce forces thinly to cover commitments to defend Hawaii and the sea routes to Australia and New Zealand. Japan consolidated its hold on the East Indies by February 1942, brushing aside the initial Allied opposition in the short-lived ABDA (American British Dutch Australian) command. The US Navy dared not risk naval battles which endangered its few remaining aircraft carriers until others could be built to join them in the Pacific. The first American actions of the war were cautious raids on Japanese positions. In the unaccustomed and unenviable role of underdog, the navy overcame its distaste for deception. Fleet Admiral Chester W. Nimitz, Commander-in-Chief, Pacific Fleet and Pacific Ocean Areas (CinCPAC-CinCPOA) based in Hawaii, favored any means that promised to augment his meagre resources and get results.[3]

The chance for Nimitz's first engagement with the Japanese Navy, which would become the Battle of the Coral Sea occured in April 1942. Using intelligence reports from the navy's small but excellent cryptanalytic staff, the Combat Intelligence Unit at Oahu, Nimitz built deception into his plans. He would send a light cruiser, the *Nashville*, to the rich Japanese fishing grounds off the Kamchatka peninsula early in May. There he hoped the cruiser's presence would distract the Japanese as they prepared their projected operations against Tulagi and Port Moresby on the southern coast of New Guinea. The *Nashville* would be instructed to attack fishing boats and, more to the point, send radio signals designed to simulate a carrier task force. However, as the real US task force under Vice Admiral Frank Jack Fletcher prepared to intercept the Japanese invasion fleet in the Coral Sea early in May, the Kamchatka diversion did not materialize: the *Nashville* ran aground at Midway en route to the North Pacific, and had to return to Pearl Harbor for repairs.[4]

Warnings by navy cryptanalysts reading parts of the Japanese Navy's secret code in May tipped off Nimitz that the next Japanese target would be Midway.[5] He took the opportunity afforded by several weeks of advance warning to initiate various deceptions to guard his preparations to meet the Japanese at Midway. In the weeks before the battle the cryptanalysts put together where and when and even how many ships Japan would send to attack Midway early in June; Nimitz gradually had solid information and enough precious time to mass his three remaining carriers. The disparity of forces at Midway could have been crushing: Japan had 11 battleships, five carriers, 16 cruisers, and 49 destroyers against American forces with no battleships, three carriers, eight cruisers, and 14 destroyers.[6] Nimitz's carefully planned surprise and several pieces of luck gave the US its strategic victory over long odds.

Before he ordered Vice Admiral William (Bull) Halsey's Task Force 16 to leave the Solomon Islands in mid-May and to return at once to Pearl Harbor, Nimitz apparently sent Halsey secret orders: he should deliberately reveal himself to Japanese patrol planes attached to an invasion force headed toward Nauru and Ocean Islands, east of the Solomons in the Pacific. The Japanese duly reported sighting Halsey's two carriers on 15 May, after which the American task force turned and withdrew. The unexpected sighting provoked a flurry of response from the Japanese, who ultimately decided to call off their invasion of Nauru and Ocean as too risky with American carriers in the vicinity. Four days later Halsey set off for Pearl Harbor and then Midway, leaving the South Pacific temporarily without a US Naval presence. The sighting on the 15th, however, kept the Japanese nervous about the whereabouts of Halsey's carriers, in effect covering his departure, and leading into a radio deception scheme which began later in May.[7]

This plan called for simulating the presence of the carrier task force off the Solomons by using the communications and ships of the Commander of Naval Forces, Southwest Pacific, Vice Admiral Herbert Leary, based in Melbourne, Australia. Beginning on 25 May, two ships, a cruiser and a sea-plane tender, sent dummy radio traffic patterned on the recent battle in the Coral Sea as they steamed back and forth off the Solomons. Shore traffic was returned from Melbourne.[8] To Japanese traffic analysts, limited to radio as their only channel of information, it appeared that Halsey's carrier task force remained east of the Solomons where it had so recently been seen. This relatively simple deception paid handsome dividends a week later, 4 June, at Midway. Intelligence analysts of the Naval General Staff in Tokyo took note of the American radio traffic coming from the Solomons. They advised their invasion fleet commanders that US carriers remained in the South Pacific, a fact which implied, they claimed, that the Americans remained ignorant of Japan's intention to attack Midway. Even in the face of contrary evidence, after the Japanese forces already in the Pacific had monitored the tell-tale increase in radio traffic as Fletcher's task force left Pearl Harbor for Midway on 30 May, the Naval

General Staff in Tokyo remained convinced that two American carriers were still off the Solomons and that the surprise attack on Midway was secure.[9]

In the event, Nimitz's ambush north-east of Midway caught the Japanese striking force by surprise. They lost four aircraft carriers and the battle of Midway. The United States gained one of its most important victories in the Pacific. According to Japanese historians of Midway, the US won as much because of Japanese intelligence failures as because of the admittedly brilliant American intelligence achievements. The sighting of Halsey's carriers, when 'confirmed' by the subsequent simulated radio traffic, misled intelligence analysts in Tokyo.[10] Even when Admiral Isoroku Yamamoto's main force, lurking west of the Midway striking force, caught the suspicious increase in radio traffic from Pearl Harbor on the 30th, they did not warn the commander who led the strike. They maintained their own radio silence instead. Admiral Nagumo was lulled to disaster at Midway, according to Fuchida and Okumiya, by the

commander who led the strike. They maintained their own radio silence instead. Admiral Nagumo was lulled to disaster at Midway, according to Fuchida and Okumiya, by the

> Naval General Staff's persistent misestimate, maintained until the very eve of the battle and communicated to the advancing Japanese forces, that an American task force was operating in the Solomons area, strongly implying that the enemy had no suspicion of the impending Japanese attack on Midway.[11]

The deception supporting Midway was an excellent example of how a small-scale, locally planned and implemented deception could materially contribute to a victory which, in this case, had great strategic significance. After losing four carriers at Midway, Japan never resumed the offensive in the Pacific.

NOTES

This paper was originally presented at the US Army War College Conference on Intelligence and Military Operations, Carlisle Barracks, Pennsylvania, 22–25 April 1986. The views expressed in this article are those of the author and do not reflect the official policy or position of the Department of Defense or the US government.

1. Roberta Wohlstetter, *Pearl Harbor, Warning and Decision* (Stanford, 1962), pp.379–80; and 'Cuba and Pearl Harbor: Hindsight and Foresight', *Foreign Affairs*, 43 (July 1965), 691–707, 704.
2. JSC 79/4/D, 'Directive: Security Control for Military Operations', 26 August 1942, CCS 334 JCS (8–4–42) section 1, Record Group (RG) 218, Modern Military Records (MMR), National Archives (NA), Washington, D.C.; Vernon E. Davis, *The History of the Joint Chiefs of Staff in World War II*, vol. II: *Development of the JSC Committee Structure* (Washington D.C.: Historical Division, Joint Secretariat, Joint Chiefs of Staff, 1972,) pp. 330–31.
3. John B. Lundstrom, *The First South Pacific Campaign: Pacific Fleet Strategy, December 1941–June 1942* (Annapolis: Naval Institute Press, 1976), p. 84.
4. Ibid.
5. W.J. Homes, *Double-Edged Secrets* (Annapolis: Naval Institute Press, 1979), pp. 85–95.
6. Thaddeus Tuleja, *Climax at Midway* (New York: W.W. Norton & Co., 1960), pp. 51–52,62.
7. Ludstrom, *The First South Pacific Campaign*, pp. 156–9.
8. Ibid., p. 180.
9. Mitsue Fuchida and Masatake Okumiya, *Midway: The Battle that Doomed Japan* (Annapolis: Naval Institute Press, 1955), pp. 129–30.
10. Ibid., pp. 109–10.
11. Ibid., pp. 232–3.

CHAPTER 24

A Pentagon Proposal to Continue a Deception Organization after WW2 (1945)—General Conclusions[1]

Colonel Newman Smith

Prior to World War II, Newman Smith was a combat veteran and a successful international banker who spoke French, German, and Spanish. He took a position in charge of strategic deception in Joint Security Control. Colonel Smith was one of only four American officers to work on strategic deception during the war.

When the United States entered World War II, the use of military cover and deception on a systematic theater and world-wide basis was unknown to the United States Army. Its value as a military weapon was conclusively proved during the various campaigns of the war. This is generally confirmed by senior Army commanders.

Despite British experience from 1940 until the United States' entry Into the war and the fact that cover and deception had been very successfully practiced by them during that time, the United States authorities failed to profit by this experience and did not fully grasp its import and value in winning battles until two years later.

While a deception organisation finally was set up within the United States Joint Chiefs of Staff organisation, its directives were restricted through a lack of understanding of deception as it was then being practiced, and no firm deception policies were formulated until near the end of the war.

Finally after senior commanders were convinced beyond question of the value of theater and world-wide cover and deception, no steps were taken for its vigorous development in theaters of primary United States responsibility until it was too late to realise its full effect.

United States Pacific theater commanders, unfamiliar with results obtained by cover and deception in the Mediterranean and European theaters, were indifferent to its use, except on a local tactical scale, and initially appeared reluctant to attempt its overall employment.

A preponderance of reserve officers were assigned to cover and deception duties, thus failing to provide an adequate nucleus of Regular Army officers for its future peacetime development and use.

1. Colonel Newman Smith, "A Pentagon Proposal to Continue a Deception Organization after WW2 (1945)," Informal Memorandum on the Origin, Development and Activities of the Special (Deception) Section, Joint Security Control, 1942–1945, Washington, DC: The Pentagon, November 20, 1945, p. 37, Paragraphs 125–134.

The organisation initially created for implementing deception failed to meet the requirements of the highly specialized and complex character of the mission on a theater and world-wide scale. By splitting functions between different groups, progress in correcting deficiencies was further retarded.

While the principle of theater and world-wide cover and deception was finally accepted as a weapon of war, only a piecemeal wartime policy grew out of it. No formalized peacetime doctrine and policy with respect to cover and deception now exists, nor is any agency of the War Department charged with its future development. The Navy Department has established an organization responsible for the peacetime study and development of cover and deception.

General Recommendations

It is urgently recommended:

a. That an interim peacetime deception organization be established within the War Department at the earliest practicable date to accomplish the missions outlined in Section XVII, in collaboration with the Navy deception organization.

b. That emphasis be placed upon a thorough study of the experience gained in World War II, with a view to the establishment of a permanent peacetime War Department and inter-service organization charged with the making of plans and keeping them current and in a ready-for-use status.

The Pentagon
Washington, D.C.
November 20, 1945

Newman Smith
Colonel, MI
Chief, Army Special Section

A Russian Tradition of Stratagem: 1939 to the Present

CHAPTER 25

Catching NATO Unawares: Soviet Army Surprise and Deception Techniques

C. J. Dick

Mr. Charles Dick, educated as a historian and trained as an intelligencer, was a Research Associate at the Soviet Studies Research Centre, Royal Military Academy, Sandhurst. He was the Centre's Director from 1989, including after its renaming in 1993 as the Conflict Studies Research Centre, until his retirement in 1995.

Marx and Lenin teach that victory in war goes to the side with superior resources. This is a disheartening notion for the Soviet leadership as the Warsaw Pact is markedly weaker than NATO in almost every measure of military potential – size (and reliability) of population, wealth, industrial power and technological progress; only in currently deployed military strength is the Pact stronger. General Secretary Mikhail Gorbachev and the Soviet leadership well know that, in 1941, the Germans came within measureable distance of destroying Soviet communism, and that in order to preserve their power and system, the Soviets must avoid the repetition of such a war. They realize, as well, that a strategic nuclear exchange would make the last war look like a skirmish.

The Soviets do not, however, seem to conclude from this that a war against NATO cannot be won. Historically they have been conservative in their use of military force, and it would be difficult, in 1986, to postulate a plausible scenario for a decision to attack NATO, but they have by no means adopted the fashionable Western notion that war can no longer serve as an instrument of policy.

Surprise in Its Strategic Context

The Soviets conclude instead that a war with NATO must be won very quickly, in its initial period: this is defined as "the period of time which elapses between the start of hostilities and the completion by the combatants of their mobilization, concentration and deployment." Of course, as the aggressor will have a considerable advance in preparations and mobilization over the defender, the concept of the initial period really applies only to the latter. This means that the Soviets must seize a vital area of NATO and destroy key combat groupings before NATO can either complete its defensive preparations or agree on the use of nuclear weapons.[1]

1. This section owes a great deal to P.H. Vigor's admirable *Soviet Blitzkrieg Theory*, London, 1983.

The federal German Republic, the economic and military heart of NATO Europe, represents, in the Soviet view, just such a small and vulnerable area, which could be swallowed up in the course of a single, strategic offensive operation. The distance from the inner German border (IGB) to the Ruhr, across the North German Plain, is only 300-400km. The distance to Mainz through the Fulda Gap is only about 150km. In a significant comparison, the Soviets point out, that, in the Vistula-Oder operation during World War II, they advanced 500km on a front about 500km wide in 17 days – an average of 30km a day for combined-arms (infantry) armies and up to 70km a day for tank armies. In the process, 35 German divisions were totally destroyed and another 25 suffered 50-70% casualties. The Soviets see no reason why today's entirely mechanized army could not be at least as effective under favourable circumstances.

The Soviets identify five essentials for a quick victory. These are surprise, a heavy blow, a rapid advance, simultaneous attacks throughout the enemy's depth and air superiority.

The Advantages of Surprise

- Recent Soviet writings stress that "the element of surprise has long been the most important principle of military art... The employment in modern combat of nuclear weapons, highly effective conventional weapons and highly mobile forces has drastically increased the role and significance of surprise strikes."[2] Surprise is seen to confer five advantages.
- NATO's reinforcement plans will be pre-empted and rendered largely unworkable. Given the alliance's maldeployment, it may even be possible to prevent some NATO corps from occupying FEBA positions; certainly surprise will thwart NATO's rather extensive obstacle and field fortification plan which could threaten the viability of Soviet tactical and even operational techniques. Instead of having to conduct difficult and costly breakthrough operation, the Soviets will impose their style of war, clashing with NATO formations while they are on the move, in a series of meeting engagements – a form of combat for which they train intensively, but which is ignored by most of the alliance armies. The Soviets believe that the meeting engagement will be the typical battle in the next war, and their tactics (so often derided by Western officers) are perfectly suited to such combats.[3]
- Surprise, and the initiative it confers, is seen to be force multipliers making it possible to achieve at least a limited strategic objective with much smaller forces than would be required against a prepared enemy. A particularly significant example is the Soviet attack on Manchuria in 1945 when the Red Army chose to strike with a superiority of only 1.2:1, but with the advantage of surprise.[4] The benefits of surprise in a future war are seen similarly to outweigh the disadvantages of incomplete mobilization. Since surprise obviates the need for breakthrough operations, there is no need for large, vulnerable

2. Lt.Gen. V.G. Reznichenko, *Taktika* (Moscow, 1984).
3. For a further examination of this important point, see C.J. Dick, "Soviet Battle Drills," *IDR* 5/1985, and "Soviet Operational Concepts," *Military Review*, October and November, 1985.
4. For the importance of the Manchurian operation in Soviet thinking, see J. Despres, L. Dzirkais and B.Whaley, *Timely Lessons of History—the Manchurian model for Soviet strategy* (Santa Monica, 1976).

concentrations and strong second echelons at every level.[5] In short, it undermines the rationale of both NATO's operational nuclear force and the Deep Strike plan, at least in its deeper (and more expensive) versions.

- The concept of operational manoeuvre groups (OMGs) seems to have become an essential feature of contemporary Soviet operational planning (see C.N. Donnelly in IDR 9/1982 and C.J. Dick in IDR 6/1983). Surprise makes it much easier to insert major groupings into the enemy rear, thus prolonging the effects of surprise.
- By ensuring that there will be a series of meeting engagements, surprise lessens the logistic burden and the number of casualties in offensive operations. Soviet research shows that, in 1944-45, tank armies involved in fast-moving, manoeuvre warfare and advancing at 16-45km a day suffered only one-third the loss in men and two-thirds the tank losses (mostly easily repairable mechanical breakdowns) of tank armies advancing 4.5-13km a day. They also used only one-third the amount of fuel and one-fourth the quantity of ammunition of the armies involved in fighting through a prepared, balanced defense.
- It will be just as important for the USSR to surprise the Warsaw Pact as to catch NATO unawares. Otherwise, some reluctant Pact allies might attempt to opt out and leak Soviet intentions to the West. It would be as well to give the ordinary soldiers and populations of these allies little time to reflect on the need for and desirability of war.

A Heavy Blow

The weight of the blow must be great enough to carry it to the strategic objective (here postulated as the Rhine) in a single, non-stop offensive operation. Moreover, it must be delivered by the forces in place in peacetime, as a major mobilization and forward deployment would prejudice surprise. As Marshal V.D. Sokolovski has written: "It is necessary to keep deployed in peacetime sufficient armed forces to reach at least the nearest strategic objectives before successive echelons are mobilized and sent into action ... so that the main aims of the war can be attained in the initial period, without additional mobilization ... He who, right from the start, can get his troops deepest into enemy territory will be best able to exploit his nuclear strikes and prevent the enemy from mobilizing. This will be of great importance in Europe because distances are so small."[6]

Herein, doubtless, lies an explanation for the steady growth of the Group of Soviet Forces in Germany (GSFG), and the modernization of the East German Army. Over the last decade, GSFG has expanded by approximately 10% in tanks, 20% in combat APCs/MICVs, 35% in artillery and over 50% in attack helicopters. These increases take into account the much publicized Brezhnev troop withdrawals of 1979, and the trend shows no sign of changing.

5. See C.N. Donnelly, "The Soviet Concept of Echelonning," *NATO Review*, December 1984, and Dick, "Soviet Operational Concepts," *op. cit.*
6. V.D. Sokolovski, *Military Strategy*, 3rd edition, translated in London, 1975.

Speed and Flexibility in the Advance

- A high tempo of operations has a paralyzing effect on enemy commanders, who become disoriented when prevented from implementing long prepared defensive plans and faced with uncertainty about the attacker's progress and intentions. They are limited to reacting belatedly to the latter's initiatives, making proper exploitation of his weaknesses impossible.
- The psychological impact of a rapid advance on enemy commanders, troops and civilians alike can quickly induce a climate of collapse, not least in that most important military objective of all, the minds of the enemy's political decision-makers. The Soviets believe that NATO, being an alliance of disparate, bourgeois states, riddled with disaffected groups and with no real cohesion under the surface, is prone to this sort of collapse.[7]
- Speedy progress can negate the enemy's nuclear weapons, reducing his decision-making time and making targeting difficult. Friendly and enemy units will become so intermingled that nuclear strikes will become impossible in many cases.

Simultaneous Attacks Throughout the Enemy's Depth

The enemy must be attacked throughout his entire deployment. Much investment has been poured into Soviet forces for this purpose; the operational–tactical missile troops now being equipped with missiles accurate enough to deliver improved conventional munition (ICM) warheads effectively, long-range strike aviation, seven airborne divisions, at least eight air-assault brigades (a growth area) and many independent battalions, a massive special-purpose forces effort, as well as OMGs and forward and raiding detachments from the ground forces. Apart from spreading confusion and panic, these attacks in the enemy's depth have four principal missions:

1. The destruction, or at least disruption, of enemy nuclear delivery means and associated command, control and logistics.
2. The disruption of the enemy's command, control and communications.
3. The destruction and disruption of enemy air and air-defense forces as part of the counter-air operation.
4. The obstruction and delay of enemy reserve and troop deployments and of the withdrawal of defeated enemy elements, as well as disruption of the logistic system. In this way, if the defense crumbles from within, it will be possible to dispense with a strong second operational echelon, the mobilization of which would prejudice surprise and which is vulnerable to nuclear, or even conventional interdiction.

Air Superiority

This is an obvious precondition both for rapid advance and for deep operations. The counter-air operation has now assumed an importance almost equal to the ground operation, as has the air-defense operation. The type of aircraft being supplied to the modern-

7. See B. Byely, et. al., *Marxism-Leninism on War and Army* (Moscow, 1972).

ized Soviet Air Force, its reorganization and its new operational concepts demonstrate the Soviet awareness of this point.

The Possibilities of Strategic Surprise

Strategic surprise is defined as concealment of the intention to launch an offensive and/ or its timing.[8] It is achieved through "large-scale deceptive actions, regroupings and concentrations, concealing troops and installations, and misinforming the enemy."[9] Historically, the Soviets have been very successful in implementing such measures. Perhaps the most pertinent example is the war with Japan in 1945. The Japanese knew they were going to be attacked, but believed the blow would not fall before March 1946; they thought that the bad weather combined with the urgent Soviet need to increase force levels in the Far East would preclude an offensive in 1945, while the government was taken in by Soviet diplomatic deception.

The Limits of Modern Surveillance

Many experts reject past examples of successful surprise as irrelevant, arguing that modern means of surveillance make surprise impossible. In doing so, they ignore three basic points.

1. If the aggressor's objective is a limited one, as has been postulated for the Soviets in a war with NATO, then his tell-tale preparations will be correspondingly small (given as high a level of peacetime preparedness as that of the GSFG and of selected non-Soviet Warsaw Pact formations).
2. Surveillance merely provides data. Technical progress in intelligence gathering has not made interpretation any easier nor lessened the possibilities for deception and disinformation. This is particularly important because the people that the Soviets have to surprise are NATO's political leaders, who will make the crucial decisions about mobilization and deployment. Political leaders, at least in a democracy, may be overly cautious, concerned that a hasty reaction could provoke the very war they wish to avoid.
3. Surprise does not have to be total to impart of substantial advantage to an aggressor. This is particularly true if the objective is a limited one and victory can be achieved before its effects have worn off.

It must be remembered that, while progress in the surveillance field has indeed amounted to a "transparency revolution", other technological advances have actually made surprise easier to achieve. The increased range of modern missiles and aircraft makes it possible to deliver a massive, deep, highly accurate blow (conventional as well as nuclear) without warning. The time required to move fully mechanized, combat-ready formations in the forward area from their peacetime garrisons to their deployment areas is now to be reckoned in hours, rather than the days or weeks that used to

8. This section owes much to Michael I. Handel, "Intelligence and Deception" and "Intelligence and the Problems of Strategic Surprise," *Journal of Strategic Studies*, 1/1982 and 3/1984, respectively.
9. Quotation from Major General Matsulenko, "Operational Maskirovka in the Third Period of the Great Patriotic War," the *Great Soviet Encyclopaedia*, Vol.15.

provide ample warning. Moreover, the equipment, structures and training of many contemporary armies make it very difficult to distinguish between offensive and defensive preparations and exercises. This is particularly true if the potential aggressor is wont to use highly visible, large-scale military activity for the purpose of applying political pressure, or if he desensitizes the intended victim by accustoming him over time to a high level of military activity.

Thus, for example, the Egyptians made warlike preparations twenty times between January and October 1973 and actually cancelled three cross-canal attacks at the last minute because of Israeli counter-mobilizations; they would not have struck in October had the Israelis reacted then, but would have continued their game of cat and mouse until an occasion when their enemy remained supine.[10] In the Polish crisis of 1981, on the other hand, when Soviet and Warsaw Pact military preparations were undertaken which would normally have been considered alarming, NATO did not react because these preparations could be explained in terms of the Polish situation.11

Assessing Intentions

Clearly, as war-like preparations can be made ambiguous, an assessment of an enemy's capability to attack is an insufficient basis on which to reach the decision to counter-mobilize. Such precipitous counter-mobilization could appear threatening to the enemy. At best, it would lead to a series of futile and expensive mobilizations which would soon become politically unacceptable (the reason for the Israeli failure to react in October 1973). It is essential, therefore, to divine the enemy's intentions before preparing for war and this puts a premium on political over military indicators.

- *Disinformation:* A major problem facing intelligence staffs in this effort is sorting genuine signals from the background "noise" in time. Before initiating preparations for war, NATO governments are going to demand very strong evidence of the USSR's intentions to attack. Given systematic Soviet deception and disinformation, intelligence will be unable to supply it. The Soviets will do everything they can to complicate and confuse the picture and it will be extremely difficult to penetrate a well prepared, long-term disinformation plan. Analysts are therefore generally gloomy about the prospects of avoiding surprise at the outset of the next war.
- *Surprise in Europe:* It is worth noting that the Soviet Union has achieved strategic surprise, as to intention and/or timing, three times in recent history—in the 1968 invasion of Czechoslovakia, the 1979 occupation of Afghanistan and the 1981 events in Poland. If the Kremlin leaders cold-bloodedly decided that an invasion of Western Europe was necessary, they would desensitize NATO, accustoming the alliance to a threatening military posture and level of preparedness, while using every available conduit to assure the West that war was un-thinkable. They could then manufacture some crisis (such as Poland in 1981) as a cover for the final, unconcealable military preparations.

10. See Janice Stein, "Military Deception, Strategic Surprise and Conventional Deterrence—a political analysis of Egypt and Israel, 1971-1973," Journal of Strategic Studies, 1/1982, for an excellent analysis of this crisis and its wider implications.
11. See C.N. Donnelly, "The Military Significance of the Polish Crisis," RUSI and *Brassy's Defence Yearbook*, 1983.

If, in a less likely scenario, there was no pre-conceived plan to go to war, but that decision was only taken during the course of a crisis, the Soviets would endeavour to convince the West that it was "just another crisis," solvable like all its predecessors. Surprise, at least as to timing, would then be achieved by striking before the crisis peaked (as in the Polish case) or after it appeared to be resolved (as with Czechoslovakia in 1968).

In any case, to steal a march of even a few days could give the Soviets a decisive edge over NATO, given the alliance's dependence on several days of preparation to establish a viable defense. SACEUR's intelligence staff may warn of the probability of attack for some time, but the doubts of many of NATO's leaders will delay the decision to institute countermeasures until it is too late. Thus, SACEUR is unlikely to be astonished when the Soviets cross the IGB, but he is likely to be surprised in the military sense, in that he will not have been allowed to deploy and prepare defenses in good time.

Operational Surprise

Operational surprise is defined by Major General Matsulenko as stemming from successful concealment of the timing, strength, direction and/or mode of the offensive.[12] Obviously, it will be most easily achieved if preceded by strategic surprise. Indeed, the two in combination can be devastating, particularly if the objective is limited.

Eight Methods

Soviet sources insist that operational surprise can also be achieved on its own. Although modern surveillance means make it difficult fully to conceal preparations for a large-scale offensive, Soviet Major General S.P. Solov'ev states that concealment of the true scale, and especially the direction and timing, of the main attack is "a quite achievable task which should always occupy the centre of attention ... Innovative search in solving the problem of gaining the element of surprise in military operations is one of the most important tasks of military cadres."[13] He identifies eight ways in which operational surprise can be achieved.

1. By forgoing detailed preparations and attacking "from a flying start", surprise as to timing can be achieved. The Soviet invasion of Manchuria provides a good example. The Soviets launched a sudden onslaught with only a 1.2:1 superiority in men, 4.8:1 in tanks and artillery and 2.5:1 in aircraft – figures not too dissimilar to the force ratios pertaining to the Central Region today.
2. The enemy can be confused as to actual intentions by misleading behaviour, such as a low level of activity or obviously defensive preparations.
3. Attacking at an unlikely time usually catches the enemy badly off balance. In both the Manchurian and Vistula-Oder operations, for instance, the defense was caught off guard by the Soviets' decision to attack in unfavourable weather.
4. Time after time in history, an attacker has gained a decisive advantage by striking along an unexpected axis terrain dismissed by the enemy as unsuitable for a major

12. Major General Matsulenko, op cit. The term "operational" is used in the Soviet sense of warfare conducted at the level of front (army group) and army (corps).
13. Maj Gen S.P. Solov'ev, "Strategic Surprise," Voennaya Mysl. 1 /1979.

offensive; in Soviet eyes, the best terrain for tanks is any area where there are no enemy anti-tank weapons.

5. Unexpected operational methods can have a paralyzing effect on the enemy. The German blitzkrieg is the best known example. It is less commonly realised that the Soviets more than repaid them in their own coin with the imaginative use of "mobile groups".
6. A considerable shock can be delivered by the unexpected employment on a large scale of new, or greatly improved weapons. Soviet T-34s and Katyushas came as an unpleasant technological surprise to the Wehrmacht, as did more recently (in 1973) the SA-6/ZSU-23-4 combination to the Israelis. Soviet writings lay great emphasis on the impact this aspect of surprise will have in the next war.
7. Strict security is, of course, essential for surprise, and Soviet secrecy is legendary. Their security was never penetrated by the Germans, but this was understandable. In the Vistula-Oder operation, which involved 2.2 million men, 34,500 guns and mortars, 6,500 tanks and SP artillery guns and 4,800 aircraft, only oral instructions were issued until D-day minus four, when only four written operation orders were issued.
8. The last method, deception, is so important that it will be treated separately.

Operational Deception

Despite the "transparency revolution," Western armies remain vulnerable to operational level deception for three reasons.

The overconfidence of some intelligence officers renders them fallible. This is especially true where oracular faith is placed in SIGINT, the most easily spoofed of all intelligence-gathering means.

Closely related to the first point is the pronounced tendency to disparage Soviet operational flexibility. Soviet lack of tactical imagination is taken to imply a lack of operational imagination too, despite all the historical evidence to the contrary. Knowing this, the Soviets can tailor their deception to confirm the enemy view of their stereotyped practices. US forces may be particularly vulnerable to this, because of an often dubious picture of their enemy and because of Air Land Battle's reliance at times on templating the Soviet forces in the absence of hard intelligence.

Modern war has seen a simultaneous massive increase in the volume of data available to intelligence staffs and a drastic shortening of the time available for analysis, dissemination and subsequent decision-making. This must increase the possibilities of surprise through deception. The Soviets will try to allow the enemy enough time to notice and act on their deception, but insufficient time to analyze the information in detail.

The mobility and flexibility of modern mechanized and air-mobile formations are such that the deception plan does not have to work for long. As with strategic deception, it does not have to be totally successful either. An ambiguous picture will often be enough to prevent the enemy from reacting in time. (It is also worth mentioning that the very nature of manoeuvre warfare, with its sudden, rapid and radical changes in the situation and frequent shifts of emphasis between axes will generate so much "noise" that, even without hostile deception to complicate matters, intelligence will have the greatest difficulty in divining enemy intentions.)

Figure 1 Vistula-Order Operation

Examples of Maskirova

Soviet military writings use the term "maskirovka", an amalgam of concealment, camouflage, simulation and disinformation which combines security and deception. Though at unit and sub-unit levels, maskirovka may, because of time constraints, consist of little more than the use of dead ground, coupled with a misleading smoke concentration on a false axis, it can become very sophisticated at higher levels. During World War II, massive concentrations were achieved, totally undetected by the Germans (often in very bare terrain). Prior to the battle of Kursk, the Central Front alone was covertly reinforced by ten rifle divisions, seven tank regiments, 23 tank-destroyer brigades, over 9,000 guns and mortars and 1,300 aircraft.

In offensive operations, simulation took place on a grand scale. Thus, before the Vistula-Oder operation, over 1,000 dummy tanks and SP artillery guns were deployed with scores of real ones (to create tracks), and with the real communications nets of the 1st and 2nd Tank Armies, on a false axis; all the German intelligence combat indicators for each stage of an offensive in preparation were duly supplied. Meanwhile, the true concentration (on an axis through difficult terrain) was successfully disguised as a false one. The Soviets were also prepared to devote very substantial forces to feint attacks on false axes. For example, the 4th Guards and 5th Shock Armies were used in this way in the Iassi-Kishinev operation, as was the reinforced 36th Army in Manchuria. These troops, who took very heavy casualties, were not considered wasted. They were not only misleading the enemy, but also fixing enemy formations (and drawing in reserves) and killing enemy soldiers.

Deception is not a forgotten art, as it tends to be in the West. Rather it is stressed as a mandatory component of all strategic, operational and tactical plans. In fact, the Soviets have formed well-equipped specialist units at operational level purely to implement deception. Sophisticated modern surveillance techniques permit the design of dummies to emanate "heat, light and electro-magnetic energy."[14]

Tasks of Maskirovka

Maskirovka is pervasive in Soviet military literature and has a myriad of uses, but in the operational sphere, eight basic tasks are listed:

1. Masking an increase in or redeployment of forces and/or weapons which the enemy has spotted.
2. Blocking the enemy's perception and/or identification of new weapons.
3. Very importantly, distracting the enemy's attention from other activities.
4. In high-intensity combat, effectively overloading the enemy intelligence's analytical capability.
5. By presenting an illusion of strength where there is weakness, inspiring awe in an opponent or focusing his attention on an area where no real threat exists.
6. In a complementary fashion, lulling the enemy into a false sense of security by presenting an image of weakness where there is really strength.
7. Accustoming the enemy to particular patterns of behaviour so that he does not recognize offensive preparations.
8. Confusing enemy expectations, leading him to misunderstand Soviet actions so that he fails to find the correct response to them.

Principles of Deception

It is obviously impossible to list the ingredients of Soviet deception as one would for the recipe for a cake. The first principle, as Soviet writers never tire of reiterating, is creativity and originality; stereotype is fatal to any prospect of success. Scrupulous adherence to the following basic principles is also stressed:

A. Any deception plan must have a clearly defined aim which supports the real operational scheme. It must be directed at a specific target, usually the enemy commander, and be based on his known prejudices and likely reactions; ideally, it will confirm his preconceptions (as known, for instance, from a study of his training manuals and exercise play, from his reaction to previous moves and from the nature of his deployment).

B. Control is always exercised at the highest possible level. Otherwise, various uncoordinated lower-level deception schemes can compromise each other, and the main plan, through revealing anomalies.

C. Preparation and execution must be thorough. Sufficient resources must be allocated to ensure credibility, and their activities constantly monitored to eliminate errors and breaches of security. There must be close co-ordination between the actual operational plan and the deception scheme (preferably passing the former off

14. Colonel V. Shchedrov, "Camouflaging Troops During Regrouping of Troops and Manoeuvre," Voennaya Mysl. 6/1966.

as the latter). Each stage of the deception scheme is implemented when it appears that the enemy has fallen for the previous one.

D. Credibility is, of course, a *sine qua non*. The deception plan must be a plausible alternative to the real one, and it must unfold in logical and realistic fashion, feeding the enemy with the combat indicators he would expect to see, yet not so readily as to excite suspicion; the enemy intelligence must be made to work for its material so that it believes it has penetrated security (and so that, as a consequence, it defends its interpretation with conviction).

E. It is important to feed as many of the enemy's intelligence collection sources as possible; a picture painted by one alone will arouse suspicion. There must, too, be the normal minor inconsistencies and ambiguities. Timing is also important. Intelligence must be given enough time to collect and interpret the false information, but not enough for thorough analysis.

F. Flexibility in execution is a vital attribute of deception. Enemy reactions, both predicted and unforeseen, must dictate the unfolding of the plan; radical modifications may become necessary if, for instance, the enemy misses some signals, misinterprets some and does not buy others.

G. Security is as vital to the deception plan as it is to the operational one.

Conclusions

The current debate about improving NATO's conventional defense tends to concentrate on "bean counts" and the relative merits of rival items of hardware for scarce funds. Perhaps both military and political leaders need to spend more time on the intangibles, beginning with the problem of how to cope with the danger of surprise without, at the same time, running the risk of alarming the Soviets in period of crisis, or of giving them an excuse for pre-emptive action.

- *Field fortifications:* One useful measure, which could help to lessen the pact of surprise, might be the construction of a belt of field fortifications, manned by infantry, along the IGB. Soviets have considerable respect, born of painful wartime experience, for a combination of anti-tank ditches, minefields and strong points in depth backed by strong, armoured reserves.[15] Such mobility denial would, in effect, act as a force multiplier for the defense, cancelling out the advantage surprise confers on the attacker. The Soviets would be forced to echelon their offensive forces in greater depth to ensure penetration. This would compel them either to deploy substantial extra formations, compromising surprise or to plan a much weaker initial blow on a narrower front. In either case, NATO's task would be greatly simplified and the prospects of "deep strike" having worthwhile targets dramatically improved. All this would be achieved without impairing crisis stability, given the unambiguously defensive nature of such defenses.
- *Accepting surprise:* NATO's senior commanders would do well to plan on the basis that they will (not may) be the victims of strategic surprise. To count on several days for preparation is to confuse warning time, which may exist, with preparation time,

15. Major J.B.A. Bailey, in "The Case for Preplaced Field Defenses," IDR 7/1984, makes a persuasive case for field fortifications which need not reflect either the expense or the mentality of a "Maginot Line".

which almost certainly will not. It is more than likely that little, if any, augmentation of force levels will be possible prior to hostilities, and obviously reinforcement schedules will be most unreliable in the face of Soviet deep attacks. As things stand at present, the Soviets will be able to impose their style of warfare on NATO. This means that the alliance will have to accept the inevitability of meeting engagements and battles. Western forces need to accept the logic of this and train for them.

CHAPTER 26

The Red Mask: The Nature and Legacy of Soviet Military Deception[1]

David M. Glantz

David M. Glantz is an American military historian and the editor of the "Journal of Slavic Military Studies." Glantz received degrees in history from the Virginia Military Institute and the University of North Carolina. He entered active service with the United States Army in 1963 and retired at the rank of Colonel.

In August 1945 the Soviets conducted their geographically most challenging and extensive strategic operation of the Second World War. In response to the Allied request for Soviet assistance in the war against Japan, first made in 1944 and repeated in 1945, the Soviets at the Potsdam Conference (May 1945) committed themselves to operations against Japanese forces in Manchuria and on the Northern island possessions of Japan (the Kuriles and Sakhalin). There were about 700,000 Japanese troops in Manchuria facing an almost equal number of Soviet troops. An additional 300,000 to 400,000 Japanese in Korea might be expected to figure in the operation, with another 100,000 on Kurile Islands and Sakhalin. A well-orchestrated Soviet attack would probably negate the requirement to engage those Japanese forces in Korea. On the basis of comparative strength, the Manchurian operation would be somewhat larger than that in Belorussia. Using Belorussia as a guide, Soviet forces in the Far East would have to be at least doubled in strength to more than 1.5 million. The Soviets did realize that Japanese armored weakness would permit Soviet armor superiority, making up for part of the lack in manpower. Thus, initial Soviet planning assessed the need for about 1.5 million men and at least one tank army and one to two tanks or mechanized corps to defeat Japanese forces in the requisite time frame.

Several conditions delayed Soviet preparations for the operation, increased the importance of the Manchurian operation, and placed major significance on the factor of surprise and the role of *maskirovka*. The first reality was the size of the projected theater of operations (roughly 600 by 1,000 kilometers) and more important, the distance of the theater from European Russia. To raise Soviet force strength to over 1.5 million men would require moving almost 700,000 men over 9,000 kilometers along the umbilical cord of the Trans-Siberian Railroad from the European theater to the Far East. Even more significant was the necessity of moving large armored forces (one tank army and one tank corps) and the massive quantity of necessary equipment and supplies. To

1. David M. Glantz, "The Red Mask: The Nature and Legacy of Soviet Military Deception," in *Strategic and Operational Deception in World War Two*, Michael Handel, editor, London: Frank Cass & Co., 1987, pp. 231, 234–239. Reprinted by permission of Frank Cass & Co. (Taylor and Francis Group).

maintain strategic surprise this movement would have to be kept as secret as possible. Once deployed for the offensive, those forces would have to advance through the difficult and varied terrain of Manchuria to a depth of almost 900 kilometers. Terrain problems also necessitated deceiving Japanese forces about actual routes of the Soviet advance.

The second reality was the degree of fanatical resistance the Japanese had displayed in earlier operations against US forces. There was no reason to assume more passive Japanese reception of a Soviet advance into Manchuria. This made it necessary for skillful operations, using the best operational and tactical techniques developed in the western war. Thus, the Soviets had to send experienced forces and commanders to the Far East. The final reality confronting the Soviets on the very eve of the offensive was that of time constraints. American use of the nuclear bomb, and the possible ensuing Japanese collapse, made it imperative that the offensive should achieve its goals in a matter of days, rather than weeks or months. The Soviets determined that Manchuria had to be secured within 30 days and the main entrances into central Manchuria within one week. This was as much a political as a military necessity.

From virtually every perspective, *maskirovka* would make the difference between success and failure. To a greater extent than in the west, successful strategic *maskirovka* was imperative, while operational and tactical *maskirovka* had to be perfect if the overall operational plan was to be realized.

The Soviets understood the difficulty in masking their overall offensive intentions. General S. M. Shtemenko noted:

> Our striving for surprise in our operation was complicated very much by the fact that the Japanese had long and steadfastly believed in the inevitability of strategic surprise was hardly a practicable matter. Nevertheless, after pondering this problem, we returned more than once to the first days of the Great Patriotic War. Our country had also expected the war and prepared for it; however, the German attack proved unexpected. Consequently, it was not necessary to prematurely repudiate surprise in the current case.

Surprise in the Manchurian operation depended first on maintaining the secrecy of the operational plan and the nature and scope of Soviet offensive preparations. Also, to an extraordinary degree, surprise would depend on the form of the attack, which had to be conducted in a manner *guaranteed* to pre-empt Japanese defenses or paralyse the Japanese command and control structure. This meant the use of unorthodox operational and tactical techniques, tested earlier but never before relied upon to such a major extent.

The first task was to secure planning for the operation. The Soviets followed procedures developed in previous operations and severely restricted the number of planners and of planning documents. A Far East Command directive restricted full participation in planning to the *front* commander, the member of the *front* military council (political officer), the chief of staff, and the chief of the operational directorate. *Front* chiefs of troop branches and services participated in developing only their functional portions of the plan and were not familiar with overall *front* missions. The directive read:

> The army commander will be assigned missions personally and verbally, without the delivery of written directives from the front. The participation in the development of the army operational plan will be established on the same basis as within the front. All documentation

concerning the plans of troop operations will be kept in the personal safes of the front and army commanders.

The most sensitive aspect of offensive preparations was the problem of masking the extensive force and *materiel* build-up in the Far East. This effort would take considerable time to complete, and would rely on a limited transport network already highly vulnerable to Japanese observation. This concentration, regrouping and deployment of forces and *materiel,* and the movement of commanders and staffs required to lead the forces were carried out under the cloak of a host of *maskirovka* measures. High-level commanders and their staffs traveled to the Far East (normally by air) under assumed names, wearing bogus badges of rank, branch and service. Movement by rail of *materiel* began as early as January 1945; and, whenever possible, the Soviets relied upon indigenous stocks and production to equip their forces. Movement of men and *materiel* occurred only at night under strict camouflage conditions. Hundreds of kilometers of artificial covers were built to physically mask the rail line from Japanese observations (especially along the Amur and Ussuri Rivers where the rail line was under direct observation of Japanese forces on the far bank).

During June and July the Soviets used 136,000 rail cars to move the requisite number of men and *materiel* to launch the offensive (22-30 trains per day). While the Japanese noted the increased volume of traffic, Soviet camouflage probably concealed from Japanese eyes 50 per cent of the volume of men and *materiel* moved.

Once in the Far East, the *fronts* placed arriving units and *materiel* in widely dispersed camps and depots in positions distant from the front, but accessible for quick deployment forward, needing only limited use of forward assembly areas and jumping-off positions. During these movements troops conducted exercises which made it possible for units to train, adapt to local conditions, and prepare for the attack while maintaining their concealment. One to two days before initiation of the attack, where necessary, forces moved into their initial positions, in which movement of any kind, food preparation, and wood-cutting were strictly forbidden. All newly arrived units maintained strict radio silence. Similar measures covered the extensive redeployment of forces already stationed in the Far East and Trans Baikal regions.

The Far East command and the *front* headquarters published detailed directives and instructions concerning general and specific *maskirovka* plans to govern their forces. For example, the 1st Far Eastern Front (previously the Primor'ye group and 25th Army) issued extensive instructions calling for the creation of false troop concentrations, a decoy offensive sector, active defense preparations in other sectors, and activation of bogus reconnaissance. Troops performing these missions were not informed about the false nature of their activities.

Meanwhile, all activity in front sectors was kept at normal levels. The population carried on normal activities, military units followed usual routines, and radio traffic continued with no change. Newly arriving units moved only at night and rested in forests and villages out of the enemy's view. Where cover was unavailable (as in the steppes), tanks, vehicles and personnel dug in and concealed themselves with camouflage materials or nets. The Soviets prepared unit command posts in advance, either digging them in or using existing structures which blended in with the surroundings. All communications nets lines were dug in as well as artillery firing positions prepared in advance by specially designated engineer units. Though these and other measures did little to alter Japanese judgements about ultimate Soviet offensive intentions, they did conceal the timing, scope and location of the impending attack.

The extensive secrecy and *maskirovka* measures were but one aspect, albeit an important one, concerning preparations for the operation. Equally important were techniques used by the Soviets to surprise the Japanese regarding the form of the attack. Most of these techniques were by no means new, having been used on various occasions in the German war, when the Soviets had been able to judge their effectiveness. However, in Manchuria the Soviets used a wide array of these techniques on a grander scale than ever before.

As in 1939, Japanese attitudes toward likely Soviet performance conditioned the Japanese for being surprised. The Japanese apparently stereotyped Soviet combat performance on German reports of how the Soviets had operated early in the war (1941-42). Thus, they underestimated the ability of the Soviets to employ armor skillfully and co-ordinate combined arms forces in poor terrain. They underestimated their logistical capability to effect and support large troop concentrations in terrain they (the Japanese) considered unsuited for large-scale military operations, particularly the areas adjacent to western Manchuria. The skill of Soviet combat performance far exceeded Japanese expectations.

The form and location of the Soviet attack at the strategic, operational and tactical levels also surprised the Japanese. The Soviet's two-front strategic envelopment through both eastern and western Manchuria contradicted Japanese expectations and deployments. Soviet units routinely crossed terrain the Japanese considered impassable, leaving it virtually undefended. Moreover the Japanese were unable to counter the Soviets' conscious decision to use every possible avenue of approach in their attack. Soviet use of armor in first echelon at every level of command—initially, or shortly after the beginning of the attack—also caught the Japanese off guard. Having discounted the threat of armor in such difficult terrain, they could not deal with it. The Soviet tendency to bypass fortified positions confounded Japanese commanders and rendered most Japanese defenses useless.

Tactically as well, the Japanese were unprepared to deal with Soviet combat techniques. The Soviet use of small, task-organized assault groups with heavy engineer and firepower support clashed with the image of human waves of infantry in the assault. Perhaps most unexpected for the Japanese was Soviet reliance on forward detachments to probe Japanese defenses, bypass them, and attack deep into the Japanese operational rear. The Soviet commanders' display of initiative at all levels did not fit Japanese preconceptions of Soviet performance. In fact, the scope of Soviet use of rapid maneuver surprised and confused the Japanese, resulting in a general paralysis of their command and control. To a greater extent than at any time before, in Manchuria the Soviets followed the advice found in all of its earlier regulations, specifically to achieve *maskirovka* by 'employing methods of fighting that are new for the enemy and weapons unknown to him'.

Soviet *maskirovka* measures achieved marked success. Certainly in mid-1945, the Japanese anticipated that the Soviet Union would ultimately join the war against Japan. The Soviet attack when it did occur, however, 'caught the Japanese totally unprepared for an invasion they expected'. In late January 1945, the Japanese Army Vice Chief of Staff predicted the Soviets would abrogate the Japanese-Soviet Neutrality Pact and enter the war against Japan in the latter half of 1945. It happened in April. After the German surrender in May, the Japanese High Command (JSHQ) assessed that Soviet entry into the war was unlikely until the spring of 1946, when Japan would be closer to surrender. They conceded, however, that Soviet attack preparations could be complete by August or September 1945; but thought weather conditions in that period were not conducive for an attack. All these assessments were based on a political judgment that Soviet war losses would force them to await an American invasion of Japan before they would join the war. Moreover, in August Manchuria was subject to heavy, often torrential monsoon rains which turned its few roads into morasses.

After Yalta, the Japanese estimated it would take at least three months after the German surrender for the Soviets to move necessary forces to Manchuria for an offensive. By July Japanese estimates of Soviet reinforcements put Soviet strength at about 1.3 million men. By August or September the Japanese estimated the Soviets would have 40-50 divisions.

The Japanese did detect large-scale Soviet rail movements into the Far East. From February to August the Kwantung Army counted 800-1,000 trains of 40 cars each and deduced the movement of 20-25 Soviet divisions. Thus, Japanese intelligence 'underestimated between 30 to 50 per cent the forty Soviet divisions actually deployed eastward', which caused them to believe 'that they still had time to prepare their defenses against the Soviet invasion in the spring of 1946'. Interestingly enough, while undercounting military units and incorrectly estimating deployment regions, the Japanese kept a fairly accurate count of overall Soviet manpower strength in the Far East.

Based on these optimistic estimates, the Japanese were slow to implement changes in force disposition associated with their 1944 alteration of overall Kwantung Army planning from an offensive to a defensive posture. Many units had not completed their redeployment by the time of the Soviet invasion.

As in Belorussia, there was a dichotomy between intelligence assessments at lower level and high-level headquarters, with the lower-level taking a more realistic view. At a 4 August conference at 3rd Area Army the Kwantung Army operations section said no invasion could be expected until September. If the attack did not begin then, the Kwantung Army believed it would probably begin in the spring. High-level optimism muffled warnings from below, starved lower level units of meaningful intelligence information on the Soviets, and stifled the war preparations of low-level units. 'In short, the Japanese were unprepared strategically, operationally, or tactically for the massive Soviet blow which fell on 9 August 1945.' The captured deputy chief of staff of the Kwantung Army, General M. Tomokatsu, declared that the specific dates of the entry of the USSR into the war remained unknown, despite the fact that the Kwantung Army had detected the Soviet build-up. Hence, he declared that the declaration of war by the Soviet Union on 8 August was completely unexpected. General Semizu, commander of Japanese Fifth Army, stated, 'We did not expect that the Russian army would go through the taiga, and the offensive of Russian forces from almost inaccessible regions proved a complete surprise for us'. To a great extent, the combat performance of the Kwantung Army during the Manchurian operation attested to the degree of surprise the Soviets achieved by using new operational and tactical techniques enabling them to achieve objectives in half the time. An operation optimistically judged to last at least 30 days was in fact completed within 15 days.

Manchuria represented a sterling effort for the Soviets regarding the wartime use of *maskirovka*. More important, it was one of their few experiences of employing *maskirovka* in the initial stages of war and has therefore become a major focal point of study. Specifically, the Soviets have investigated the role of *maskirovka* in that increasingly sensitive period of the transition from peace to war.

Soviet *maskirovka* measures at the strategic, operational and tactical levels achieved greater success in Manchuria than in any previous operation. They were able to confuse the Japanese regarding the timing, scope, location and form of the offensive. Hence, Manchuria has become a textbook case for the Soviets on how to conduct strategic *maskirovka*. The time and special considerations of the offensive seemed to create conditions and requirements somewhat analogous to theater offensives in a modern nuclear context. That fact has prompted even more intense Soviet study.

The Chinese Tradition of Stratagem: circa 350 BC to the Present

CHAPTER 27

Foreword to General Griffith's Sun Tzu[1]

B.H. Liddell Hart

B. H. Liddell Hart was a British soldier and military historian. Liddell Hart became an officer in the Kings Own Yorkshire Infantry at the outbreak of World War I. Liddell Hart, after retiring from the military, became a writer and set out to explain the high casualty rates of the war.

SUN TZU's essays on 'The Art of War' form the earliest of known treatises on the subject, but have never been surpassed in comprehensiveness and depth of understanding. They might well be termed the concentrated essence of wisdom on the conduct of war. Among all the military thinkers of the past, only Clausewitz is comparable, and even he is more 'dated' than Sun Tzu, and in part antiquated, although he was writing more than two thousand years later. Sun Tzu has clearer vision, more profound insight, and eternal freshness.

Civilization might have been spared much of the damage suffered in the world wars of this century if the influence of Clausewitz's monumental tomes *On War*, which moulded European military thought in the era preceding the First World War, had been blended with and balanced by a knowledge of Sun Tzu's exposition on 'The Art of War'. Sun Tzu's realism and moderation form a contrast to Clausewitz's tendency to emphasize the logical ideal and 'the absolute', which his disciples caught on to in developing the theory and practice of 'total war' beyond all bounds of sense. That fatal development was fostered by Clausewitz's dictum that: 'To introduce into the philosophy of war a principle of moderation would be an absurdity—war is an act of violence pushed to its utmost bounds.' Yet subsequently he qualified this assertion by the admission that 'the political object, as the original motive of the war, should be the standard for determining both the aim of the military force and also the amount of effort to be made'. Moreover, his eventual conclusion was that to pursue the logical extreme entailed that 'the means would lose all relation to the end'.

The ill-effects of Clausewitz's teaching arose largely from his disciples' too shallow and too extreme interpretation of it, overlooking his qualifying clauses, but he lent himself to such

1. B. H. Liddell Hart, in Foreword to Sun Tzu, *The Art of War, 2nd Edition*, General Griffith, Sun Tzu, translated by General Griffith, MA: Oxford, 1963, pp. v-vii. By permission of Oxford University Press, USA.

misinterpretation by expounding his theory in a way too abstract and involved for concrete-minded soldiers to follow the course of his argument, which often turned back from the direction which it seemed to be taking. Impressed but bemused, they clutched at his vivid leading phrases and missed the underlying trend of his thought—which did not differ so much from Sun Tzu's conclusions as it appeared to do on the surface.

The clarity of Sun Tzu's thought could have corrected the obscurity of Clausewitz's. Unfortunately, Sun Tzu was only introduced to the West, by a French missionary's summary translation, shortly before the French Revolution, and although it appealed to the rational trend of eighteenth-century thinking about war its promise of influence was swamped by the emotional surge of the Revolution and the subsequent intoxicating effect of Napoleonic victories over conventional opponents and their too formalized tactics. Clausewitz began his thinking under the influence of that intoxication, and died before he could complete the revision of his work, so that this lay open to the 'endless misconceptions' he had foreseen in his testamentary note. By the time later translations of Sun Tzu were produced in the West, the military world was under the sway of the Clausewitz extremists, and the voice of the Chinese sage had little echo. No soldiers or statesmen heeded his warning: 'there has never been a protracted war from which a country has benefited'.

There has long been need of a fresh and fuller translation of Sun Tzu, more adequately interpreting his thought. That need has increased with the development of nuclear weapons, potentially suicidal and genocidal. It becomes all the more important in view of the re-emergence of China, under Mao tse-tung, as a great military power. So it is good that the task should have been undertaken, and the need met, by such an able student of war and the Chinese language and thought as General Sam Griffith.

My own interest in Sun Tzu was aroused by a letter I received in the spring of 1927 from Sir John Duncan, who was commanding the Defence Force which the War Office had dispatched to Shanghai in the emergency arising from the advance of the Cantonese armies under Chiang Kai-shek against the Northern war lords.

Duncan's letter began:

> I have just been reading a fascinating book 'The Art of War' written in China 500 B.C. There is one idea which recalled to me your expanding torrent theory: 'An army may be likened to water: water leaves dry the high places and seeks the hollows; an army turns from strength and attacks emptiness. The flow of water is regulated by the shape of the ground; victory is gained by acting in accordance with the state of the enemy.' Another principle contained in the book is acted upon by Chinese generals of today; it is 'the supreme art of war is to subdue the enemy without fighting'.

On reading the book I found many other points that coincided with my own lines of thought, especially his constant emphasis on doing the unexpected and pursuing the indirect approach. It helped me to realize the agelessness of the more fundamental military ideas, even of a tactical nature.

Some fifteen years later, in the middle of the Second World War, I had several visits from the Chinese Military Attaché, a pupil of Chiang Kai-shek. He told me that my books and General Fuller's were principal textbooks in the Chinese military academies—whereupon I asked: 'What about Sun Tzu?' He replied that while Sun Tzu's book was venerated as a classic, it was considered out of date by most of the younger officers, and thus hardly worth study in the era of mechanized weapons. At this, I remarked that it was time they went back to Sun Tzu, since in that one short book was embodied almost as much about the fundamentals of strategy and tactics as I had covered in more than twenty books. In brief, Sun Tzu was the best short introduction to the study of warfare, and no less valuable for constant reference in extending study of the subject.

B. H. LIDDELL HART

CHAPTER 28

Stratagem: The Chinese View in the Sun Tzu[1]

Scott Boorman

Scott A. Boorman has been a junior fellow of the Harvard University Society of Fellows and a member of the Harvard Sociology Department. He is the author of The Protracted Game: A Wei-ch'I Interpretation of Maoist Revolutionary Strategy *and various articles on Chinese military history and comparative strategic theory.*

The ancient Chinese strategist Sun Tzu is the only Chinese strategic thinker who has ever gained much cognizance in the West, with the latter-day exception of Mao Tse-tung. This recognition, though not based in most cases on deep historical evaluation, is well merited, since Sun Tzu is also cited among military thinkers with overwhelming frequency in Chinese historical and political literature, as well as being studied by most Chinese statesmen with pretensions toward strategic competence and by most officials confronted with the handling of military situations. When one gets deeper into the historical source material, to be sure, one wonders whether statesmen and soldiers have paid much more than superficial respect to the writings of Sun Tzu (or any other theoretical exponent of strategic matters); but there is no doubt that this brief book (in present form, it is only a few thousand characters) has had a far more controlling influence on Chinese strategic behavior than any comparable work, including that of Clausewitz, has had in the West. The main point, in any case, is the guiding picture of strategy through stratagem. A remark by a Western Sinologist on a classic of Chinese rhetoric was not *directly* intended for this context, but bears verbatim quotation: "It would not matter whether [that book] were one-fifth or four-fifths fictional, this would remain the central political fact of the age."

Although, as previously indicated, the *Sun Tzu* text is at least tolerably well known in contemporary Western strategic circles, in the Western literature there is no interpretation of its significance that is very helpful to understanding its central themes. To evaluate the picture of the world presented in the *Sun Tzu*, several concrete aspects of the organization and style of the text should be discussed at the outset. As in most Chinese social-theoretical and philosophical works of the period, the style is terse and epigrammatic and consists of a sequence of maxims strung together, roughly organized by topic, but offering no consistent sequential argument. Metaphors and analogies are frequent; historical examples are lacking, though the standard commentaries fill in many examples from later epics. The commentary interpretations are of great impor-

1. Scott Boorman, "Stratagem: The Chinese View in the Sun Tzu", William W. Whitson, editor, *The Military and Political Power in China in the 1970s,* New York: Praeger Publishers, 1972, pp. 318–323. Reproduced with permission of Praeger Publishers via Copyright Clearance Center.

tance, since for the most part they have heavily influenced the interpretations of the text for at least the last millennium.

Following is an excerpt from a standard translation [by Giles] of the first chapter of the text, which follows some introductory remarks stressing the importance of the study of war and the basic aspects of the military art. The relevance to the topic of stratagem needs no comment:

> All warfare is based on deception. Hence, when able to attack, we must seem unable; when using our forces, we must seem inactive; when we are near, we must make the enemy believe we are far away; when far away, we must make him believe we are near. Hold out baits to entice the enemy. Feign disorder, and crush him. If he is secure at all points, be prepared for him. If he is in superior strength, evade him. If your opponent is of choleric temper, seek to irritate him. Pretend to be weak, that he may grow arrogant. If he is taking his ease, give him no rest. If his forces are united, separate them. Attack him where he is unprepared, appear where you are not expected.

These military devices, leading to victory, must not be divulged beforehand.

It is clear that Sun Tzu is here stressing primarily that level of stratagem that consists of attempting to induce the enemy to make a wrong decision because he is misled as to one's own situation and strategy. The possible devices for implementation of this aim are numerous, and it is to the credit of the text that it does not bog down in an attempted catalog of the mechanisms (by contrast to certain much later Chinese strategic works, which, as Herbert Franke has pointed out, fairly reek of armchair stratagems). In a later chapter (Chapter 9), however, there is a very interesting and quite concrete discussion of counterstratagems: the problem of filtering one's intelligence with the aim of disclosing the enemy's true situation when he is trying to present it deceptively. For example, "Humble words and increased preparations are signs that the enemy is about to advance. Violent language and driving forward as if to the attack are signs that he will retreat."

For a modern Chinese, perhaps the most familiar collection of stratagems and counterstratagems of this nature is that contained in the fictionalized *San-kuo chih yen-i* (Chronicle of the Three Kingdoms). This is a rather romanticized, and hence very popular, narrative account in the colloquial style of a famous period of factionalism and dynastic decay in the third century A.D., known to have made a deep impression on the young Mao Tse-tung. Carried to its romantic limit, this aspect of Chinese warfare leads to the near-paranoid view of Chinese conflict behavior advocated many years ago [1900] by a former French military attaché [G. de Cotenson] stationed in Peking. He argues,

> Leur diplomatie consiste ordinairement à faire croire à l'ennemi qu'on veut s'entendre avec lui, ou aux rebelles que grâce leur sera faite, à leur promettre des honneurs pour les attirer dans son camp et les massacrer à loisir.

Ruses of this kind are deeply bound up with Western perceptions of Asian strategic subtlety. From a less culturally chauvinistic viewpoint, stratagem of this nature is scarcely profound; in fact, to quote Balzac, "Il n'y a pas de théorie, il n'y a que de partique dans ce métier."

It should be noted, however, that a certain element of the last, and highest, form of strategem—that which seeks to manipulate the opponent's utility function itself—appears in Sun Tzu's prescriptive list, where he speaks of angering the enemy and encouraging his arrogance. This subtle, but significant, distinction is supported by various standard commentaries on this particular passage. Li Ch'uan (himself an author of a well-known military text of the eight century A.D., the so-called *T'ai-po yin-ching* (Secret Classic of Venus) remarks, for example, that the purpose of angering the enemy commander is to cause his spirit to vacillate—i.e., to lose his head. Another commentator, Chang Yu, makes the point even more strongly. By irritating the enemy, he remarks, the commanding general can cause the former to "advance lightly, without stratagem." Even more extensive commentary has been generated by the remark on encouraging the arrogance of the enemy, with many historical examples being cited on instances where wars were lost because a commander was induced by the enemy to be overconfident of his own capabilities and thus to have a distorted view of the over-all situation.

A few isolated quotations, to be sure, prove nothing. But the image of manipulating the enemy's view of the world is pervasive. To quote the much later (eleventh century A.D.) *Tzu-chih t'ung-chien* (Comprehensive Mirror for Aid in Government):

> Now, the Way of War is this: attacking the heart is the best, attacking walls is the worst: battle launched at the heart is the best, battle launched at soldiers is the worst. I would wish that Your Excellency subdue their minds only.

Carried to its logical conclusion, manipulation of the mind of the enemy implies perturbation of his utility function to the point where he does not fight at all. The real development of our highest level of stratagem occurs only in the third chapter of the Sun Tzu, entitled "Attack by Stratagem". In this section, the emphasis shifts from merely outwitting the enemy (deception in the limited sense) to attempting to break his will without fighting. This is not a humanitarian concept, any more than is the indirect approach of Liddell Hart, though one can indeed trace a continuum between the type of bloodless victory imagined by the Sun Tzu and (for example) the almost totally ethical and nonstrategic remarks in such texts as the Huai-nan tzu. The objective is to hold one's own intent invariant, while simultaneously maneuvering the enemy into a position where he is willing to grant it.

Hence to fight and conquer in all your battles is not supreme excellence; supreme excellence consists in breaking the enemy's resistance without fighting. . . . The general, unable to control his irritation, will launch his men to the assault like swarming ants, with the result that one-third of his men are slain, while the town still remains untaken. Such are the disastrous effects of a siege. Therefore the skilful leader subdues the enemy's troops without any fighting; he captures their cities without laying siege to them; he overthrows their kingdom without lengthy operations in the field. With his forces intact he will dispute the mastery of the Empire, and thus, without losing a man, his triumph will be complete. This is the method of attacking by stratagem.

The picture that emerges from this passage is very different from that suggested by a contemporary Western treatise [by Whaley]: "The ultimate goal of stratagem is to make the enemy quite certain, very decisive, and wrong." The example of the town and countryside [given by Shakespeare, "If we lose the field we cannot keep the town"] provides a modern paradigm case of the kind of stratagem Sun Tzu is talking about in the above passage. In point of fact, large numbers of KMT military units scattered in cities and Schwerpunkte across the North China plain eventually surrendered quite passively without a fight to encircling Chinese Communist forces based in the countryside. Chinese Communist guerrilla warfare effectively played upon the rigidities of thinking always present in the Nationalist command to maneuver the Nationalists into an inflexible strategic policy inimical to their own best interests. This kind of stratagem does not fit naturally into the disjunctive framework provided by Liddell Hart's concept of "alternative objectives."

Having once introduced the Chinese Communist revolutionary movement, it is relevant to suggest a view of the interrelations among the triad formed by the concept of stratagem in the abstract, Sun Tzu, and Maoist strategy. In another work, the present author has argued for a structural analogy between Maoist revolutionary strategy and the game of wei-ch'i (Japanese go). Essentially, the role of wei-ch'i was to provide for Chinese Communist doctrine a concrete, representational model of what have been well-labeled as the geometrical and physical aspects of warfare. There remains, however, a considerable realm of strategic action in revolutionary warfare with which no analogy to a complete-information game can cope. Broadly speaking, this is the realm of incomplete information and imprecise objectives. Although various analogies have been drawn between Sun Tzu and Maoist strategy as a whole, it is this latter, highly psychological, realm of strategic behavior that furnishes most of the common ground between Sun Tzu and Mao Tse-tung.

In other words, the main influence of Sun Tzu on Mao has been in the area of strategy centering around deception and stratagem. Analogies between Maoist insurgency and the thought of Sun Tzu are much weaker in the sphere of warfare more geometric—the realm of strategy customarily associated in the West with the name of Antoine Jomini. In no sense does Sun Tzu recommend protracted guerrilla warfare with widely dispersed base areas of the type essential to Maoist thinking and—under suitable structural transformations—to the game of wei-ch'i. Eventually, with more historical and theoretical work, one may hope (in a felicitous phase of Benjamin Schwartz) to analyze the organic chemistry of the Chinese strategic tradition in much greater detail: but some of the distinctions here suggested may provide a crude beginning.

CHAPTER 29

Chinese Deception Doctrine: A View from Open Sources[1]

Michael Pillsbury

Michael Pillsbury is a defense policy adviser, former government official and author of books and reports on China. In 1975–76, while an analyst at the RAND Corporation, Pillsbury published articles in Foreign Policy and International Security recommending that the United States establish intelligence and military ties with China. He served on the staff of four U.S. Senate Committees from 1978–1984 and 1986–1991.

Preface

The Introduction and Overview of this report provides a number of preliminary insights about how Chinese open sources deal with the issue of military deception. It is based on a survey of 300 Chinese books and journals published since 1992. Drawing from several Chinese journal articles about deception, Part One suggests several tentative propositions about what Chinese deception doctrine may be, including the proposition that China's approach to deception may be quite different from other nations.

Book One is a 60 page translation of excerpts from a book entitled *Military Deception* published in 1992 and written by a senior colonel who currently serves at China's National Defense University. The style of this book resembles a textbook or comprehensive survey. It provides a theoretical schema for using three main types of deception at three levels of national strategy, military campaign, tactical and single soldier deception. To illustrate these theoretical concepts of deception, the book describes in detail about 150 examples of deception in practice, 100 of which are from Chinese history, fifty are from deceptions practiced by Western nations, and only about 15 are deceptions practiced by contemporary Chinese, of which only 6 are from the post 1949 period. In my view, the book's theoretical concepts are highly valuable, but it is unfortunate that the author could provide so few illustrations since 1949. He does, however, stress the importance of multi-channel deception using SIGINT, imagery, and HUMINT, as well as the need for several techniques of counter deception analysis and collection.

Book Two is a 40 page translation of excerpts from a set of books entitled *Deceptive Strategy* published in 1996 by a senior colonel who serves in the PLA General Staff Department as Director of Military Research. This book also provides a set of theoretical concepts about how to use deceptive strategy, as well as more than 200 examples. To a greater degree than the first book, most of the examples are drawn from Chinese history. This book's preface contains calligraphy congratulating the author on the book from many of China's highest ranking general officers and the defense minister. It also

1. Michael Pillsbury, "Chinese Deception Doctrine: A View from Open Sources," Washington, DC: FDDC, c.2003, pp. 1–4.

reveals the existence of a national commission for research on deceptive strategy on which many of these Chinese generals serve. In my view, this book's importance stems in part from its effort to examine "deceptive strategy" in a comprehensive manner, beginning in China 3,000 years ago, and its assertions that Chairman Mao developed a unique deception style for China during the Chinese civil war that remains relevant for the future.

These books were acquired in China for the Pentagon's Office of Net Assessment project on Chinese military open sources.

Introduction and Overview

This report fulfills a request for a condensed translation of two Chinese books on deception theory. The military authors of the two books seem to suggest that China's military possesses its own unique theory of deception. They believe that China can compensate for military inferiority in wartime by employing this deception doctrine. These two books are not unusual. They fit into a larger context of Chinese military writings of the past decade. Two important Chinese military themes have been the revival of ancient Chinese history as a guide to current military issues, and a search for strategies and military technologies suited for use by inferior forces [China] against superior forces [the United States.] Chinese writings by more than 200 authors about these themes have been translated in *Chinese Views of Future Warfare* and *China Debates the Future Security Environment* [National Defense University Press]. It will be difficult for any reader who is unfamiliar with these Chinese strategic debates to plunge into these two Chinese books about deception. At least two assumptions are understood by the Chinese senior officers who are the intended audience of these two books on deception. The authors appear to assume, contrary to common US attitudes:

** Deception by an inferior power of a superior can be decisive, and is not a mere force multiplier.

** Extremely useful models for future military deception may be found in ancient Chinese history.

Implications for the U.S.

If Chinese views are so unfamiliar, why should American readers make any effort to understand them? There are several implications for the US contained in these two books about the future potential role of military deception:

** The potential of successful deception against the United States appears to include:

- Increasing US forces' vulnerability to surprise,
- Inducing US forces to make mistakes,
- Minimizing warning time from the US intelligence community to US policymakers,
- Reducing any the role the US could play, either diplomatically or militarily,
- Affecting domestic US policy debates,

- Inducing US policy paralysis or at least causing extended delays in crisis decisionmaking,
- Shaping the battlefield through directed information without its source becoming known,
- Exporting Chinese deception techniques to other nations hostile to the US,
- Countering the effects of US weapons systems and US operational concepts.

The scope of these two books is broad, and the authors appear to define "deception" to include measures to deceive with respect to:

1. China's intentions,
2. China's programs, and
3. China's existing capabilities.

Both books draw upon American and Soviet writings about deception, but both authors state that China understands deception theory better. The US is credited with excellent deception techniques, as is the former USSR. An article recently credited the Soviet Union with aiding China in 1957 by providing a useful manual of military deception.

Limits of Open Sources

The two authors provide no concrete examples of any contemporary Chinese deception practices. Instead, they carefully restrict their books to over one hundred historical examples of deception. It is well known in China that military authors cannot obtain security approval for open publication if they discuss sensitive matters about Chinese armed forces. Thus, it cannot be expected that open sources will reveal specific and concrete applications of deception. There appears to be no similar taboo, however, on discussing deception theory and doctrine in the abstract, as long as the materials used are purely historical. However, both authors make clear that they are advocates of the use of deception. They claim the mantle of authority of both Chairman Mao's writings and the classic *Art of War* by Sun Tzu. To underline that they are addressing the future employment of the deception doctrine, one of the books does go so far as to offer the full text of a sample deception operation order that a military region commander could issue. Both books seem to be supplemental reading for a classified training course that PLA officers have stated is conducted in various staff schools in China. Presumably, the classified courses use more specific manuals and real world examples, rather than the hundreds of historical case studies used in these two open source books. It is possible that the classified military training courses use the same approach to deception theory and doctrine as these two books. After all, one of the books is endorsed by a dozen senior generals, and its author serves on the General Staff. The other book has been cited approvingly by additional authors.

The condensed, translated version of these two books amounts to 100 pages. A full, verbatim translation would exceed 1500 pages. A full translation would also have to explain literally hundreds of Chinese historical references that would be obscure without detailed annotations. Yet even a condensed version of Chinese deception doctrine and theory as discussed in these two books may be of interest for three reasons:

**First, as a great civilization with thousands of years of its own political and military history, contemporary Chinese military authors may have important ideas to contribute to those in the West with an interest in the general history and theory of deception.

**Second, and more narrowly, China may have developed its own unique deception doctrine that merits study as part of any effort to understand Chinese national security strategy.

**Third, the use of open source Chinese materials can often usefully supplement other US government research and analysis. Obviously, in the modern world, planning for military deception would be a sensitive and secret matter for any nation. It could not be addressed in detail in open source books. Indeed, open sources could even be a channel to practice deception on the unwary foreign reader. So, readers should be alert to such issues in making any interpretation of Chinese [or any other nation's] writings on deception theory.

Background of the Two Authors

Before turning to the description of the main features of Chinese deception theory put forward in these two books, some background to put them into context may be useful. The authors of both books occupy important positions, but apparently do not meet with foreign military delegations and are not known outside China. Only a few dozen Chinese officers visit the US annually, out of a universe of many hundreds of Chinese published military authors, so the fact that they are not known may be insignificant. Some believe all important Chinese military authors are known in the US. However, others fear that this is not so yet. It is important to try to understand whether these two authors have any authoritative status, or whether they are representative of mainstream views.

Several Chinese sources have helped to provide background about the author of the more important of the two books, Colonel Ma Jinsheng. Colonel Ma is not further identified in his book Military Deception. However, several PLA generals have told me this year that Colonel Ma is not a pen name, and that he is a real person working in Beijing. A second confirmation about Colonel Ma was found in the comments of a PLA author of another book entitled Battlefield Deception published in 1999, who expressed his appreciation to Colonel Ma and identified him as the Director of the National Defense University Press. A third confirmation of Colonel Ma's background has come from the quotation of one of Colonel Ma's basic concepts about deception theory by yet another PLA author from the nuclear rocket forces in 1999 in a collection of articles about information warfare.

Unlike Colonel Ma, the author of the second book Colonel Chai Yuchiu is identified briefly on the book jacket as a member of the General Staff Department of the PLA. Other PLA officers have acknowledged to me that they know him. His position title is stated to be Director of Military Studies of the General Staff. Unfortunately, no PLA officer I have interviewed has been willing to offer details of what this vague job title entails. Requests by several Americans to meet him have been declined, with the explanation that Colonel Chai does not ever meet with foreigners. However, Colonel Chai, like Colonel Ma, has also been quoted in books by other PLA authors. One of a series of six books published by military authors in 1999 and 2000 about Chinese defense strategy referred to Colonel Chai's study. Thus, both Colonel Ma and Colonel Chai appear to be respected, serving Army officers who are senior members of what might be called the defense intellectual community in Beijing, with Colonel Ma at the National Defense University [which trains future Chinese general officers] and Colonel Chai at the General Staff headquarters. Their Chinese military rank of "senior" colonel means that they are roughly equivalent to a brigadier general in other national systems. If promoted, they would become major generals.

A Unique Chinese Deception Theory?

These two books are not the sole important works on Chinese deception doctrine. At least five other books have been acquired as well as a dozen articles. Some of these materials [not translated here] focus on narrow problems. Two example are ways to use deception to neutralize US aircraft carriers and ways to use deception to attack Taiwan. All these books and articles, however, imply as a theoretical point that deception operations are not the responsibility of a special unit. Rather, the most senior commanders are responsible for deception plans and operations. This approach differs from other nations. The US-UK deceptions for the Normandy landing in 1944, the Soviets in the 1980s , and the Iraqis in the 1990s all had special organizations for deception. Iraq, for example, had extensive organizational arrangements for concealment. According to Scott Ritter's *Endgame* [p. 126]: "A "Concealment Operations Room" of the Special Security Organization "coordinates with the SSO to implement concealment measures for past programs, covert procurement, and covert manufacturing. . . . The concealment mechanism was the most difficult issue facing UNSCOM in its work in Iraq."

This difference between Chinese deception theory and that of other nations raises the question:

** Are there different "styles" of national deception doctrine or practice? If so, what does Chinese deception theory suggest would be the "signatures" of attempted Chinese deception?

** Logically, countering Chinese deception would have to begin with a search for such "signatures."

Examples of Chinese Deception Aspirations

** Colonel Ma presents a categorization of deception into 4 levels, 3 types of goals and by duration. He emphasizes that successful deception can endure months or years. "Chains" of deception operations can be attempted, each of which is examined or tested through a kind of feedback loop. Then the deceiver may adjust both his "message" and the multi channels in use. SIGINT, HUMINT, and other channels must all be coordinated, then whatever channel the opponent trusts the most must be used the most, whatever "message" is most credible to achieve the goal must be repeated and reinforced. Less credible messages or distrusted channels need to be downgraded.

** Both Colonel Ma and Colonel Chai appear to believe that successful military deception can be "decisive"—not just a force multiplier, but a show-stopping, war-winning weapon that requires appropriate attention and investment by the commander. Failed deceptions may be discovered.

** Deception includes not just "moving" the enemies' forces but pre-war development of secret weapons that are called "assassin's clubs." These deception weapons create a kind of technological surprise that is revealed only at decisive movements to win a war.

** Both Colonel Ma and Colonel Chai appear to consider deception more important than Information Warfare, and even define IW as a part of a larger deception doctrine.

** Colonel Ma and Colonel Chai state that the best models for deception can be found in Chinese ancient history, especially the founders of the six greatest Chinese dynasties. Both use two or three time more Chinese examples than Western ones to illustrate their deception theory.

** Colonel Ma defines the most successful deception as one that makes the enemy actually "move" his forces and commit decisive, fatal errors. Lesser deceptions just confuse or deny information to the opponent. He labels this "seduction" deception because it lures the enemy commander to do something. Colonel Chai argues that Chairman Mao created this concept of luring or seducing the enemy to make fatal errors during the Civil War in the 1940s, and that Mao borrowed it from ancient history.

** Colonel Ma defines deception as an integral part of strategy planning. It is not a separate and specialized activity. He states,

> Any deception process is the organic combination of strategy planning and preparation, and the application of skills. Any effective military deception does not deviate from the guidance of military strategy. Only under the guidance of military strategy can military deception be correctly carried out, otherwise it will act in a way that defeats one's purpose [to try to go south by driving the chariot north], and things will go contrary to one's wishes.

Before formulating deception plans, commanders must first consider and assess the following six issues:

1. What is the "shi" configuration or propensity of our side and the possible enemy trends?
2. How can we use our power (plans one, two and three)?
3. Is it possible to create a "shi" [propensity] that can cause a doubling effect to occur in one's own capabilities?
4. If this "shi" can be created, what kind of changes must the enemy make in deployment?
5. Will the enemy be willing to make this kind of change?

If the enemy is not willing to do so, should coercion methods or deception methods be used to cause the enemy to make changes beneficial to our side?

After the above questions have been resolved one by one, the deception issue can be put on the agenda. The process of resolving the above issues is the process of planning and preparing strategy. The planning and preparation of strategy is the core issue of military deception."

** To implement deception, Ma and Chai both state requires extensive use of special operations forces [qi bing] coordinated with regular troops, both to deceive the enemy and to collect intelligence for the feedback loop to determine how well the deception's message and channels are working, and what adjustments are needed.

** Low tech deception appears to be favored. So called *tu ban fa, or* "the country bumpkin style" deception can defeat high technology weapons, as a part of deception doctrine. A PLA author wrote in the May 2000 issue of Military Digest that the Yugoslav Army brilliantly used eight simple and backward means of deception:

- Last minute anti-air defense system changes to deny the US time to input missile targeting;
- Deploying old planes to attract NATO bombers, while concealing newer fighters in bunkers;
- Changing radio frequencies so that the enemy could not locate positions;
- Using smoke screens and radar reflectors against lasers and infrared to "trap" laser bombs;
- Burning tires near airports to confuse the enemy's cruise missiles;
- Covering targets with special materials to confuse reconnaissance;
- False SIGINT to seduce missiles to explode prematurely;
- Delays in attacking stealth aircraft until the moment bomb bay doors opening.

** Both Colonel Ma and Colonel Chai emphasize that intensive intelligence collection and analysis about the enemy leader's perceptions and character and his advisory system must be conducted. In this way, his perceptions can be fostered to seduce him to commit fatal errors. Colonel Ma offers a detailed checklist for how to do this. Ma states that deception requires intelligence collection as follows:

1. Find out the main interests and characteristics of the target of deception, for frequently this is more important than a general study of his life.
2. Find out what wars the target of deception fought in his military career, as well as the reasons and process of his successes and failures. Often this is more significant than understanding the enemy's combat principles and military theory.
3. Find out what people are always around the target of deception, who is most in charge when speaking to him, who most easily arouses his dislike when speaking to him. Frequently this is more critical than understanding the general officer roster and chain of command relations.
4. Find out what the target of deception needs the most and is most concerned with (including personal, family, work, and combat), for often this is more vital than understanding the enemy 's general situation and terrain.
5. Find out what major setbacks and good fortune the target of deception has encountered throughout his career, and what person or thing created these setbacks and good fortune. Frequently, this is more important than understanding his general resume.

** Military exercises and military equipment can be observed by satellites. Therefore, military technological surprises ["assassin's clubs"] and military training must be subordinate to deception doctrine. Colonel Ma praises Egypt for deceiving US satellites in its 1973 war against Israel. The doctrine appears to be that China must not reveal all its capabilities in military exercises. On the other hand, both Colonel Ma and Colonel Chai discuss the need for military "displays" to intimidate or to coerce. Colonel Ma provides a sample "deception directive" from a theater commander at the Military Region level:

> Three days before the attack is launched, serious deception action must be taken to show the enemy our military will attack "A" area:
>
> 1. Unload tanks at rail stations toward "A" area and set up decoy artillery.
> 2. Organize engineer troops to clear roads to "A" area.
> 3. Organize part of forces around the clock to move to "A" area and to construct field command centers and fortifications, then have them move back and forth many times.
> 4. Extensively use mobile artillery or "floating" artillery companies and battalions to show "A" area has re-enforcements.
> 5. Set up mobile radios to pretend forces are being mobilized in the Corps group army headquarters.
> 6. Deliberately have recon teams in enemy areas lose some maps with important markings about "A" area.
> 7. 12 hours before the attack is launched, implement combat recon in enemy forward lines around "A" area.

This sample deception instruction clearly tells us the general purpose, the tasks, and the timing so that the Group Army can then make their own deception planning.

It is very important that Headquarters must have non-stop inspections of deception preparations.

Chinese History as a Source of Future Deception Models

This introduction and overview does not provide a complete summary of the many points these two books explore. Chinese deception theory is little known in the West, perhaps because none of the major academic books on deception discuss China. Indeed, a common theme in Western books on deception is that it was invented in the West. Yet perhaps a unique view of the Chinese military is that China's own rich history contains all the main types of deception, and therefore the most valuable lessons for the future. In this translation, a circle has been inserted before all Western examples used by the authors, and a square before all Chinese examples. The page numbers to the original Chinese books are shown in brackets on each page for the convenience of readers who wish to refer to the Chinese text.

Section VI: Tactical Camouflage: Now You See It, Now You Don't

Introduction

Visual camouflage has been practiced in war and hunting throughout history. The term itself first came into use during World War I, and is derived from the French verb camoufler, "to make up for the stage."[1] Camouflage in war involves the use of materials and techniques to hide, blend, disguise, decoy, or disrupt the appearance of military targets and their backgrounds. The use of these techniques helps prevent an enemy from detecting or identifying friendly troops, equipment, activities, or locations. Properly designed, these techniques take advantage of the immediate environment and natural and artificial materials.[2]

Camouflage varies based on the surroundings in which the object will be viewed (air, land, sea), the climate, and the terrain. Camouflage on the ground is relatively simple and has involved no more than the use of local vegetation and nets. Camouflaging large industrial targets or large sensitive military sites may require the expertise of engineers and chemists because of their knowledge of structures and materials. At sea, scientists with knowledge of optics and how light and color affect visibility have to inform the camoufler, along with experts who can minimize the knowledge gained from satellites, thermal cameras, and radar. Making aircraft less visible presents perhaps the greatest challenge. How do you effectively camouflage an aircraft when it is on the ground or in the air, under varying lighting conditions, and based on the surroundings at both the departure airfield and the at the target location? New theories of camouflage were invented to deal with the challenges of aircraft camouflage.[3]

Finally, the use of decoys is integral to camouflage. While camouflage aims to hide what is real, decoys attempt to show what is false. Decoys can give the enemy an inflated estimate of strength or can draw the enemy's attention away from a real attack. The use of decoys complements camouflage.

1. Guy Hartcup, *Camouflage: A History of Concealment and Deception in War,* New York: Charles Scribner's Sons, 1980, p. 7.
2. HQ, Department of the Army. FM 20-3, *Camouflage, Concealment and Decoys,* Washington, DC, 1999.
3. Hartcup, 8, 9, 14.

CHAPTER 30

Now You See It, Now You Don't: Camoufleurs, Conjurers and Pickpockets[1]

Roy Behrens

Simply the best all-round introductory book on camouflage. A brilliant fulfillment of the promise of the author's previous 1981 book on the subject. This book not only gives us the theory that ties deception to art, design, and war, but extends its reach to encompass con games, magic, and perceptual illusions in general. Consequently, it is perfect for beginners and a finishing school for veteran camoufleurs.

Mr. Behrens (1972 MA from the Rhode Island School of Design), a Vietnam War veteran, was an Associate Professor of Art at the University of Wisconsin (Milwaukee).

In 1842, the British novelist Charles Dickens visited the United States, a trip he described in *American Notes*. That same year, having returned to England, he purchased the tricks, equipment and props of a retiring stage magician, and began a tradition of putting on shows for his family and friends on birthdays and other occasions.

> [Magic] takes us for a time into an unreal world.
>
> —Warren E. Steinkraus "The Art of Conjuring" in the *Journal of Aesthetic Education* (1979).

Dickens as a magician, writes Phyllis Rose, "turned watches into tea caddies, made pieces of money fly through the air, burned up pocket watches without burning them. He caused a tiny doll to disappear and then to reappear with little messages and pieces of news for different children in the audience. But his greatest trick, the climax of it all, was his manufacture from an ordinary gentleman's hat of a plum pudding."[7.1]

◄ **FIGURE 7.A** John Tenniel Illustration [detail] for Lewis Carroll's *Alice's Adventures in Wonderland*, 1865.

The "first fundamental principle" of magic, claims the Great Merlini, a fictional magician in a mystery novel by Clayton Rawson, is

1. Roy Behrens, *False Colors: Art, Design and Modern Camouflage*, Iowa: Bobolink Books, 2002, pp. 151-171.

misdirection. "The other two—and they are used by magicians, criminals, and detective story authors alike—are *imitation* and *concealment*. Understand how these principles operate, and you should be able to solve any trick, crime, or detective story."[7.2]

Of course, Merlini might also have listed camouflage artists, since camouflage is governed by the exact same principles as magic—regardless of whether we call them "unit-forming factors" or "laws of disguise"—one proof of which is offered by the contributions of a British stage magician named Jasper Maskelyne to Allied combat efforts in World War II.[7.3]

> At that point, early in '42, anyone in the arts was invited to apply for camouflage work—and we all rushed out there very green; not even knowing how to salute.
>
> —William Pahlmann quoted in Reit, p. 79.

> Personally I dreaded going into the army. I mean, who wants to go into the army? That's how I got involved in the Camouflage Society, I figured if I had to go into the army, I might as well do something behind the lines.
>
> —Sam Leve quoted in Naverson, p. 43.

Maskelyne was a grandson of one of the most famous stage magicians of the 19th century, John Nevil Maskelyne, whose career as a magic performer began by duplicating on stage the "inexplicable manifestations" of spiritualists. He went on to develop one of the most enduring magic acts in history, an endeavor which finally ended in 1917, when, at age seventy-seven, he became ill during a performance and died a few weeks later. His son Nevil took over the act, but, seven years later, when Nevil died, the firm was inherited by his three sons, the youngest and most notable of whom was Jasper.

So this family of performers was widely known in England when Jasper Maskelyne interrupted his stage career in 1939 in order to offer his services to the Royal Engineers as a camouflage expert. Soon after, he was directed to the Camouflage Training Center at Farnham Castle for six weeks of instruction in how to conceal, deceive and distract. As it turns out, he was not the first Maskelyne to participate in a British war effort: As explained in his autobiography, his famous grandfather had contributed to military balloon research during the Boer War, while his father had shared information on artillery and gunpowder during World War I.

As you know, a conjurer gets no credit when once he has explained his trick and if I show you my method of working, you will come to the conclusion that I am a very ordinary individual after all.

—A. Conan Doyle
A Study in Scarlet

The camouflage training at Farnham, recalls Maskelyne, "almost drove me out of my mind." For six weeks, he listened to lectures in which "I learned how Artic rabbits suffer a change of color when snow falls, and why tigers hang about in tall grass," whereas his experience as a stage magician "had taught me more about the subject than rabbits and tigers will ever know. I could, in fact, have hidden myself and most of the rest of the class so efficiently that the lecturers would never have found them in the duration of the war, but that would only have caused trouble."[7.4]

Crouched in a corner of that imaginary railway compartment, he [E.M. Forster] would have worn a kind of protective coloring, like an oak-egger or a stick insect—or, rather, like a retired booking-office clerk from a station on a branch line.

—William Plomer,
The Autobiography of William Plomer, p. 304.

Among Maskelyne's lecturers and classmates were people who would later be recognized for a variety of achievements, if rarely in the area of camouflage. One of the chief instructors, for example, was Hugh B. Cott, the British zoologist, who had been a student of John Graham Kerr, and who had just written a classic book on biological camouflage, with notes on its military applications, titled *Adaptive Coloration in Animals*.

Julian Trevelyan, an artist with the unit, recalls that he and others laughed at Cott's "passionate addiction to countershading" although they acknowledged its value. Cott was himself an artist, a scientific illustrator, who made exquisite pen-and-ink drawings for his own book.[7.5] When he and others from Farnham were sent to the war zone in North Africa, he continued his scientific research with a menagerie of snakes, lizards and beetles that he kept in empty gas cans and studied in breaks between duties.

Trevelyan was a printmaker and the brother-in-law of Robin Darwin (a descendant of Charles Darwin), an artist who also contributed to military camouflage. Before enlisting, Trevelyan had founded a company called the Industrial Camouflage and Research Unit with three other artists: Stanley William Hayter, John Buckland-Wright and Roland Penrose. As independent consultants, they claimed to have knowledge of how to create camouflage designs that would protect factories from bombardment by enemy aircraft. In truth, as Trevelyan admitted later, "we in our camouflage unit knew very little more about it than the man in the corner garage. We had none of us done much flying, and when we had flown we had not addressed ourselves particularly to the problem of what makes things conspicuous from the air. Had we done

[Shortly before World War II, Sybil Moholy-Nagy received a book in London from Bauhaus designer Marcel Breuer in Germany] which, when opened, was found to be [Hitler's autobiography] *Mein Kampf*. It was worse than a poor joke, she and Carola [Giedion] were furious and threw it away with the rubbish. Breuer arrived soon after, apparently happy at being away from Nazi Germany, only to find two furious dames attacking him with no mercy. When he could get a word in he explained that, in order to get some of his money through German Customs, he thought it would be a bright idea to interleave their leader's great book with bank notes. They would surely not examine it with any great care. There was immediate pandemonium, all rushed down, hoping the rubbish had not yet been taken away. When they found the book, all was forgiven.

—Jack Pritchard
View From a Long Chair
p. 111.

so we would soon have realized that a lot of our assumptions were false, and that the pattern of the world from above is read very differently from the way in which we had supposed."[7.6]

Nevertheless, he continues, "it was easy to sell any kind of camouflage" in the early months of the war, especially greenish squiggly shapes, applied indiscriminately, in part because citizens seemed to believe "that the green stripes were a charm that somehow bought them immunity from the unknown hazards of war."[7.7]

After Trevelyan had become a camouflage officer, he would occasionally "be asked to give a demonstration of how to paint some piece of equipment so as to merge it with the broken country around. I would arrive on the barrack square with pots of paint and brushes, and set to work daubing the shield of some anti-tank gun with spots of different greens and browns, touching in the underside of the barrel itself with pure white on the principle of Cott's gazelles. Against the dreary barrack walls it looked an unholy mess, but when it was wheeled out into the country and placed against a hedge, there were cries of astonishment at my magic. This role, half-clown, half-magician, was one that I found camouflage officers were more or less expected to fill."[7.8]

A needle is much simpler to find in a haystack than in a bin of other needles.

—Colin Watson

For some time I have been trying to find the right word for the shimmering glancing twinkling movement of the poplar leaves in the sun and wind. This afternoon I saw the word written on the poplar leaves. It was "dazzle." The dazzle of the poplars.

—Francis Kilvert
in Plomer (1960), p. 91.

At Farnham at the same time was Trevelyan's former partner, Roland Penrose, a Surrealist

Even the face of the soldier is streaked with bars of color like that of an Indian and he thus becomes quite invisible and is as effectively concealed as any wild animal lying under cover in the woods.

—Ernest Peixotto
"Special Service for Artists in War Time," p. 6.

▲ **FIGURE 7.B**
In his civilian guidebook, Roland Penrose included drawings of disruptively camouflaged clothing, much like this example from World War I.

▶ **FIGURE 7.C**
World War II instruction sheet for a camouflage face painting kit for soldiers, using techniques like those Lee Miller used.

painter who later produced a pioneering biography of Picasso. In fact, he was one of the first to point out the significance of Picasso's recognition of the resemblance between Cubism and World War I camouflage. Penrose was not in the military, but sat in on the training as a civilian camouflage instructor. In 1941, he wrote and illustrated a guidebook for civilians called the *Home Guard Manual of Camouflage*.[7.9]

It is hard to believe that a Surrealist as puckish as Penrose could lecture about trickery without introducing pranks as part of the lecture. Thus, Penrose found a clever way of demonstrating the conspicuousness of even the slightest movement: He would, Trevelyan remembers, "stand his audience round a piece of grass in which a small button was lying, which was of course invisible. He would then pull it rapidly along with a thread, and immediately all eyes were fixed on it."[7.10]

Penrose's wife was Lee Miller, the American model and photographer who had earlier been the wife of the Surrealist photographer Man Ray. In his notebook, Penrose provides a written account and a photograph of an afternoon tea in an outdoor garden with friends during which, as a "camouflage experiment," Miller removed her clothing and smeared herself with a matt dull greenish camouflage cream developed by a cosmetics company. It was Penrose's theory, he writes, "that if you could hide such eye-catching attractions as hers from the invading Hun, smaller and less seductive areas of skin would stand an even better chance of becoming invisible."[7.11]

Among those trained as World War II British army camoufleurs were the painters Robert Medley, Edward Seago, Patrick Phillips, Blair Hughes-Stanton and Frederick Gore;

designers Victor Stiebel, Steven Sykes, James Gardner, Jack Keefer, and Ashley Havindon; sculptor John Codner; film producer Peter Proud; theater set designer Oliver Messel; and a host of others, among them stained glass artisans, a *Punch* cartoonist, a circus manager, an art restorer, and art connoisseurs.

Monday, 7th April [1941] This afternoon reported to Victory House, along with forty other miserable recruits, all clutching their suitcases. We were conveyed to Paddington in camouflaged cattle-trucks, hurling from one wall to another as we went round corners, and arrived in Gloucester at six p.m.

—Joan Wyndham *Love Lessons: A Wartime Journal*, p. 187.

Like Penrose, Seago found it hard to be totally serious when designing camouflage. He was, for example, once asked to provide hundreds of life-sized human dummies, each of which was mounted on a remote controlled turntable and placed beside a dummy anti-aircraft gun on the south coast of England. As a joke, Seago made all of the dummies appear to be imitations of the discredited British prime minister Neville Chamberlain, but dressed in a helmet and battle fatigues.[7.12]

Reading between the lines of Trevelyan's autobiography, one senses that he was not fully convinced of Jasper Maskelyne's extraordinary command of camouflage, or, at any rate, not as convinced as the magician himself. Maskelyne, recalls Trevelyan, "entertained us with his tricks in the evenings [at Farnham], and tried, rather unsuccessfully, to apply his techniques to the disguise of the concrete pill-boxes [in which anti-aircraft guns were hidden] that were then appearing everywhere overnight. He was at once innocent and urbane, and he ended up as an Entertainments officer in the Middle East."[7.13]

By Maskelyne's account, which he recorded later in a book titled *Magic—Top Secret*, nearly all the entertainments devised by himself and the team that he fondly referred to as the "Magic Gang" were directed toward Adolf Hitler. "I and others made the mailed fist [of the Third Reich] strike wildly against empty air, with considerable loss to its sense of balance," recalls Maskelyne. "We made it draw timidly back when nothing but cardboard and canvas opposed it; and, at other times, crash home upon knobby obstacles from which it recoiled hurt, surprised and bleeding."[7.14] As proof of his effectiveness, he continues, there were no more satirical drawings of him in German newspapers after 1941: "Instead, Hitler did me the honor of placing my name on his dreaded 'Personal Black List,' and a reward was offered for my capture, perhaps because it was hoped that the Gestapo 'magic' might make me talk."[7.15]

Many of the Magic Gang's projects were tricks that might best be described as mimetic distractions and decoys, objects and events that not only called attention to themselves but also appeared to be some other thing than what they actually were. In one case, for example, he and his soldiers used scrap metal to build dozens of tanks and artillery guns, then filled the gun barrels with a stage magician's recipe for a harmless but conspicuous flash that looked and sounded like real artillery fire. Then, to create the effect of a

A town occupied by an army looks like a deserted shell of grey inhabitants watched over by khaki ghosts.

—Stephen Spender in Goldsmith (1986), p. 62.

Papa was an airplane watcher in south Georgia; he even had this little hat to wear. He'd count every plane that came over and check to see if it was an enemy plane. They made camouflage nets in the gym at the high school.

—Bunny Johnson in Sutton and Waite (1992), p. 22.

column of tanks moving across the desert, he hired North Africans to drive their camels across the dunes, each animal dragging behind it a dust-producing wooden frame. In another incident, he and his crew constructed "a fleet of dummy submarines, full size, able to float

▲ **FIGURE 7.D**
DAN SCHRAD
Hand-drawn metamorphosis in which Beethoven becomes a clarinet.

like real submarines, but also able to be folded up by a few men so as to travel in a five-ton truck."[7.16]

> Black, slack, stack, stark, stare, stale, shale, whale, while, white.
>
> —E.J. KAHN, JR

IN DISCUSSIONS OF magic, the inevitable explanation is that "the hand is quicker than the eye." But that itself is a diversion, since the success of most magic depends not on extraordinary swiftness and manual dexterity but rather, as advised earlier by the Great Merlini, on the misdirection of attention, or distraction. It is to the performer's advantage that we, as vigilant and sophisticated audience members, should follow very closely the actions pointed out to us (figure), while ignoring the seemingly marginal acts—the trifles, Sherlock Holmes would say—that really enable the trick to occur (ground). The magician, writes Warren Steinkraus, "requires that his viewers watch every overt move diligently, thereby missing secretive moves."[7.17]

While the viewer attends—as directed—mainly to the figure, the magician must follow both figure and ground. This requires a sort of double consciousness, explains Steinkraus, with the result that "there are always two performances going on, the one we obviously see and the one carefully screened from our view. A conjurer must be double-minded as no other performer dares to be. He can never lose himself totally in the effect he is presenting, but at the same time he must appear so intently involved in doing what we observe that he gives us not the slightest hint that he is doing quite otherwise."[7.18]

▲ **FIGURE 7.E**
Tom Grothus
Adjusting the Antenna (1984).

[So-and-so] would not have thought of concealing his whereabouts, being a person who considers himself a "tough" soldier.

—John Gaitha Browning
quoted in Toliver, p. 34.

Those two performances, one manifest, the other latent, are essentially two constellations or two Gestalts. They are comparable to the alternating interpretations of the rabbiduck, although in magic, as in dreams, one of them is deliberately made visible while the other remains inconspicuous. The primary tactic of magic, writes Max Dessoir, "lies in the power to direct the thoughts of the audience into such a groove [the explicit Gestalt] that a solution of the trick seems for a moment the natural result of the artificially underlying causes."[7.19]

In his biography of William James, the scholar Jacques Barzun describes an amusing moment in the life of that philosopher and psychologist when he and a student were taking a stroll near Harvard University. As they walked, they encountered on the street an elderly gentleman who was talking to himself and who appeared to be oblivious to the world, so much that they might collide if they had not stepped aside. When the stranger had

All the arts seem to have gravitated to the camouflage unit—even musicians.

—Charles Burchfield
His Golden Year, p. 24.

passed, the student turned to James and said, "Whoever he is, he's the epitome of the absentminded professor," in reply to which her teacher said, "What you really mean is that he is present-minded somewhere else."[7.20]

Likewise, it is not that magicians persuade an audience to be absentminded, but rather that they cleverly prompt them to be present-minded toward other events that are of little or no consequence. Misdirection means simply directing the mind toward diversionary aspects; and magicians insure that this happens by methods that everyone uses daily. However, unless we are employed in sales, we probably use them more often to be clearer, not misleading, in the process of communicating. "My dentist informs me," writes Nathaniel Schiffman, "that dentists and doctors use the same [misdirection] trick when administering shots to patients. They first use the fingers to pinch the skin, rendering the patient a bit less sensitive to the needle to come."[7.21]

When I was in the Air Corps during World War II, I tried to get into camouflage, but they needed radio operators at that time. Because I was a drummer I passed the Morse Code aptitude test 100% and as it turned out my entire platoon consisted of jazz musicians...A less military group I never saw.

—Byron Burford
in a letter to the author (August 26, 1998), p. 62.

Perhaps the most familiar means of misdirection is the simple act of pointing. Magicians direct the attention of an audience by giving them verbal instructions about where to look, by pointing with the hands, fingers, thumbs, elbows, and feet, by motioning with ones

▲ **FIGURE 7.F**
JOHN DOPITA
A hand-drawn metamorphosis in which Beethoven becomes a piano, c.1995.

head, by facing something, or simply by staring at it.

Another common way of calling attention to an object is through contrast (color, size, sound, movement, position, and so on), by creating what Gregory Bateson would call "news of difference," a technique that is often effectively used by pickpockets.

Cock-cook-cool-fool-foul-soul-sour-slur-slum-glum-glim-frim-gram-cram-craw-crow.

—E.J. KAHN, JR

The pickpocket's use of misdirection is mentioned in Hugh B. Cott's lecture to the Royal Engineers: "When a pickpocket intends to rob you of your watch or wallet," he writes, "he, or his confederate, takes care to distract your attention from what he intends to do by creating a diversion. He draws your eyes from what is really happening to what seems to be happening."[7.22]

He compares pocket-picking to the confusing effects of disruptive coloration in animals. "Distributed over the body are irregular patches of contrasted colors and tones," continues Cott, which "tend to catch the eye of the observer and to draw his attention away from

the underlying form of the animal which exhibits them. The patterns themselves may be conspicuous enough, but since they contradict the form of the animal on which they are superimposed, they pass for part of the background, in the same way that the pickpocket's tactics of bluff pass for a commonplace incident."[7.23]

In books about magic, dreams, camouflage, and other forms of misinformation, it is customary to include an inventory of methods, a list of the kinds and varieties of deceptive tactics. Rarely are two of these lists the same, but they typically include such strategies as condensation, substitution and displacement (described earlier as Freudian dream techniques), or, in the lingo of stage magicians, such categories as *production*, *restoration*, *modification*, *transposition*, *relocation*, and so on. In general, one might simply say that all of these are variations on one overriding strategy, the tactic of doing the opposite from what is really going on, or what would normally be expected.

That exact same principle was recommended in 500 BC by Sun Tzu, a Chinese army tactician, who said: "All warfare is based on deception. When able to attack, we must seem unable; when using our forces, we must seem inactive; when we are near, we must make the enemy believe that we are far away; when far away, we must make him believe we are near.

Hold out the bait; entice the enemy. Feign disorder and crush him."[7.24]

In magic performances, a thing appears where nothing was. Or, where something was only a moment ago, there is suddenly nothing. The parts of a thing are restored to a whole. One thing becomes some other thing, or exchanges places with another. That which is solid and heavy becomes insubstantial and weightless. The impenetrable becomes penetrable; the inanimate animate. That which is vulnerable becomes invulnerable. The mental becomes physical; the physical mental. Large becomes small; front becomes back; top becomes bottom; in becomes out, and so on. The list could continue indefinitely, in the sense that so long as an attribute can be defined and identified, the opportunity exists for a magic trick (or dream technique or camouflage strategy) to counteract that attribute.

Art historian Wylie Sypher said that "the great Cubist achievement was camouflage." With that in mind, it is of value to compare the above roster of magic categories with Sypher's surprisingly parallel list (cited in an earlier chapter) of techniques used by the Cubists. Among these, he writes, are "a breaking of contours, the passage, so that a form merges with the space about it or with other forms; planes or tones that bleed into other planes and tones; outlines that coincide with other outlines, then suddenly reappear in new relations; surfaces that simultaneously recede

Plagiarists are always suspicious of being stolen from—as pickpockets are observed commonly to walk with their hands in their pockets.

—Samuel Taylor Coleridge
Table Talk (1836)

◄ **FIGURE 7.G**
Victorian-era metamorphosis of a musician and his cello.

When it was dark we stood on the balcony and leant back looking at the sky and the barrage balloons, and G said, "I've never known it so warm in the evenings, not in three years. All the blimps are up tonight, they should paint them blue like the sky for camouflage."

—Joan Wyndham
Love Lessons, p. 68.

and advance in relation to other surfaces; parts of objects shifted away, displaced, or changed in tone until forms disappear behind themselves."[7.25]

Suzi Gablik has compiled a similar list for her book about the life and work of the Surrealist artist René Magritte, an account of the primary methods by which Magritte's paintings "disrupt any dogmatic view of the physical world."[7.26] She also reveals that his interest in art began during his childhood, when, as he was playing in an abandoned cemetery, he happened upon a painter who (in Magritte's words) "seemed to me to be performing magic."[7.27]

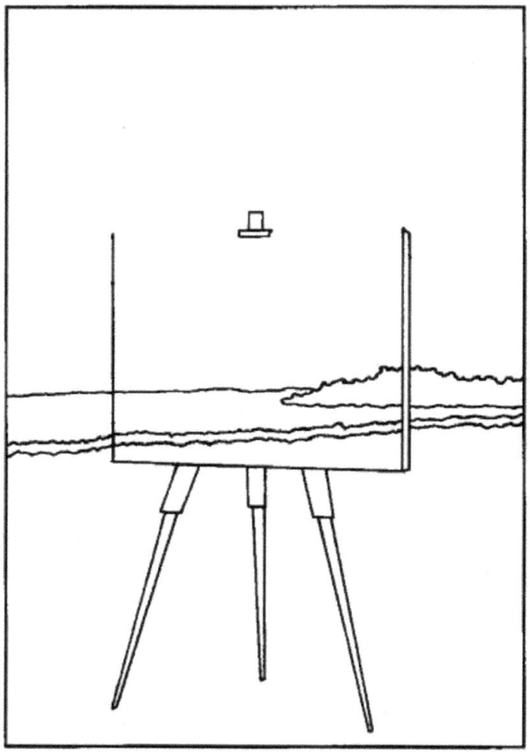

▲ **FIGURE 7.H**
Author's diagram, loosely based on paintings by the Belgian Surrealist RENÉ MAGRITTE, in which the boundary is blurred between a painting and the thing painted.

In various of Magritte's paintings, a carrot turns into a bottle. A bird becomes a leaf. A man's head is a light bulb. Leather shoes are also feet. A rock floats. A figure is a hole through which it is also the background. An image of a landscape is itself the thing it represents. A painting of a pipe is labeled *Ceci n'est pas une pipe* ("This is not a pipe"). A flying bird is made of stone. A single apple fills a room. The smoke from a man's pipe becomes his nose. Hundreds of businessmen float in air, while precisely aligned in perspective. And a steam locomotive juts out of a fireplace wall, its smoke rising into the chimney.

By day I wore a dashing uniform, and Flossie dressed in an immaculate waitress outfit that could not camouflage her glorious body.

—Art Buchwald
Leaving Home

There was a sensation in London in 1875 when Jasper Maskelyne's famous grandfather, the magician John Nevil Maskelyne, announced a new addition to his stage performance. It was a robot, a cross-legged seated mechanical man called "Psycho" who was good at playing whist, a forerunner of bridge. This Oriental manikin, which was only twenty-two inches high, was seated on a small platform which was in turn supported by a transparent glass cylinder. Having withstood an inspection by members of the audience, it moved its head to look at cards (both playing cards and alphabets), and used its arm to pick them out as a means of responding to questions about mathematics, or to spell out words. It also took an occasional puff from a cigarette.[7.28]

▲ **FIGURE 7.1**
Les Coleman
Dislocation (c. 1996)

Show business is dog eat dog. It's worse than dog eat dog. It's dog doesn't return dog's phone calls.

—Woody Allen

At a performance by Psycho on June 24, 1876 (which was, incidentally, the day before the Battle of Little Bighorn), one of the audience members was the Rev. Charles Lutwidge Dodgson, who was an Oxford College don and a teacher of mathematics. A prolific writer, within the previous decade he had published two children's classics, *Alice's Adventures in Wonderland* and *Through the Looking-glass and What Alice Found There*, which he signed with the now famous pseudonym of Lewis Carroll (in which he hid his given names, Charles Lutwidge, by translating them into Latin as Carolus Ludovicus, then reversing the order and changing them back into English again). He had specifically come to observe Maskelyne's amazing mechanical man, after which he concluded (incorrectly) that "I have no doubt [that it] has a dwarf in it."[7.29]

> [Camouflage violated] the military obsession with neatness, spit-and-polish, drill-field regularity... [It] was considered careless and "undisciplined." It made a virtue of dispersion, irregularity, and improvisation; and like modern ecologists, camoufleurs tried to blend with the natural landscape instead of exploiting and dominating it.
>
> —Seymour Reit
> *Masquerade*, pp. 65-66.

But Carroll was also presumably there for another, more general reason, which was to satisfy his lifelong fascination with magic. A generation younger than Charles Dickens, he had practiced magic all his life. As a child, Martin Gardner writes, Carroll "dabbled in puppetry and sleight of hand, and throughout his life enjoyed doing magic tricks, especially for children. He liked to form a mouse with his handkerchief then make it jump mysteriously out of his hand. He taught children how to fold paper boats and paper pistols that popped when swung through the air."[7.30]

> We [military camoufleurs] actually helped the soldiers' morale. They said they slept better at night knowing we had camouflaged the barracks...Many of the officers I worked for thought of us as dreamers with magic paint buckets and expected far more from us than we could possibly deliver.
>
> —Harper Goff
> quoted in Naverson, p. 98.

Lewis Carroll's *Alice* books are exemplars of the adroit use of ambiguity—they are, as one critic befittingly said, "an infinite onion, with many layers"—in the sense that they lend themselves willingly to a wide variety of complex interpretations, from psychoanalytic to literary to religious and political, on a level for children as well as adults. Since at least the 1930s, there have been frequent, repeated attempts to dissect both the writer's psyche and his writings (which he claimed to have been "nothing but nonsense") through filters as heavily tinted as Freud's, or that of his disaffected student, Carl Jung. It was in reply to such "psycho-analyses" of Carroll's writings, as if they were transcripts from his therapy sessions, that James Joyce wrote in *Finnegans Wake* that "We grisly old Sykos have done our unsmiling bit on 'alices, when they were jung and easily freudened."

Ignoring for a moment their enormous differences, the *Alice* books might have been written by Magritte, while Lewis Carroll might have made Magritte's paintings. Or maybe the Magic Gang could have produced the dreamlike magic works of both.

In various episodes of Carroll's books, a little girl fills up a room. A poem is printed backwards. There is a grin without a cat, and an insect with bread for wings and a sugar cube for a head. Memory is discredited since it "only works backwards." The text of a tale about a mouse has been typeset to resemble the shape of a tail. A wooded landscape is a giant chessboard. The sea is hot, and pigs have wings.

The *Alice* books are the literary equivalent of a Victorian magic show, like those that Charles Dickens gave for his family and friends. "Like all great magicians," writes John Fisher, Carroll "needed only simple, everyday objects such as a ball of wool, a kitten and some chessmen, together with the almost

▲ **FIGURE 7.J**
JOHN TENNIEL
Illustration of the bread-and-butterfly from Lewis Carroll's *Through the Looking-glass and What Alice Found There*, 1871. This paradoxical creature has wings of bread-and-butter, a body of crust, and a lump of sugar for a head. It lives on weak tea with cream. If it doesn't eat, it will of course die; but if it eats it will also die, since its head will dissolve while drinking the tea.

obligatory mirror and the assistance of a little girl to produce a spectacle unrivalled in later years by Houdini or Disney."[7.31]

"The time has come," the Walrus said,
"To talk of many things:
Of shoes—and ships—and sealing wax—
Of cabbages and kings—
And why the sea in boiling hot—
And whether pigs have wings."

—Lewis Carroll
Through the Looking-glass and What Alice Found There.

The *Alice* books are said to rely on the language of dreams, which explains why they often gets targeted for literary psychoanalysis, and why Carroll's writings were praised and promoted by the Surrealists as the antecedents of their tradition. But it is also camouflage, in the sense that its enigmatic author—for whatever reason—made clever and often astonishing use of "the guise of nonsense" to demonstrate "the ephemerality and unimportance of our most cherished categories."[7.32] ✂

He [Pablo Picasso] was thought to be dead at birth in Málaga on October 15, 1881. Then his uncle Salvador Ruiz, a celebrated Spanish physician who had delivered the boy, calmly puffed cigar smoke up the baby's nose, provoking howls of protest. Thus did Picasso embark on 91 years of rugged life.

—Robert Hughes
Time magazine (May 26, 1980), p. 79.

THE ETYMOLOGY OF "CAMOUFLAGE"

THE WORD "CAMOUFLAGE" can be traced back to the 16th century to the French term for a practical joke (a variation on a hotfoot) by which a person who fell asleep in a chair could be abruptly and rudely aroused. It was called a *camouflet*, a term that may have been derived from *chault mouflet*, an archaic expression for "hot face." The joke was accomplished by lighting the tip of a hollow paper cone, then holding the opposite smoldering end under the victim's nose. The stupefied sleeper would spring to his feet as soon as a noseful of smoke was inhaled.

Three centuries later, *camouflet* had come to mean a small but lethal powder charge by which a tunneling enemy troop could be entrapped beneath the ground, buried and suffocated. From that came *camoufler*, a verb that means "to get made up," to put on cosmetics and costumes in order to play a theatrical role—or, in a somewhat more sinister sense (it was an underworld slang term), to be deceptive, to blow smoke, or, more specifically, to disguise oneself for illicit purposes.

It was from this lineage of root words that "camouflage" became the term for the purposeful use of a visual disguise as a military stratagem. During World War I, the term was exported to English from French. According to the *Oxford English Dictionary*, it first appeared in British print on May 25, 1917, in the *London Daily News*, where it was used in a sentence that said "The act of hiding anything from your enemy is termed 'camouflage.'"

A few months later the same newspaper announced that the King of England had "paid a visit to what is called a camouflage factory" where he was introduced to "all the latest Protean tricks for concealing or, as we all say now, 'camouflaging' guns, snipers, observers."

By the end of World War I, "camouflage" had amassed so many figurative, nonmilitary connotations that *The Anthenaeum* observed that the word "has met with more wear and tear in a few months than many receive in a century." During the same period, one finds reference to a "camouflage autocracy," to eggs "camouflaged" in a scramble, and even to a telephone as having been "camouflaged" by its design. "Instead of saying 'He is a bluff,'" wrote one journalist in 1917, "we say 'He is nothing but camouflage.'"

Section VII: Strategic Deception

Introduction

Strategic Deception

Strtegic deception focuses on the *big picture*. At its center are the policies or even ideology of a state or significant nonstate actor. Accordingly, the target of the deception effort must be the people and offices that can change and influence policy. These targets have traditionally been heads of state, ministers or cabinet members, and senior military commanders.[1] Today, this target list should be expanded to include others who are in a position to sway the actions and decisions of state and nonstate actors. To be sure, identifying the players on this expanded list presents a challenge.

The risky and expensive nature of strategic deception inevitably confronts what Clausewitz called the *friction of the whole machine*. Strategic surprise is very difficult to achieve because of the possibilities of disclosure associated with a protracted, slow-moving operation and the bungling that inevitably ensues when a full bureaucracy is involved. Despite his reservations, Clausewitz ultimately tells us that high level deception, designed to achieve surprise, is "never wholly unproductive" and is always worth the effort because even small effects at the strategic realm can have significant ramifications.[2]

This section looks at the heady domain of military deception on a grand scale. This, as we'll see in the next three cases, is the stuff of Pearl Harbors and D-Days. Note, however, that exactly the same set of psychological principles apply here as throughout the spectrum from grand strategy to minor tactics.

1. Roy Godson and James J. Wirtz, eds., *Strategic Denial and Deception,* London: Transaction Publishers, 2002), p. 2.
2. Bart Whaley, *Stratagem: Deception and Surprise in War,* Norwood, MA: Artech House, 2007, p. 71.

CHAPTER 31

Biggest Hoax of the War: Operation FORTITUDE: The Allied Deception Plan that Fooled the Germans about Normandy[1]

Wentworth Eldredge

At the start of World War II, Wentworth Eldredge joined the United States Army Air Force. Assigned to intelligence work, Eldredge helped distract German agents from the Allies' plans for an invasion of France through Normandy.

The following article is a brief but accurate summary covering the essential elements of the British and American FORTITUDE deception.

1. H. Wentworth Eldredge, "Biggest Hoax of the War: Operation FORTITUDE: The Allied Deception Plan that Fooled the Germans about Normandy," *Air Power History*, Vol. 37, No.3, Fall 1990, pp. 15–22.

Biggest Hoax of the War

Operation FORTITUDE: The Allied deception plan that fooled the Germans about Normandy

by Dr. H. Wentworth Eldredge, Major, USAF in WWII

Editor's Note: General Omar Bradley called FORTITUDE "the biggest hoax of the war." That successful scheme to deceive Germany about the date and place of the invasion of France in 1944 made an immense contribution to the success of the landings.

But because the British take security very, very seriously there has never been a definitive book devoted to the subject, though many historians and novelists have written about one or another aspect of the operation. Not until four decades had passed did the public learn about it when the Associated Press revealed the main outlines, citing the recollections of Colonel Arthur A. McCrary, Ret., of the 3103rd Signal Service Battalion, who remarked on the 40th anniversary of D-Day, "I was surprised to find out many of the things that were going on. At the time the information was passed out strictly on a need-to-know basis."

Another participant, far higher up and at the policy level, was the only Army Air Forces officer involved in the planning and execution of FORTITUDE — Capt. (later Major) H. Wentworth Eldredge, on leave from Dartmouth College, from which he retired in 1974, having been head of its Sociology Department. A frequent lecturer at the NATO Defense College and the French, German, and Belgian General Staff Schools, he is now writing his memoirs. Here are portions of his chapter on FORTITUDE.

I was getting increasingly bored with being Assistant Historian of the Eighth Air Force—not what I had joined up for! My boss, Bruce Hopper, was a distinguished Harvard professor, and writing the history of the Eighth was an important task. But it was sedentary, and I craved something more active. So I was intrigued by what Dennis Wheatley (whose beautiful step-daughter, Diana, I was squiring and would later marry) was up to. Obviously, it would not do to ask, but I gathered it was some sort of deep and devious planning. It sounded "just my cup of tea."

Dennis and wife Joan laid on a delightful party full of brass hats downstairs in Chatsworth Court in December 1943. They had a flat there, having been bombed out of their St. Johns Wood house. Dennis was a splendid host, having once been a successful wine merchant. In England a wine merchant has the status of beloved friend of the upper classes. Dennis had been much too lavish and went broke in the Depression. For example, Tallulah Bankhead went to her grave owing him 3,000 pounds sterling; today, with a purchasing power of about $20,000, quite a bit to drink up! Later Dennis became the best-selling author of 55 books, chiefly adventure novels but with others documenting his WW II experiences. Published in 15 languages, sales ran to 60,000,000 copies.[1]

Dennis had been brought up at the Royal Naval School, served in WW I in the London Artillery and was gassed, transferred to the RAF, becoming a Pilot Officer and eventually a Wing Commander, thus having the unusual distinction of serving in all three military branches. He landed his final job as one of Winston Churchill's confidants and advisors by unofficially concocting a variety of "future possibilities" that Germany or England might carry out. Dennis had a vivid imagination and was a colorful character who dressed the part, lining his blue RAF overcoat with scarlet silk and carrying a leather swagger-stick with concealed sword—both strictly non-regulation.

He trusted his step-daughter and asked Diana if she knew any bright American who did not have the national failing, "a big mouth." Diana said she knew one, a college professor who might fit the bill. I suspect that one of the goals of that Chatsworth party was to have some of the important soldiers, sailors and airmen there "vet" me. The Brits at that point knew they had to open up the FORTITUDE Plan to the Yanks. Next to ULTRA, it was the most carefully kept secret of the European war. But, after all, two-thirds of the troops fighting in Europe would be American, so Amercians should be part of the FORTITUDE inner circle.

Invited Into the Circle

After the party Dennis suggested, if I was interested, that I invite Lt. Col. William Baumer, a U.S. Army General Staff officer in London, to a good lunch and see what happened. Dennis alerted Baumer, and I duly asked him to join me at the Coq d'Or, a very good spot. We hit it off at once, easily comparing thoughts on how careers intersect

Shoulder patches of three notional (i.e. imaginary) Divisions, U.S. Army 1944-45. (the author)

and other topics. He had graduated from West Point in 1933 but left the Army to make money as a writer, returning to uniform upon the outbreak of WW II. Did I know Ralph Ingersoll? No, but I knew about his complicated venture with *PM*, an unorthodox daily newspaper he had started pre-war in New York City. I thought he was very bright and very tough. Baumer said, "Good," he'd make an appointment for me to see Ingersoll, who was at ETO Hq. in Grosvenor Square.

I went to his office two days later. He was an Ordnance captain then, though he rose to Lt. Col. later. In those days I was quite "liberal," as was *PM*, so we got along fine. He said he would send up a request for me and blast me loose from Hopper, adding that he would tell me what he did when I was formally transferred. It turned out to be FORTITUDE.

Backgrounding

Perhaps a definition of "deception" in military parlance is needed at this point:

A "Cover Plan" is a false one attempting to hide the real operational intention. It is aimed at the enemy commander through his own intelligence facilities of all sorts. It is a dangerous and subtle thing from start to finish because, if it is discovered by the enemy, it can point to the real plan. Obviously, security must be very tight. Though deception in war goes back as far as the Trojan Horse, in WW II the British were forced to amplify it after Dunkirk because they did not have nearly enough trained troops or weapons to stand up to the expected Nazi invasion. So the Brits began "manufacturing" Divisions out of thin air, a practice they continued after the threat of German invasion ceased. Even without that subterfuge it was difficult enough to ascertain just what forces they had since each Division was habitually formed from bits and pieces of the yeomanry, County Regiments, Guards Regiments and new recruits. FORTITUDE was mostly a British show, with some help from "the bloody Yanks." The main responsibility on the American side was carried by three men—Col. William ("Billy") Harris, a G-3 officer assigned by ETO as its controlling voice; Ralph Ingersoll, elevated to Major; and Captain Eldredge. We were wonderfully lucky to share this unique assignment and lucky also to have sufficient sophistication to work with the very, very bright Limeys who ran it. The deception "scenario" (Ingersoll's word) was this:

1) **Location**: *Main attack* to be on the Pas de Calais with preliminary diversion in Normandy shortly before.

2) **Date**: *Later* than the real time.

3) **Time**: *Not* at first light of day, which would be the real intent.

The idea sounds very simple. But to carry it out was complicated as hell, with a whole nation as a base; a second nation which had the reputation of the biggest mouths in history; a third nation, the French, who leaked worse than the Americans and so were told nothing; millions of soldiers, sailors and airmen; plus numerous friendly and some not-so-friendly diplomats with ears at many keyholes.

Elements of the Plan

Here is a summary of the seven major ingredients in Operation FORTITUDE with Col. Johnnie Bevan, British Army, as the master manipulator of what was officially dubbed The London Controlling Section.

1) Fake Army. Lt. Gen. George S. Patton, an officer well known and respected in the German military, was appointed and publicly announced as commander of FUSAG (First Army Group). FUSAG was both a real entity (converted after the Normandy landings to 12th Army Group) and a phoney—the cover plan headquarters.

To create the image in German minds of a huge invasion force building up in East Anglia to as many as 50 "notional" (the British variant of "imaginary") Divisions required a lot of showmanship—especially since the fake FUSAG actually mustered only 1,500 men.[2]

So we made a big thing of Georgie Patton and his white bull-terrier attending local town meetings. Extensive lighting arrangements were set up to indicate, even in a black-out, the existence of sprawling military establishments. Notional Divisions, both British and American, were given not only fictitious names but appropriate shoulder patches and other identifying gear. Most of the names and patches were hatched in Roger Fleetwood Hesketh's fertile mind; my only contribution was in naming the fake 59th Infantry "The Rattlesnake Division," as symbolized on its patch (see picture). Soldiers wearing such patches were "seen" and reported to Germany by spies who had been "turned."

2) Wireless networks. Military assemblies are more than large groups of men and weapons sitting quietly before battle. There is also a huge volume of routine radio and other forms of communication. This was simulated in escalating volume by two American signals units which infiltrated false traffic into FUSAG, both real and fake headquarters. This could be overheard across the Channel. The Brits had a similar outfit mimicking traffic between their real and phoney forces. Every now and then transmission would be arbitrarily halted to alarm and confuse the Germans by implying a sudden attack.

3) Dummy landing craft, planes, tanks, etc. were constructed of rubber and wood by the hundreds and positioned in harbors or on sham airfields in areas clearly threatening the Pas de Calais.[3] German recon did not spot many of the dummies, if at all, for the RAF Spit/Hurricane interceptors were too good, maintaining almost complete control of the air in eastern and southern England.

4) Air action. Both the AAF and RAF bombed and otherwise attacked the Pas de Calais and Normandy areas with about the same volume of bombs and bullets. Targets included artillery and buzz-bomb sites, radar stations, railroad yards, bridges and canals as well as storage areas and selected factories.

Blasting V-1 and Radar Sites

The "ski-sites," as the V-l buzz-bomb launching pads were called because they looked from above like skis lying on their sides, were significant targets entirely aside from deception planning. There were 96 of them, chiefly in the Pas de Calais area since it was the nearest to Hitler's major target, London.

Air attacks on them began in late 1943 and mounted steadily in size, causing much anguish to United States Strategic Air Force commanders who thought oil refineries deep in Germany were more important objectives. But more knowing heads prevailed, Eisenhower among them, and CROSSBOW, as the anti-buzz-bomb-site campaign was dubbed, was given "over-riding priority" on December 15, 1943. Eighth AF Mission #164, its largest to date, deployed more than 1,300 aircraft dropping 1,700 tons of bombs. During March 1944 the AAF and RAF dispatched 2,800 CROSSBOW sorties, dropping 4,150 tons. The figure rose to 4,140 sorties and 7,500 tons in April and leveled off in May with 4,600 sorties and 4,600 tons. Of the May figures, all from U.S. planes, 800 sorties and 2,600 tons were dropped by VIII Bomber Command heavies, with 3,800 sorties and 2,000 tons from Ninth AF mediums. Deliberately obscured in all these statistics is the tonnage dropped on interdiction targets like those Normandy area, but the message conveyed to the German High Command was clear.[4]

A similar campaign against German radar stations was executed by the RAF under the direction of the brilliant scientist Dr. R. V. Jones, who had discovered during the Blitz in 1941 how the Germans used intersecting radar beams to guide their night attacks on targets as small as Coventry. There were, Jones recorded in his superb *The Wizard War*, some 600 individual radar stations along the coast from Denmark around France and down to Spain. The RAF followed a deception policy in attacking and destroying almost all of these tiny targets: For every such attack in Normandy two were carried out elsewhere. On the night of D-Day, Jones wrote, one operator did detect the approach of the Allied armada but his reports were "discounted up the chain of command because they were not confirmed by other radar stations."[5]

Still a third role for the airmen in FORTITUDE was use of WINDOW, the bundles of shredded tin-foil habitually dropped by RAF night bombers because they confused German radar. The night before D-Day RAF 617 Squadron's Lancasters were schooled to fly in rectangular orbits measuring eight by two miles and moving steadily eastward up the English Channel. The effect on the remaining radar stations would be that of an approaching fleet of ships. But bad weather and the day's delay in executing OVERLORD caused cancellation.[6]

The near-storm-level weather was one factor in convincing Rommel there would be no landings that week, and he was on vacation. None of his troops in Normandy were on alert. Only by pure chance was a good German Division (352 Inf.) moved into line to practice for the invasion. That was what caused the mess at Omaha Beach, where the U.S. forces were nearly pushed into the sea.

Other Elements of the Plan

5) Resistance groups in France and the Low Countries. They were told only that the main attack would be in the Pas de Calais—this on the assumption that it would be leaked to the Krauts. Rumors were floated that British and American teams would be sent in as decoys to be captured.

6) Double agents. The British had not only broken the main German codes but also captured all the spies infiltrated clumsily into Great Britain. Some of these were executed or imprisoned, others "turned"—i.e., converted into mouthpieces for their British masters and communicating with Germany by radio, by mail through neutral Spain and Portugal and by assorted other devices such as invisible ink and diplomatic pouches. A total of 120 double-agents, real and notional, operated at one time or another under British control. All students of FORTITUDE agree that the skill with which these strange men and women were puppeteered by their masters was the most important single contribution to the plan's complete success. Several books have detailed their work, notably in the case of GARBO, whose real name was Jean Pujol. His own book, somewhat immodestly but accurately titled, *Operation Garbo: The Personal Story of the Most Successful Double Agent of World War II*, begins with a tribute to Lt. Col. Roger Fleetwood Hesketh, British Army. Of all the brilliant Brits in FORTITUDE, Roger was the most inventive and the officer I recall most keenly and admiringly. An Eton and Oxford grad, he took care never to meet GARBO. The intermediary was "Tommy" Harris, civilian son of a Scotch father and Spanish mother and no relation to Billy. Tommy spoke Spanish fluently and was GARBO'S "spy-master." Such indirect linkage was one more way by which FORTITUDE's secrecy was preserved.

Title page of this article reproduces a German high command map of Great Britain as of May 15, 1944. It reflected the German assessment of the locations of major American, Canadian, and British ground combat units at the time. The map came into the hands of Allied intelligence, and showed graphically the success of Operation FORTITUDE. Map 1 on this page is a simplified version of the German map. It depicts the German holdings of US and Canadian units of division, corps, army, and army group level. Beside it on Map 2 are the actual locations of the real units on May 15, 1944. The Germans were fooled into thinking that the potential invasion force was twice as strong as actual.

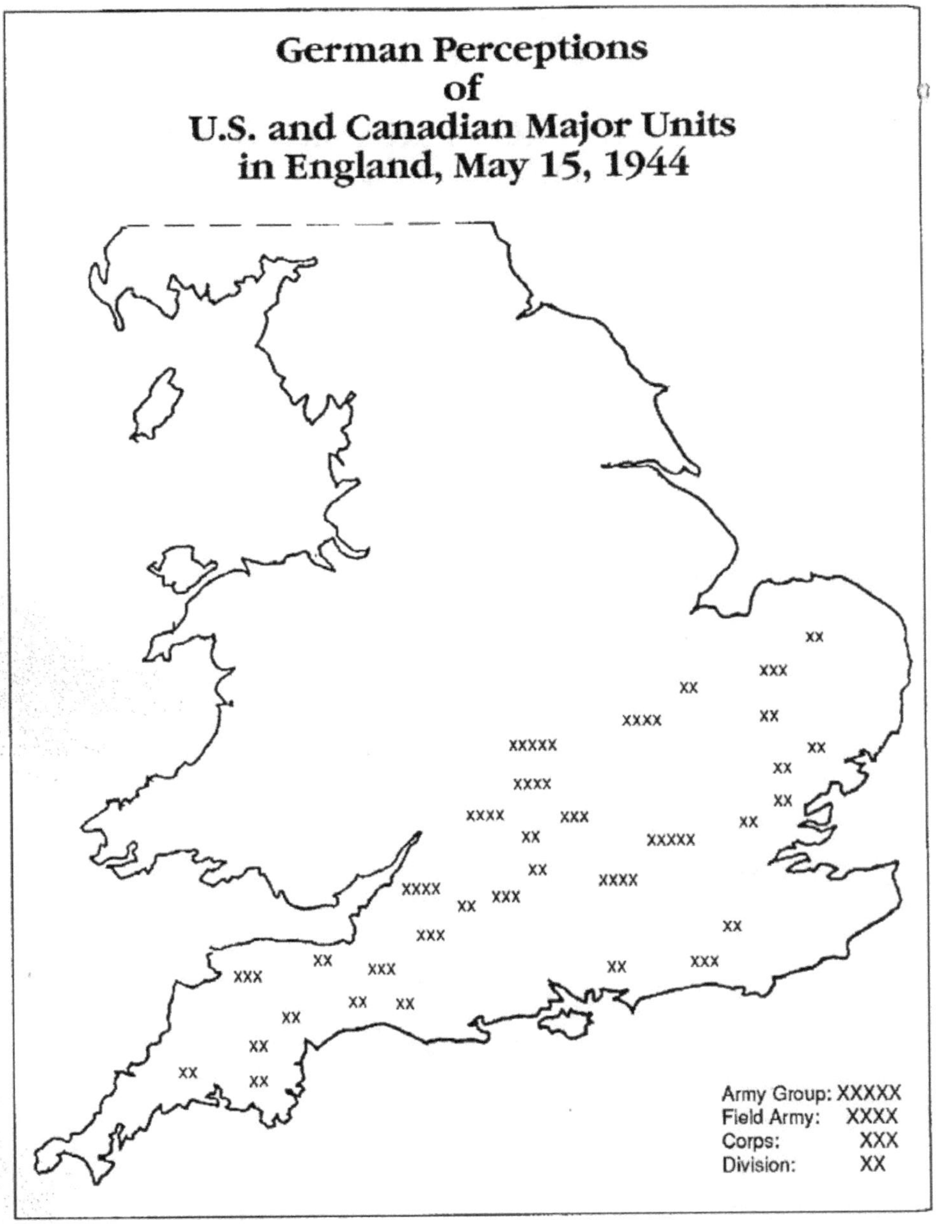

Actual Locations of U.S. and Canadian Major Units in England, May 15, 1944

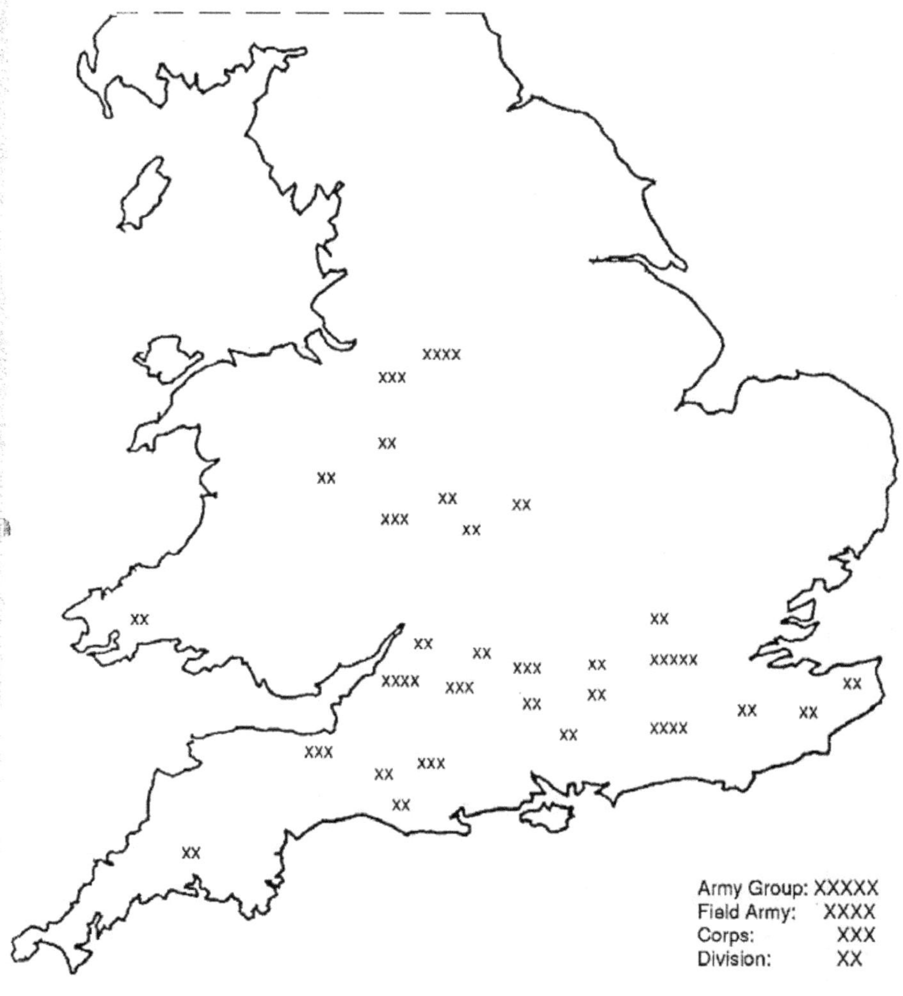

Washington 1982 "Lessons Learned" seminar on "Cover Planning" by the CIA and Joint Chiefs of Staff. L to R: Lt.Col. Roger Fleetwood Hesketh, British Army (Ret.), Prof. H. Wentworth Eldredge; Prof. R.V. Jones; William A. Casey, Director, Central Intelligence; Capt. Douglas Fairbanks, Jr., U.S. Navy (Ret.); Maj.Gen. William Baumer, U.S. Army (Ret.); Maj.Gen. William A. Harris, U.S. Army (Ret.) (the author)

GARBO's greatest feat was his long transmission (122 minutes) sent on June 9, three days *after* the landings, in which he insisted that the landings in Normandy were indeed a feint and that FUSAG, now enlarged, he said, to 75 Divisions, remained the true threat. Post-war study of the German records reveals that General von Rundstedt as well as Hitler himself remained convinced that the Pas de Calais was the objective. Several new Divisions were sent there instead of to Normandy, and the existing force of Panzers held in place. All told at least 15 German Divisions were held back for a fortnight or longer.[7]

7) Miscellany. A ten-mile strip along the entire southern coast of England was closed to civilian traffic. Airmail between England and Lisbon was halted in March, travel to Ireland banned in April, and all Service leaves stopped. Civilian mail was arbitrarily delayed. Rumors were floated through the diplomatic community, and all outgoing mail from foreign embassies was ostentatiously censored, causing much gnashing of teeth among the "cookie-pusher" set.[8]

Filling in the Blanks

I was, of course, directly involved in only a fraction of these actions, but picked up a lot of data by hearsay at the time and through books and articles written subsequently by some of my FORTITUDE associates, many of whom, such as Lt. Commander (later Captain) Douglas Fairbanks, Jr., of the U.S. Navy, were unknown to me at the time. He was brought in as American liaison for several feinting maneuvers by the Royal and U.S. Navies in the English Channel.[9]

My digs in London were in the DeVere Hotel on Kensington High Street, where I had one scruffy little room too high up (bomb bait). In the "little blitz" of early 1944 I once got under my bed to get out of the way of flying glass. During my first days on the job nothing happened at all as I slowly got my bearings. Ingersoll took me to Norfolk House, where the basic work on OVERLORD planning itself was carried on and which had important links with FORTITUDE. For FORTITUDE itself, British Lt. Col. Jervis Reade was in charge of the dummy physical things, while his compatriot Lt. Col. Roger Fleetwood Hesketh was in charge of "Special Means" (double agents). Reade and I had little in common but I got along famously with Roger who was about my generation and died two years ago. His "history" of FORTITUDE, still unpublished though widely quoted, is the best single summary. Shortly I was taken to the "holy of holies" where the very, very secret London Controlling Section (LCS) was housed underground at Great George Street—15 feet from Churchill's war room and emergency cot.

Next to the Manhattan Project (the making of the atom bomb), the actual story of FORTITUDE was the best-kept secret of WW II. The story of the Manhattan Project is well-known, with Soviet espionage filling in the few official gaps. On the other hand the British have slowly buried the records of cover planning, even with the full story of ULTRA pretty much revealed. I know, as I tried to get some information in 1986 from the British Cabinet Secretary and was told very politely but firmly that "in its present state" the manuscript on the London Controlling Section was not available. Its author, the noted war historian and Oxford don Sir Michael Howard (formerly a Coldstream Guards officer), suggested that I add my voice to those who want it released. He said I would have no luck, but he would be grateful if I tried, adding that he himself could not have a copy of his own manuscript. I have heard since that the barrier is being relaxed and some portions of his study will soon be published.

Col. John H. Bevan, British Army, was a most remarkable fellow. "Johnnie," as he was finally known to me, had been in WW I and afterward spent a year in banking in Germany. Later he prospered as a very high-level broker in London (it was rumored that he invested most of the Royal Family's holdings). You did not ask that sort of question from Johnnie; in fact you did not ask any questions at all if you were smart. In WW I Johnnie had been part of a pioneering British intelligence program. At a big house in the country a complete German Hq. staff was formed of Englishmen who spoke German and understood Germany very well. They wore German uniforms and decorations, atc German food and spoke only in German. And they had a complete German staff to serve them. Their task was to puzzle out what would be the next German move. I do not know how long this lasted or how successful it was. Like all mysterious things British it seems to have slipped down through the cracks of history.

Making it Convincing

But now it was WW II and I was trying to tell Johnnie and Roger juicy and true-sounding things to enhance their understanding and impression of the American army now growing rapidly in Britain. Roger wrote in English while drafting what the double agents would transmit to Germany. Tommy Harris and GARBO would turn Roger's messages into Spanish for transmission via Portugal or Spain. I was worried how I could continue to be effective with no way to check the results. Similarly, the officers running BRUTUS, a Polish flying officer, and TRICYCLE, an elegant Yu-goslav, came to check with me about their uncertainties. Eventually I went to Bevan and said I could not go on in good conscience without some concrete evidence that this "chicken feed," as the Brits called it, was believed by the German General Staff. Johnnie tried to placate me by saying the Krauts were running aerial recon and had a pretty good idea from that. I replied that German aerial recon was almost non-existent and the explanation would not wash, adding that I would quit unless I were shown some real evidence. Johnnie gave me a stern look but said to come back in a couple of days, which I did. He told me he had arranged for me to see Col. Telford Taylor, U.S. Army (actually OSS) at 20 Grosvenor Square, where G-3 Special Plans was located.

I went to see Taylor, who said pleasantly enough, "Hello, Eldredge," then stopped and started again when all the curtains were drawn. His manner changed and he said very seriously that *I would be shot if I revealed what he was going to tell me.* This of course brought me up sharp and I replied, "Yes, Sir." He then told me the main German codes had been broken and the information gained regularly thereby was known as ULTRA. I thought "No More Drinking" and hoped I did not talk in my sleep! Taylor went on to be the Assistant Prosecutor at the Nuremberg Trials and I did not see him again until two or three years later when we met eating spare-ribs in the Shoreham Hotel in Washington. He smiled broadly as he said, "Hello, Eldredge."

Maj. H. Wentworth Eldredge, USAF in London, 1945. (the author)

So I went to Johnnie and asked to see some recent traffic. Roger, who of course was already in the know about ULTRA, then showed me the traffic that was relevant to my job. "Need to know" was the watch word and I did not ask questions that did not clearly concern me (except now and then!).

One early assignment involved a deception known as FORTITUDE NORTH—a fake army in Scotland aimed at Norway to pin down the German troops there. To inspect it I went to Edinburgh with several other officers from Norfolk House. Next I hopped across the Irish Sea in a British C-47 to talk with an American major general who was forming a corps which later became part of Patton's real army. This corps had as its basis parts of the 5th Infantry Regiment, previously in Iceland. I told him this was carried by the Krauts as the 55th Infantry Division, which did not exist. He readily agreed when I explained that we would send some signal corps men to build up the normal radio traffic of a division settling down. Later we would have two fake Divisions follow it; he was not to use radio traffic until he heard from us and he was not to tell his staff anything about this deception. FORTITUDE NORTH, though dropped from the main FORTITUDE plan quite early, kept 300,000 Krauts deployed on the Norwegian Coast who could have been transferred to France.[10]

Late in the spring of 1944 deep gloom suddenly fell over all the Brits. No one volunteered much to me, but I gathered that one of our most important double agents (TRICYCLE) might be compromised, threatening the whole cover plan. It turned out that TRICYCLE, whose real name was Dusko Popov, came under German suspicion but was cleared. Not so ARTIST, a young, disillusioned German officer named Johann Jebsen operating for the British out of Lisbon: He was spotted by the Nazis and slain.[11] To play things safe, TRICYCLE was taken off FORTITUDE entirely, and we relied more and more on GARBO.

It is now clear to all students of the complex and enigmatic course of FORTITUDE that GARBO's long message of June 9, in Hesketh's own words, "tipped the balance"[12]

That same pregnant June 9, as discovered by MAGIC (the U.S. breaking of the Japanese PURPLE code), the Japanese Ambassador, who was more Nazi than the Nazis themselves, had a long talk with Hitler. The Japanese Military Attache then reported to Tokyo, saying in part ". . . because one separate Army Group is stationed on the southeast coast of Britain, it is expected that plans will be made for this to land in the Calais and Dunkirk areas."[13]

Another bit of convincing evidence that FORTITUDE had worked was the Allied capture in Italy of a German map dated May 15 and showing FUSAG heavily deployed in East Anglia and threatening the Pas de Calais (see Map).

On May 25, 1945, after the German surrender, Billy Harris, Ralph Ingersoll and I, speaking for "Special Plans Branch, G-3, 12th Army Group," submitted a seven-page Top Secret "Informal Report" on FORTITUDE to Joint Security Control. One paragraph read;

The success of the FORTITUDE plans is a monolithic fact in any appreciation of the effectiveness of Cover and Deception. There is no doubt whatever that enormous losses were avoided by the successful achievement of surprise, which was the deception's first objective. The defeat and capture of a large part of the 15th German Army while retreating from the Pas de Calais rests on the military foundation of having been pinned down, out of the battle, by wholly imaginary assault forces dreamed up by the deception planners and implemented by deception means on the east and northeast coasts of England. Enemy forces were also held in Norway and Southern France.

Bibliography & Notes

FORTITUDE was only one segment of an over-all deception program against Germany established by the U.S. Joint Chiefs of Staff and code-named BODYGUARD. Books on the subject tend to include FORTITUDE as only one chapter. The two such books whose treatment of FORTITUDE comes closest to being "complete" are Charles Cruikshank's *Deception in World War II* (Oxford University Press, London 1979), and *Strategic and Operational Deception in the Second World War* (Frank Cass, London 1987), containing a chapter by T. Cubbage entitled, "The Success of Operation FORTITUDE: Hesketh's History of Strategic Deception," which quotes extensively from Hesketh's own unpublished book. Other books and sources are also cited below.

1. Wheatley, Dennis. *The Deception Planners, My Secret War.* (Hutchinson, London 1980), pp. 180-81 & 188-191.
2. Cruikshank & Cubbage, *op cit.*
3. In Roy Hoopes's *Ralph Ingersoll, A Biography* (Atheneum, NY 1985) he is quoted (p. 278) as claiming to have persuaded American industry to make rubber tanks and planes to replace wooden ones fabricated in Britain. This is by no means certain since the British were already using rubber tanks as decoys in North Africa. Eldredge is quoted on the same page as saying, "Ingersoll was the trickiest, most elusive person I've ever dealt with. I've never met anyone who was such a bright guy who was such a goddamned liar. He'd say anything to get what he wanted."
4. Statistics cited come from Craven & Cate, *The Army Air Forces in World War II*, Vol. III, pp. 85-106. Greatly more detailed figures and background are available at the Air Force Historical Research Center, Maxwell AFB, Alabama, including records of all missions by the Eighth and Ninth Air Forces as well as documents about FORTITUDE targets of interdiction in the Pas de Calais (#519.323-22) and FORTITUDE plans and directives (#505.61-21).
5. Jones, R. V. *The Wizard War*, British Scientific Intelligence, 1939-1945 (Coward, McCann & Geoghegan. NY 1978) pp. 400-405, 411 & 415.
6. Jones. *op cit*, pp. 405-406.
7. Besides GARBO's own book, *Operation GARBO: The Personal Story of the Most Successful Double Agent in WW II*, co-authored by Nigel West (Random House, NY 1985), see Wheatley, *op cit*, pp. 181 and 188; Cruikshank, op cit, p 217; F. W. Winterbotham, *The ULTRA Secret* (Harper & Row, NY 1974) pp. 121-137; and Sir J. C. Masterman, *The Double Cross System in the War of 1939 to 1945* (Yale University Press, New Haven, CT 1972) pp. 146-158.
8. Wheatley, *op cit*, p. 190.
9. Cruikshank. *op cit*, p. 214.
10. Cruikshank *op cit*, says on p. 112, "At the beginning of March there were 18 divisions in Scandinavia . . . in the middle of May an additional first-class division was sent to Norway."
11. GARBO. *op cit*, p. 122.
12. *Strategic and Operational Deception in World War II*, *op cit*, p. 340.
13. Masterman. , p. 156.

CHAPTER 32

Soviet Deception in the Cuban Missile Crisis—Learning from the Past[1]

James H. Hansen

Hansen was a 30-year veteran of DIA, serving as a senior intelligence analyst until 1990. He then joined the CIA as an analyst and operations officer.

Moscow's surreptitious dispatch of nuclear-capable SS-4 and SS-5 surface-to-surface missiles to Cuba in 1962 upset the strategic balance in an alarming way.[1] The resulting showdown—which the Russians call the "Caribbean Crisis" and the Cubans call the "October Crisis"— brought the world to the brink of nuclear war. From its inception, the Soviet missile operation entailed elaborate denial and deception (D&D) efforts. The craft of denying the United States information on the deployment of the missiles and deceiving US policymakers about the Soviet Union's intent was the foundation of Nikita Khrushchev's audacious Cuban venture. Piecing together the deception activities from

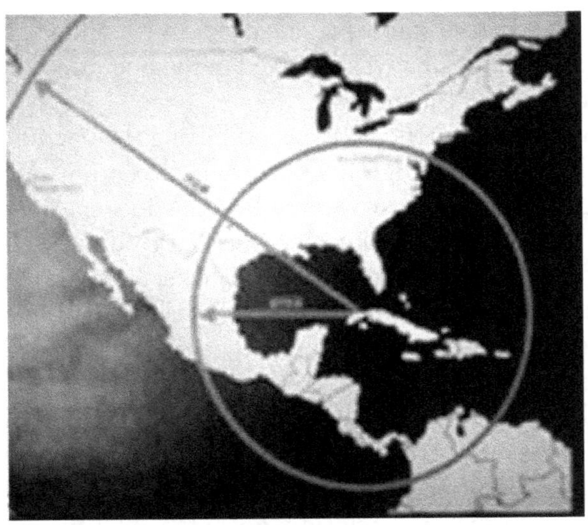

Figure 1 Range of Soviet SS-4 medium-range ballistic missiles and SS-5 intermediate-range ballistic missiles, if launched from Cuba.

1. James H. Hansen, "Soviet Deception in the Cuban Missle Crisis," *Studies in Intelligence*, Vol. 46, No.1, 2002, pp. 49–58. Retrieved at: https://www.cia.gov/library/center-for-the-study-of-intelligence/csi-publications/csi-studies/studies/vol46no1/article06.html.

declassified US, Russian, and Cuban accounts yields insights that can help us anticipate and overcome the D&D efforts of a growing number of foreign adversaries today.

Maskirovka

Moscow has always had a flair for D&D, known in Russian as maskirovka. Its central tenet is to prevent an adversary from discovering Russian intentions by deceiving him about the nature, scope, and timing of an operation. Maskirovka covers a broad range of concepts, from deception at the strategic planning level to camouflage at the troop level.[2] Russian military texts indicate that maskirovka is treated as an operational art to be polished by professors of military science and officers who specialize in this area.

DIA analysis preceding the missile crisis noted that the Soviet Army had probably employed large-scale battlefield deception "more frequently and with more consistent success than any other army."[3] The Soviets practiced extensive maskirovka before their move into Czechoslovakia in 1968. Moscow also trained foreign forces to apply deception, including North Vietnamese units before the Tet offensive in 1968 and Egyptian forces before crossing the Suez Canal in 1973.

Close-hold Planning

Gen. Anatoli Gribkov—then a senior member of the Soviet General Staff—provides revealing insights into the early planning of the operation. He says that, after Nikita Khrushchev decided to emplace the missiles in Cuba in the spring of 1962, the General Staff detailed only five officers—four generals and a colonel—to serve as the center of the military planning apparatus. Col. Gen. Semyon Ivanov, chief of the General Staff's Chief Operations Directorate, was in overall charge.[4] During that summer, the circle of collaborators and contacts expanded to include members of each of the relevant service branches, but secrecy and need-to-know prevailed. The most senior officers brought into the plan were at least told that Cuba was involved in the operation, but only a few were informed of the exact nature of the mission.[5]

The top civilian and military officials conceptualizing the operation did not see eye-to-eye about the likelihood of pulling off a successful deception. At the very center of those making the decisions stood First Deputy Prime Minister Anastas Mikoyan, Presidium member Frol Kozlov, Defense Minister Marshal Rodion Malinovsky, and Marshal Sergei Biryuzov, commander of the Strategic Rocket Forces.[6] Alternate Presidium member Sharaf Rashidov was brought in as well, possibly for the cover that he later provided for traveling delegations.[7] Biryuzov and his experts believed that the deployment could be made expeditiously and secretly, without the US discovering the missiles. Mikoyan was surprised at this judgment and believed the marshal to be a fool.[8] Rashidov was confident that the missiles could be hidden, claiming that they could be placed so as to blend in with the palm trees. Gribkov held that only somebody inexperienced in military matters could reach such a conclusion, given the extensive preparations needed for each missile site.

Throughout the early planning stage, no secretaries were used to prepare final typed texts. A colonel with good penmanship wrote the proposal that the Defense Council adopted. It grew into a full-fledged plan, still handwritten, which was approved by Malinovsky on 4 July and Khrushchev on 7 July.[9] From May through October, for reasons of security, no communications about the proposed, planned, and actual Soviet deployments in Cuba were sent, even by coded messages. Everything was hand-carried by members of the small coterie of senior officials who were directly involved.[10]

Developing a Cover Story

The General Staff's code name for the operation—ANADYR—was designed to mislead Soviets as well as foreigners about the destination of the equipment. Anadyr is the name of a river flowing into the Bering Sea, the capital of the Chukotsky Autonomous District, and a bomber base in that desolate region. Operation ANADYR was designed to suggest to lower-level Soviet commanders—and Western spies—that the action was a strategic exercise in the far north of the USSR. Promoting the illusion, the troops that were called up for the Cuban expedition were told only that they were going to a cold region. Those needing more precise instructions, such as missile engineers, were informed that they would be taking ICBMs to a site on Novaya Zemlya, a large island in the Arctic where nuclear weapons had long been tested.[11]

To strengthen the concealment, many units were outfitted with skis, felt boots, fleece-lined parkas, and other winter equipment.[12] Moreover, perhaps to further backstop the cover plan, Moscow tapped four ground forces regiments from the Leningrad Military District in the north for dispatch to Cuba. The deception was so thorough that it fooled even senior Soviet officers sent to Cuba. One general there asked Gribkov why winter equipment and clothing had been provided. The general admonished him to "think like an adult," and explained, "It's called ANADYR for a reason. We could have given away the game if we had put any tropical clothing in your kits."[13]

Getting the Cubans on Board

Secrecy surrounded the first Soviet delegation that went to propose the audacious plan to Fidel Castro and other Cuban leaders. The officials arrived in Havana with little fanfare on 29 May, amidst a delegation of agricultural experts headed by Rashidov. The group included Col. Gen. Ivanov and several missile construction specialists and other military experts, whose job it was to determine whether the missiles could be deployed in secrecy.[14] Ambassador Aleksandr Alekseev took Cuban Defense Minister Raul Castro aside to explain that "Engineer Petrov" in the group actually was Marshal Biryuzov, and that he needed to meet with el lider maximo without delay. Only three hours later "Engineer Petrov" was shown into Fidel Castro's office.[15] The Cuban leadership unanimously and enthusiastically gave its approval in principle.[16]

Soviet maritime policy began to shift in accordance with these first trips. In June and July, the USSR began to charter Western ships to carry general cargo from the Soviet Union to Cuba, reserving its own freighters for carrying military cargo.[17]

During 2-17 July, a Cuban delegation led by Raul Castro traveled to Moscow to discuss Soviet military shipments, including nuclear missiles. Khrushchev met with the Defense Minister on 3 and 8 July. Raul Castro initialed a draft treaty with the Soviet Defense Minister that governed the deployment of Soviet forces to Cuba. This pact was not to be publicly revealed until a visit that Khrushchev planned to make to Cuba in November.[18]

The Russians began to dispatch officers and specialists covertly to Cuba by air. On 10 July, Gen. Issa Pliyev, traveling under the name "Pavlov," arrived in Cuba to command the Soviet contingent.[19] Two days later, 67 specialists touched down. They journeyed as "machine operators," "irrigation specialists," and "agricultural specialists." Their covers, however, could not have withstood probing—they had been assigned to occupations about which they knew nothing. They were urged to consult the few genuine specialists traveling with them to gain some rudimentary knowledge of their ostensible jobs.[20] On 17 July, Havana announced that Cuba and the USSR had signed an agreement establishing a regular Moscow-Havana civil air route. US intelligence analysis at the time speculated that the new Tu-114 flights were bringing Soviet military officers and sensitive electronic and signal-monitoring equipment to Cuba.[21]

Then-Minister for Industry Ernesto "Che" Guevara and the head of the Cuban militia led another delegation to Moscow during 27 August-2 September. The purpose was to introduce Fidel Castro's revisions into the draft treaty. The Cubans proposed that the deployment be made public in order to head off any American overreaction; Khrushchev, however, successfully argued for continued secrecy.

Keeping the Secret at Home

In the Soviet Union, the men and equipment destined for Cuba were assembled, loaded, and moved by rail at night under reinforced guard. The train routes and final destinations were kept secret. Mail and telegrams along the way were strictly prohibited.[22]

To mask the immensity of the overall effort, the shipments to Cuba left from eight ports—four in the north (Kronstadt, Liepaya, Baltiysk, and Murmansk) and four on the Black Sea (Sevastopol, Feodosiya, Nikolayev, and Poti).[23] Western access to these ports was closed off. It was normal for the Soviets to close ports when munitions were being loaded, but this time the surface-to-surface missiles were being put on the ships under tight security and cover of darkness.[24]

The troops were housed at nearby military facilities during the two or three days required to load a ship. Guards were posted to prevent anyone from leaving the area. No letters, telegrams, or telephone calls were permitted, a rule that also applied to the officers.[25] The ships' crew members, some of whom made more than one run to Cuba, were forbidden shore leave and correspondence.[26] Secrecy was so strict that couriers carried all messages between the ports and the Defense Ministry in Moscow.[27]

On board, the Soviets applied the same maskirovka measures that they had adopted when they first began to send weapons to Cuba. Packing crates or special shipping containers concealed and protected weapons carried as deck cargo. Certain telltale military equipment was boarded up with planks to make it look like the ship's superstructure. Even on-deck field kitchens were disguised.[28] The Soviets shielded crated military hardware—such as missiles and launchers—with metal sheets to defeat infrared photography.[29] They stored other combat and specialized equipment below, out of sight. Ordinary automobiles, trucks, tractors, and harvesters were placed on the top deck to convey the impression that only civilian and agricultural gear was being transported.

The freighter Poltava, which sailed to Cuba in September, was a good example. No external signs indicated that it was carrying missiles. On deck were cargo trucks, none of which were associated with the missiles. Nonetheless, some US experts speculated that the ship might be carrying ballistic missiles deep in its hold, because the Soviets tended to use large-hatch ships of the Poltava and Omsk classes to deliver such missiles.[30]

The ship captains were not told where their cargoes were to be delivered. Before casting off, the captain and the troop commander jointly received a large sealed envelope. Unfastening it, they found a smaller envelope to be opened only at a certain set of geographic coordinates in the Atlantic Ocean. When they reached the designated point, an officer from the KGB's Special Department joined them for the opening of the envelope. The instructions told them to proceed to a Cuban port and authorized them to inform the ship's company of the destination.[31] The concern for secrecy permeated the process. The last sentence of the captain's letter read: "After familiarizing yourself with the contents of this document, destroy it."[32]

Every ship involved in Operation ANADYR carried thick folders, prepared by Defense Ministry staff officers, which contained background information on a number of countries with which the USSR had good relations. The study materials on Cuba were buried in these packets, so that not even the compilers would know the real focus of the operation.

Secrecy in Transit

The Soviet ships made false declarations when they exited the Black Sea and the Bosporus. Cargo records were altered and the tonnage declared was well below what was being carried.[33] The ships would declare from Odessa, although they had loaded at other ports. Often ships going to Cuba listed Conakry, Guinea, as their destination. When the volume of traffic increased, a number of ships did not give their destinations but simply stated that they were carrying "general cargo" and "awaiting orders."[34]

Transit through the Bosporus and the Dardanelles Straits presented a special challenge. Not only were the soldiers kept below decks, but the captains were under orders to prevent any foreigners from boarding, even the Turkish pilots who usually guided civilian ships through those tricky waters. Whenever the pilots approached the Soviet ships, the Soviet crews would lower bulging parcels of vodka, brandy, caviar, sausages, and other delicacies. Gribkov noted that this transparent bribery worked well: "Everyone likes to get presents, even pilots."[35]

The captains were instructed to take all possible evasive action in the event of attacks or an effort to board their ships. Should evasive action fail, they were to "destroy all documents with state and military secrets," take measures to protect the personnel, and sink the ships.[36] Should their vessels experience mechanical failure en route, the captains were to explain to ships offering assistance that they were exporting automobiles.[37] Had this occurred, it might have provided clues—the USSR had few cars of any kind and was not recognized as an automobile exporter.

Moscow also resorted to diplomatic means to reduce US reconnaissance of the ships en route. In July 1962, the Soviets described US reconnaissance missions in international waters as "harassment," and requested through their GRU officer in Washington, Col. Georgi Bolshakov, that these flights be stopped for the sake of better bilateral relations.[38] In retrospect, this overture clearly appears to have been an effort by Khrushchev to delay the discovery of weapons related to Operation ANADYR. Bolshakov met with Attorney General Robert Kennedy more than a dozen times.

Most of the voyages lasted from 18 to 20 days. Due to strict maskirovka measures, the troops were kept below decks except for a few minutes at night when small groups were allowed to exercise and get some fresh air.[39] During the tropical days, heavy tarpaulins covered the hatches to the lower decks where the troops were berthed. With little air circulation, the inside temperature climbed to 120 degrees Fahrenheit or higher. Rations were issued twice a day and only in darkness.[40] Many of the troops on board swore that they would never again set foot on a ship.

Although the restrictions made conditions on board nightmarish, the deceptions worked. Gen. Gribkov states that "US intelligence discovered neither the true significance of the surge in Soviet shipping to Cuba nor the mission of our troops on the island until nearly all the men had come ashore and, still moving in large numbers only by night, had been deployed to their assigned positions."[41]

Unloading in Cuba

As the Soviet troops arrived, Cuban officials took steps to support Moscow's maskirovka plan. In early fall, they began to exert control over the movements of all foreigners on the island. News reporters and foreign embassy personnel were forbidden to travel outside Havana. In the city, Cuban agents surveilled and harassed foreigners, especially British embassy officials.[42]

The planners had selected 11 Cuban ports to receive the Soviet ships: Havana, Mariel, Cabanas, Bahia Honda, Matanzas, La Isabella, Nuevitas, Nicaro, Casilda, Cienfuegos, and Santiago de Cuba.[43] They earmarked three of them—Bahia Honda and Mariel on the northwest coast and Casilda on the south coast—to receive the surface-to-surface missiles and nuclear warheads.[44]

Even before the Soviet ships approached Cuban ports, a number of maskirovka precautions had been implemented. At Mariel, for example, the Soviets built a large cinder-block wall around the unloading area so that none of the port activity could be observed by land-based agents.[45] As the ships lay in port, KGB officers kept watch on deck. All Cubans, even militiamen, were barred from the port areas.[46] Local inhabitants within a mile of the waterfront in Mariel had to evacuate their homes.[47]

The first SS-4 missiles arrived in Mariel on board the Omsk on 8 September. The Indigirka brought the initial shipment of nuclear warheads on 4 October.[48] According to one source, this ship carried 99 nuclear charges—some two-thirds of all nuclear weapons sent to Cuba and over 20 times the explosive power dropped by all Allied bombers on Germany throughout World War II.[49]

Most of the military technicians also came ashore at Mariel. Deception activities throughout the transit stage and the strict security measures at Mariel hindered the ability of US intelligence agencies to estimate the number of Soviet troops. The plan for ANADYR that was approved in early July had called for moving 50,874 men. That total included personnel for field hospitals, bakeries, mechanical workshops, and other support units, all with a three-month supply of food and fuel. During September, the plan was revised to eliminate submarine and surface ship squadrons, due to potential resupply problems and concerns that their presence might sound an alarm bell in Washington. By late October, the size of the contingent in Cuba had reached about 41,900 personnel—quadruple the size that US intelligence agencies figured.[50]

Nonetheless, in the hectic initial days in Cuba, secrecy created more than a few glitches. The General Staff had neglected to provide passwords to facilitate communication between the arriving transport ships and the Cuban greeting parties. Accordingly, some ship captains and on-board troop commanders had difficulty accepting orders to reroute their ships from their originally assigned ports. The captain of one ship even turned back out to sea rather than allow a Cuban patrol boat crew to come aboard to guide him to his anchorage.[51]

Usually two or three days were required to unload a ship with military cargo, and maskirovka requirements invariably complicated the work. Equipment that had at least a superficial resemblance to agricultural machinery was unloaded in broad daylight, but weapons and other military equipment could be unloaded only at night. From the docks, specialized equipment was stored in sheds or moved directly to designated bases along back roads at night.[52]

All this time, Radio Moscow was claiming that the USSR was only giving Cuba "machine tools, wheat, and agricultural machinery," along with "some 7,000 tons of various fertilizers."[53] This description was consistent with the false identities provided to many of the Soviet military specialists and also with the daytime unloading activity.

Movement to Field Sites

The maskirovka measures were not air tight. In the initial stages of the operation, the United States received reports from friendly nations, newspaper correspondents, and other sources indicating that hundreds of Russian troops in fatigues had been seen in Havana and in seemingly endless convoys along Cuba's main highways. Many young Russian men also had been observed sightseeing in the Cuban capital in checked, cotton shirts and cheap trousers.[54] Although the Soviets and Cubans took extra precautions to keep gawkers away from the wharves and moved the nuclear cargoes away under black canvas and escorted by heavy guard, the chatty Cubans gave a steady stream of clues to US SIGINT collectors.[55]

At the same time, the Soviets and Cubans mounted a major campaign using HUMINT channels to bolster the overall deception effort. The planners leaked accurate information about the deployment so as to mask it. The information was funneled through counterrevolutionary organizations and their press in the United States, especially in Miami. The CIA discounted the information, because it did not consider the groups and people peddling it to be credible. This strategy was highly effective, according to a former Cuban intelligence officer.[56]

The deception campaign that exploited the émigrés' lack of credibility was unwittingly backstopped by correspondence between Cubans and their friends and relatives in the United States. From June to September, Cuban intelligence intercepted some 17,000 letters that had something to say about the deployment of Soviet troops and missiles in Cuba. In late September, Cuban authorities permitted those letters to arrive in Miami as part of the deception campaign. Just as Havana expected, the CIA paid no attention to these letters.[57]

For US intelligence analysts, the amount of "noise" from Cuba grew deafening. Reports flooded in from Cubans, tourists, foreign diplomats in Cuba, and newspaper officials reporting in a private capacity. At the CIA focal point at Opa-Locka, Florida, intelligence officers screened countless reports and debriefed Cubans who had fled the island.[58] Most of the reports from Cuba were exaggerated or imaginary—some were so outrageous that they were laughable and made all the others suspect. There were far-fetched tales of African troops with rings in their noses, lurking Mongolians, and even Chinese troops. These accounts followed earlier erroneous reports of Soviet military equipment secreted away in caves, underground hangers, and concrete domes. The previous reports had cast doubt on the reliability of sources, so US analysts found it easy to dismiss the stream of reports of Soviet missiles.[59]

From the port areas, the canvas-covered SS-4 missiles were moved in night convoys, under tight security, to sites in the interior of the island. Security was tightened so that the troops disembarked dressed as civilians, and their escorts—Soviet personnel who had arrived earlier—were required to wear Cuban military uniforms and issue commands along the convoy routes only in Spanish.[60] On the march or bivouacked, Soviet military men remained dressed in civilian clothing and were forbidden to mention their military designations or the ranks of their commanders. Moreover, all communications between the Soviet military headquarters in Havana and units in the field had to be made in person, not written or sent by radio. Except for very brief hookups and equipment tests, Soviet troops maintained total radio silence in order to mask their identity, location, and troop strength from US intelligence.[61]

In retrospect, some Soviet and Cuban officials found it remarkable that the operation remained secret for a full month after the missiles arrived in Cuba.[62] The missile carriers were too big to go unnoticed on the back roads of the island for long. As they rumbled through the little Cuban towns, they left a trail of downed telephone poles and mailboxes. When a peasant's shack had to be moved or knocked down to allow a missile carrier to turn a tight corner, those who witnessed the event were bound to talk. Soviet and Cuban efforts to discredit such anecdotal accounts paid off.

Disingenuous Diplomacy

Soviet spokesmen kept up a steady stream of denials and disinformation in September. On 4 September, Ambassador Anatoli Dobrynin sought out Robert Kennedy and stated that he had received instructions from Khrushchev to assure the President that there would be no surface-to-surface missiles or offensive weapons placed in Cuba. Dobrynin also added that the Attorney General could assure his brother that the Soviet military buildup was not of any significance.[63] On 6 September, Theodore Sorenson, special counsel to President Kennedy, met with Dobrynin, who reiterated his assurances that Soviet military assistance to Cuba was strictly defensive in nature and did not represent a threat to American security. The following day, Dobrynin assured US Ambassador to the United Nations Adlai Stevenson that the USSR was supplying only defensive weapons to Cuba. On 11 September, TASS announced that the USSR neither needed nor intended to introduce offensive nuclear weapons into Cuba.[64]

In late September, Khrushchev embarked on a barnstorming tour in the Turkmen and Uzbek republics. This high-profile trip, which extended into the first week of October, emphasized agricultural themes. In none of Khrushchev's many speeches during his travels was there any reference suggesting aggression or threats to the United States.[65]

The pattern continued. On 13 October, a high State Department official, Chester Bowles, questioned Dobrynin on whether Moscow intended to put offensive weapons in Cuba; the Ambassador denied any such intention. On 17 October, GRU Col. Bolshakov brought Robert Kennedy a personal message directly from Khrushchev to President Kennedy that "under no circumstances would surface-to-surface missiles be sent to Cuba." The next day Foreign Minister Gromyko met with President Kennedy for two hours. Gromyko assured him that the Soviet aid to Cuba "pursued solely the purpose of contributing to the defense capabilities of Cuba and to the development of its peaceful economy."[66]

Denouement

The missile sites themselves could never have remained hidden for long. They were constructed in areas expropriated from Cuban landowners, had no fences or walls, and

Figure 2 U2 photograph of SS-4 missile site in Cuba.

were exposed to aerial observation.[67] Standard maskirovka doctrine gave preference to deployment in wooded areas, yet Cuba's forests were generally sparse, consisting of a few clusters of palm trees or a thick undergrowth of bushes. Such vegetation could not cover all of the missile equipment.[68] SS-4 launchers are anchored to large concrete slabs and surrounded not only by the missiles, but also by multiple buildings, fuel trucks and tanks, and hundreds of meters of thick cable. To try to maintain secrecy, Soviet commanders forbade their troops from taking any leave from their deployment sites and ruled out using Cuban labor. Nonetheless, Soviet commanders and planners knew that although the tractor-trailers and associated large objects could be covered by canvas, their masses could not be shrunk.[69] Heavy equipment might obscure part of the missile site signature from ground-level, but from above it stuck out markedly.[70]

On 14 October, a U-2 aircraft photographed the area of San Cristobal, where the first missile unit was being deployed. In only six minutes, US Air Force Maj. Richard Heyser snapped 928 photographs that yielded the first confirmation of offensive missiles in Cuba.[71] Washington stepped up intelligence collection of all kinds, readied massive air attack and invasion plans—including sending nuclear-armed B-52s aloft—and engaged in extensive policy deliberations in the Executive Committee. On 22 October, President Kennedy revealed the missile buildup to the world. Confronted with the photographic evidence, the Russians informed Raul Castro that more attention would have to be paid to concealing the site work and camouflaging the missiles and other heavy equipment.[72] The Soviet units stretched tarpaulins and nets over the missiles, and daubed paint or mud across the canvases. This marked the first time that they tried to conceal their missiles from the air, probably hoping to mask the total number of missiles and protect against sabotage.[73] By 28 October, however, the confrontation, including Kennedy's imposition of a naval and air quarantine on the shipment of offensive military equipment to Cuba, led Khrushchev to agree on a formula to end the crisis. The Russians began to dismantle their bases.

In Conclusion

The Soviet deception effort was comprehensive, but not free from shortcomings. The early, overly optimistic assessments by Marshal Biryuzov and Rashidov evidently went unchallenged. Some cover arrangements were slapdash. Many of the slips occurred not in the USSR but in Cuba, when Soviet units had to unload their weapons, transport them to the field, and set them up. The operation might have been enhanced by the presence of maskirovka specialists in all Soviet units in Cuba. In the 1980s, a special maskirovka directorate was created within the General Staff. Such an organization would have played a vital role had it existed in 1962.[74]

On 4 February 1963, the President's Foreign Intelligence Advisory Board issued a major postmortem report over the signature of its chairman, James R. Killian, Jr. The Killian report described the introduction and deployment of Soviet strategic missiles in Cuba as a "near-total intelligence surprise."[75] It concluded that the Intelligence Community's analysis of intelligence indicators and its production of current intelligence reports "failed to get across to key government officials the most accurate possible picture of what the Soviets might be up to in Cuba" during the months preceding 14 October. The report took the Community to task for inadequate early warning of hostile intentions and capabilities; failure to provide senior policymakers with meaningful, cumulative assessments of the available intelligence indicators; and failure to produce a revision of the erroneous National Intelligence Estimate (NIE 8-3-62) of 19 September 1962.[76]

Nowhere does the 10-page Killian Report mention adversarial denial and deception. Within US intelligence organizations, the awareness and systematic study of foreign D&D had not been developed, and would not emerge until some 20 years later. It is likely that with a trained, well-staffed, and deception-aware analytic corps, the United States could have uncovered Khrushchev's great gamble long before Maj. Heyser's revealing U-2 mission.

Only now, four decades later, can we uncover the extent of the use of deception in the events leading to the Cuban missile crisis. To paraphrase Sir Winston Churchill, perhaps the least-explored aspect of the crisis was the Soviet effort to cloak the truth of its strategic missile deployment within a body-guard of lies, on a scale that most US planners could not comprehend.

Footnotes

1. Cuba is approximately 145 kilometers from US shores. The SS-4 medium-range ballistic missiles, which were deployed first to Cuba, had a range of up to 2,500 kilometers. The SS-5 intermediate-range ballistic missiles had a range of up to 5,000 kilometers.
2. Jennie A. Stevens and Henry S. Marsh, "Surprise and Deception in Soviet Military Thought," Military Review, July 1982, pp. 25-35.
3. Defense Intelligence Agency (DIA), Soviet/Warsaw Pact Ground Forces Camouflage and Concealment Techniques, DDI-1100-161-78, January 1979, p. vii.
4. Gen. Anatoli I. Gribkov and Gen. William Y. Smith, Operation ANADYR: US and Soviet Generals Recount the Cuban Missile Crisis (Chicago, Berlin, Tokyo, and Moscow: edition q, inc., 1994), p. 24. This is a vital source, given Gen. Gribkov's role in planning and implementing the operation.
5. *Ibid.*
6. Raymond L. Garthoff, Reflections on the Cuban Missile Crisis, revised edition (Washington, DC: The Brookings Institution, 1989), pp. 12-13.
7. Dino A. Brugioni, Eyeball to Eyeball: The Inside Story of the Cuban Missile Crisis (New York: Random House, 1991), p. 84. The author was a key figure at the National Photographic Interpretation Center in 1962.
8. Garthoff, p. 17.
9. Gribkov and Smith, p. 24.

10. Garthoff, p. 17.
11. Aleksandr Fursenko and Timothy Naftali, "One Hell of a Gamble": Khrushchev, Castro, and Kennedy, 1958-1964 (New York and London: W. W. Norton & Company, 1997), p. 191. This valuable source relies on material from Soviet/Russian archives.
12. Gribkov and Smith, p. 15.
13. *Ibid.*
14. Garthoff, p. 15; also, Gribkov and Smith, p. 14.
15. Fursenko and Naftali, p. 186.
16. James G. Blight, Bruce J. Allyn, and David A. Welch, with the assistance of Davis Lewis, Cuba on the Brink: Castro, the Missile Crisis, and the Soviet Collapse (New York: Pantheon Books, 1993), p. 8. This source draws extensively on input from key Soviet/Russian and American officials who had a hand in the crisis.
17. Brugioni, p. 92.
18. Gribkov and Smith, p. 21.
19. Fursenko and Naftali, p. 192.
20. Gribkov and Smith, pp. 37-38.
21. Brugioni, p. 93.
22. Gribkov and Smith, p. 56.
23. *Ibid.*, p. 29.
24. Brugioni, p. 149.
25. Gribkov and Smith, p. 30.
26. *Ibid.*, p. 56.
27. *Ibid.*, p. 29.
28. *Ibid.*, p. 30.
29. *Ibid.*, p. 56.
30. Brugioni, pp. 149-150.
31. Gribkov and Smith, p. 31.
32. *Ibid.*
33. US SIGINT revealed that Soviet vessels were making false port declarations and listing less than their known cargo-carrying capacity. By late August, the National Security Agency noted that 57 voyages to Cuba had taken place in a little over a month and that some ships were on their second voyage in that period of time. See: National Security Agency, NSA and the Cuban Missile Crisis (Fort Meade, MD: NSA Center for Cryptologic History, 1998), pp. 3-4.
34. Brugioni, p. 149.
35. Gribkov and Smith, p. 57.
36. Fursenko and Naftali, p. 192.
37. *Ibid.*
38. *Ibid.*, p. 193.
39. Gribkov and Smith, p. 35.
40. *Ibid.*, p. 56.
41. *Ibid.*
42. Brugioni, p. 148. The Cubans probably knew that the British were helping the United States, which did not have formal representation in Havana.
43. Gribkov and Smith, p. 38. Gribkov's map puts Nicaro in the wrong location, apparently confusing it with Niquero on the Southeast coast.
44. Fursenko and Naftali, p. 216.
45. Brugioni, p. 150.
46. David Detzer, The Brink: Cuban Missile Crisis, 1962 (New York: Thomas Y. Crowell, 1979), p. 69.
47. *Ibid.*, p. 57.
48. Gribkov and Smith, p. 52.

49. Fursenko and Naftali, p. 217. There is conflicting source information on the number of warheads specifically for the SS-4 missiles. Gribkov states that 36 such warheads were introduced. This issue cannot be resolved based on current evidence, but 36 appears to be a likely figure as that tracks with Soviet doctrinal requirements for refire missiles.
50. Gribkov and Smith, p. 28.
51. *Ibid.*, pp. 38-39.
52. *Ibid.*
53. Detzer, p. 57.
54. Brugioni, p. 101.
55. NSA, op. cit., pp. 2-3. The US SIGINT ship Oxford was hugging the Cuban coastline at that time.
56. Domingo Amuchastegui, "Cuban Intelligence and the October Crisis," Intelligence and National Security, Volume 13, Number 3, Autumn 1998, p. 101. This is a special issue on intelligence and the missile crisis, edited by James G. Blight and David A. Welch. It is a unique collection of articles on the roles played by different intelligence services.
57. *Ibid.*
58. Detzer, p. 59.
59. *Ibid.*, p. 60.
60. Gribkov and Smith, p. 39. This maskirovka requirement is comparable to that used during the Korean conflict, when Soviet pilots were instructed to speak in Chinese while flying missions to try to fool US SIGINT units.
61. *Ibid.*
62. *Ibid.*, p. 52, and Amuchastegui, p. 101. A cautionary note here is that this article is the only published account by a former Cuban intelligence officer thus far.
63. Brugioni, p. 115.
64. Blight, Allyn, and Welch, pp. 463-464.
65. Brugioni, pp. 157-158.
66. Blight, Allyn, and Welch, pp. 465-466.
67. Brugioni, p. 150.
68. Gribkov and Smith, p. 40.
69. *Ibid.*, p. 55.
70. *Ibid.*, p. 40.
71. The details of the U-2 mission are found in Volume XI: Foreign Relations of the United States, 1961-1963, Cuban Missile Crisis and Aftermath, edited by Edward C. Keefer, Charles S. Sampson, Louis J. Smith, and David S. Patterson (Washington, DC: US Government Printing Office, 1996), p. 29. Although uncertain about the status of the weapons in Cuba at the time of discovery, we now know that only some of the nuclear-capable delivery systems were ready for action in late October. Of the 36 SS-4s deployed, for example, only about half were ready to be fueled—an 18-hour process—and not one had been programmed for flight. See Gribkov and Smith, p. 63.
72. Gribkov and Smith, p. 53.
73. Detzer, p. 194.
74. This revelation comes from a former GRU officer who wrote under the pen name of "Viktor Suvorov" and produced books and articles on the Soviet military and intelligence forces in the 1980s.
75. Central Intelligence Agency, CIA Documents on the Cuban Missile Crisis (Washington, DC: CIA, 1992), p. 367.
76. *Ibid.*, pp. 367-368.

CHAPTER 33

The Barry and Thomas Critique of the Pentagon Report: In the Lessons and Non-Lessons of the Air and Missile Campaign in Kosovo[1]

Anthony Cordesman

Anthony Cordesman was a member of the strategic assessment group that assisted General Stanley McChrystal in developing a new strategy for Afghanistan. Before joining CSIS, Cordesman served as director of intelligence assessment in the Office of the Secretary of Defense and as a civilian assistant to the deputy secretary of defense.

The Kosovo air campaign shows that the most sophisticated weapons of war are easily fooled by relatively primitive decoys. So, as weapons systems become more sophisticated, the opportunities for deception remain the same, or perhaps greater.

The most controversial critique of the Pentagon report is contained in an article in Newsweek by John Barry and Evan Thomas.[4] According to this report, the Secretary of Defense suppressed an Air Force report that showed that the number of targets verifiably destroyed was only a "tiny fraction" of the US claims: 14 tanks, not 120; 18 armored personnel carriers, not 220; 20 artillery pieces, not 450. The article states that Air Force investigators, spent weeks combing Kosovo by helicopter and by foot, only found evidence of just 58 successful strikes out of the 744 "confirmed" strikes by NATO pilots during the war.

The article states that General Clark was under constant pressure to produce positive bombing results, and that, "The surgical strike remains a mirage. Even with the best technology, pilots can destroy mobile targets on the ground only by flying low and slow, exposed to ground fire." It also says that the, "Pentagon essentially declared victory and hushed up any doubts about what the air war exactly had achieved."

According to Barry and Thomas, Clark doubted the resulting Air Force claims and initially tried to obtain accurate assessments. At the end of the war Clark publicly scoffed at claims by General Nobojsa Pavkovic, the Serbian ground commander, that Serbia had lost only 13 tanks. However, Clark's staff reported that Pavkovic might be right, and Clark dispatched a team into Kosovo in June to do an on-the-ground survey. It had 30 experts; some from NATO but most were from the Munitions Effectiveness Assessment Team, or MEAT. Barry and Thomas report that the NATO bombing was accurate against fixed targets, like bunkers and bridges, although it was able to use deception. For example, one bridge was protected from the

1. Anthony Cordesman, *The Lessons and Non-Lessons of the Air and Missile Campaign in Kosovo,* Westport, CT: Greenwood Press, 2001, pp. 159–162.

high-flying NATO bombers by constructing, a fake bridge 300 yards upstream, made of polyethylene sheeting stretched over the river. As a result, NATO "destroyed" the phony bridge many times.

The Serbian deception effort proved to be much more effective in protecting ground force targets. Mock artillery pieces were made of long black logs stuck on old truck wheels. A two-thirds scale SA-9 antiaircraft missile launcher was fabricated from the metal-lined paper used to make European milk cartons. They report that the MEAT team found dozens of burnt-out cars, buses and trucks—but few tanks.

According to Barry and Thomas, General Clark reacted by ordering the team to drive to the sites where Serbian weapons were supposed to be damage and to walk the terrain. In early August 1999, the MEAT team returned to Air Force headquarters at Ramstein air base in Germany with sine 2,600 photographs of the sites. Both General Walter Begert, the Air Force deputy commander in Europe, and General Clark are reported to have rejected their conclusions and Clark is said to have insisted that the Serbs had concealed their damaged equipment or that the MEAT team had failed to properly survey the sites. The MEAT team, insisted, however, that it had conducted proper surveys and saw no physical evidence that heavy equipment had been lifted by cranes or dragged away.

Again according to Barry and Thomas, the USAF was ordered to prepare a new report. In September, Brigadier Gen. John Corley produced a survey that showed that NATO had successfully struck 93 tanks, close to the 120 claimed by General Shelton at the end of the war, and 153 armored personnel carriers, not far from the claim of 220. Corley's team did not do new field research, however but rather looked for indicators that would the pilots' claims. The article quotes "senior officer" at NATO headquarters who examined the data as stating that more than half of the hits the Corley report declared to be "validated kills," had only one piece of supporting evidence such as a blurred cockpit video or a flash detected by a spy satellite. The latter evidence is virtually meaningless since satellites usually can't discern whether a bomb hits anything when it explodes.

The Newsweek article also states that NATO sources reported that, British Gen. Sir Rupert Smith -- Clark's deputy -- and his chief of staff, German Gen. Dieter Stockman privately cautioned Clark not to accept Corley's numbers, and that the U.S. intelligence community also questioned the Corley report. It states the CIA put far more credence in the results of meeting of U.S. and British intelligence experts in November 1999, which determined that the Yugoslav Army after the war was only marginally smaller than before the conflict.

It is painfully clear from both the Barry and Thomas article, and from private interviews by the author with very high level US officers and defense officials, that Defense Secretary Cohen and General Shelton played a major role in ensuring that none of these issues were dealt with the report on the lessons of war that was sent to Congress. The report did provide a chart from the Corley report showing that NATO killed 93 tanks but as Barry and Thomas note the text effectively disowns the chart without providing any alternative information, "the assessment provides no data on what proportion of total mobile targets were hit or the level of damage inflicted.".

Off the record discussions with senior US Air Force officers also indicate, however, that they feel the Barry and Thomas article unfairly discounts a great deal of evidence obtained using intelligence sources, and that the MEAT team relied far too heavily on direct, after action inspection. They note with accuracy that such methods did not prove reliable in Vietnam, or in the Gulf War. The author also found that they did not prove accurate in the October War between Israel, Egypt, and Syria – a war where the author wrote the report that the Secretary of Defense sent to Congress and had to coordinate much of the US-Israeli analytic effort. On the basis of these discussions, the author believes that the NATO statistics are more likely to be correct than those of the MEAT team are, although any such conclusion is highly uncertain and most of the material evidence is classified and cannot be discussed.

The Fog of War, Surgical Bombing, and the Revolution in Military Affairs

It would take a major team effort with full access to classified material to resolve these uncertainties, however, and it is brutally clear that it would have to be independent of the Office of the Secretary of Defense and Joint Staff to have any credibility. As has been note earlier, Kosovo is the third consecutive major action in which the Office of the Secretary of Defense and Joint Staff have demonstrated they lack the professional integrity to be trusted in carrying out honest and objective studies of the effectiveness of US military forces in combat. The reporting and non-reporting by these groups on the Gulf War and Desert Fox was no better.

At one level, a key lesson seems to be that the assessment of war can no longer be trusted to the Office of the Secretary of Defense and Joint Staff. It is also unclear that it can be trusted to the services. For example, the US Army suppressed a self-critical report on the lessons of the Gulf War and then circulated a large book filled with self-praise to the Congress. The US Navy issued a preliminary report with considerable self-criticism, but classified its detailed report and did not circulate it outside the Navy. The only service to attempt an independent and objective analysis, to declassify the results, and give it broad circulation was the US Air Force, which reported in its Gulf War Air Power Survey. While the historical sections of all four services have since done excellent work, one lesson of the war in Kosovo is that the analysis of lessons must be done by an independent body, that it must report to Congress or some authority outside the Department of Defense, and that its results must be a public as possible – both to ensure that US military personnel and planners are properly informed and to establish checks and balances on the kind of public relations-oriented reporting that is now common place.

These issues also have grave policy implications. The "fog of war" remains the reality, and precision is not omniscience. Concealment, dispersion, deception, and the other aspects of asymmetric warfare adapt to defeat many of the advances in sensors. The uncertainties created by the real-world imperfections in targeting are compounded by the fact that advances in precision often lead to small visible damage areas that make observable damage far less apparent than the impact of area bombing. As Serbia showed during and after the war, the manipulation of the media and the physical phenomena left after an attack are also a key form of information warfare.

Unfortunately, this reality seems to go so deeply against the grain of the military and public information officers, that the end result is that the "fog of war" becomes a "fog of deception" in which self-deception plays no little part. The end result is that policy makers, military officers, and strategic analysts are almost certainly being given a grossly exaggerated picture of the ability to target mobile forces in the field, to inflict damage with modern airpower, and to do so in ways that minimize losses and collateral damage. Both the current success of the "revolution in military affairs" and the ability to improve it are almost certainly being exaggerated, and so is the ability to achieve decisive results with limited force. As a result, the US and its allies cannot learn the right lesson from their military experience, and the cost of political posturing by of the Office of the Secretary of Defense and Joint Staff is almost certain to be paid for in blood.

CHAPTER 34

Denial and Deception Practices of WMD Proliferators: Iraq and Beyond[1]

David A. Kay

David A. Kay is best known for heading the Iraq Survey Group and acting as a Weapons Inspector in Iraq after the 2003 U.S. invasion. Dr. Kay served as the UN Chief Weapons Inspector from 1991–1992. Following that, he was Vice President of Science Applications International Corporation (SAIC) from 1993–2002.

WITH THE END of the Cold War and the Soviet threat, a broad consensus seems to be forming that U.S. foreign and defense policies and budgets can now be premised on the absence of any major threat to the United States or its allies and on such timely detection of any newly developing threat that defensive measures can be deferred until after a threat emerges. The case of Iraq should offer a warning against too much optimism.

The failed efforts of both the International Atomic Energy Agency (IAEA) safeguards inspectors and national intelligence authorities to detect prior to the Persian Gulf War a nuclear weapons program of the magnitude and advanced character of Iraq's should stand as a monument to the fallibility of on-site inspections and national intelligence when faced by a determined opponent. The Iraqi military buildup, as well as the multiple failures of its timely detection, is an experience rich in lessons that, if correctly understood, may help in detecting other covert weapons programs and, equally important, U.S. understanding of the limits of its ability to guarantee timely detection.

During the inspections in Iraq after the Gulf war, an immense military production establishment was found that was producing or striving to produce a broad range of chemical, biological, and nuclear weapons and missiles capable of delivering them. The nuclear program alone involved investments amounting to around $10 billion. Inspectors and those who monitor proliferation were surprised by the magnitude and advanced state of Iraq's efforts to obtain nuclear explosive devices.[1] At the time of the Gulf war Iraq was probably only 18 to 24 months away from its first crude nuclear device and no more than three to four years away from more advanced, deliverable weapons. Moreover, the amount of foreign assistance and technology that

David A. Kay is assistant vice president of Science Applications International Corporation. He was chief inspector of the three early UN nuclear weapons inspections in post–Gulf war Iraq that uncovered the Iraqi nuclear weapons program.

The Washington Quarterly • 18:1 pp. 85–105

1. David A. Kay, "Denial and Deception Practices of WMD Proliferators: Iraq and Beyond," *The Washington Quarterly*, Vol. 18, No. 1, Winter 1995, pp. 85-105.

had fueled the Iraqi arms program was truly staggering.

On April 18, 1991, Iraq submitted to the United Nations (UN) Special Commission on Iraq (UNSCOM) and the IAEA details of the quantities and locations of its chemical, biological, ballistic missile, and nuclear materials stockpiles. This listing was required by the terms of Security Council resolution 687 and was designed to provide the baseline for the inspection activities that were to lead to the dismantling of Iraq's weapons of mass destruction (WMD). It was signed by none other than President Saddam Hussein and accompanied by a letter from Iraq's foreign minister, Ahmed Hussein. This declaration acknowledged nearly 10,000 nerve-gas warheads, 1,000 tons of nerve and mustard gas, 1,500 chemical weapon bombs and shells, and 52 SCUD missiles with 30 chemical and 23 conventional high explosive warheads. Iraq, however, denied that it had any nuclear materials that fell under the resolution or any biological weapons.

Although this declaration was eye-opening with regard to the Iraqi chemical weapons stockpile and its acknowledgment of chemical warheads for SCUDs, it was widely acknowledged at the time—and subsequently confirmed by on-site inspection—to be seriously misleading in the nuclear, biological, and missile fields. It also was soon discovered to be a serious misstatement of the size of the Iraqi chemical weapons arsenal as well. Rolf Ekéus, chairman of UNSCOM, informed the Security Council on July 31, 1991, that inspectors had already found four times more chemical weapons than Baghdad had declared (46,000 chemical shells as compared to 11,500, and 3,000 tons of chemical agents as compared to 650 tons declared). By October 1991, inspectors had found 100,000 chemical shells and bombs—almost 10 times the number initially declared.

On the nuclear front, the scale of deception and denial was even greater. Iraq's initial declaration of April 19, 1991, avowed that it had no proscribed nuclear materials, but this was amended on April 27 to acknowledge that it did have what had been reported to the IAEA before the war, including 27.6 pounds of highly enriched uranium and 22 pounds of low enriched uranium and a peaceful research program centered on the Tuwaitha Nuclear Research Center. Subsequent inspections found an altogether different program. Iraq in fact had, beginning in 1981, embarked on a clandestine uranium enrichment program using three different methods: electromagnetic isotope separation, the so-called calutron method; chemical enrichment; and gaseous centrifuge enrichment. At the time of the invasion of Kuwait, it had begun the start-up for industrial-scale enrichment using calutrons and had acquired the material, designs, and much of the equipment for 20,000 modern centrifuges. Design, component testing, and construction of manufacturing facilities for actual bomb production were well advanced. Over 20,000 people were found to have been employed in this clandestine nuclear weapons program.

All of this is by way of prologue and scene setting to the main thread of Iraq's deception and denial activities, which were aimed at covering up its weapons programs. For purposes of analytical clarity, I have separated these activities into four separate, but overlapping, layers: political, procurement/acquisition, facilities and activities, and inspections. My conclusions are drawn largely from Iraq's nuclear program after 1981. I believe, however, that the deception and denial activities applied in the nuclear case are largely representative of those used in other weapon areas as well.

Political

According to Jaffar dhia Jaffar, the scientific leader of the Iraqi nuclear weapons efforts, in the wake of Israel's successful attack on the Osirak research reactor in June 1981 Iraq launched an internal review of the future course of its nuclear activities. This review concluded that a major effort should be made to develop a uranium enrichment capability, and that this program should be clandestine while Iraq maintained its status as a full member of the Nuclear Non-Proliferation Treaty (NPT). In these internal deliberations, as described by Jaffar, the deciding factor in the decision to stay in the NPT was the desire of the military and security services not to attract any undue attention to Iraq's developing nuclear program that would complicate procurement and development efforts. As he described it to UN inspectors, the argument ran, "Let Israel believe it destroyed our nuclear capacity, accept the sympathy being offered for this aggression and proceed in secret with the program."[2] If one accepts this account as broadly true of the internal Iraqi discussions in fall 1981 as to how to proceed with their nuclear activities, then by their own admission their program was rooted in a fundamental deception as to their NPT undertaking. Jaffar suggested in a July 1991 meeting to several inspectors that IAEA headquarters be told that he was thinking of traveling to Vienna for the 1991 IAEA General Conference to offer a seminar on "How to beat the NPT." He added, correctly, that this suggestion when reported to Vienna would not be welcomed.

Throughout the 1980s Iraq played an active role in the IAEA's various programs, welcomed agency staff and officials as visitors to its Tuwaitha Nuclear Research Center, sought and received equipment, expert assistance, and training from the IAEA's technical assistance program, and was a frequent member of the agency's Board of Governors. Iraq was a far from silent member of the IAEA, leading the demands for international condemnation of Israel's attack on its safeguarded reactor and pushing for international protection for peaceful nuclear facilities against military attack. Iraq repeatedly emphasized the peaceful and safeguarded nature of its nuclear facilities and contrasted its openness to inspection with Israel's refusal to sign the NPT.[3] In an irony that I am sure that I appreciate more than the Iraqi government, Iraq played a leading role in board meetings in 1985 in praising, and ensuring the continuance and eventual expansion of, a program I was leading to introduce more rigorous evaluation of IAEA programs. The praise was so extravagant that a member of the director general's staff asked me during the discussion whether I had an Iraqi relative. The ultimate irony is that Iraq's IAEA governor had in fact, as chairman of the Iraq Atomic Energy Commission (IAEC), presided over the initiation and large expansion of their nuclear weapons program in the early 1980s.

The strength of the Iraqi denial and deception program is that it was anchored in a real fact—*the Israeli destruction of the Osirak nuclear reactor.* This piece of reality was reinforced by the accompanying sympathy that this attack engendered in the West and the desire of those who carried out the Osirak attack, as well as those who admired the technical skills it demonstrated, to believe that it had dealt a decisive blow to the Iraqi nuclear program. The extent to which the rooting of the Iraqi deception activities in the reality of the destruction of their Osirak reactor assisted the deception effort can best be captured in a study prepared in 1982 at the request of the Senate Foreign Relations Committee by the Congressional Research Service (CRS) of the U.S. Library of Congress. The CRS concluded that:

> At the time of writing, Iraq's long term potential had been reduced by Israel's bombing of Iraq's large research reactor. The assistance which that reactor might have provided now will not be available until it is rebuilt. . . . Assuming that the research reactor were to be rebuilt and put into operation by the late 1980s, it seems unlikely that Iraq's intellectual and industrial base can grow enough to materially reduce its dependence upon foreign imports and technical assistance during this time. It seems unlikely too that Iraq's own industries will be able to design and build a reprocessing plant, an enrichment plant, or a nuclear power station for many years to come. . . . This assumes the undertaking could be kept secret for several years so that other countries would not have an opportunity to take countermeasures. While the success of such an undertaking cannot be ruled out as impossible, it nonetheless appears to be improbable. In summary, Iraq's potential nuclear weapons capability at the outset of the 1990s would still be small, assuming that it does not get access to engineers and technicians experienced in the production of nuclear weapons materials.[4]

Outside the nuclear arena, Iraq skillfully played throughout the 1980s on the fear in the West and Persian Gulf of Iranian fundamentalism. Iraq's portrayal of itself as a secular bulwark against the wild mullahs of Tehran provided an effective rationale for its large arms buildup. The fact that much of this took place during the course of Iraq's war with Iran, in which most Western governments provided assistance to Iraq despite their own professed neutrality and often despite domestic legislative prohibitions against military assistance, provided protection against too rigorous scrutiny of Iraq's arms buildup.

Procurement/Acquisition

The physical scale of Iraq's arms program is what has astounded most observers. The facilities were large, over scale even for Iraq; well designed; amply furnished with the latest technologies; and often constructed with foreign assistance. For the nuclear program alone it is easy to identify more than 25 major facilities. What is less clear, at least to most outside observers, is the multilayered deception efforts that Iraq applied to avoid the unmasking or accurate piecing together of this program by outsiders as it was being assembled. These deception efforts consisted of, at least, nine major thrusts. Each of these nine areas is worthy of a much fuller treatment. I hope that as pending criminal actions are completed and some serious security concerns become less pressing, a fuller treatment of these techniques will be possible.

Compartmentalization and Multiple Project Codes

The Iraqi nuclear weapons program was heavily compartmentalized and was managed through a complex system of multiple and shifting project codes. Not only did various elements of the project teams generally not know of the total program, they often did not know of the real purpose of their own work. Firms—international as well as Iraqi manufacturing facilities—would be approached to manufacture components without being told the end use for these components. In the case of Iraqi manufacturing facilities, plans and material would be delivered by the IAEC with instructions to produce a set of components. After production, the items and all scrap or unutilized material would be taken back. With regard to international firms, again the end use would not be specified. As one example, many of the large pole magnets of the calutrons used to enrich uranium were produced in Austria by an Austrian state-owned firm that shipped the finished magnets to Iraq—half by trucks through Turkey and half by ship through Hamburg. The Austrians did not ask the purpose of these magnets and the Iraqis certainly did not volunteer this information. Much the same story applies to the high-quality copper that was used to wrap these magnets. It was produced in Finland to Iraqi specifications and went by ship to Iraq. In both these cases—and in many others—the Iraqis did not even have to go through the motions of creating a cover story, because the operating rule of the commercial world is "Don't ask, just sell."

Project codes, another example of simply good tradecraft, were an integral part of the Iraqi deception effort. PC-3 or Petrol Chemical Project 3, the overall project code for the Iraqi nuclear weapons program, became widely known after the Gulf war, but this misses the reality of the Iraqi use of project codes. Iraq used hundreds of multiple, constantly shifting, and overlapping codes for individual components of all of its arms program. It became clear during the course of the post–Gulf war inspections, where these codes were a constant source of confusion, that these codes had presented a real obstacle to various national intelligence services trying to track the Iraqi program. On more than one occasion, during the course of an on-site inspection, several of the governments that were providing intelligence support to the inspections would openly disagree on which element of the Iraqi weapons program a particular code related to. It is important to remember that the codes were used not only for the Iraqi nuclear program but for all of its most sensitive weapons programs. During the inspections, the project codes were a nightmare as inspectors tried to correlate documents, invoices, shipping crates, Iraqi declarations, and actual physical facilities. It is reasonable to assume that this relatively simple deception technique must have proved equally frustrating to analysts trying to decipher the Iraqi weapons program before the war.

Misdirection as to Purchaser and End Use

Most export control regimes do not attempt to block all exports to a given country, but rather to deny benefit from exports to specific institutions (e.g., the military), or to inhibit potentially harmful uses (e.g., chemical weapons programs). The guidelines of the London Suppliers Group identify items related to nuclear weapons that should not be exported to non-nuclear weapon states. In practice, however, nuclear export controls have proved to be vulnerable to deception. One major problem area is so-called dual-use exports. A high-speed, computer-numerically-controlled (CNC) machine tool may be used to produce truck engines for a civilian automotive industry or gas centrifuge rotors for a nuclear weapons program. The more modern and industrialized the economy the wider the possible permitted uses for almost any import. In a case that came to light during the inspections, Iraq ordered preforms for centrifuge endcaps from the Swiss firm Schmiedemeccanica SA, a metal forging company, and declared that the end use was for automotive part forgings. In 1992 the Swiss decided to drop prosecution after determining that these preforms so closely resembled the end use declared by the Iraqis that the Swiss firm could not be reasonably expected to know otherwise.

To further complicate matters, it is up to the importer to identify the user and uses for any imports and to a lesser extent to the exporting firm or its agents to confirm these facts.[5] Very seldom are on-site checks on such declarations made either before or after sales—all in all a system asking for abuse and deception. Iraq clearly understood the weakness of national export control regimes.

During the course of the UN inspections after the Gulf war it became clear that the University of Baghdad, the Ministry of Industry, petrochemical projects, and many other seemingly innocent purchasers had been nothing more than shells for the Iraqi nuclear weapons program. Falsified end uses were routinely declared, and in most cases the stories did not have to be terribly clever because—until Iraq lost the Gulf war—they never had to withstand any type of check after the item was received in Iraq. This stratagem worked particularly well for Iraq during the 1980s because parallel with a clandestine weapons program there existed a very large conventional military program that, by and large, was not subject to Western export controls and also a booming civilian economy based on Iraq's oil revenues and loans from Gulf states.

Buy-in to Foreign Manufacturers

Iraq also used another deception strategy that is particularly hard to monitor in a world that places a high premium on international freedom to trade and invest, and indeed, a world where governments compete to attract foreign investment. Iraq invested in legitimate foreign firms that either manufactured equipment useful for the Iraqi clandestine weapons program or that could legitimately order such equipment purportedly for their own use. The most publicized case of this type was the Iraqi acquisition of the British firm, Matrix Churchill. Matrix Churchill was a small machine tool producer that in the mid-1980s was facing severe financial problems. When the Iraqi investors arrived on the scene, they were welcomed as financial rescuers. Over the years Matrix Churchill provided not only a valuable source for illegal exports of high-performance machine tools, but also a useful cover for ordering products from other firms for onward transshipment to Iraq.

Middlemen and Front Companies

Iraq benefited in building up its weapons program from its ability to use a complex network of middlemen and front companies. Some of these front companies and middlemen were purely creatures of the Iraqi weapons program, but others were associated with the shadowy world of smugglers, traders, and fast-buck artists that swarm to wherever there is money to be made. Modern electronic communications and the opening up of so many borders has made this a particularly powerful deception technique when money or material needs to move across borders without undue notice of its origin or destination. The Gulf region is particularly fertile ground for such operations because its new wealth reinforces its old cultural practices of smuggling and the predisposition to avoid too much attention from governments.

Multiple Purchases

The Iraqi weapons program is also striking in terms of the amount of money it spent—and seemed to overspend—in procuring foreign material. This "shop until you drop" mentality had several origins. As a deception strategy, it worked on a rather profound understanding of the way export controls are administered and how intelligence services interpret information. In seeking to purchase export control–sensitive items, the Iraqis would often place orders in quantities that were below the size that triggered controls. Their calculation was that this would avoid controls, and, even if detected, such small requests would signal that their program was in the early stages of development and not well thought through and hence not of great concern. Indeed, during much of the 1980s the threat estimate assigned to Iraq's nuclear program was downgraded because analysts could not find a pattern in Iraq's procurement that indicated a well-organized program moving toward early production of nuclear weapons. The Iraqis also seem to have understood that the discovery of one procurement channel would not trigger an intensive search for other pathways to obtain the same item. Iraq seems to have learned early that an effective deception can be based on a defeat or loss—much faster than the victors learned that every victory contains the seeds for self-deception.

Multiple purchases of the same items were also central to the very nature of the Iraqi nuclear weapons program and if correctly understood could have raised serious alarm flags about that program well before the Gulf war. Iraq was engaged in a crash program to acquire nuclear weapons, roughly modeled on the U.S. Manhattan project. Research and development activities for uranium enrichment were paralleled by construction of large production facilities. For example, the large facility at Tarmiya designed to produce highly enriched uranium using the electromagnetic isotope separation (EMIS) process was constructed in parallel with Iraq's efforts to build their first small EMIS machines. Similarly, a very large facility to produce thousands of gas centrifuges at Al Furat was being constructed before Iraq had frozen the design on the type of centrifuge to be built. In such a crash program—particularly where the government rules by terror—any material bottlenecks would have very serious consequences. Iraqi officials attempted to insure against such bottlenecks by multiple purchases on the theory that even if several procurements failed for whatever reason the weapons program would go on unchecked. It was better, from the Iraqi perspective, to spend a lot of money for supplies, than to lack required supplies when needed and have a critical program grind to a halt.

System Integration in Iraq

Many of the equipment items required for the production of WMD and advanced conventional arms are legally exportable so long as they do not contain particular attachments (e.g., laser alignment systems or flow forming mandrels of the specific dimensions associated with gas centrifuge rotors). The principle involved is that without such attachments the equipment can only be used for less capable tasks either associated with permitted military applications or civilian uses. In addition, export controls operate on the assumption that even if states like Iraq manage to obtain the restricted attachments separately, they would be unable to successfully integrate them with the legally obtained parts to form the more capable equipment. This assumption reflects the view that countries like Iraq lack the system integration skills needed to merge research and development work on a process like calutrons or centrifuges with the civil construction of a production facility, the equipment installation, and the process engineering—all of which are critical steps in getting a uranium enrichment facility up and running.

Iraq successfully exploited the weakness of these assumptions. CNC machine tools that could not be exported legally from the United States with laser alignment systems were exported without them only to be mated later in Western Europe and Iraq with such systems. Iraq also demonstrated considerable experience with project management and system integration, particularly in their EMIS program.

Systems integration is a powerful deception technique. The focus of collection efforts inevitably has to be sharp enough to exclude most of the large trade and technology flows that go into even a mid-level industrializing economy, and yet most weapons programs are composed of a lot of very innocuous items that only take on their deadly character when combined in very specific ways. If collection efforts are unable to detect the purpose of imports at their most vulnerable stage—when borders are being crossed—then a very heavy burden is placed on national technical means (NTM) and human intelligence (HUMINT) to detect the actual integration of these items into weapons production systems—and this occurs mostly inside buildings in secured areas.

"Simply Good Tradecraft"

Iraq applied a lot of what can best be described as "simply good tradecraft" to its efforts to disguise the clandestine weapons program. Invoices routinely had their supplier and banking transfer agent blacked out or completely cut out. Wooden shipping crates had the shipper and destination blacked out or physically chipped out as soon as they arrived in Iraq. Scientists who traveled abroad used false identity and travel documents. All of these techniques, and many similar ones, compounded well beyond the usual fog impeding analysis the problem of detecting the true purpose, size, and state of development of the Iraqi programs.

Competitive Commercial Negotiations

Iraq was able to use the strong desire of Western providers of technology to make sales both to conceal the true purposes of Iraqi efforts and to extract a considerable amount of proprietary information from Western companies without any compensation. A classic example of this ploy lay at the heart of Iraq's efforts to obtain technology for the chemical enrichment (Chemex) of uranium. There are two suppliers in the world of chemical enrichment technology, one is Japanese and the other is French. In the mid-1980s Iraq initiated preliminary discussions with both and indicated a desire to acquire a uranium enrichment capability using the Chemex technique. After preliminary exploration the Iraqis decided to concentrate on the French process. Inspectors learned during the course of the postwar inspections that a strong factor in motivating Iraq to abandon discussion concerning the Japanese process was that it requires a set of exotic resins that have few other uses and Iraq was worried that the large-scale importation of such resins would signal the scale of their uranium enrichment program.

Iraq engaged the French firm in negotiations stretching out over several years. The course of these soon took on a familiar pattern. Iraq would indicate it needed only a little more technical information on which to make a decision; the French would reveal more proprietary data and the cycle would begin again. The French firm was particularly anxious to make a sale as the process had yet to find its first customer. Finally, after the French firm had revealed essentially all of the technical details concerning its Chemex process, Iraq announced that it had concluded that the process was too expensive and was abandoning all interest in pursuing it. The French seem to have bought this story and went away believing that Iraq had given up on this process; Iraq went away and started clandestine development of Chemex.

Many of the natural forces of the commercial world—competition between suppliers, protection of proprietary customer information, reluctance to admit that a deal has been lost, and a lack of a basic counterintelligence perspective in most commercial enterprises—can actually help a proliferator's deception effort.

Large Industrial and Conventional Military Programs

The scale of the industrial and conventional military programs in Iraq in the 1980s—the conventional military program alone is estimated by Iraq to have cost $110 billion—must have made it almost impossible for any intelligence service to keep abreast of all the commercial discussions, imports of material and technology, and new construction that were under way and that might have had some relevance to clandestine weapons capability. Although it would be wrong to assert that the conventional military program and the large civil economy investments were part of an Iraqi deception effort, it would be equally wrong to dismiss the fact that they served that purpose. It is essential to recognize that in all but a few countries, such as isolated North Korea, the "noise" of normal commercial activities and conventional military programs may provide inherent deception for clandestine programs. This is particularly true for chemical, biological, and missile programs, which share many of the characteristics of innocent undertakings.

Facilities and Activities

The Iraqis paid as much attention to concealing the actual purposes of facilities and activities as they did to deception in their purchasing and acquisition efforts.

Facilities

Iraq well understood the deception advantage of hiding a clandestine program in plain sight. Their major declared nuclear research center at Al Tuwaitha was also the initial center of their clandestine nuclear program. Al Tuwaitha was visited every six months by IAEA safeguard inspectors who announced that there was no sign of diversion of nuclear material, thus giving a general clean bill of health to the facility. Two deputy directors general of the IAEA and many staff members visited Al Tuwaitha and reported no suspicious activity. Also selected Western scientists were invited to the facility to reinforce the impression that it was an open center of peaceful research.

This openness, however, was carefully controlled. The IAEA inspectors were only allowed to visit portions of three of the almost 100 buildings at Al Tuwaitha, and Iraqi escorts guided the IAEA staff and Western scientists through chosen laboratories and conducted meetings in the center's conference building. Although the restrictions on the movements of these visitors in retrospect seem like a warning sign, they were not recognized as such by most of the visitors at the time. The safeguard inspectors were used to such restrictions in every country they visited, and it is only since the experience in Iraq that this has begun to slowly change. The other visitors were controlled by "busy" schedules, generous hospitality, and cultural tourism. One of the IAEA deputy directors general, a former defense official of an Asian country, assured the UN inspectors before their inspection after the Gulf war that he had toured all of Al Tuwaitha and there was nothing at all there related to uranium enrichment or a weapons program. Although self-delusion was no doubt involved in some of the assessments of the peaceful nature of Al Tuwaitha, the Iraqis deserve credit for the skill with which they controlled visitors and the clever layout of the facility itself. The distribution of the buildings, the visual screening provided by trees, and the careful routing of the internal road system made it very difficult for any visitor without access to overhead intelligence to accurately understand the size of the center and the relationship of the buildings to each other.

Understanding the Algorithms of the Photointerpreters/Analysts

The Iraqis demonstrated on numerous occasions that they had a rather accurate understanding of the limitations of U.S. technical collection systems and of how data gathered by such systems were interpreted.

The catalogue of techniques is long but includes construction of buildings within buildings (Tuwaitha); deliberately constructing buildings designed to the same plans and for the same purposes to look different (Ash Sharqat and Tarmiya); hiding power and water feeds to mislead as to facility use (Tarmiya); disguising operational state (Al Atheer); diminishing value of a facility by apparent low security and lack of defenses (Tarmiya); severely reducing off-site emissions (Tuwaitha and Tarmiya); moving critical pieces of equipment at night; and dispersing and placing facilities underground.

All of these techniques, and others, vastly complicated the job of understanding the nature of the Iraqi nuclear weapons program and provide at least a partial answer as to how a program so large could have been so incompletely understood by the world's intelligence communities.

Personnel

The Iraqi nuclear program involved at least 20,000 personnel, many of whom had training and contacts abroad. This was a potentially large source of leakage of information on the aims and direction of these activities. Iraq faced this problem and adopted a series of deception practices designed to limit any such loss.

First, Iraq managed its flow of personnel to ensure that students were not all sent to the same universities and countries. This had several advantages. Training in most scientific disciplines follows somewhat different approaches in different countries and provides access to multiple networks of information. This is particularly true in the various engineering and science disciplines that most concern a nuclear weapons program. For example, information concerning the ability and techniques involved in focusing X-rays was classified in the United States long after it was part of the general physics literature in Japan, Germany, and Britain. Also, by dispersing students, Iraq made it more difficult for any one country to fully appreciate the breadth of technical skills being built up in Iraq. And this dispersal of students certainly made it more difficult to track individual Iraqi scientists. Concerns with privacy and academic freedom as well as a low collection priority have meant that systematic data on foreign students are collected in few countries and seldom shared with other countries.[6]

Second, Iraq moved from an almost total dependence upon foreign training in the 1970s to almost totally indigenous undergraduate and master's level training, and even credible doctorate-level training in some fields, in the 1980s. This vastly complicated the task of accurately accessing the talent and direction of Iraq's scientific infrastructure. As we examined Iraq's weapons laboratories it became clear that the staff kept abreast of the latest scientific work in English, French, Russian, and German. On the other hand, few resources seem to have been devoted in the West to tracking the theses and other work coming out of Iraqi institutions.

Third, Iraq purposely followed strategies designed to ensure that the purposes of scientific visits would not be easily discerned. False names and passports were used, the purpose of visits and requests for training were misidentified to sound innocuous, and the employer was misidentified to avoid any connection with the IAEC or the military.

A final barrier to leakage of information from the staff of the nuclear program was a cleverly crafted set of fringe benefits and punishments that discouraged those who knew important details of the Iraqi program from divulging information. First, the staff of the IAEC received an automatic draft exemption from the Iraqi army—and this was at the time of the bloody Iran–Iraq War. Second, salaries were consistently kept above the announced government salary scale. Other benefits, including bonuses and access to travel and foreign goods, were also substantial. Although the benefits were good, the severance package was horrible in its brutality. Those who attempted to defect and were caught or who evidenced any sign of lack of loyalty to the regime—including not reporting on family members who were disloyal—were pitilessly executed, in many cases in front of their family.

Inspections—Prewar and Postwar

As a subscriber to the NPT, Iraq was subject prior to the Gulf war to routine—roughly once every six months—IAEA safeguard inspections of its declared nuclear material. These inspections took place only at the Tuwaitha Nuclear Research Center—the only site for which Iraq declared nuclear materials. As Tuwaitha was the initial center for Iraq's clandestine nuclear program, it was important both that these inspections not inadvertently stumble upon weapons activities and that Iraq appear to be fully cooperating with the IAEA and not trying to conceal some activities. Inspections were also the key to the post–Gulf war unraveling of Iraq's WMD program. In both the prewar and postwar phase, deception and denial activities played a key role in trying to limit the impact of inspections. The deception and denial activities can be separated into six major thrusts.

Fat and Happy Inspectors

Under the IAEA safeguard procedures, a country is allowed to designate the inspectors that it will accept to inspect its facilities. This provision, which had originally been insisted upon by some West European states to limit what they thought would be commercial espionage by inspectors from other states, has turned out to be a major brake on the integrity and efficiency of the safeguard system. Iraq consistently accepted only inspectors from the then Soviet Union, its allies in the Eastern bloc, and Africa. Not only were Soviet inspectors presumed to be more controllable, given their nation's close military arrangements with Iraq, but they were also the beneficiaries of a quirk in the way they were paid that made inspection missions highly profitable. Soviet inspectors had to turn over their salary to their government, which kept most of it to cover the cost of providing their housing in Vienna and "other expenses" and deposited the balance into a ruble account of no use to the inspectors while in the West. But there was one exception. Soviet inspectors were allowed to retain all of their per diem allowance received while on official travel. Thus the less a Soviet inspector spent on lodging and food while on inspection missions, the more he had in hard currency to spend on goods unobtainable at home. Iraq was a particularly profitable mission for inspectors so inclined because the official conversion rate of the Iraqi pound before the Gulf war was artificially high and consequently the UN per diem rate was substantial.

Soviet and Eastern bloc inspectors and many from the developing world were widely known to seek out the cheapest available accommodations and to enjoy official entertainment as a money-saving strategy. The Iraqis exploited this desire by providing cheap accommodations, at a considerable distance from the nuclear research center at Tuwaitha, and by hosting late evening official dinners for inspectors. The result was that the inspectors were often weary, but happy at the money that they were saving and not inclined to do more than the limited and minimal official inspection tasks. To do more would have been to risk that Iraq would refuse to accept a particular inspector with the consequent loss for that inspector of valuable income.

During the first round of post–Gulf war inspections, the Iraqi authorities continued to try to delay and reduce the coverage of inspections by suggesting long and pleasant lunches, late dinners, and sightseeing visits to some of the interesting archeological sites several hours from Baghdad.

Threats and Intimidation

Iraq during the post–Gulf war inspections moved from positive inducements for easy inspections to physical and verbal threats and intimidation. Inspectors were awakened with telephoned threats; obscene and threatening notes were slipped under hotel doors; hotel rooms were ransacked; verbal abuse on the street and at inspection sites became common; on several occasions inspectors were physically attacked by outraged Iraqi "civilians"; UN vehicles were bombed and tires slashed; and shots were fired over the heads of inspectors as a team photographed Iraq's secret uranium enrichment equipment. The Iraqi authorities made it clear to UN and IAEA authorities which inspectors they found acceptable and which were not to be sent to Iraq again. Formal letters of censure were sent by Iraq to the UN and the IAEA concerning some inspectors, and the Iraqis demanded that no Americans be allowed to lead inspection missions. It would be easy to dismiss most of this except that it did have an impact. For a while Americans were not put in charge of missions, and some inspectors reported that the IAEA leadership sought to soften the tone of its inspection presence.

Study and Monitor Inspectors and Procedures

Iraq both before the war and during the postwar inspections carefully monitored inspection procedures and adapted its deception activities accordingly. During the course of the fourth postwar inspection the Iraqi national safeguard head—responsible for ensuring that national nuclear safeguards were applied—appeared and repeatedly bragged that he had applied everything he had learned from his earlier job as an IAEA safeguards inspector to ensure that his former IAEA colleagues would not detect the clandestine nuclear weapons program during their pre–Gulf war inspections. In great detail he described the dual books, limited access, and keep-the-inspectors-happy strategy that Iraq had used.

During the postwar inspections the Iraqis engaged in active bugging of hotel rooms, meeting rooms, and office spaces used by the IAEA and the UN. Local Iraqi staff who were the drivers and maintenance personnel were often candid in their admission that the Iraqi security authorities required that they report daily on what they observed. Neither the UN nor the IAEA provided their staff with any training in counterintelligence or communications security. Although vigorous attempts were made to show how such lapses were seriously hindering effective inspections—and several embarrassing cases where IAEA "surprise" inspections became known in advance—this remained an uphill battle against an entrenched mind-set that still does not recognize the problem.

Misdirection

The postwar inspections began with the old IAEA cooperative inspection model of asking the Iraqi authorities to take the inspectors where they wanted to go. It should be no surprise to anyone that during the first inspection Iraq used this arrangement to carefully keep the inspectors away from sensitive sites. In the second inspection, when this same tactic was tried, the inspectors replied by producing handheld Global Positioning Satellites (GPS) instruments and current intelligence data, including very accurate line drawings of Iraqi weapon sites. This hindered but did not stop the Iraqi efforts to seek to misdirect the inspection effort at every stage.

Cover Stories

Another form of misdirection was the cover stories, some quite elaborate, that Iraq brought out in an attempt to shield its clandestine program. Tarmiya, one large site of Iraq's massive uranium enrichment effort, was said to be only a high voltage transformer testing facility; the Palm building where much of the testing for the Iraqi calutrons was done was said to be only an air force truck repair and radar testing facility; and the first calutrons seen by inspectors were described as only scientific mass spectrometers. Some cover stories were quite clever and required a great deal of scientific and inspection effort to penetrate, and some may never have been identified as false.

When cover stores were penetrated, Iraq would abandon them in a controlled fashion that tried to indicate that minimal disclosure was complete disclosure. The first admitted separation of plutonium was said to be the only separation—until another larger separation was uncovered by inspectors. Denial of any centrifuge activity was followed finally by an admission of a small centrifuge enrichment program that was said to be all there was—until a massive program was discovered by inspectors. Even a penetrated cover story served a useful purpose of allowing another opportunity for Iraq to claim that it had turned over a new leaf and was making a full and final disclosure of its secret activities. More than 10 such "full and complete" disclosures have been made, and yet the UN has remained unconvinced that the various clandestine programs have been fully disclosed.

Destroy/Shield

Before the first inspectors entered Iraq after the war and later when the inspectors began to close in on important elements of the Iraq nuclear program, Iraq began to systematically destroy facilities so that their true purpose and state of development could not be identified. At the Tuwaitha Nuclear Research Center, the large calutron test facility (80 by 120 meters) had escaped all damage from the coalition bombing, but it was completely leveled by Iraq and covered with dirt before the first UN inspector entered the site. Laser and centrifuge test facilities also at Tuwaitha were similarly destroyed, leaving unresolvable uncertainty as to how much progress Iraq had made with these programs. In the North near Mosul at Al Jesira where UCl_4 and UF_6 were produced to feed the uranium enrichment program, two substantial buildings were leveled by Iraq and one was covered with thick layers of epoxy paint to make all sampling efforts impossible.

One reason why many of the early inspections in Iraq became confrontational is that Iraq made it impossible to obtain easy answers as to how far its program had proceeded. Faced with a massive denial effort and forceful deception, the inspectors had little choice but to adopt a search strategy that would take account of Iraq's moves. These counters required the inspectors to move to sites without prior notice, go after documents, make greater use of technical intelligence methods, and assert a healthy skepticism about all Iraqi claims.

Implications for Intelligence of Denial and Deception Activities beyond Iraq

In the wake of the discoveries of the extent of the Iraqi clandestine weapons programs and the role that organized denial and deception played in shielding these programs, the important question that remains is whether Iraq is only an exceptional case or whether it is the prologue to a period in which denial and deception activities gain even greater importance in shielding clandestine programs. Two major areas of concern that deserve greater attention are the implication of denial and deception for collection systems and for the intelligence analysis process itself.

Intelligence Collections Systems

Multilateral Information Sharing—the Downside. The UN inspection efforts in Iraq marked a watershed in the willingness of the U.S. intelligence community to share with an international inspectorate its high-quality intelligence product, including a broad range of NTM, HUMINT, and measurement intelligence (MASINT), and equally the willingness of the UN system openly to receive and act upon such information.[7] This was a marriage of immediate necessity, without a long courtship and a careful weighing of the long-run implications of the union. Once it became clear to U.S. policymakers that the Iraqi WMD program was much larger than estimated before the Gulf war and that much of it had survived—and the intelligence community was the first to perceive this—then the only instrument for eliminating Iraq's WMD capability became the UN inspectors. On the other hand, Iraq had demonstrated during the first IAEA inspection in May 1991 that if the inspectors did not have current intelligence Iraq would be able to continue to deceive the IAEA and to obtain a clean bill of health as to its nuclear program. The second IAEA inspection team that entered Iraq, in June 1991, was provided with a broad range of high-quality intelligence, both before and during its mission, and this directly led to the stunning revelation of the previously unknown Iraqi uranium enrichment program. This initial success in intelligence sharing led to greater collaboration and was the essential element in the inspection successes in the second half of 1991 that led to the rapid unraveling of the Iraqi nuclear program.

Intelligence sharing between the United States and the UN that began in mid-1991 was not always a natural and riskless undertaking. The U.S. intelligence community was faced with a situation for which there were no parallels or rules. The UN system was riddled with cold war rivalries, nationals of many states had dubious records of being trustworthy, the UN and IAEA were totally unprepared at the basic mechanical level for handling confidential information,[8] and UN/IAEA personnel were not subjected to security clearances nor generally had they any real appreciation for the importance of protecting sources and methods. In the face of this, decisions had to be made about sharing the product of some of the very best collection methods available to the United States—and these were methods that would have to be continued to be relied on to monitor situations around the world. Even today—three years after the inspections began—the speed at which decisions were made to share intelligence information and the range of support provided to inspectors in the field is still amazing. In the clash of cultures represented by the open world of the UN/IAEA and the closed, secret world of the intelligence community, it needs to be recognized that the intelligence community took the first step toward sharing and ensuring that the UN system would have the information required to meet this first cold war challenge.

By the end of 1991, the lesson of Iraq that multilateral inspections can be made genuinely effective only if they are coupled with accurate and timely intelligence had been taken to heart by both the UN and the intelligence community. Although many problems remain in this relationship—and some of these deserve greater attention than they have been receiving—the sharing of U.S. intelligence with multilateral efforts is clearly now officially sanctioned.

The initial decisions to support the Iraq inspections with national intelligence represented a clear example of enlightened leadership. By 1993, however, other factors were also at play and beginning to lead to even greater efforts to share national intelligence information. With the disappearance of the Cold War, the scramble by intelligence community components to acquire customers and to appear useful began to lead to a fundamental shift in the philosophy of sharing sources and methods with multilateral organizations. Methods, including some of the best available MASINT, are now being shopped to multilateral institutions as vital enhancements to efforts to halt nuclear, chemical, and biological weapon proliferation as well as to thwart international crime, the drug trade, and money laundering.

In this rush to provide new tools for important missions—and to justify old programs in a new environment—probably too little attention is being paid to one of the most important lessons from Iraq. This lesson is that determined proliferators pay close attention to sources and methods with a view to countering them. Iraq's deception practices grew out of close scrutiny by Iraq and states that supported its efforts of U.S. collection methods. As the United States shares its collection methods more widely with the international community to support worthwhile international efforts, the underlying algorithms and weaknesses of these methods will become much more widely understood. It is inevitable that these "best available" collection methods are not going to be as good five years from now as they were five years ago when they face a determined nation that wants to avoid detection of its activities.

New Investment Requirements. If existing U.S. intelligence sources and methods are being degraded by sharing—and there seems little doubt that they are—then new investments in upgraded technical means are necessary to maintain a qualitative edge.[9] These investment decisions should also pay considerable attention to developing technical systems that have pathways by which to release information that will not degrade core collection capabilities. During the Cold War little attention had to be given in assessing investment decisions on collection technologies as to whether the release of their product would degrade the underlying collection method. The rule was to limit the release and thereby protect the method. In the post–cold war world intelligence sharing will become the norm, and consequently technical selection criteria used in new investment decisions must reflect this choice.

Analysts and Deception

The front line in the struggle to penetrate denial and deception efforts will remain the intelligence analysts. The challenge they face is becoming much more difficult.

Virtual Proliferation and Virtual Deception. The scale of the Iraqi program—large and expensive, with a dedicated weapons cycle—may turn out to be a thing of the past and not representative of the character of future weapons programs. The industrial base of many middle-size, developing countries is such that a modest chemical and biological weapons program could be easily accommodated within their open civilian infrastructure, without the construction of significant dedicated facilities and with only trivial and short fused adjustments associated with final weaponization. The same peaceful petrochemical projects that make agricultural pesticides can make advanced nerve gases; a well-equipped university or hospital biochemical laboratory is quite adequate for the production of many biological warfare agents. Even in the nuclear arena, the civil and nuclear infrastructure of countries such as Taiwan and the Republic of Korea, and most of Western Europe, could easily and quickly produce a small number of simple nuclear weapons with little new construction and few warning signatures. In sum, the United States is entering a period of virtual proliferation in which capabilities are generally available and the real question becomes one of motivation and intentions. In such an environment the task of intelligence collection and analysis becomes harder as the requirements for denial and deception become easier. Discerning capabilities will become less significant than discerning intentions absent any detectable change in capabilities.

Spread of Deception Technology. Technologies that are capable of supporting robust deception efforts are now rapidly making their way into the civil economy, among them pollution control technology, data encryption, and electronic emission control. The most revolutionary advances are in the area of pollution control technology, which offers the capability to substantially reduce and even eliminate off-site emissions of particles, effluents, and thermal and electromagnetic signatures. Such technologies are generally not subject to export controls or even tracking of foreign sales, and yet they strike at the heart of MASINT capabilities. Even newer technologies that, for example, are able to actively adjust their emissions to emulate various structures and processes, are emerging in the commercial entertainment market.

When the Standard of Success Becomes Not Conviction but Uncertainty' OR Why "Shopping for Intelligence" Makes Deception Easier. A deception campaign that results in bureaucratic paralysis through uncertainty or disagreements as to capabilities and intentions will often be sufficient to obtain a state's objective—it will not need to reach the higher "gold" standard of widespread belief in the deception. Governmental authority, including the intelligence community, is becoming more fragmented both within states and between states. This fragmentation in authority very often gives an effective veto to the unconvinced. It is the rare policymaker who, if he or she desires, cannot find some element of the intelligence community that will either support his or her view with intelligence or question the foundation of the views of others. The objective of deception that seeks to delay or avoid action or to place a benign explanation in play has only to be to convince one and not the many—a far easier standard than in even the recent past.

In the case of WMD, the period of maximum vulnerability of a proliferator is precisely during the period of development of the first few weapons. Once, for example, one or two nuclear weapons are developed and hidden, the attraction of preemptive military actions or economic sanctions diminishes very rapidly. Yet perversely, deception during the development phase is the easiest. Analysts tend to look for reprocessing facilities, uranium enrichment facilities, biological warfare sites that look like those of the United States. The challenge to disguise such sites to look benign or to adopt an obsolete technology with which the analysts may be unfamiliar is not too difficult. This challenge becomes even easier when only one or two analysts need be convinced that the reprocessing site might be a rayon factory and they are able to shop this view to policymakers who will often be seeking to avoid or delay hard choices.

Countering Proliferation Deception

This article could easily appear to be a paralyzing depiction of the possibilities of denial and deception operations. This is not my intent. The outlook for the success of deception is depressing and the problem will become tougher, but there are actions and management steps that, if taken, can provide a far better chance of successfully defeating deception attempts.

Three "principles" should be borne in mind.

- Niches or the "it's not your account" type of thinking/management make successful deception more likely. Intelligence analysis is inherently a compartmentalized and specialized activity and this is never more so than when significant downsizing looms. On the other hand, successful detection of deception operations rests fundamentally on the ability to question the accuracy of observable facts and interpretations. Something does not "seem right," it "does not feel right," it "fits together too well," or "can that sensor be spoofed by . . .?" These are all questions and surmises that can easily be rebuffed by organizational and turf barriers. Other damaging attitudes are, "you don't need to worry about my platform," "the PIs [photo interpreters] are the best in the business," or the ultimate put-off, "that's my job to worry about that, you just do yours."
- Detection of deception should not be ghettoized in the realm of counterintelligence or those who run collection systems. Obviously counterintelligence and managers and developers of collection systems have to worry about deception and can be a valuable resource in its detection. As deception technologies have improved, however, the detection of deception is less likely than in the past to result from a failure in the basic deception, that is, for example, a failure to mask all emissions. The keys in the future will more likely come from evidence of motivations and intentions or an unease about an overall pattern of activity rather than any specific, observable deception. If this is true, then all analysts need to be aware of deception possibilities and able to question all "facts."
- Believe deception can happen and study how it has and can happen. The message that needs to be at the forefront of every analytical activity is that *when the observable facts can be altered by deception measures, then reason is not a guarantee against deception.* This is a difficult message to get across in operations where the worth of the analytical product tends to be judged by the analyst's logical rigor, and it becomes even more difficult to communicate this message of caution at a time of generational

change in a bureaucracy. Deception operations, such as Iraq's, and new deception possibilities need to be studied and not just by specialists. For example, it would be interesting to know how many analysts have participated in reviews of the impact of environmental control technologies on their ability to understand what is happening in their region of concern or how many analysts track the movement of such technologies into their regions. Analysts need to be given new tools to aid in the identification of deception activities.

Policy Nexus: What U.S. Interests Are Affected by Deceptions?

There is a tendency to view the growing likelihood of successful deceptions as only an intelligence problem. Surrounded by a fog of uncertainty, intelligence assessments have never been easy. Active deception has only increased the inevitable ambiguity that is associated with intelligence. Although this reasoning is no doubt soundly based, it misses the new context of the post–cold war world. Deception threatens two important aspects of U.S. national security policy for coping with emerging threats.

Ability to Build and Maintain Coalitions and Control Regimes

The proliferation of WMD and means to deliver them is now recognized as one of the highest-priority threats to the United States. Although there is no single answer as to how to meet this threat, it is clear that a major thrust will have to be to maintain effective technology control regimes that limit the further spread of such technologies. Technology control regimes (e.g., the IAEA, the Missile Technology Control Regime, the Australia Group, and the Chemical Weapons Convention [CWC]), to be effective, increasingly depend upon reliable, shareable intelligence. If deception activities make this intelligence appear to be less reliable or fear of deception makes it less shareable, then these regimes will lose their effectiveness. Even if the U.S. intelligence community is not deceived, but allied services are, then their willingness to support multilateral steps or even unilateral national controls will be lessened. Treaty compliance regimes such as the IAEA and CWC inevitably will operate with sources and methods that are less robust against deception than those of the United States. The willingness to share U.S. knowledge of deceptions with such regimes will have to struggle with the reluctance arising from loss of sources and methods. This means that such regimes will often (usually) operate with less than the best available intelligence and may not be able to detect or even confirm violations—even those that the United States may have a high confidence exist. The potential for such a process to unleash a corrosive type of relationship, both with the international control regimes and in bilateral relationships, has already been foreshadowed to some extent in efforts to keep WMD material out of the hands of Iraq and Iran.

Beyond multilateral technology controls are the issues of effective national export controls. General national export controls are rapidly disappearing and if replaced at all will only be replaced by limited controls on a few target states of notoriously bad character. Deception practices will at a minimum delay the targeting of such offenders and may in many cases make it impossible to build an effective case for controls until it is too late to check the proliferation.

Military Planning and Action

It is a military axiom that you cannot target what you cannot find, and that you only target what you find. Programs to develop WMD are most vulnerable to military and political actions before any weapons have ever been produced. At such times there are only limited concerns over collateral damage from preemptive military attacks and no concern that a state might be able to reply with a WMD. On the other hand, absent compelling evidence of such a weapons program, most Western governments will have considerable difficulty in embarking upon preventive military action. At this stage the role of deception is to raise doubt as to the true nature of activities of concern.

Although greater attention is now being given to the use of deception to actively shield a program, less attention has been given to deception that attempts to raise concern that a state may already have acquired such weapons. The most effective purpose of a small number of first generation nuclear weapons (a small number is probably any number more than 0 and less than 10) is intimidation and breaking up coalitions. U.S. military forces are rapidly changing from a forward deployed force to a power projection force that must depend upon forward bases and long logistical trains to operate. If those whose interest the United States is defending and those upon whose territory it relies for operational support believe that they are threatened by a nuclear adversary, they may well see political compromise and negotiations as superior to combat. Even the United States is likely to show less relish for military options that raise the possibility that it may actually have to face a nuclear or biological attack. In such cases deception may be used effectively to raise the serious possibility either that a country has a nuclear weapon where no weapon at all may exist or that it is likely to use a WMD if forced into combat. If deception practices make it impossible for the intelligence community to give confident answers as to whether a country has a nuclear weapon or is likely to use one, and the location of such weapons cannot be given with confidence, then military and political action will become very unattractive. The intelligence community will then be the certain target of the charge that a "failure" of intelligence has left the policymakers with no choices except accommodation or unacceptable risks.

The views expressed in this article are solely those of the author.

Notes

1. "The point is not how wrong the United States was about Iraq's timetable for acquiring a bomb, but rather how greatly the United States underestimated the magnitude of the Iraqi covert effort. As it stands, such a massive miscalculation of a nation's capability, high or low, can surely happen again." John M. Deutch, "The New Nuclear Threat," *Foreign Affairs* 71 (Fall 1992), p. 128.

2. Conversation with the author, when the latter was serving as chief inspector on the Sixth UNSCOM/IAEA Nuclear Inspection Team in July 1991.

3. Mr. Al-Ashimi, Iraq's delegate to the IAEA's General Conference in 1981, declared that Iraq "had been one of the first countries to ratify the Treaty on the Non-Proliferation of Nuclear Weapons (NPT), despite the discrimination introduced by that Treaty . . . the Iraqi Government had always given its full support to the Agency's activities with respect to the peaceful uses of nuclear energy and the control of proliferation." IAEA, 24th Regular Session, GC (XXIV) OR.220, Vi-

enna, February 1981. Iraq was represented at the IAEA at the time the Gulf war began by a former chairman of the Iraq Atomic Energy Commission who had held that office when Iraq's program gained speed in the early 1980s.

4. Congressional Research Service, Library of Congress, "Analysis of Six Issues about Nuclear Capabilities of India, Iraq, Libya, and Pakistan" (Washington, D.C.: GPO, January 1982), pp. 37–39.

5. The newsletter *Nuclear Fuel* reports that several shipments of preformed tubes for scoops in gas centrifuges from the German metalworking firm Team GmbH were shipped to Pakistan after being declared as bodies for ballpoint pens. *Nuclear Fuel*, June 20, 1994, pp. 9–11.

6. I dealt for many months with a senior Iraqi scientist whose entire university training, from undergraduate to doctorate, had been in the United States and whose first job had been at a U.S. nuclear power plant, and yet basic data on or pictures of this key individual could not be found.

7. Although this was not the first time that intelligence information had been made available to the IAEA—intelligence-based briefings had been supplied earlier on several suspect nuclear programs—the breadth of information and the openness of the sharing during the Iraqi inspections has been so great as to mark a distinctly different approach.

8. The IAEA had not a single secure telephone or fax when the inspections began nor a single secure conference room.

9. It is striking that, at least in the unclassified testimony, U.S. national security officials have not shared with Congress this concern as they try to develop broader support for continued investment in technical collection methods.

Section VIII: Irregular Warfare and Non-State Actors

Introduction

Terrorists and Other Non-State Actors—Constraints and Opportunities

Warfare has often been described as symmetric, where belligerents have similar power and available resources and generally use similar strategies and tactics. World Wars I and II were mostly fought symmetrically. Asymmetric warfare describes conflicts where absolute power and resources are significantly different among warring factions. Accordingly, for the weaker belligerents, strategies and tactics must be sufficiently different to offset their quantitative and qualitative disadvantages. To be sure, the stronger actor must also make adjustments to prevail. The American Revolution was mostly fought asymmetrically, especially in the south. The symmetries of warfare have generally focused on power and tactical differences but have included disparities in the interests of the belligerents as well.[1]

Contemporary terrorist and insurgent threats have forced the academic and military communities to breathe new life into the study and practice of asymmetric warfare in general and warfare against nonstate and illicit actors in particular. Previous research attempted to explain how weaker actors prevail over stronger opponents, the nature of the strategies used by successful weak and strong actors, and occasionally, but not sufficiently, the nature of the weaker actor. There has been little research conducted on the relationship between stratagem and asymmetric warfare.

Weaker actors, whether they are insurgents, terrorists, state proxies or sympathetic groups, are forced to operate in the shadows, even covertly, to survive. This point is non-trivial and has important ramifications for denial, deception and ultimately for the scope of political-military strategy and operations. For nonstate and illicit actors, an additional geometric element, an imbalance, exists between the requirement for deniability and the need to conduct operations. The revelation of any substantive information about a nonstate or illicit actor can be lethal to their very existence.

Accordingly, illicit groups and nonstate actors must practice denial as a strategic necessity, not as a choice. If these actors trifle with this requirement they risk being rolled up and destroyed by the state. J. Bowyer Bell clearly points out that the greater the need for denial, the greater the cost in operational efficiency. The practical implication of devoting extensive resources towards concealing one's existence and activities is that there are minimal residual resources for strategic deception or the strategic offense. In other words, the greater the need for denial, the less the likelihood of deception. The

1. Andrew Mack, "Why Big Nations Lose Small Wars," *World Politics*, Vol 27, No. 2, January 1975, p. 181.

occurrence of tactical surprise is much more likely and is necessary to advance, or at least sustain, an illicit actor's cause.[2]

Correcting this imbalance between the requirement for deniability and the need to conduct operations is fundamental to success for illicit groups. This correction must involve increasing the legitimacy or justice of the cause within the geographic bounds of the conflict as well as internationally. As the perception of legitimacy increases, the struggle can slowly move out of the shadows and operate more overtly while maintaining sufficient security for organizational survival.[3]

For illicit groups, the future of strategic deception may be directly linked to the perception of legitimacy or justice of the cause. In fact, deception can have a significant role in defining the nature of the struggle itself.[4] Illicit organizations often have affiliated legitimate or semi-legitimate groups operating to inform target audiences and frame the struggle in terms of justice. Affiliated radical groups can easily operate in open societies outside the geographic area of the fight. In other words, strategic deception in the modern information environment can disguise the true nature of a political-military struggle. Strategic deception can, therefore, help correct the imbalance between the requirement for deniability and the need to conduct operations for illicit groups. The greater the perceived legitimacy, the less the need for strategic denial.

In conclusion, strategic deception by illicit groups who do not have significant perceived legitimacy is highly unlikely; the requirement for strategic denial is simply too great. Establishing legitimacy by deceptive means can offer illicit groups the freedom of maneuver necessary to redirect resources for the offense, shored up by strategic deception. For the state, delegitimizing the illicit actor by every means available, including deception, can ultimately render the illicit actor irrelevant, or even better, result in his destruction. However, one should never forget that self-deception is also a significant threat if one unequivocally believes that illicit groups are incapable of strategic deception.

2. J. Bowyer Bell, "Conditions Making for Success and Failure of Denial and Deception: Nonstate and Illicit Actors," in *Strategic Denial and Deception*, Roy Godson and James Wirtz, editors, London: Transaction Publishers, 2002, p. 129.
3. Ibid., 134.
4. Ibid., 141.

CHAPTER 35

Conditions Making for Success and Failure of Denial and Deception: Nonstate and Illicit Actors[1]

J. Bowyer Bell

J. Bowyer Bell received his doctorate from Duke University and went on to teach at M.I.T., Harvard and Trinity Universities. While studying the Middle East, Bell became interested in the tactics and techniques of the IRA, and subsequently published his book, "The Secret Army: The IRA 1916–1970." Continuing his work in war and peace studies, Bell founded the International Analysis Center, whose clients included the United States Department of Justice and the Central Intelligence Agency. Also an artist, the author painted scenes from the conflicts he witnessed. Since his death in 2003, his paintings continue to be shown around the world.

There are many kinds of nonstate actors: transnational companies, failed states, revolutionary organizations, or religious orders. Many operate secretly and some illegally. Organizations and movements, defined by the recognized and legitimate as illicit, must seek cover to operate. Some organizations are thus partly covert while others, particularly the most illicit, are apt to be mainly if not entirely, covert. For these latter groups to operate, denial is so vital that it becomes a strategic necessity achieved by the creation of a real and conceptual underground. Both rebels and criminals, just like lethal cults or business agents stealing patents, depend upon cover to operate illicitly; but the clandestine world of the terrorist or the criminal is an alternative reality rather than a mere matter of hiding the operations of a journalist or a banker. This denial through the fabrication of a clandestine underground world is a strategic construct for the most illicit organizations. Such strategic denial permits operations to go forward, although at a high cost in efficiency.

Those operating most covertly, in fact, seldom have the resources, time or opportunity for strategic deception planning. Instead they achieve organizational goals by relying on operational surprise, effective denial and the vulnerabilities of the opponent. Only rarely does strategic denial allow strategic deception planning, and then most often to groups with some legitimacy, with havens, with conventional resources, or to those not assumed dangerous by the le-

1. J. Bowyer Bell, "Non State and Illicit Actors," in *Strategic Denial and Deception: The Twenty-First Century Challenge*, Roy Godson and James J. Wirtz, editors, NJ: Transaction Publishers, 2002, pp. 129-131, 159-161.

gitimate. Thus, the more illicit and hidden the group, the more difficulty members have organizing strategic deception. A movement may be assumed to be benign, thereby maintaining denial and deceiving the perceived enemy. But once an asymmetrical conflict arises, the resources demanded by denial, maintenance of the organization, or normal operations absorb most organizational resources.

As a result, the threat of strategic deception rather than strategic denial by illicit organizations is real but rare except on a tactical level. Regimes that seek to co-opt such organizations often find the capacities and discipline of the underground less than anticipated and control elusive. State-sponsored terrorists have state assets while a revolutionary terrorist relies on perceived truth, will and history. Yet, strategic deception does exist and often does threaten national stability, global order, and the innocent—and such deception is hidden by the strategic denial of illicit organizations.

Strategic denial is assured by the creation of a parallel world that both exhilarates and protects, offers haven and enhances the life of the committed. This is as true for the Sicilian Mafias who seek respect, power, money, and at times display as it is for the terrorist who seeks to change the direction of history through recourse to violence. The criminal flourishes in the hidden world protected by denial while the mobster uses denial to get out of the hidden world into legitimate power. In both cases denial is a strategic necessity, not a tactical choice nor a cluster of ruses. A cult may be a secret world of the chosen that is perceived as benign by outsiders until its members commit murder. The known criminal and the potential gunman who focus public attention are the types of individuals who require denial to operate on a day-to-day basis. The Islamic fundamentalists operating in New York in the nineties offer an insight into the dynamics of strategic denial: the creation of a parallel reality as cover and haven.

Strategic deception is rare because of a lack of resources and talent. In the cases of the 1993 Holy Jihad World Trade Center bombing in New York, the Irish Republican Army's preparations for a three-stage campaign in January 1970, and the Sicilian Mafia's covert participation in the drug trade after 1957, the requisite talent and means were available. Most illicit criminal and political organizations, covert and limited, conservatively pursue advantage with violence rather than cunning, with technical and tactical deceit rather than elegant strategic planning. For most illicit organizations, denial is a strategic necessity that can afford protection, continuity, a base and reassurance but rarely tactical assets. Severely limited in most cases by states, illicit organizations can rarely afford time dedicated to strategic deception, are inclined to reveal the truth as they perceive it, and tend to rely on techniques and tactics to purse advantage. Time, tangible assets, and technological skills often will allow strategic deception; but few criminals and fewer rebels have the resources to plan strategically. Yet, potential targets would be deceived if they imagine that such deception is never a threat.

Conclusion

Deception is integral to illicit organizations. Cover is a strategic necessity that, at a great cost in efficiency, allows illicit organizations the opportunity to achieve their objectives. The revolutionary organization is, in fact, an alternative reality that seeks to be invisible except when it needs to propagate a version of the truth and to bolster the impact of its armed operations. In the case of the truth, remarkably little deception is deployed. The revolutionaries consider their faith to be inviolate and convincing, without need of apology or equivocation. Reality is often visible. Islamic Jihad preached the truth, the IRA hid it for two years and the Sicilian Mafia makes the hidden truth yield public respect. In operational matters, much that takes place underground is focused on maintenance, supply and distribution, command and control, and communication that needs to be hidden but need not otherwise be duplicitous. Secrecy apart, most operations focus on obvious targets. Strategic deception is rare.

An organization about to embark on an armed struggle usually does so after much contemplation. At the beginning, there is time and motive for deception. Most rebels tend to be truthful about their intentions, means and ultimate aspirations. Most criminals have no choice and most cults are benign. Thus, the big strategic questions usually have been narrowed to the date to begin operations. Once committed, the organizations rarely have the resources or time to make strategic deception planning possible. By contrast, deception is readily found in tactical operations. Rebels have threatened what they did not plan to do or used false warnings and fake devices as ruses to pave the way for the real operation. Many groups seek to project greater assets or greater legitimacy, but most deception is apt to be an operational, technical tool. Such tactical deception, over time, can become quite sophisticated. What is most often sought, however, is simple surprise rather than combining a series of ruses in a grand tactical campaign.

The prime reason that deception planning is scanty and strategic deception operations are rare underground is that time and resources are lacking. The operational leadership is often unable to meet. Planning is by rote, the desire for momentum overwhelming the opportunities for novelty. Outsiders assume that the underground is simply a secret version of everyday life. They assume that in the underground planning and freedom to act are institutionalized. This is almost never the case. Nothing is easy underground. Planning can be halting and communications are always faulty. Faith supplies the energy for an armed struggle, but always at the cost of efficiency. Deception is perceived as a luxury while denial becomes second nature as part of the underground reality. Deception in operational matters is deployed to assure secrecy and at times as an integral part of tactics.

Denial makes everything possible, and on occasion tactical deception makes the spectacular possible for the revolutionary. Unexpected violence can have an exaggerated effect on events. The revolutionary often can be assured that the initial attack or operation will be successful.

Illicit organizations and movements must seek cover to operate effectively. Some organizations are thus partly covert while others, particularly criminals and revolutionaries, are entirely covert. For them denial is so vital that it becomes a strategic necessity achieved by the creation of an alternative reality. Denial is achieved not through planning but mutual possession of revealed truth. Tactical deception arises from the cover so provided as well as necessity, habit and custom, experience, and luck. Those operating covertly seldom have the resources, time or opportunity for strategic deception planning, and instead rely on operational deception to achieve surprise. Such tactics are made possible by effective denial that focuses on the opponent's vulnerabilities. Every armed struggle employs technical and tactical deception as well as strategic denial. With time, a safe haven, and sufficient assets, strategic deception planning is more likely. Thus, the more illicit and hidden the group, the more difficult it is for them to organize strategic deception.

CHAPTER 36

The Evolution of a Revolt[1]

Lieutenant-Colonel T.E. Lawrence

In this too often overlooked landmark paper, Lawrence of Arabia presents the first theory of what he calls "irregular war," namely asymmetrical combat between regular military units and "irregulars." It is based on his closely involved experience with the Arab Revolt against the Turkish Army occupiers in the Arabian Peninsula, from its inception in June 1916 until victory in Palestine and Syria in 1918. Although Lawrence exaggerated the effectiveness of his guerrilla operations in diverting substantial Turkish strength from the Palestinian front, his theory stands on its own.

Lawrence had been invited by his former colleague, Col. Guy Dawnay, to contribute an article to the inaugural edition of editor Dawnay's "Military Quarterly." Accordingly, Lawrence drew from the sections on military strategy in the second and later destroyed draft of his monumental "Seven Pillars of Wisdom," which wouldn't be published in its world-famous third draft until 1926. There he would recycle his "The Evolution of A Revolt" article into Chapters 33 and 59. Still later, with Lawrence's permission, Capt. Liddell Hart lightly edited this piece for his article, "Science of Guerrilla Warfare," which he published above Lawrence's initials in the Encyclopedia Britannica, 14th Edition (1929). Liddell Hart sent the 15 guinea fee along to Lawrence, who in delight wrote back, "The cheque delights me. I fancy Guy Dawnay only paid £10 for the original article."

Lawrence, a most unusual Englishman, had graduated Oxford in 1910 with a degree in history and a deep interest in archaeology. He served in Military Intelligence during World War I, during which he rose to the rank of lieutenant colonel. Afterwards he championed Arab rule in the Near East.

A major part of Lawrence's operations was ambushing the Turkish trains moving on the Hejaz Railway. However, he had deliberately left out the one detail that explains how he could know when and where to hit these trains, or even how he knew which kinds of supplies or troop reinforcements each train carried. How could he hit them with minimal chance of being outgunned, much less counter-ambushed? He was assured of this crucial intelligence through a cunning deception operation that he either devised himself or, more likely, inherited. Lawrence, not a modest man, kept this secret during his lifetime only because it involved one of the deepest secrets of British Military Intelligence—namely, the British were freely reading all Turkish codes except two, the Ildirim Radio Code and the Telegraph Code. By constantly cutting the railway telegraph lines, Lawrence forced the Turks to switch all railway communications to radio, thereby unintentionally but automatically sharing this private information with British Intelligence. Although Lawrence's exploitation of this closely-held British SIGINT

1. Lieutenant-Colonel T.E. Lawrence, "The Evolution of a Revolt," *Army Quarterly*, Vol.1, No.1, London: October, 1920, pp. 55-69.

secret was published by Liddell Hart in 1938,2 its significance wasn't noticed until recently through the research of Yigal Sheffy (1998), an Israeli military historian and former IDF intelligence officer.

The Arab Revolt began in June, 1916, with an Arab offensive, a surprise attack by the half-armed and inexperienced tribesmen upon the Turkish garrisons in Medina and about Mecca. They had no success, and after a few days' effort they withdrew out of range of the fort artillery, and began a blockade. This method forced the early surrender of Mecca, whose road communications were too long and rough to be held by the Turks. Medina, however, was linked by railway to the Turkish main Army in Syria, and, thanks to their superior numbers and equipment, the Turks were able in a week's fighting to restore the line and reinforce the temporarily-besieged garrison there. The Arab forces which had attacked it fell back gradually as the Turks became more offensive, and at last moved fifty miles south-west into the hills, and there took up a position across the main road to Mecca. At this point the campaign stood still for many weeks, while both sides breathed, and the Turks prepared to take the initiative, by sending an expeditionary force to Mecca, to crush the revolt where it had started. They moved an army corps to Medina by rail, and strengthened it beyond establishment with guns, cars, aeroplanes, machine guns, and quantities of horse, mule and camel transport. Then they began to advance down the main western road from Medina to Mecca. The total distance was about two hundred and fifty miles. The first fifty miles were easy: then came a belt of hills twenty miles wide, in which were Feisal's tribesmen standing on the defensive: after the hills was a level stretch, for seventy miles along the coastal plain to Rabegh, rather more than half-way. Rabegh is a little port on the Red Sea, with good anchorage for ships. In it was Sherif Ali, Feisals eldest brother, with more tribal forces, and the beginnings of an Arab Regular Army, recruited from officers and men of Arab Blood, who had served in the Turkish Army, and were now willing to fight against their old masters for their national freedom.

Our military advisers had told us that Rabegh was the key of Mecca, since no hostile force could pass along the main road without occupying it and watering at its wells under the palm trees. Its defence was therefore of the main importance. The Navy could cooperate effectively from the harbour, and the circle of the palm-groves must be laid out as an entrenched position, and held by regular troops. They thought that Beduin tribesmen would never be of any value in a fixed position, and that therefore an Arab regular force must be formed and trained as soon as possible to undertake this duty. If the Turks advanced before the new force was ready, the British would have to lend a brigade, of British or Allied troops, to save the Sherif in his extremity, by maintaining this stop-block.

A personal reconnaissance of the Arab positions, here and in the hills where Feisal was, caused me to modify the views of the experts slightly. Feisal had some thousands of men, all armed with rifles, rather casual, distrustful fellows, but very active and cheerful.

2. *T. E. Lawrence to His Biographer, Liddell Hart* (New York: Doubleday, Doran & Co., 1938), 115, reprinting Liddell Hart, *T. E. Lawrence in Arabia and After* (London: Cape, 1935), 313. Lawrence's note reads: "[P]art of the trouble was that we did not use W/T for 'operation' messages. Allenby every morning for breakfast had the log of Turkish signals over the preceding 24 hours: we read their every message —and I presume they read all of ours. To keep our moves secret we used air-mail or word of mouth. To keep the Turks public, one of my cares was to distribute wirecutters over their rear, and cut their telegraph at least daily."

They were posted in hills and defiles of such natural strength that it seemed to me very improbable that the Turks could force them, just by their superior numbers: for in some ways it is easier to defend a range of hills against nine or ten thousand men than against nine or ten. Accordingly, I reported that the tribesmen (if strengthened by light machine guns, and regular officers as advisers) should be able to hold up the Turks indefinitely, while the Arab regular force was being created. As was almost inevitable in view of the general course of military thinking since Napoleon, we all looked only to the regulars to win the war. We were obsessed by the dictum of Foch that the ethic of modern war is to seek for the enemy's army, his centre of power, and destroy it in battle. Irregulars would not attack positions and so they seemed to us incapable of forcing a decision.

While we were training the regulars (of course not sending officer or light machine guns to Feisal in the hills meanwhile), the Turks suddenly put my appreciation to the test by beginning their advance on Mecca. They broke through my "impregnable" hills in twenty-four hours, and came forward from them towards Rabegh slowly. So they proved to us the second theorem of irregular war—namely, that irregular troops are as unable to defend a point or line as they are to attack it.

This lesson was received by us quite without gratitude, for the Turkish success put us in a critical position. The Rabegh force was not capable of repelling the attack of a single battalion, much less of a corps. It was nearly impossible to send down British troops from Egypt at the moment: nor do I think that a single British brigade would have been capable of holding all the Rabegh position: nor was the Rabegh position indispensable to the Turks: nor would a single Arab have remained with the Sherif if he introduced British troops into the Hejaz.

In the emergency it occurred to me that perhaps the virtue of irregulars lay in depth, not in face, and that it had been the threat of attack by them upon the Turkish northern flank which had made the enemy hesitate for so long. The actual Turkish flank ran from their front line to Medina, a distance of some fifty miles; but, if we moved towards the Hejaz j railway behind Medina, we might stretch our threat (and, accordingly, their flank) as far, potentially, as Damascus, eight hundred miles away to the north. Such a move would force the Turks to the defensive, and we might regain the initiative. Any-how, it seemed our only chance, and so, in January, 1917, we took all Feisal's tribes-men, turned our backs on Mecca, Rabegh and the Turks, and marched away north two hundred miles to Wejh, thanks to the help of the British Red Sea Fleet, which fed and watered us along the coast, and gave us gun-power and a landing party at our objective.

This eccentric movement acted like a charm. Clausewitz had said that rearguards modulate the enemy's action like a pendulum, not by what they do, but by their mere existence. We did nothing concrete, but our march recalled the Turks (who were almost into Rabegh) all the way back to Medina, and there they halved their force. One half took up the entrenched position about the city, which they held until after the Armistice. The other half was distributed along the railway to defend it against our threat. For the rest of the war the Turks stood on the defensive against us, and we won advantage over advantage till, when peace came, we had taken thirty-five thousand prisoners, killed and wounded and worn out about as many, and occupied a hundred thousand square miles of the enemy's territory, at little loss to ourselves.

However, we were not then aware that Wejh was our turning point. We thought we had come to it to cut the railway, and I was at once sent up country to do this, as a means to take Medina, the Turkish headquarters and main garrison. On the way up I fell ill, and spent ten days on my back in a tent, without anything to do except to think about war and analyse our hitherto empirical practice for its real import.

I was unfortunately as much in charge of the campaign as I pleased, and had had no training in command to fit me for such a work. In military theory I was tolerably read, for curiosity in Oxford years before had taken me past Napoleon to Clausewitz and his school, to Caemmerer and Moltke, Goltz and the recent Frenchmen. These had seemed very partial books, and after a look at Jomini and Willisen I had found broader principles in the eighteenth century, in Saxe, Guibert and their followers. However, Clausewitz was intellectually so much the master of them all that unwillingly I had come to believe in him. Tactically the only campaigns I had studied step by step were the ancient affairs of Hannibal and Belisarius, Mohammed and the Crusades! My interests were only in pure theory and I looked everywhere for the metaphysical side, the philosophy of war, about which I thought a little for some years. Now I was compelled suddenly to action, to find an immediate equation between my book-reading and our present movements.

However, the books gave me the aim in war quite pat, "the destruction of the organized forces of the enemy" by "the one process battle." Victory could only be purchased by blood. This was a hard saying for us, as the Arabs had no organized forces, and so a Turkish Foch would have no aim: and the Arabs would not endure casualties, so that an Arab Clausewitz could not buy his victory. These wise men must be talking metaphors, for we were indubitably winning our war . . . and as I thought about it, it dawned on me that we had won the Hejaz war. We were in occupation of 99 percent of the Hejaz. The Turks were welcome to the other fraction till peace or doomsday showed them the futility of clinging to our window pane. This part of the war was over, so why bother about Medina? It was no base for us, like Rabegh, no threat to the Turks, like Wejh: just a blind alley for both. The Turks sat in it on the defensive, immobile, eating for food the transport animals which were to have moved them to Mecca, but for which there was no pasture in their now restricted lines. They were harmless sitting there; if we took them prisoner, they would cost us food and guards in Egypt: if we drove them out northward into Syria, they would join the main Army blocking us in Sinai. On all counts they were best where they were, and they valued Medina and wanted to keep it. Let them!

This seemed unlike the ritual of war of which Foch had been priest, and so I began to hope that there was a difference of kind between us and him. He called his modern war "absolute." In it two nations professing incompatible philosophies set out to try them in the light of force. A struggle of two immaterial principles could only end when the supporters of one had no more means of resistance. An opinion can be argued with: a conviction is best shot. The logical end of a war of creeds is the final destruction of one, and Salammbo the classical textbook-instance. These were the lines of the struggle between France and Germany, but not, I thought, between Germany and England, for all efforts to make our men hate the enemy just made them hate war, and later on by the Armistice we made the Great War *fall* short of the Foch ideal. To me it seemed only a variety of war: and I could then see other sorts, as Clausewitz had numbered them, personal wars for dynastic reasons, expulsive wars for party reasons, commercial wars for trading reasons.

Then I thought of the Arab aim, and saw that it was geographical, to occupy all Arabic-speaking lands in Asia. In the doing of it we might kill Turks: we disliked them very much. Yet "killing Turks" would never be an excuse or aim. If they would go quietly, our war would end. If not, we would try to drive them out: in the last resort we would be compelled to the desperate course of blood, on the maxim of "murder" war, but as cheaply as possible for ourselves, since the Arabs were fighting for freedom, a pleasure only to be tasted by a man alive.

My own personal duty was command, and I began to unravel command and analyse it, both from the point of view of strategy, the aim in war, the synoptic regard which sees everything by the standard of the whole, and from the point of view called tactics, the means towards the strategic end, the steps of its staircase. In each I found the same elements, one algebraical, one biological, a third psychological. The first seemed a pure science, subject to the laws of mathematics, without humanity. It dealt with known invariables, fixed conditions, space and time, inorganic things like hills and climates and railways, with mankind in type-masses too great for individual variety, with all artificial aids, and the extensions given our faculties by mechanical invention. It was essentially formulable.

In the Arab case the algebraic factor would take first account of the area we wished to conquer, and I began idly to calculate how many square miles . . . perhaps a hundred and forty thousand . . . and how would the Turks defend all that . . . no doubt by a trench line across the bottom, if we were an army attacking with banners displayed . . . but suppose we were an influence (as we might be), an idea, a thing invulnerable, intangible, without front or back, drifting about like a gas? Armies were like plants, immobile as a whole, firm-rooted, nourished through long stems to the head. We might be a vapour, blowing where we listed. Our kingdoms lay in each man's mind, and as we wanted nothing material to live on, so perhaps we offered nothing material to the killing. It seemed a regular soldier might be helpless without a target. He would own the ground he sat on, and what he could poke his rifle at.

Then I estimated how many posts they would need to contain this attack in depth, sedition putting up her head in every unoccupied one of these hundred thousand square miles. I knew the Turkish Army inside and out, and allowing for its recent extension of faculty by guns and aeroplanes and armoured trains, still it seemed it would have need of a fortified post every four square miles, and a post could not be less than twenty men. The Turks would need six hundred thousand men to meet the combined ill wills of all the local Arab people. They had one hundred thousand men available. It seemed the assets in this part of command were ours, and climate, railways, deserts, technical weapons could also be attached to our interests, if we realized our raw materials and were apt with them. The Turk was stupid and would believe that rebellion was absolute, like war, and deal with it on the analogy of absolute warfare. Analogy is fudge, anyhow, and to make war upon rebellion is messy and slow, like eating soup with a knife.

So much for the mathematical element, which I annoyed the others by calling hecastics. The second factor was biological, the breaking-point, life and death, or better, wear and tear. Bionomics seemed a good name for it. The war-philosophers had properly made it an art, and had elevated one item in it, "effusion of blood," to the height of a principle. It became humanity in battle, an art touching every side of our corporal being, and very war. There was a line of variability (man) running through all its estimates. Its components were sensitive and illogical, and generals guarded themselves by the device of a reserve, the significant medium of their art. Goltz had said that when you know the enemy's strength, and he is fully deployed, then you know enough to dispense with a reserve. But this is never. There is always the possibility of accident, of some flaw in materials, present in the general's mind: and the reserve is unconsciously held to meet it. There is a "felt" element in troops, not expressible in figures, guessed at by the equivalent of sσεa in Plato, and the greatest commander is he whose intuitions most nearly happen. Nine-tenths of tactics are certain, and taught in books: but the irrational tenth is like the kingfisher flashing across the pool, and that is the test of generals. It can only be ensured by instinct, sharpened by thought practising the stroke so often that at the crisis it is as natural as a reflex.

Yet to limit the art to humanity seemed to me an undue narrowing down. It must apply to materials as much as to organisms. In the Turkish Army materials were scarce and precious, men more plentiful than equipment. Consequently our cue should be to destroy not the Army but the materials. The death of a Turkish bridge or rail, machine or gun, or high explosive was more profitable to us than the death of a Turk. The Arab Army just now was equally chary of men and materials: of men because they being irregulars were not units, but individuals, and an individual casualty is like a pebble dropped in water: each may make only a brief hole, but rings of sorrow widen out from them. We could not afford casualties. Materials were easier to deal with and put straight. It was our obvious duty to make ourselves superior in some one branch, gun-cotton or machine guns, or whatever could be made most decisive. Foch had laid down the maxim, applying it to men, of being superior at the critical point and moment of attack. We might apply it to materials, and be superior in equipment in one dominant moment or respect.

For both men and things we might try to give Foch's doctrine a negative twisted side, for cheapness' sake, and be weaker than the enemy everywhere except in one point of matter. Most wars are wars of contact, both forces striving to keep in touch to avoid tactical surprise. Our war should be a war of detachment: we were to contain the enemy by the silent threat of a vast unknown desert, not disclosing ourselves till the moment of attack. This attack need be only nominal, directed not against his men, but against his materials: so it should not seek for his main strength or his weaknesses, but for his most accessible material. In railway cutting this would be usually an empty stretch of rail. That was a tactical success. We might turn the average into a rule (not a law—war is antinomian, said Colin), and at length we developed an unconscious habit of never engaging the enemy at all. This chimed with the numerical plea of never giving the enemy's soldier a target. Many Turks on our front had no chance all the war to fire a shot at us, and correspondingly we were never on the defensive, except by rare accident. The corollary of such a rule was perfect "intelligence," so that we could plan in complete certainty. The chief agent had to be the general's head (de Feuquiere said this first), and his knowledge had to be faultless, leaving no room for chance. We took more pains in this service than any other staff I saw.

The third factor in command seemed to be the psychological, that science (Xenophon called it diathetic) of which our propaganda is a stained and ignoble part. Some of it concerns the crowd, the adjustment of spirit to the point where it becomes fit to exploit in action, the prearrangement of a changing opinion to a certain end. Some of it deals with individuals, and then it becomes a rare art of human kindness, transcending, by purposeful emotion, the gradual logical sequence of our minds. It considers the capacity for mood of our men, their complexities and mutability, and the cultivation of what in them profits the intention. We had to arrange their minds in order of battle, just as carefully and as formally as other officers arranged their bodies: and not only our own men's minds, though them first: the minds of the enemy, so far as we could reach them: and thirdly, the mind of the nation supporting us behind the firing-line, and the mind of the hostile nation waiting the verdict, and the neutrals looking on.

It was the ethical in war, and the process on which we mainly depended for victory on the Arab front. The printing press is the greatest weapon in the armoury of the modern commander, and we, being amateurs in the art of command, began our war in the atmosphere of the twentieth century, and thought of our weapons without prejudice, not distinguishing one from another socially. The regular officer has the tradition of forty generations of serving soldiers behind him, and to him the old weapons are the most honoured. We had seldom to concern ourselves with what our men did, but much with what they thought, and to us the diathetic was more than half command. In Europe it was set a little aside and entrusted to men outside the General Staff. In Asia we were so weak physically that we could not let the metaphysical weapon rust unused. We had won a province when we had taught the civilians in it to die for our ideal of freedom: the presence or absence of the enemy was a secondary matter.

These reasonings showed me that the idea of assaulting Medina, or even of starving it quickly into surrender was not in accord with our best strategy. We wanted the enemy to stay in Medina, and in every other harmless place, in the largest numbers. The factor of food would eventually confine him to the railways, but he was welcome to the Hejaz railway, and the Trans-Jordan railway, and the Palestine and Damascus and Aleppo railways for the duration of the war, so long as he gave us the other nine hundred and ninety-nine thousandths of the Arab world. If he showed a disposition to evacuate too soon, as a step to concentrating in the small area which his numbers could dominate effectively, then we would have to try and restore his confidence, not harshly, but by reducing our enterprises against him. Our ideal was to keep his railway just working, but only just, with the maximum of loss and discomfort to him.

Accordingly, I put in a few damages to the line, enough to annoy the enemy without making him fear its final destruction, and then rode back to Wejh, to explain to my chiefs that the Arab war was geographical, and the Turkish Army for us an accident, not a target. Our aim was to seek its weakest link, and bear only on that till time made the mass of it fall. Our largest available resources were the tribesmen, men quite unused to formal warfare, whose assets were movement, endurance, individual intelligence, knowledge of the country, courage. We must impose the longest possible passive defence on the Turks (this being the most materially expensive form of war) by extending our own front to its maximum. Tactically we must develop a highly mobile, highly equipped type of army, of the smallest size, and use it successively at distributed points of the Turkish line, to make the Turks reinforce their occupying posts beyond the economic minimum of twenty men. The power of this striking force of ours would not be reckoned merely by its strength. The ratio between number and area determined the character of the war, and by having five times the mobility of the Turks we could be on terms with them with one-fifth their number.

Our success was certain, to be proved by paper and pencil as soon as the proportion of space and number had been learned. The contest was not physical, but mineral, and so battles were a mistake. All we won in a battle was the ammunition the enemy fired off. Our victory lay not in battles, but in occupying square miles of country. Napoleon had said it was rare to find generals willing to fight battles. The curse of this war was that so few could do anything else. Napoleon had spoken in angry reaction against the excessive finesse of the eighteenth century, when men almost forgot that war gave license to murder. We had been swinging out on his dictum for a hundred years, and it was time to go back a bit again. Battles are impositions on the side which believes itself weaker, made unavoidable either by lack of land-room, or by the need to defend a material property dearer than the lives of soldiers. We had nothing material to lose, so we were to defend nothing and to shoot nothing. The precious element of our forces were the Beduin irregulars, and not the regulars whose role would only be to occupy places to which the irregulars had already given access. Our cards were speed and time, not hitting power, and these gave us strategical rather than tactical strength. Range is more to strategy than force. The invention of bully-beef has modified land-war more profoundly than the invention of gunpowder.

My chiefs did not follow all these arguments, but gave me leave to try my hand after my own fashion. We went off first to Akaba, and took it easily. Then we took Tafileh and the Dead Sea: then Azrak and Deraa, and finally Damascus, all in successive stages worked out consciously on these sick-bed theories. The process was to set up ladders of tribes, giving us a safe and comfortable route from our sea-bases (Yenbo, Wejh or Akaba) to our advanced bases of operation. These were sometimes three hundred miles away, a long distance in lands without railways or roads, but made short for us by an assiduous cultivation of desert-power, control by camel parties of the desolate and unmapped wilderness which fills up all the centre of Arabia, from Mecca to Aleppo and Bagdad.

In character these operations were more like naval warfare than ordinary land operations, in their mobility, their ubiquity, their independence of bases and communications, their lack of ground features, of strategic areas, of fixed directions, of fixed points. "He who commands the sea is at great liberty, and may take as much or as little of the war as he will": he who commands the desert is equally fortunate. Camel raiding-parties, as self-contained as ships, could cruise without danger along any part of the enemy's land-frontier, just out of sight of his posts along the edge of cultivation, and tap or raid into his lines where it seemed fittest or easiest or most profitable, with a sure retreat always behind them into an element which the Turks could not enter. We were fortified in our freedom of movement by an intimate knowledge of the desert-front of Syria, a country peculiarly and historically indefensible against attack from the east. I had traversed most of it on foot before the war many times, working out the movements of Saladin or Ibrahim Pasha, and, as our war-experience deepened, we became adepts at that form of geographical intuition, described by Bourcet as wedding unknown land to known in a mental map.

Our tactics were always tip and run, not pushes, but strokes. We never tried to maintain or improve an advantage, but to move off and strike again somewhere else. We used the smallest force, in the quickest time, at the farthest place. If the action had continued till the enemy had changed his dispositions to resist it, we would have been breaking the spirit of our fundamental rule of denying him targets.

The necessary speed and range were attained by the extreme frugality of the desert men, and their high efficiency when mounted on their she-riding-camels. The camel is an intricate animal, and calls for skilled labour in the handling: but she yields a remarkable return. We had no system of supply: each man was self-contained and carried on the saddle from the sea base at which the raid started, six weeks' food for himself. The six-weeks' ration for ordinary men was a half-bag of flour, forty-five pounds in weight. Luxurious feeders carried some rice also for variety. Each man baked for himself, kneading his own flour into unleavened cakes, and warming it in the ashes of a fire. We carried about a pint of drinking water each, since the camels required to come to water on average every three days, and there was no advantage in our being richer than our mounts. Some of us never drank between wells, but those were hardy men: most of us drank a lot at each well, and had a drink during the intermediate dry day. In the heat of summer Arabian camels will do about two hundred and fifty miles comfortably between drinks: and this represented three days' vigorous marching. The country is not so dry as it is painted, and this radius was always more than we needed. Wells are seldom more than one hundred miles apart. An easy day's march was fifty miles: an emergency march might be up to one hundred and ten miles in the day.

The six weeks' food gave us a range of over a thousand miles out and home, and that (like the pint of water) was more than ever we needed, even in so large a country as Arabia. It was possible (for me, the camel-novice in the Army, "painful" was a better word) to ride fifteen hundred miles in the month without re-victualing, and there was never a fear of starvation, for each of us was riding on two hundred pounds of potential meat, and when food lacked we would stop and eat the weakest of our camels. Exhausted camel is poor food, but cheaper killing than a fat one, and we had to remember that our future efficiency depended on the number of good camels at our disposal. They lived on grazing as we marched (we never gave them grain or fodder), and after their six weeks on the road they would be worn thin, and have to be sent to pasture for some months' rest, while we called out another tribe in replacement, or found fresh riding-beasts.

We did not hamper ourselves with led-camels. The men carried with them a hundred rounds of ammunition and a rifle, or else two men would be an "automatic" team, dividing the gun and its drums between them. They slept as they were, in their riding cloaks, and fared well enough till the winter of 1917–1918, which caught us on the five-thousand foot hills of Edom behind the Dead Sea. Then we lost many men and camels frozen to death, or trapped in the snow, which lay over all the highlands in deep drifts for weeks, while we vainly appealed to Egypt for tents and boots and blankets. In reply we were advised that Arabia was a tropical country!

The equipment of the raiding parties aimed at simplicity, with nevertheless a technical superiority over the Turks in the most critical department. We had great quantities of light machine guns, used not as machine guns, but as automatic rifles, snipers' tools, by men kept deliberately in ignorance of their mechanism, so that the speed of action would not be hampered by attempts at repair. If a gun jammed, the gunner had to throw it away, and go on with his rifle. We made another special feature of high explosives, and nearly everyone in the revolt was qualified by rule of thumb experience in demolition work. We invented special methods of our own, for rapid work under fire, in the course of our months of practice, and before the end were dealing with any quantity of track and bridges economically and safely.

On some occasions we strengthened tribal raids by armoured cars, manned by Englishmen. Armoured cars, once they have found a possible track can keep up with a camel party. They are, however, cumbrous and shorter-ranged, because of the difficulty of carrying petrol. Therefore we seldom used them more than a hundred miles from home. On the march to Damascus, when we were nearly four hundred miles off our base, we first maintained them by a baggage train of petrol-laden camels, and afterwards by the help of the Air Force were able to give them further supplies by Handley-Page. Cars are magnificent fighting machines, and decisive whenever they can come into action on their own conditions. But though each has for main principle that of "fire in movement," yet the tactical employments of cars and camel-corps are so different that I do not recommend their being used in joint operations, except in very special circumstances. We found it demoralizing to both, to use armoured and unarmoured cavalry together.

The distribution of the raiding parties was unorthodox. It was impossible to mix or combine tribes, since they disliked or distrusted one another. Likewise we could not use the men of one tribe in the territory of another. In consequence, we aimed at the widest distribution of forces, in order to have the greatest number of raids on hand at once, and we added fluidity to their ordinary speed, by using one district on Monday, another on Tuesday, a third on Wednesday. This much reinforced their natural mobility. It gave us priceless advantages in pursuit, for the force renewed itself with fresh men in every new tribal area, and gave us always our pristine energy. Maximum disorder was in a real sense our equilibrium.

The internal economy of the raiding parties was equally curious. We aimed at maximum articulation. We were serving a common ideal, without tribal emulation, and so we could not hope for any *esprit de corps* to reinforce our motives. Soldiers are made a caste either by being given great pay and rewards in money, uniform, or political privileges; or, as in England, by being made outcasts, cut off from their fellows by contempt. We could not knit man to man, for our tribesmen were in arms willingly, by conviction. There have been many armies enlisted voluntarily: there have been few armies serving voluntarily under such trying conditions, for so long a war as ours. Any of the Arabs could go home whenever the conviction failed him. Our only contract was honour.

Consequently we had no discipline, in the sense in which it is restrictive, submergent of individuality, the lowest common denominator of men. In regular armies in peace it means the limit of energy attainable by everybody present: it is the hunt not of an average, but of an absolute, a 100-percent, standard, in which the ninety-nine stronger men are played down to the level of the worst. The aim is to render the unit a unit, and the man a type, in order that their effort shall be calculable, their collective output even in grain and in bulk. The deeper the discipline, the lower the individual efficiency, and the more sure the performance. It is a deliberate sacrifice of capacity in order to reduce the uncertain element, the bionomic factor, in enlisted humanity, and its accompaniment is compound or social war, that form in which the man in the fighting line has to be the product of the multiplied exertions of the long hierarchy, from workshop to supply unit, which maintains him in the field.

The Arab war was simple and individual. Every enrolled man served in the line of battle, and was self-contained. We had no lines of communication or labour troops. The efficiency of each man was his personal efficiency. We thought that in our condition of warfare the sum yielded by single men would be at least equal to the product of a compound system, and it was certainly easier to adjust to tribal life and manners, given elasticity and understanding on the part of the commanding officers. Fortunately for our chances nearly every young Englishman has the roots of eccentricity in him, and so we got on well enough. Of course we used very few Englishmen in the field, not more than one per thousand of the Arab troops. A larger proportion would have created friction, just because they were foreign bodies (pearls if you please) in the oyster: and those who were present controlled by influence and advice, by their superior knowledge, not by an extraneous authority.

In practice we did not employ in the firing line the greater numbers which the adoption of a "simple" system put theoretically at our disposal. We preferred to use them in relay: otherwise our attack would have become too extended. Each man had to have liberal work-room. In irregular war if two men are together one is being wasted. The moral strain of isolated action makes this simple form of war very exacting on the individual soldier, and demands from him special initiative, endurance and enthusiasm. Our ideal was to make action a series of single combats. Napoleon, in his pregnant valuation of the Mamelukes in terms of French soldiers, first gave me the idea: Ardant du Picq widened its application: the prejudices of historians are generally the richest part of their histories. Our value depended entirely on our quality, not on our quantity. We had to keep always cool, for the excitement of a blood-lust would impair the science of our combatants, and our victory depended on our just use of speed, concealment, accuracy of fire. Irregular war is far more intellectual than a bayonet charge.

The illiteracy of our forces was not harmful, since we worked intentionally in these small numbers and explained our plan verbally to everyone. Their very illiteracy has trained them to a longer memory and a closer hearing of the news. Nor were our tactics too subtle, for they had to be translated into independent action through the heads of our followers, and success was impossible unless most of them used their intelligence to forward our conception against the moral and material accidents of the path. This dilution of tactical ability to the level of the lowest interpreter was regrettable, but not all loss. The only alternative would be independent enterprise, and a mediocre design, persisted in, is grander than a series of brilliant expedients and will overcome them in the end.

By careful persistence, kept strictly within our strength and following the spirit of our theories, we were able eventually to reduce the Turks to helplessness, and complete victory seemed to be almost within our sight when General Allenby by his immense stroke in Palestine threw the enemy's main forces into hopeless confusion and put an immediate end to the Turkish war. We were very happy to have done with all our pains, but sometimes since I have felt a private regret that his too-greatness deprived me of the opportunity of following to the end the dictum of Saxe that a war might be won without fighting battles. It was an irony of fate to entrust this side-show of a side-show, with its opportunity of proving or disproving the theory, to an outsider like myself, not qualified technically to make the best of it. I would have given so much to show that Saxe was the greatest master of his kind of war, but now all I can say is that we worked by his light for two years, and the work stood. This is a pragmatic argument that cannot be wholly derided.

Unfortunately our campaigns lacked a historian as much as an executant. Now that I try to write down what we did, and why, some of our principles look truisms (mankind would so rather believe a sophism) and some look contradictory. The fault must be either in my exposition or in my observation. Savage warfare seems never to have been thought out in English from the savage point of view, and the Arab revolt would have been a great opportunity for a thinker to test its possibilities on a grand scale. Our war was so odd and so far away that coy Authority left us to ourselves. We had no base machinery, no formal staff, no clerks, no government, no telegraphs, no public opinion, no troops of British nationality, no honour, no conventions. The experiment was a thrilling one, which took all our wits. We believed we would prove irregular war or rebellion to be an exact science, and an inevitable success, granted certain factors and if pursued along certain lines. We did not prove it, because the war stopped: but here the thesis is:

It seemed that rebellion must have an unassailable base, something guarded not merely from attack, but from the fear of it: such a base as we had in the Red Sea Ports, the desert, or in the minds of the men we converted to our creed. It must have a sophisticated alien enemy, in the form of a disciplined army of occupation too small to fulfill the doctrine of acreage: too few to adjust number to space, in order to dominate the whole area effectively from fortified posts. It must have a friendly population, not actively friendly, but sympathetic to the point of not betraying rebel movements to the enemy. Rebellions can be made by 2 percent, active in a striking force, and 98 percent, passively sympathetic. The few active rebels must have the qualities of speed and endurance, ubiquity and independence of arteries of supply. They must have the technical equipment to destroy or paralyse the enemy's organized communications, for irregular war is fairly Willisen's definition of strategy, "the study of communication" in its extreme degree, of attack where the enemy is not. In fifty words: Granted mobility, security (in the form of denying targets to the enemy), time, and doctrine (the idea to convert every subject to friendliness), victory will rest with the insurgents, for the algebraical factors are in the end decisive, and against them perfections of means and spirit struggle quite in vain.

CHAPTER 37

The Inherent Vulnerabilities of Technology: Insights from the National Training Center's Opposing Force[1]

Colonel John D. Rosenberger

Colonel Rosenberger was the Chief of Staff of the First Cavalry Division at Ft.Hood and the Commander of the 11th Armored Cavalry Regiment (ACR). At the U.S. Army's National Training Center, Rosenberger found that the Red team (OPFOR) could consistently overcome the Blue teams (BLUFOR) high tech systems with low tech camouflage and deception measures. This finding validated what the Serbs did against NATO during the Kosovo War in 1999.

The following is an excerpt from a speech that was given by COL John Rosenberger, USA, at the annual U.S. Marine Corps Command and Control Symposium at Quantico, Virginia, in May 2000. He was the Chief of Staff of the 1st Cavalry Division at Fort Hood and the Commander of the 11th Armored Cavalry Regiment (ACR). The 11th ACR performs the mission of the Opposing Force (OPFOR) by acting as aggressors against U.S. Marine and Army units often equipped with the very latest in advanced technologies who build up, train, and maneuver at the National Training Center (NTC) at Fort Irwin.

To the 2,500 troopers of the 11th Armored Cavalry Regiment who make up the Opposing Force (OPFOR) at the U.S. Army's National Training Center (NTC), it came as no surprise to watch the 3rd Serbian Army march back into Serbia virtually unscathed by the relentless attacks of Global air power during the Kosovo conflict. It also came as no surprise to see the Serbian Army employ a wide variety of physical and electronic deception techniques, remain tactically well dispersed, and hide their combat systems in the infrastructure of cities and villages to preserve their combat power. Such tactics are old news to the combined arms team of the NTC OPFOR. The same strategies used by the Serbian Army have been learned and employed successfully by the OPFOR at the NTC since 1994—adaptive countermeasures critical to preserving combat capability at the tactical level of war against the impressive array of intelligence collection and attack technologies employed by America's joint team. This example is only one of several insights the OPFOR can provide into the limitations and vulnerabilities of current war fighting technology

1. Colonel John D. Rosenberger, "The Inherent Vulnerabilities of Technology: Insights from the National Training Center's Opposing Force," Speech at the annual USMC Command and Control Symposium, Quantico, Virginia, May 2000.

that underpins America's style of warfare in the 21st century. All of these insights and lessons learned are directly relevant to the pending combat operations in Iraq.

In the past 10 years, NTC OPFOR has exposed many limitations and vulnerabilities inherent to the war fighting technologies U.S. joint services are currently pursuing. Moreover, they have learned to defeat them just like any adaptive and savvy opponent will do—exactly as the Serbian Army did awhile back. These exposed vulnerabilities are compelling on several fronts, not simply to ensure we make smarter technological investments in the years ahead but, equally important, to ensure that we do not forfeit combat effectiveness, the ability to deter, or the ability to quickly defeat our enemies at both the operational and tactical levels of war. As a start, we at NTC have learned that active and passive force protection measures are vital to preserving combat power against asymmetric technologies—asymmetric, in this case, meaning some technological capability that provides a decisive advantage over an opponent in combat. For example, cruise missiles, laser-guided bombs, satellite reconnaissance systems, high-altitude reconnaissance aircraft, and unmanned aerial vehicles have provided the United States an asymmetric combat advantage over all opponents in the past decade. In response to these capabilities, opposing forces, at least at NTC, have learned that thermal deception, vehicle and unit dispersion, decoys of all types, camouflage, concealment, and electronic deception are vital means and ways to protect and preserve ground combat power.

Furthermore, the OPFOR has learned that air power and overhead intelligence acquisition systems have significant limitations and are inherently vulnerable to deception—even in desert and mountainous terrain and, by extension, certainly so in densely forested areas and jungles, not to mention complex urban terrain. We have learned that if we limit our movement, do not create dust clouds, remain tactically dispersed, use camouflage, and employ decoy equipment, we will absorb few losses. The Serbian Army and paramilitary forces employed the same methodology of force protection in the dense forests, cities, and villages in Kosovo. By using a combination of these force protection techniques, the effectiveness of attacks against ground forces can be limited and thereby endured.

We have learned how to deceive the operators and analysts behind the intelligence acquisition screens and leverage them to set conditions for success. Techniques in offensive operations and infiltration can be used to create a weakness in the enemy's defense, permitting rapid penetration and exploitation. Employment of these techniques set conditions for OPFOR tactical success several times in the past. The Serbs used similar techniques to preclude effective air attacks against their ground combat forces and deceive NATO forces about their actual strength, disposition, and location. Even more ingenious, the Serbs used appreciation of this vulnerability to lure NATO attack aircraft into attacking organized columns of civilian vehicles, then exploiting the scenes of carnage via the international media. Such a strategy, designed to attack the solidarity of the NATO coalition, is an extreme example of information warfare. In short, against a savvy opponent, acquisition systems have little intelligence value to tactical and operational commanders unless the data or images are confirmed quickly by another real-time imagery system or a well-trained reconnaissance team with the capability and optical resolution to discern the exact composition and types of vehicles acquired.

Another important lesson we have learned is that the key to defeating forces equipped with sophisticated collection, targeting, and situational awareness technologies is to gain information dominance quickly in the initial phase of the operation. If we can disrupt the enemy's ability to move information across the battlefield, then we can quickly level the playing field and negate the asymmetric advantage.

The location of stationary and relatively immobile communication node centers is easy to predict, given a line-of-sight analysis within an area of operations. There are a limited number of accessible positions where comprehensive line-of-sight communications can be established and sustained. Accordingly, the OPFOR tasks both its division and regimental reconnaissance teams to find these large, easily identifiable communication sites during the reconnaissance phase of an operation. Once found, and they always are, we attack the sites with accurate long-range artillery, rockets, or fixed-wing assets during the first phase of offensive or defensive operations. This strategy stops the flow of digits, quickly levels the playing field, and eliminates the asymmetric advantage afforded by the technology.

Line-of-sight technologies are easily disrupted by hills and mountainous terrain, unless continually supported by multiple aerial or ground retransmission stations positioned within the brigade's area of operations. Furthermore, this type of technology is even more limited, if not ineffective, when fighting in cities, a lesson painfully learned by the Russians in Groznyy, Chechnya, in 1996 and again this past year. While the Russians struggled to maintain FM communications to control operations, the Chechens used cellular telephones and commercial satellite communications to coordinate their defensive operations within the city.

In summation, from the experience from the OPFOR and the study of actual battlefield maneuvers, we have learned that there is no substitute for well-trained ground reconnaissance teams in war fighting at the tactical level of war. Despite all the intelligence and information technology provided to brigade task force commanders over the past 6 years, the OPFOR regimental commanders, using 1960s–1970s technology and unaided by overhead reconnaissance systems, have always had better, near-perfect information about the strength, composition, location, and disposition of their opponents. Their opponents, on the other hand, have remained and continue to remain relatively blind, despite the bloom of technology. This ability to see the battlefield better than their opponents, despite the introduction of sophisticated technologies, is provided by our division and regimental reconnaissance teams, undoubtedly some of the best-trained tactical reconnaissance teams in the world.

The indisputable fact is that well-trained observers (reconnaissance teams) in sufficient number to establish observation throughout the depths of the battlefield and armed with effective, secure communication, easily offset the supposed asymmetric advantages of overhead reconnaissance platforms in the business of close combat at brigade level and below. Moreover, from a practical perspective, overhead reconnaissance platforms cannot classify a bridge and determine if it will support the movement of forces, find and determine feasible fording sites across rivers or streams, locate minefields or bypasses, or provide any accurate information about enemy strength and dispositions within cities, the most likely battlefields in our future.

In conclusion, if the insights provided cause you to question the direction, design, and investments we have made in trying to create information dominance at the tactical level of war, that is good. If these insights foster a change in your perspective about the practical value and utility of technology by exposing its limitations and vulnerabilities, that is good, too. If these same insights drive our joint team to pursue more prudent technological investments in the future or drive the creation of better organizations, equipment, doctrine, tactics, and techniques for employing technology in the future, then that is even better. If they convince you that we should keep teaching our soldiers, sailors, and Marines how to read a map and navigate with compass in hand and keep teaching artillerymen how to survey their firing positions and teach our staff personnel what to do when the screens go blank, that is icing on the cake.

Finally, if these arguments convince you that OPFOR at our combat training centers can provide critical insights into the limitations and vulnerabilities of technology, informing our judgment to ensure that the Department of Defense wisely adapts to and dominates threats in the 21st century, then my objective has been accomplished. One thing is for certain: If we ignore the lessons and successful countermeasures that our Opposing Force has made and continues to make against technology, then we ignore the work of these great soldiers at our peril.

CHAPTER 38

Tactical Deception and Strategic Surprise in al-Qai'da's Operations[1]

Richard H. Shultz, Jr. and Ruth Margolies Beitler

Richard H. Shultz, Jr. is the director of the International Security Studied Program and professor of International Politics at the Fletcher School of Law and Diplomacy. Ruth Margolies Beitler is a Professor of Comparative Politics in the Department of Social Sciences at the United States Military Academy.

This article accurately portrays al-Qai'da as a nonstate equivalent of a counter-intelligence (CI) state. It also highlights the time, preparation and foresight necessary for strategic surprise. Finally, it shows how denial and cover protect and ultimately enhance military capability.

1. Richard H. Shultz, Jr. and Ruth Margolies Beitler, "Tactical Deception and Strategic Surprise in al-Qai'da's Operations," *Middle East Review of International Affairs (MERIA)*, Vol. 8, No. 2, June 2004, pp. 56–81.

How do Middle East terrorist organizations build deception tactics into their training to exploit the vulnerabilities of militarily superior opponents? This paper argues that al-Qa'ida has developed into the non-state equivalent of the counterintelligence (CI) state, using CI principles and practices in its strategy. The article examines how al-Qa'ida employs counterintelligence tools to organize, deploy, and fight.

In this era of asymmetric warfare, Middle East terrorist organizations have become increasingly adept at manipulating information about their operational plans and actions to exploit the vulnerabilities of militarily superior opponents. The surprise attack by al-Qa'ida against the United States on September 11, 2001 is the quintessential case in point. By adopting counterintelligence methods, including denial and deception, al-Qa'ida was able to camouflage its operational units inside the United States.

To understand al-Qa'ida's ethos and operational profile, it is necessary to discern how it has built deception, denial, and other counterintelligence (CI) practices into the training of its operatives, and planning and execution of its operations. This has significant implications for the United States' war against al-Qa'ida.

THE COUNTERINTELLIGENCE IMPERATIVE

Leading specialists of international terrorism consider al-Qa'ida markedly different from its counterparts in the 1980s.(1) What past experiences and new developments are central to grasping al-Qa'ida's operational distinctiveness?

In the 1980s, specialists in intelligence proposed a theory of counterintelligence (CI) focused on regimes whose principal concern was an obsession "with enemies, internal and external. . . that security and the extirpation of real or presumed threats." This perception generated the creation of unprecedented security and intelligence organizations to fight these foes.(2) A characteristic of these states included a millenarian ideology preaching fundamental change through the destruction of all existing evils and pursuing a totally new way of life.(3) The consummate example of the counterintelligence state was the Soviet Union.(4) Looking at the methods employed by al-Qa'ida over the past decade, it is clear that it has taken on most of the main attributes of such counterintelligence actors.

Of course, all terrorist groups have to pay considerable attention to security and counterintelligence matters in terms of defensive CI measures to protect it against penetration and its members from apprehension. Yet, al-Qa'ida also employs an array of offensive CI instruments as well, which makes it quite unique among terrorist organizations.

Fourth-Generation Warfare and al-Qa'ida

In the 1990s, al-Qa'ida carried out an organizational and operational transformation in the way terrorist movements functioned. In effect, it initiated a revolution in terrorist affairs (RTA).(5) Some military analysts suggest this RTA by al-Qa'ida is a key development in the transformation of warfare into its 4th generation.(6) Martin van Creveld, in his 1991 book, The Transformation of War, was among the first to propose that a new form of war was emerging. He argued, "The modern paradigm for warfare, in which nation-states wage war for reasons of state, using formal militaries. . .[is] being eclipsed by a post-modern approach . . National sovereignties are being undermined by organizations [terrorists and other non-state armed groups] that refuse to recognize the state's monopoly over armed violence."(7)

How were non-state armed groups, including terrorist groups, adapting to globalization, network-based organization, and information age technologies? How would these developments affect the terrorist's capacity to execute unconventional attacks? How would terrorist groups cooperate among themselves and with state sponsors in order to operate and fight globally?

The response can be defined as 4th Generation Warfare (4th GW), which al-Qa'ida appears to have emulated:

- 4th generation warfare is highly irregular, unconventional and decentralized in approach.
- Asymmetrical operations are employed to bypass the superior military power of nation-states to attack and exploit vulnerable political, economic, population, and symbolic targets, thus demoralizing both government and its populace.
- Both organization and operation are masked by deception, denial, stealth, and related techniques of intelligence and counterintelligence tradecraft. They wear no uniforms and infiltrate into the populations of the nation-states they seek to attack.
- Information age technologies profoundly influenced terrorist organizations and operations. The development of network-based terrorist organizations with transnational connections through cell phones, fax machines, e-mail, and websites provide these non-state armed groups with global reach.
- Modern communications and transportation technologies greatly complicate this new battlefield. Not only are there no fronts, but also the old distinctions between civilian and military targets become generally irrelevant.
- Laws and conventions of war applied to nation-states do not constrain terrorists (and their state sponsors) as they seek new and innovative means, including the use of weapons of mass destruction, to attack civilians and nonmilitary targets and inflict terrible carnage.
- 4th generation warriors, frequently in the name of religious-based millenarian ideologies, are remorseless enemies. Their operations are marked by unlimited violence, unencumbered by compassion or constraints.

Since the September 11 attacks, much has been learned about al-Qa'ida regarding how it bears an unmistakable resemblance to this template of 4th Generation Warfare (4th GW). It adapted itself to globalization, network-based organization, and information age technologies while relying on deception, denial, stealth, and related techniques of counterintelligence tradecraft.

Indeed, in one piece that appeared in an al-Qa'ida internet magazine following September 11, one operative described how the organization had studied, learned from, and adapted 4th GW concepts to its war fighting doctrine.(8) The author, Abu Ubeid al-Quarashi, began by discussing the origins of those ideas as conceptualized in U.S. defense journals, citing key references. Next, he noted that since the late 1980s, "The Islamic nation has chalked up the most victories, in a short time, in a way [the world] has not known since the rise of the Ottoman Empire. These victories were achieved. . . against the best armed, best trained, and most experienced armies in the world (the USSR in Afghanistan, the United States in Somalia, Russia in Chechnya, and the Zionist entity in southern Lebanon)."

How did they do it? According to al-Quarashi, "The mujahideen proved their superiority in 4th Generation Warfare using only light weaponry. They are part of the people, and hide amongst the multitudes." These victories established "precedents for world powers and large countries being defeated by [small] units of mujahideen . . despite the great difference between the two sides." The ultimate example, he underscored, was on September 11, when al-Qa'ida "dealt Americans the most severe blow ever to their morale." The key lesson learned was that "the time has come for the Islamic movements facing a general crusader offensive to internalize the rules of fourth-generation warfare. They must consolidate appropriate strategic thought, and make appropriate military preparations."(9)

Counterintelligence Principles

How does al-Qa'ida employ counterintelligence tools in support of its terrorist operations? Intelligence analysts must be able to distinguish between real signals indicating an attack, and "noise" which refers to irrelevant messages or those sent intentionally to deceive. True signals are "always embedded in the noise and irrelevance of false ones."(10) There was considerable "noise" prior to the September 11 attacks that allowed al-Qa'ida to achieve strategic surprise. Through increased communications among terrorist cells, the intelligence community knew that an attack was imminent, yet did not know where or how the terrorists would strike.(11)

Preconceptions can lead to self-deception and also play a role in strategic surprise. Likewise, tactical or technological innovation may impede an accurate assessment of intelligence. Prior to September 11, the intelligence community was cognizant that an attack was imminent, yet few expected hijacked airliners to be used as weapons. For the most part, analysts defined and linked the concept of "hijacking" to the traditional understanding that protagonists seize planes and make demands in exchange for releasing the plane and hostages. Worse, the intelligence community did not understand the extent to which al-Qa'ida had deployed units to the United States to prepare for and carry out the strategic surprise of September 11.

Islam and Denial and Deception

Al-Qa'ida's use of denial and deception operational techniques akin to those employed by intelligence services is also shaped by related concepts found in Islam. Two in particular are relevant here--*taqiyya* and *kitman*. Broadly speaking, *taqiyya* means precautionary dissimulation and keeping one's convictions secret, while *kitman* constitutes mental reservation or concealment of malevolent intentions. For the most part, the concepts refer to Shi'a practices that have become tenets of Shi'a theology.(12) *Taqiyya*, the practice of hiding one's true belief under duress, can be found in the Koran. For example:

> Let not the believers take the disbelievers as guardians instead of the believers, and whoever does that will never be helped by Allah in any way, unless you indeed fear a danger from them (illa an tattaqu minhum tuqat). . . Whoever disbelieved in Allah after his belief--except him who is forced thereto and whose heart is at rest with faith.(13)

These verses point to the permissibility of concealing one's ideology and opinion when in danger.(14) Other cultures have similar dissimulation concepts to protect themselves from harm.(15) Yet for al-Qa'ida members, the Koran provides examples to be emulated. One important illustration stems from Ammar ibn al-Yasir's experience in the seventh century. Persecuted by non-Muslims for his belief in Islam, he renounced his faith but then went to Muhammad to repent. Muhammad explained that as long as he renounced his faith with his lips and not with his heart, he would not be perceived as a traitor. Magnus Ranstorp argues that the concept of *taqiyya* is an important element of al-Qa'ida operations.(16)

Further, theological doctrines often turn operational and this can be seen in the interpretation and utilization of *taqiyya* and *kitman*. For example, during the Spanish inquisition, "[i]n operational terms, *taqiyya* and *kitman* allowed the 'mujahadeen' to assume whatever identity was necessary to fulfill their mission; they had doctrinal and theological, and later jurisprudential sanction to pretend to be Jews or Christians to gain access to Christian and Jewish targets: 'the mujahadeen can take the shape of the enemy.'" Moreover, since the 7th century, "deceptive triangulation" was a favored tactic used by Muslims to convince their enemies that jihad was not aimed at them but at another enemy, or simply used to deny that there was jihad at all.(17)

In the context of modern terrorism, these concepts have been interpreted and used for operational purposes. Indeed, they are considered important tools for radical Islamic terrorists. As one study points out, for those faced with asymmetrical terrorism:

> Al-Qaeda training manuals. . . carry detailed instructions on the use of deception by terrorists in Western target countries. . . The study of *taqiyya* and *kitman* is crucial to an understanding of Islamic fundamentalism and terrorism ranging from the issuing of false terrorist threats, operational and strategic disinformation issued by al-Qaeda in the form of 'intelligence chatter' to the use of *taqiyya* and *kitman* by terrorists during interrogation and the use of systematically misleading expressions concerning Islam and terrorism by Muslim spokesmen.(18)

The use of these traditional concepts by terrorist organizations like al-Qa'ida was confirmed by Shaykh Taj al-Din al-Hilaly, the grand mufti of Australia: "Al Taqiyya is with tongue only; not the heart. A believer can make any statement as long as the 'heart is comfortable.' The 9/11 terrorists lived and visited in the United States for two years before the 9/11 attacks. How did they acculturate? By the use of taqiyya. Meaning: I hate you but I smile at you in public."(19)

Evidence presented at the trial of men accused of bombing the American embassies in Kenya and Tanzania in 1998 revealed that "al-Qa'ida members were told to mingle with Westerners, and conceal their extremist beliefs... lead a double life--not atypical for members of [a]l Qaeda."(20)

AL-QA'IDA CASE STUDIES

To what extent has al-Qa'ida trained for and employed denial, deception and related CI tools as part of how it organizes and deploys units globally to execute surprise attacks against the United States? This can be determined by: 1) examining the place of these CI tools in al-Qa'ida doctrine and training; 2) scrutinizing the trial testimony of al-Qa'ida operatives involved in the East Africa embassy bombings; and 3) appraising the use of denial, deception, and other CI instruments by the September 11 cells.

Doctrine and Training

Consider, for example, a manual entitled "Military Studies in the Jihad Against the Tyrants" (hereafter "Military Studies") designed for training al-Qa'ida members.(21) Comparing the operational principles laid out in the document against two terrorist operations--the 1998 East Africa embassy bombings and September 11--provides a useful framework for anticipating future operations by this terrorist network.

The authorship of "Military Studies" remains unclear. However, the investigation into the terrorist bombings in East Africa provided clues. The British police found the manual in the home of Anas al-Liby. He was an associate of Ali A. Muhammad, a former sergeant in the U.S. Army from 1986 to 1989.(22) Muhammad admitted that he trained Usama bin Ladin's disciples in Afghanistan.(23) He also translated U.S. military manuals from English into Arabic. A raid on Muhammad's California home uncovered a computer file that contained terrorist training methods, including the use of explosives, assassination, and surveillance techniques. Muhammad subsequently pleaded guilty.(24)

"Military Studies" begins with a chilling declaration:

> The confrontation that we are calling for does not know Socratic debates. . . Platonic ideals. . . nor Aristotelian diplomacy. But it knows the dialogue of assassination, bombing, and destruction . . Islamic governments have never and will never be established through peaceful solutions and cooperative councils. They are established as they [always] have been: by pen and gun, by word and bullet. . . [and] by tongue and teeth.

The document reads like a typical military field manual. It sets the stage with the rationale for action (the repression of the faithful by apostate regimes and unbelievers), strategic objectives (to overthrow godless regimes and replace them with Islamic ones), and operational requirements (membership qualifications). It then delves deeper into the tactical realm, carefully structured along functional tasks. The manual is primarily concerned with one main goal: "to avoid detection at all costs when infiltrating an enemy state." The manual contains an abundance of tactical lessons on how to use disguises, elude authorities, and execute a wide range of missions. Given the covert nature of the work involved, secrecy and trickery are core strategies.

The introductory section recites stories from the Koran and ancient proverbs that extol deception as a virtue. Lesson two of the manual, "Necessary Qualifications for the Organization's Members," discusses the importance of keeping secrets and concealing information.

According to the manual, Muhammad once advised, "Seek Allah's help in doing your affairs in secrecy." "Military Studies" also recounts old proverbs on secrecy: "The hearts of free men are tombs of secrets," and "Muslims' secrecy is faithfulness, and talking about it is faithlessness." This section also includes two tales of

Muhammad ordering his men to deceive and one story of how Muhammad himself kept secrets. The manual contains other historical lessons that the author draws from the Roman era, World War I, and the Israeli Mossad.

Lesson three examines the issue of counterfeit currency and forged documents. It states, "All falsification matters should be carried out through the command and not haphazardly" though it never states how these actions fit into operations.

Lesson four discusses the correct use of apartments and hiding places. It advises that the member must avoid being conspicuous and should blend into society by, for example, shaving beards. It recommends, for instance, that a physician working undercover should have proper documentation in his apartment to include an actual medical diploma or government permit. It cautions, "a common laborer should not enter a fancy hotel because that would be suspicious and draw attention."

Lesson five focuses on the importance of communications and transportation. The introduction notes: "It is well known that in undercover operations, communication is the mainstay of the movement for rapid accomplishment. However, it is a double-edged sword: It can be to our advantage if we use it well and it can be a knife dug into our back if we do not consider and take the necessary security measures." The section details methods required to prevent discovery by the security apparatus of enemy states. One passage explicitly refers to deception as a counterstrategy: "When the command is certain that a particular telephone [line] is being monitored, it can exploit it by providing information that misleads the enemy and benefits the work plan." It then turns to tools of the information age such as wireless communications. The manual advises, "The device should be placed in a location with high wireless frequency, such as close to a TV station, embassies, and consulates in order to prevent the enemy from identifying its location."

The ninth lesson--security plans--is particularly noteworthy. A security plan is defined as "a set of coordinated, cohesive, and integrated measures that are related to a certain activity and designed to confuse and surprise the enemy." For the plan to succeed, it must be 1) "realistic and based on fact so it would be credible to the enemy before and after the work," and, 2) "coordinated, integrated, cohesive, and accurate, without any gaps, to provide the enemy [the impression of] a continuous and linked chain of events." In case of capture and interrogation by the enemy, "each member should make up a fictitious commander as the mastermind that conceived, planned, trained, and executed the operation," and claim he had been sent away to another country to confuse the investigative trail.

Lesson Ten is concerned with the operatives charged with executing an assassination, assault, kidnapping, or bombing. The members responsible for such missions should be trained to 1) employ cleverness, canniness, and deception; 2) keep secrets; and, 3) remain unknown to the enemy. In terms of deception, "the members should be well disguised and placed in a location close to that of the operation." In terms of denial, not all members should know about the plan until shortly before executing it in order to avoid leaks. After the operation, all evidence must be eradicated.

Lessons eleven and twelve focus specifically on intelligence and espionage. "Military Studies" emphasizes the importance of knowing the enemy by again turning to the era of Muhammad. The manual notes that he used informants in most of his attacks. It reconciles Muslim principles and intelligence operations: "Spying on the enemy is permitted and it may even be a duty in the case of war between Muslims and others. Winning the battle is dependent on knowing the enemy's secrets." The lessons then describe ways to disguise operations by studying enemy vulnerabilities. For example, the manual counsels that a search through open source materials could produce valuable results. In fact, bin Ladin's operatives had apparently relied on publicly available reports by the General Accounting Office to study the security weaknesses of sensitive U.S. facilities.(25) The rest of the manual gets down to specific tactics in great detail.(26)

Another training manual is one that has been dubbed in press reports as the 11-volume "Encyclopedia of Jihad." Comprising over 7,000 pages, subjects covered include: how to spy; kinds of security;
military intelligence; sabotage; secret observation; assassination; protection of leaders; laws of sabotage; and arms use.(27)

These manuals provide a glimpse into how al-Qa'ida trained its operatives in the tools and practices of offensive and defensive counterintelligence. Along with similar documents found in Afghanistan and elsewhere, they underscore the immense challenges posed by al-Qa'ida. Its culture of secrecy, apparent proclivity for deception, diffuse organizational structure, and way of training operatives has no doubt enhanced its elusive nature as a terrorist group. These facts highlight the importance of the CI dimension in al-Qa'ida's doctrine and, as we shall see below, in its East Africa and September 11 operations.

THE EAST AFRICA EMBASSY BOMBINGS(28)

On August 7, 1998, two bombs exploded about nine minutes apart at the United States embassies in Nairobi and Dar Es Salaam. These attacks killed 224 and wounded another 4,107 people. Twelve of the dead and twenty-four of the injured were U.S. citizens, mainly Foreign Service officers. In each case, the explosive device was a truck bomb placed in close proximity to the embassy complexes.

The attacks occurred five months after Usama bin Ladin issued a fatwa announcing a "Jihad Against Jews and Crusaders" in February 1998; a fatwa he would repeat again in May 1998. Evidence from the transcript of the East Africa Embassy Bombing Trial reveals that planning for these two operations began in early 1998.(29)

The bombing of the embassy in Nairobi occurred when Muhammad Reshed Daoud al-'Awhali immobilized a Marine guard at his post at the embassy wall. Jihad Muhammad Ali (also known as Azzam) drove a truck into the embassy parking entrance. Al-'Awhali exited the truck and fled the premises seconds before the bomb exploded. A similar scenario took place in Dar es Salaam.

The base of operations for the two bombings was Kenya, where Muhammad Sadiq Odeh was the technical advisor and explosives consultant to al-Qa'ida cells located there. Both truck bombs were modified to carry a sufficient amount of TNT and mechanical detonators. Wadih al-Hajj, who had been granted U.S. citizenship in 1989, moved his family to Kenya in the late 1990s. He had previously served as an assistant to bin Ladin in Sudan and was also a veteran of the Afghan war against the Soviet Union. Al-Hajj was a key figure in the al-Qa'ida unit that carried out the bombings.

Odeh, al-Hajj, and al-'Awhali, who were convicted on all counts on May 29, 2001 and sentenced to life in prison without the possibility of parole, were among the key figures in a unit that had established itself secretly in Kenya during the 1990s. The members of the cell lived what appeared to be perfectly normal lives, waiting for the signal to go operational.

The East Africa embassy bombings were precisely coordinated and executed due, in part, to the use of basic counterintelligence tactics. Considerable evidence of the inner workings of al-Qa'ida's Kenya unit is contained in the trial transcript.(30) The archive also includes the testimony of Ali Muhammad, the only one of the seventeen defendants to plead guilty.(31)

The Kenya Cell: Key Personnel

Among the terrorists who executed the East Africa operations are Muhammad Reshed Daoud al-'Awhali, Muhammad Sadiq Odeh, Wadih al-Hajj, Khalfan Khamis Muhammad (KK Muhammad) and Ali A. Muhammad. A brief description of the background of each reveals their place in the al-Qa'ida network.

Muhammad Reshed Daoud al-'Awhali was born on January 18, 1977 in Liverpool, England but grew up in Saudi Arabia, home of his prominent family. After a rather normal upper-middle class Saudi childhood, as a teenager al-'Awhali began to read Islamist publications such as *al-Jihad*, *al-Mujahideen*, and *al-Shad.* Listening to recorded lectures by Shaykh Omar Abd al-Rahman--the blind Egyptian cleric convicted for his role in the 1993 World Trade Center bombing--furthered al-'Awhali's embrace of radical Islam and antipathy toward the United States and its presence on the Arabian Peninsula.(32)

After high school, al-'Awhali attended Muhammad bin Saud, a religious university in Riyadh. Two years before the embassy bombings, he and a friend made their way to the Khaldan Camp in Afghanistan to join the al-Qa'ida jihad. He received basic instruction in light weapons, demolitions, artillery, communications, as well as in the religious ideology of al-Qa'ida. Subsequently, upon the advice of bin Ladin, al-'Awhali went to other al-Qa'ida camps for more advanced training in intelligence, information security, and the conduct of hijackings and kidnappings. Together with Azzam, al-'Awhali received special operations and cell management training outside Kabul before proceeding to Nairobi in 1998, making lengthy preparatory stopovers in Yemen and Pakistan. On August 7, he was the passenger in the Nairobi bomb delivery truck driven by Azzam.

Muhammad Sadiq Odeh was born in Saudi Arabia to a Palestinian family. He went to high school in Jordan and enrolled in the Far Eastern University in the Philippines in 1986, where he was an architecture and engineering student. It was at that time that he became involved with a Kuwaiti Islamic center. Listening to recorded speeches by Shaykh Abdallah Azzam, bin Ladin's mentor during the Afghan war against the Soviets, Odeh became engrossed with jihad.(33)

In October 1990, instead of finishing his last year of studies in the Philippines, Odeh joined the jihad in Afghanistan. He arrived at a camp near Khost and received two months of basic training in the use of machine guns and more sophisticated weapons, such as anti-tank missiles, rocket launchers, anti-aircraft weapons, and mortars. After recovering from a head wound from an air raid in Jalalabad, Odeh attended other training camps in Afghanistan. It was at that time that he appears to have received more specialized instruction. Upon finishing a three-month religious studies course in Peshawar in 1992, Odeh, on orders from bin Ladin, spent approximately seven months in Somalia training Islamic tribesmen who opposed U.S. and UN involvement.(34)

In August 1994, Odeh left Somalia and moved to Mombassa, Kenya, where he set up a fishing business and married a Kenyan woman who was introduced to him by Mustafa Fadhl (another al-Qa'ida member). Later, he set up a furniture business with his brother-in-law. This was clearly an effort to blend into Kenyan society, becoming part of what in intelligence terminology is called a "sleeper cell."

In March 1998, the Kenya cell received orders from bin Ladin in Afghanistan to prepare the operation. On August 6, the day before the attacks, Odeh fled Kenya on a flight to Karachi. However, the Pakistani authorities discovered his fake passport and detained him.

Wadih al-Hajj was born into a Catholic family in Lebanon in 1960. He grew up in Kuwait, where his father worked for an oil company. Al-Hajj converted to Islam as a teenager, after reading the Koran. He moved to Lafeyette, Louisiana in 1978 to study urban planning at the University of Southwestern Louisiana. Because his Catholic parents had disapproved of al-Hajj's conversion, a Muslim shaykh in Kuwait paid for his university education.

Like Odeh, al-Hajj became a follower of Shaykh Abdallah Azzam, and through him received mujahadeen training to fight against the Soviet Union. He returned from Afghanistan to Louisiana in the late 1980s and married. When he graduated in May 1986, he moved to Arizona to start a family. Between 1986 and 1989 he traveled to Pakistan and Afghanistan several times.

In 1989, al-Hajj was granted U.S. citizenship and moved to Arlington, Texas. Through contacts, he was called in 1991 to support veterans of the Afghan war in New York City. At the Brooklyn-based Alkifah Refugee Center, al-Hajj became associated with operatives later convicted in the 1993 World Trade Center attack. In early 1992, al-Hajj relocated to the Sudan, where he began working for Usama bin Ladin. Although his family claims he only worked for bin Ladin's legitimate businesses, it is evident that he was also involved in terrorist operations. In 1994, al-Hajj and his family left Sudan for Kenya.

While in Kenya, al-Hajj maintained contacts with other al-Qa'ida members there, forming the clandestine sleeper cell that would be activated to carry out the Nairobi and Dar es Salaam operations. He was, in effect, the commander of the unit. In September 1997, al-Hajj and his wife and children left Kenya and returned to Texas after a Kenyan search of his house. In September 1998, one month after the bombings, he was arrested.(35)

Khalfan Khamis Muhammad was born on an island in the Zanzibar archipelago of Tanzania. From the time he was a child, he lived an unprivileged life growing up in the Kidimni area of Tanzania. Described as very religious, KK Muhammad spent more time at the madrasa than other students.(36) He did not finish secondary school, leaving during the second year of a three-year course of study.(37)

At seventeen he moved to Dar es Salaam to work for his brother in food sales. There he became more serious about Islamic studies while attending the Arusha Street mosque. Although he received al-Qa'ida military training in Afghanistan in 1995, KK Muhammad was not asked to fight with the Taliban against the Northern Alliance and returned to his brother's business in Dar es Salaam.

He was contacted by al-Qa'ida in 1998 and worked on the Tanzania bomb project without knowing the specific mission. For the operation, he rented the house in Dar es Salaam in which the bomb was constructed and the plans finalized for its delivery. He also rented the white Suzuki Samurai used to deliver the bomb to the embassy, mixed the TNT and helped load it, and rode a short distance with the driver. Just minutes after the explosion, he drove by the ruined embassy to take pictures. A few days after the bombing, KK Muhammad fled to South Africa under the alias Zahran Nasir Maulid. Once there, he worked at a fast food restaurant until his arrest.

After attending high school in his hometown of Alexandria, Egypt, Ali Abd al-Saud Muhammad received his first of two bachelor's degrees from the military academy in Cairo.(38) He served in the Egyptian army for thirteen years, from 1971 to 1984, rising to the rank of major. While in the army, he earned his second bachelor's degree in psychology and a master's degree in Islamic Studies from the University of Alexandria and later worked toward a PhD. in Islamic studies while in North Carolina.(39) Before his discharge from the Egyptian Army, he had already become a member of the Egyptian Islamic Jihad.

After leaving the Egyptian military, Ali Muhammad moved to the United States in 1985 and obtained a position as a security officer with American Protective Services in Sunnyvale, California.(40) While an employee there, he married Linda Sanchez, an American medical technician, became a naturalized U.S. citizen, and joined the U.S. Army. He served as a supply sergeant and later an Islamic politics and culture instructor for the Special Warfare Center at Fort Bragg, North Carolina. In 1988, he apparently used his leave to take an unauthorized trip to Afghanistan to fight against the Soviets. Upon achieving the rank of sergeant, he received an honorable discharge from the army three years after joining.

Having been involved with the Egyptian Islamic Jihad since the 1980s, Ali Muhammad was introduced to al-Qa'ida in the early 1990s.(41) While living in New York, he ran a leather import-export business and frequently traveled to Afghanistan to give courses on intelligence, military tactics, and explosives training.(42) He advised those involved in the 1993 World Trade Center bombing.

From 1992 to 1997 Ali Muhammad and his wife lived in California, first in Santa Clara and then Sacramento. In late 1993, he photographed and conducted surveillance of American, British, French, and Israeli targets in Nairobi, using his import-export business as a cover. It was from these photographs that bin Ladin determined where a truck bomb could best be placed in an attack on the U.S. embassy.(43) After the 1998 bombings, he was subpoenaed to appear before a grand jury where he lied under oath and was subsequently arrested. Eventually, he pleaded guilty.

Several other members of al-Qa'ida were involved in the East Africa operations. Perhaps the most important was Ali al-Rashidi, who became known as Abu Ubaidah al-Banshiri after fighting the Soviets in Afghanistan. A police officer in Egypt, he was a close associate of Ayman al-Zawahiri, the founder of Egyptian Islamic Jihad and bin Ladin's top assistant. As a member of al-Qa'ida, al-Banshiri, along with Muhammad Atef, served as the heads of al-Qa'ida's military committee. He was dispatched to Kenya in 1996 to establish the units that would eventually carry out the attacks. In Nairobi, al-Banshiri was the key military leader and the top-ranking commander until he drowned in a ferryboat accident on Lake Victoria on May 12, 1996.

Al-Banshiri appears to have been replaced by Abdallah Ahmed Abdallah (also known as Saleh, Abu Miriam, Abu Muhammad), also a member of the Egyptian Islamic Jihad and a professional soccer player. He became bin Ladin's local coordinator in Kenya for the Nairobi and Dar es Salaam bombings. On August 4, 1998, he conducted a final reconnaissance of the Nairobi embassy. Saleh assigned Hamden Khalif Allah Awad to drive the bomb truck in Dar es Salaam. He was the only one in the Kenya cell allowed, "to make overseas phone calls on behalf of al-Qa'ida." He instructed all members of his group except the bomb drivers and the locals to go to Afghanistan before August 6.(44)

Denial and Deception in the East Africa Operations

There are several examples of al-Qa'ida's use of denial, deception and other CI tactics in the events surrounding the East Africa Embassy bombings. Its operatives employed codes, false names, cover identities, disguises, sleeper cells, business fronts, and other CI tactics such as information and personnel compartmentalization.

The cells that conducted the East Africa bombings made significant use of false identities and documentation. For example, Odeh used four completely different aliases while Ali Muhammad utilized five separate identities and had appropriate documentation for each. According to trial testimony, one of the first steps in training an al-Qa'ida operative is the assignment of a new identity.(45) This was the first rule an al-Qa'ida trainee learned. When he entered the guesthouse where he would be processed, he had to put all his personal documents into an envelope to be returned to him later. In exchange, he would receive his al-Qa'ida name.(46)

Once the operative was trained and ready to deploy outside Afghanistan, his new identity or identities were appropriately documented. According to trial testimony, within al-Qa'ida there were specific sections that "specialized in obtaining travel documents for members. . . who were to deploy to foreign countries."(47) Furthermore, in Afghanistan they taught classes to operatives about how to obtain false passports and documentation when in a foreign country.(48)

Al-Qa'ida also trained operatives how to use various covers that allowed them to blend into the location to which they were deployed. In Kenya, the cell used business fronts. Ali Muhammad helped al-Hajj with a charity organization that he created to provide al-Qa'ida members with identification documents and cover.(49) Operatives went so far as to acquire extended families. For example, al-Banshiri, the top al-Qa'ida commander in East Africa until he drowned, was known to his wife and her family only as Jalal and 'Adel Habib. They never had any idea he was a member of al-Qa'ida, believing "he was an Egyptian, holding Dutch citizenship" because of his forged papers.(50)

Sleeper cells were also used extensively in Kenya to establish the operational infrastructure necessary for future surprise attacks. They created business and other fronts that made it possible to raise funds to cover the terrorists' living expenses and for the purchase of weapons and bomb-making materials needed for a future operation. Ali Muhammad admitted that he taught al-Qa'ida trainees how to create covert cells and overt fronts for operations.(51) In Nairobi, these included a car business and aiding al-Hajj in creating a charity organization.

Ali Muhammad himself is a remarkable example of the ability of an agent to penetrate the U.S. military establishment and at the same time serve as an al-Qa'ida operative. He was already associated with the Egyptian Islamic Jihad organization before he worked at the Special Warfare Center at Ft. Bragg and continued to work for al-Qa'ida while he lived in the United States.

Al-Qa'ida cells in East Africa were quite adept at creating several different business fronts. For example, al-Banshiri maintained both a business for importing gymnasium equipment and a mining business in Tanzania that he started with his brother-in-law. It was supposed to produce diamonds and gold, but never actually operated.(52) Another business venture which al-Banshiri entered into in Tanzania with his brother-in-law was a transportation company using a minibus. These businesses permitted him, as the top al-Qa'ida official in East Africa, to be seen as living a normal life and the companies gave him cover for when he traveled.

Another al-Qa'ida front was Odeh's fish business, operating out of Kilifi, Kenya. He went to Lama and Kiunga, Kenya and Kismayo, Somalia, to pick up the fish that he would offload in Kilifi. However, Odeh frequently offloaded his fish at night with his own crew (other members of the al-Qa'ida cells in Kenya) instead of during the day with Kenyan workers. This enabled him to smuggle goods from Somalia, where al-Qa'ida had an ongoing presence, to Kenya. Further, Odeh used a trucking company not associated with the Kenyan Fisheries Department to transport his cargo away from the port. In late 1997, an official of the Fisheries Department told Odeh to stop offloading his fish at night and inquired about missing statistics for his boat. He stopped offloading at night and then closed his business.(53) Until then, though, Odeh had used his fishing business as a front for several purposes: to legitimate his and other cell members' presence in Kenya, to make money selling fish, and to transport various other goods related to terrorist operations.

Of all of al-Qa'ida's business fronts, the most interesting was Help Africa People. Al-Hajj started this non-governmental charity in Nairobi in 1994. It forged identity cards for terrorists, including al-Hajj and Odeh. It was also a cover for travel to Somalia. In fact, Help Africa People even worked on a legitimate project with bona fide commercial transactions, providing aid to people with malaria in Sudan. Funded by Usama bin Ladin, it provided a legitimate reason for al-Hajj to live in Nairobi. Help Africa People worked with another NGO, the Mercy International Relief Agency (MIRA). MIRA, which worked with the UNHCR in other countries including Afghanistan, had ties to al-Qa'ida members involved in the bombing.(54) In fact, it was from the MIRA office in Nairobi that coded reports were sent to al-Qa'ida's leadership in Afghanistan detailing the activities of the Kenya cells.(55)

Members of al-Qa'ida were also obligated to perform duties they knew to be haram, forbidden by the laws and customs of Islam.(56) Such haram actions included missing prayer or only praying once a day, and drinking alcohol.

Odeh also provided a description of how al-Qa'ida cells are compartmentalized in his interrogations with the FBI. Odeh said that an operation is broken into two separate cells, the first performing infrastructure development, logistics, and operational planning. This cell conducts surveillance, tests target security, gathers intelligence, acquires the explosives, and constructs the bombs. Ideally, much of this transpires before the second cell is completely organized. The second cell carries out the act itself.(57) A letter found on al-Hajj's computer in Nairobi described how the Kenya cells stressed compartmentalization: "We, the East Africa cell members, do not want to know about the operational plans since we are just implementers. We trust our command and appreciate their work and know that they have a lot of problems."(58)

Finally, al-Qa'ida operatives also encrypted files on computer systems not only in Kenya,(59) but also on computers seized at the homes of Anas al-Liby in England,(60) Ali Muhammad in California,(61) and Khalid al-Fawwaz in London.(62) Ramzi Yousef, mastermind of the February 1993 World Trade Center bombing, left his laptop containing many al-Qa'ida documents in his apartment in Manila on January 7, 1995, after a chemical-induced explosion forced him and associates to flee. One of the associates was captured in a direct attempt to retrieve the laptop.(63) Files decrypted showed a plot to simultaneously bomb multiple U.S. aircraft. Had the terrorists used unbreakable encryption readily available on today's public market, stopping the plot and arresting those involved might not have succeeded.(64)

Based on the above, it is clear that denial, deception, and other counterintelligence tools were ingrained into the training and doctrine of al-Qa'ida cells deployed to Kenya for the East Africa bombings. The CI operational concepts laid out in "Military Studies" and other al-Qa'ida training manuals were employed by the Kenya cells to mask their activities in the period leading up to the strikes against the two embassies. A second case study of the September 11 cells provides a further look into that world.

THE SEPTEMBER 11 SURPRISE

Denial and deception were also used in each phase of the September 11 operation. In those attacks, nineteen men hijacked four commercial airliners and flew them into the World Trade Center Twin Towers, the Pentagon, and into a field in Pennsylvania en route to Washington.(65) How did they elude authorities and execute these attacks? An assessment of the September 11 cell, the operation's planning and prepositioning phase, and its implementation provides the answers.

The September 11 Cell: Key Personnel

The commander of the operation was Muhammad Atta, the pilot of American Airlines Flight 11 that flew into the North Tower of the World Trade Center. Born and raised in Cairo, Atta left Egypt in 1992 to study architecture at Hamburg University.(66) He later transferred to the Technical University of Hamburg-Harburg.

During a visit home in 1995, Atta was recruited by al-Qa'ida and by 1996 his militant views became more obvious to his acquaintances in Germany. Atta became involved in the Islamist- oriented al-Quds mosque in Hamburg.(67) During 1998, he organized elements of the cell that would later execute the September 11 attacks. One of his roommates in Germany was Ramzi bin al-Shibh who, according to reports, was to be one of the pilots on September 11. He was unable to participate in the hijackings because he was repeatedly denied visas to enter the United States.

During Atta's employment in Germany at the Hayes Computing Service, he met Marwan al- Shehhi from the United Arab Emirates and recruited him into the cell.(68) Al-Shehhi would pilot the plane that struck the South Tower. He was reportedly close to his father, a Muslim cleric, and was described as someone easily manipulated.(69)

Lebanese-born Ziad Jarrah joined Atta's cell in Hamburg, trained in Afghanistan with al-Qa'ida, and on September 11 crashed United Airlines Flight 93 into a field in Pennsylvania.(70) At the time of the attacks, he was engaged to Aysel Senguen, a German woman of Turkish origin. Jarrah had a secular Muslim upbringing in an affluent family, and attended a private Christian school.(71) In 1996, he went to Germany to study.

During his first semester, he began to read radical Islamist publications. Jarrah moved to Hamburg in 1997 to study and became involved with a radical group at the al-Quds mosque.(72) He met Ramzi bin al-Shibh who introduced him to Muhammad Atta, with whom he later traveled to Afghanistan. Jarrah's relationship with Senguen was an unstable one and between 1997 and 2000 they broke up numerous times. In February 2000, he returned to her "clean shaven and neatly dressed."(73) By that time, Jarrah had been integrated into the Atta-led cell and planned to train as a pilot in the United States. He left Germany during the summer of 2000 for Florida.

The fourth pilot, Hani Hanjour from Saudi Arabia, was not part of the Hamburg cell. He flew American Airlines Flight 77 into the Pentagon.(74) He enrolled in the University of Arizona to study English in

1991 and when he returned to Saudi Arabia the next year, his brother was surprised at his radical Islamist views. Hanjour had grown a beard, spent long hours in prayer, and rejected music and television.(75) It is not clear how Hanjour joined the Atta cell, but most likely he was recruited in a mosque in Tucson.

According to Valdis Krebs, who mapped the September 11 cell network, connections between the nineteen hijackers were kept to a minimum. Aside from Atta, who interacted with the greatest number of hijackers, even those on the same flight had little, if any, prior contact with each other in the United States.(76) Undoubtedly, the need for compartmentalization to maintain security in order to achieve surprise was seen as essential to carry out the mission.(77) To plan the operation, distant parts of the network met only when necessary to coordinate details so that intelligence organizations would have difficulty linking them together. The hijackers used internet chat rooms and e-mail to communicate. Their most sensitive communications were in person.(78) Ties among cell members remained inactive until absolutely necessary.(79)

Still, the hijackers' prior associations kept the cells interconnected even when ties were kept dormant. According to Rohan Gunaratna, al-Qa'ida is organized vertically with a loose horizontal structure of compartmentalized cells.(80) This structure has served well by allowing al-Qa'ida to accomplish missions under varying types of conditions as cells can regenerate and relocate quickly.(81) Gunaratna has also characterized al-Qa'ida's organization as "fluid and dynamic." This fact was obvious during the September 11 mission. A case in point was Ramzi bin al-Shibh. He was tasked to be the twentieth hijacker on September 11, but repeatedly failed to get a U.S. visa.(82) Al-Shibh was reassigned and became an important financial link for al-Qa'ida operatives in the United States.

Two other hijackers were principal players in the September 11 plot. Khalid al-Mihdhar and Nawaf al-Hazmi entered the United States from Bangkok on January 15, 2000. Both had trained in Afghanistan and fought in Bosnia. In January 2000, al-Mihdhar traveled from Yemen to Malaysia, where he joined al-Hazmi, who arrived from Pakistan. They met with al-Qa'ida operative Khalid bin Attash, the principal planner of the attack on the *U.S.S. Cole*. A week after their meeting, the two arrived in San Diego where they took flight lessons. In June 2000, al-Mihdhar left the United States while al-Hazmi remained and eventually joined Hani Hanjour in Arizona.(83)

It is now assumed that the hijackers carried out the September 11 attacks without a U.S.-based network in order to avoid a repetition of past mistakes that had occurred during other operations. Although it is clear that the hijackers had help from several contacts in the United States, it is doubtful that they knew about the plot. After the East Africa attacks, individuals who assisted the bombers and were cognizant of the operation remained in East Africa. Subsequently, they were arrested and provided security officials with valuable information. For this reason it appears that the September 11 attacks were planned overseas.(84) Nevertheless, the Senate and House Joint Inquiry found evidence that the hijackers received "substantial assistance" from individuals in the United States, some of whom were known to the FBI.(85)

Denial And Deception in the September 11 Operation

In February 1993, a huge truck bomb exploded in the underground parking garage of the WTC. However, American intelligence organizations did not link al-

Qa'ida to the operation at the time, though we now know they overlooked vital evidence to the contrary. The mastermind of that first attack, Ramzi Ahmed Yusef, was part of al Qa'ida's network.(86)

In hindsight, officials now believe that the scheme to create al-Qa'ida cells in the United States began in the early 1990s. One important actor in the creation of that militant network in the United States was the previously mentioned Ali Muhammad, an Egyptian, who, while serving in the Egyptian Army, joined the Egyptian Islamic Jihad and became a militant Islamist.(87)

As for the direct planning of the September 11 plot, initial preparations can be traced back to 1999,(88) a fact corroborated by Ramzi bin al-Shibh.(89) Some claim that the plan for September 11 vacillated between spreading chemical weapons using crop dusters and commandeering jets to attack the World Trade Center and the Pentagon.(90)

A key element of deception is how the hijackers entered the United States. Although some held legitimate passports and visas, their intentions were obviously concealed. For example, prior to traveling to the United States, Atta and others claimed that their passports had been stolen and so received new ones. They lied in order to conceal their travels to Afghanistan, which might have aroused suspicion.(91) Supposedly, Al-Shehhi obtained a new passport in Pakistan before he went to the UAE where he received a multiple-entry visa that was valid for ten years. Likewise, Jarrah had received a new Lebanese passport and eventually obtained a five-year entry visa to the United States.(92) For the September 11 attack, the operatives used Germany, Malaysia, and the United Arab Emirates to enter the United States.(93) The cells in each of these countries were established independently and kept separate from each other, although key players met occasionally to coordinate the mission.(94)

In March 2000, Atta sent emails from Germany to two flight schools, one in Florida and another in Oklahoma, preparing for both his mission and cover in the United States. Atta entered the United States on June 3, 2000, on a tourist visa. By July 6, he had enrolled in flight school at Huffman Aviation in Venice, Florida, with fellow hijacker al-Shehhi. That fall, Atta and al-Shehhi received their commercial pilot licenses. Eventually, Atta left the country and re-entered on January 10, 2001, on another tourist visa, again declaring to agents that he was pursuing flying lessons. Technically, Atta required a student visa to enter the United States to attend flight school. However, he convinced the inspectors to admit him after explaining that he had applied for a student visa and was awaiting approval.(95) Ironically, the authorization arrived six months after the attacks. Hanjour had been denied a visa on his initial request, but received one after he altered information on his original application.(96) He entered the United States on a student visa on December 8, 2000, claiming that he was going to study English in California. However, he never matriculated at that school.(97)

A significant problem with many of the visas stemmed not only from the hijackers' deception on the applications--including false or missing information--but also the negligence of the officials approving the visas. According to one report, fifteen of the nineteen hijackers should have been denied visas to the United States.(98) In many cases, their applications were missing information or they showed insufficient funds to support themselves for the length of time they had indicated that they would be remaining in the country.(99) Additionally, the hijackers must have known, and counted on the fact, that the United States had no specific policy to vigorously screen visa applicants from Saudi Arabia since the Kingdom is a U.S. ally. A Congressional Report confirmed, "a relatively low standard was applied in scrutinizing visa applications for accuracy and completeness in Saudi Arabia."(100)

Adding to the deception, the hijackers did not "look like" quintessential Islamic radicals. Following guidance from the al-Qa'ida training manuals, they were well-dressed, clean-shaven, lived routine American lives and even flew first class.(101) According to Ramzi Ahmed Yusef, the architect of the first World Trade Center bombing, flying first class was crucial since first class passengers are subject to less scrutiny.(102) Although many accounts paint Atta as an angry and disgruntled individual, he managed to conceal those aspects of his personality. He was known to appear polite and well-dressed. Jarrah, as mentioned previously, had a fiancée in Germany who, despite cursory glimpses of Jarrah's journey into radical Islam, assumed they would eventually live a normal, married life.(103) These measures all provided cover for the al-Qa'ida operatives.

After being taken into custody in Pakistan, Ramzi bin al-Shibh, who had roomed with Atta in Germany, discussed some of the tactics that Atta and the other hijackers used to integrate into American society.(104) When the hijackers arrived, they were required to find an appropriate "cover" to mislead and "mystify" the security services.(105) In many cases, the hijackers participated in seemingly "normal" activity, while clandestinely preparing for September 11. Ziad Jarrah took self-defense classes in which he learned to escape attacks and to counter attack.(106) Many of the terrorists joined gyms most likely to prepare for physically countering resistance on the planes. Al-Shibh also confirmed that the four hijacking crews operated as distinct units to mislead authorities and to avoid linking the hijackers together.

As dictated in the al-Qa'ida training manuals, some of the terrorists tried to avoid contact with outsiders and only interacted when necessary so as not to draw undue attention to themselves. The hijackers were described as regular neighbors who did not appear suspicious, let alone threatening.(107) They concealed their malevolent intentions and lived among Americans for extended periods of time. For example, Hanjour, who piloted the plane that rammed the Pentagon, is reported to have been in the United States, on and off, since 1991.(108)

The guidance in their training manuals also gave permission to the operatives to indulge in activities prohibited by Islam. There are several accounts of the hijackers drinking alcohol and pursuing prostitutes. Furthermore, the terrorists also misled authorities by minimizing their associations with mosques known for their radical tendencies, though there are accounts of several hijackers visiting mosques during their U.S. stay.(109) Al-Qa'ida was aware that the FBI was monitoring various Muslim communities so they deliberately relocated the September 11 cells to less conspicuous areas.(110) Several did, however, join or visit mosques in violation of al-Qa'ida instructions for operating undercover.(111) For example, Nawaf Alhazmi and Khalid al-Midhar who lived in California starting in January 2000 became members of the Islamic Center of San Diego.(112) Supposedly, Hanjour also attended a religious center in Arizona.(113) Perhaps they ignored the training manual's tenet to avoid mosques because they believed that houses of worship were off limits to FBI agents. As such, they used them as places to communicate with other al-Qa'ida members and to exchange information.(114)

Several hijackers applied to flight schools and for the most part, did not behave in ways that attracted attention. After completing flight training, the al-Qa'ida operatives also trained on simulators of jumbo jets including the Boeing 747 and 767.(115) No activity, viewed in isolation, aroused any suspicions. However, an FBI report from Phoenix alerting the government that several Middle Eastern men, possibly associated with Islamic militants, were enrolling in a flight school in Arizona, and specifically raised the question as to whether this might be occurring elsewhere(116) went unheeded.

On many occasions, the hijackers presented themselves as pilots for commercial airlines. In August 2001, Abdulazziz Alomari attempted to enter the cockpit of a flight from Las Vegas to New York. He claimed to be employed by Egypt Air and asked to ride in the jump seat as an observer. Although the pilots admitted him to the cockpit prior to the flight, he was denied access after the plane was airborne.(117) Investigators believe that Alomari was testing the security of the airlines and the ability of passengers to access the cockpit. A few days prior to the attacks, Atta and several other hijackers went to a bar in Florida. When a question was raised about the bill, Atta paid and explained that he was an American Airlines pilot.(118)

As part of the planning, several of the hijackers made "dry runs" and carried out surveillance of particular flights and safety procedures to learn more about the security loopholes on transcontinental flights.(119) Although the reasons are still unknown, each of the pilots stopped over in Las Vegas during one of their surveillance flights.(120) On May 24, al-Shehhi flew from New York to San Francisco and then immediately boarded a flight to Las Vegas where he remained until May 27 when he took a flight back to San Francisco and then to New York. Jarrah flew to Los Angeles from Baltimore and then to Las Vegas on June 7. Atta flew from Boston to San Francisco to Las Vegas on June 28 returning to Boston via Denver on July 1. He then traveled again on August 13 from Washington, D.C. to Las Vegas. On the same day, Hanjour and al-Hazmi went to Los Angeles from Washington's Dulles Airport for a final destination in Las Vegas. They returned to Baltimore the next day, stopping in Minneapolis.(121) Hanjour also conducted surveillance flights several weeks prior to the attacks by renting small planes to scout out targets or routes in the vicinity of both the Pentagon and the World Trade Center.(122)

Consistent with the instructions in their training manuals, while in the United States, several hijackers acquired fake documents, including international driver's licenses and identification cards.(123) Hanjour obtained an identification card using a fake Virginia address. Two other hijackers, Khalid al-Mihdhar and Majid Muqid also used that same address to obtain their documents.

In addition, the terrorists created a complex network of financial transactions by which they moved hundreds of thousands of dollars through U.S. accounts. The hijackers opened twenty-four bank accounts in four different banks(124) apparently with fabricated social security numbers.(125)

Outside the United States, al-Qa'ida members transferred money to the hijackers' accounts. Munir al-Mutassadiq, suspected of aiding the September 11 terrorists, admitted that he had been given a power of attorney over Marwan al-Shehhi's bank account and paid the bills when al-Shehhi was training in Afghanistan. Mutassadiq also explained that Rami bin al-Shibh contacted him from overseas to send him money from al-Shehhi's account so that he could pass the money on to al-Shehhi.(126) Al-Qa'ida operatives wanted to make tracing finances as difficult as possible and to provide cover for al-Shehhi by making it appear as if al-Shehhi had remained in Hamburg. As the judge in Mutassadiq's case noted, "This looks like you wanted to show the man is still in the country."(127) Additionally, many of the financial transactions originated from the United Arab Emirates.(128)

During the months preceding the attacks, activity accelerated to coordinate the near simultaneous hijackings on September 11. In July 2001, Atta and al-Shehhi met with Ramzi bin al-Shibh and several other Arab men in Spain, presumably to elude and confound authorities while planning the final phase of the operation.(129) During the summer of 2001, Atta also met on a monthly basis with al-Hazmi who has been described as Atta's "right hand."(130) As mentioned above, from May until a few weeks before the attacks, the hijackers took several cross-country trips, and then flew to Las Vegas. On most of the trips, the terrorists had purchased round-trip, first class tickets, but on their last trip, they purchased one-way, coach tickets.(131) Most likely, the discrepancy was an attempt to test whether or not one-way tickets--the kind the perpetrators would purchase for September 11--would arouse suspicion.

In the days preceding the September 11 attacks, the hijackers continued to deceive those around them. According to several reports, Atta visited New York, most likely to calibrate the location of the WTC into his global positioning system (GPS), which each of the pilots had acquired.(132) Atta's car was also caught by surveillance cameras moving through Boston's Logan Airport, most likely gathering last minute intelligence.(133) Although their reasons are not yet clear to analysts, Atta and Alomari drove to Portland, Maine, and flew to Boston on September 11 to connect with the plane they crashed into the North Tower. There is speculation that, based on al-Qa'ida training manuals, operatives were instructed not to travel in groups and to use "secondary stations" to avoid detection.(134)

Most of the hijackers began to buy their tickets beginning on August 26, although al-Mihdhar and Majid Muqid attempted to buy theirs on August 24. When they were unsuccessful because their address could not be verified, they went to the American Airlines counter at the Baltimore airport on September 5 and paid cash.(135) Allegedly, the terrorists bought extra seats on the aircrafts so as to lessen resistance they might encounter from passengers during the hijacking.

On the day of the attack, the hijackings were coordinated so that they would be nearly simultaneous. Supposedly, the hijackers also selected transcontinental flights with large fuel loads so as to inflict maximum damage and casualties.(136) To execute this type of coordinated attack required extensive planning, patience, and extreme secrecy. According to a cell phone call made from a passenger on one of the planes, the hijackers engaged in one last act of deception. They forced the pilots from the cockpit while ushering the passengers to the back of the plane telling them that if their demands were met, the passengers would be released without harm.(137)

In sum, as in the case of the Embassy bombings, al-Qa'ida incorporated counterintelligence tools, especially denial and deception, to execute the deadly attacks on September 11. By examining the planning, prepositioning, and execution of the September 11 attacks, it is clear that al-Qa'ida operatives internalized and implemented the tools discussed in their training manuals. Their "cover," blending in with the population, their utilization of decentralized means of communication and a sophisticated system of financial transactions all deceived U.S. intelligence authorities.

A crucial problem, which aided successes such as the East Africa Embassy bombings and September 11, was self-deception. Although officials knew that the U.S. homeland could be targeted, they were concerned with other types of attacks including truck bombs or strikes using weapons of mass destruction.(138) In 2001, the United States received information from an al-Qa'ida informant in Malaysia that the group wanted to hijack a plane and use hostages to force the release of Shaykh Omar Abd al-Rahman.(139) Perhaps the informant's admission was part of a larger scheme of disinformation to prevent U.S. security services from deciphering the plan to use planes as bombs. The CIA obviously found the information of a plane hijacking credible. On August 6, President Bush was briefed on this particular scenario.(140) Although analysts must be diligent in studying terrorist training documents, they must also be wary of self-deception if they assume that the next attack will be similar to its predecessor.

LESSONS LEARNED AND IMPLICATIONS FOR U.S. POLICYMAKERS

Understanding the philosophical underpinnings of al-Qa'ida operations, as well as its methods for training operatives is critical to U.S. national security. Misinterpreting an enemy's intentions and the manner in which they wage war is dangerous, leaving open the opportunity for disasters like that which happened on September 11. Therefore, policymakers and analysts need to become more adept at understanding al-Qa'ida's counterintelligence mindset and how it employs CI tools to include denial and deception to conduct 4th generation warfare. It is essential to comprehend both the ethos of al-Qa'ida and how it has developed as a new model for terrorist operations.

*Professor Richard H. Shultz is a professor of International Politics at the Fletcher School of Law and Diplomacy, Tufts University, and is director of the school's International Security Studies Program. His recent publications include The Secret War Against Hanoi: Kennedy and Johnson's Use of Spies, Saboteurs, and Covert Warriors in North Vietnam (1999); Security Studies for the 21st Century (co-editor and author) (1997); and War in the Information Age: New Challenges for U.S. Security Policy (co-editor and author) (1997). He is currently writing a book on Ethnic, Tribal and Religious Warriors: How Non-State Armed Groups Fight (Columbia University Press).

**Professor Ruth Margolies Beitler is an Associate Professor of International Relations and Comparative Politics at the U.S. Military Academy. Her books include The Path to Mass Rebellion: An Analysis of Two Intifadas (Lexington Press, 2004) and a monograph (co-authored with Cindy Jebb), Egypt as a Failing State: Implications for U.S. National Security, (Institute for National Security Studies, July 2003).

NOTES

1. For recent articles by Hoffman, Gunarata, and other leading specialists see Russell Howard and Reid Sawyer, (eds.), Terrorism and Counterterrorism Revised (New York: McGraw-Hill/Duskin, 2004).
2. John Dziak, "The Study of the Soviet Intelligence and Security System," in Roy Godson (ed.), Comparing Foreign Intelligence (Washington, DC: Pergamon-Brassey's, 1988), p. 65.
3. American Encyclopedia of Religion (NY: Macmillan, 1986).
4. The historical roots of such political systems were traced back to the Oriental/Asiatic despotisms of ancient times. However, those systems were said to lack certain attributes--"key ingredients"--

of their modern 20th century descendants. These missing features include a millenarian ideology and highly institutionalized security and intelligence services. The earlier despotisms, written about by Karl Witfogel and others, "were lacking in those... both in scope and intensity." It was only in the twentieth century that the counterintelligence state reached fruition. Ibid.

5. Theodor W. Galdi, Revolution in Military Affairs: Competing Concepts, Organizational Responses, Outstanding Issues (Washington, DC: Congressional Research Service, 1995), p. 4. For an online gateway to online full-text resources about the Revolution in Military Affairs, information war, and asymmetrical warfare go to The RMA Debate at <www.comw.org/rma/index.html>.

6. The 1st generation--the classical war between nation-states--was perfected by Napoleon. This was followed by the industrial age 2nd generation war of attrition based on massive firepower. It reached its apogee in WWI. Maneuver warfare, introduced by the Germans in WWII and refined by the United States in the 1980s, marked the 3rd generation.

7. Martin van Creveld, The Transformation of War (New York: Free Press, 1991), p. 224. See his retrospective in Martin van Creveld "The Transformation of War Revisited," Small Wars and Insurgency (Summer 2002), pp. 3-15.

8. For a translation, see "Bin Laden Lieutenant Admits to September 11 and Explains Al Qaeda's Combat Doctrine," The Middle East Media Research Institute (No. 344) at <www.memri.org/jihad.html>.

9. Ibid.

10. Roberta Wohlstetter, "Cuba and Pearl Harbor: Hindsight and Foresight," Foreign Affairs, Vol. 43 (July 1965), p. 707.

11. There is a debate about whether or not the intelligence community knew that terrorists would use hijackings and target the World Trade Center. See James Risen, "U.S. Failed to Act on Warnings in '98 of a Plane Attack," New York Times, September 19, 2002, p. 1. According to the article, the CIA and FBI received warnings that al-Qa'ida wanted to attack New York or Washington using airplanes.

12. The extent to which *Taqiyya* is actually accepted by many Shi'a is debatable, but the concept exists and is accepted as a legitimate tenet of Islam.

13. Koran, verses 3:28 and 16:106, as cited in Sayyid Moustafa Qazwini, "Dissimulation (*taqiyyah*)," in Inquiries about Shi'a Islam (The Islamic Educational Center of Orange County), Chapter 10. Available from <http://www.al-islam.org/shiism/10.htm>.

14. Counter Intelligence-Counter Espionage-Counter Terrorism (CI-CE-CT), "*Taqiyya* and *kitman*: The Role of Deception in Islamic Terrorism." CI-CE-CT is an Australian non-profit, educational site, dedicated to understanding and increasing capabilities to counter traditional, contemporary and emerging terrorist threats. Available from <http://www.ci-ce-ct.com/main.asp>.

15. Cyrus H. Gordon, "The Substraum of *Taqiyya* in Iran," Journal of the American Oriental Society, Vol. 97 (1977), p. 192. Gordon writes, "while Bahais, Christians, Jews and Sunnites in Iran may not foster *taqiyya* as a tenet of their faith, they resort to it as a pattern of behavior permitted by Iranian mores." He argues that it is not only a religious principle, but a social institution. Additionally, Gordon points out that in the Book of Esther, the practice of *taqiyya* was implemented, implying that the concept had pre-Islamic roots.

16. Charles Sennott, "Fighting Terror: The Investigation," Boston Globe, August 4, 2002.

17. Ibid.

18. Counter Intelligence-Counter Espionage-Counter Terrorism (CI-CE-CT), "*Taqiyya* and *Kitman*"
19. Ibid.
20. Elizabeth Neuffer, "Hijack Suspect Lived a Life, or a Lie," Boston Globe, September 25, 2001. Available from <http://web.archive.org/web/20010925123748/boston.com/dailyglobe2/268/nation/Hijack_suspect_lived_a_ life_or_a_lie%2B.shtml>.
21. United States v. Usama bin Laden, et al., S (10) 98 Cr. 1023 (LBS), Southern District of New York, 2001, government exhibit 1677-T.
22. See Judy Aita, "Ali Mohamed: The Defendant Who Did Not Go to Trial," U.S. Department of State, International Information Programs, May 16, 2001. Text online at <http://usinfo.state.gov/regionaliaf/security/a1051502.htm>.
23. Alan Feuer and Benjamin Weiser, "Translation: The How-to Book of Terrorism," New York Times, April 5, 2001, p. A1.
24. David A. Wise, "Ex-Soldier Admits Role in Embassy Blasts," Washington Post, October 21, 2000, p. A18.
25. Eric Lichtblau, "Terrorists Noted Flaws In Security, Report Says," Los Angeles Times, February 1, 2002, p. Al.
26. United States v. Usama bin Laden, et al., S (10) 98 Cr. 1023 (LBS), Southern District of New York, 2001, government exhibit 1677-T.
27. Kathy Gannon, "Jihad Manual a Testament of Death," Associated Press/Tulsa World, October 2, 2001, p. A4.
28. The primary research and initial analysis of the data for this section was undertaken by Kendall Wolf, a research assistant at the Fletcher School during the 2002-2003 academic year.
29. See <http://www.cryptome.org> and http://www.cryptome2.org>.
30. See <htttp://www.cryptome.org> and <htttp://www.cryptome2.org>.
31. See <http://cryptome.org/usa-v-mohamed.htm>; <http://www.legalservicesindia.com/Usama/bombings_trial.htm>; http://www.legalservicesindia.com/Usama/fbi.htm and <http://www.legalservicesindia.com/Usama/suspect.htm>.
32. United States of America v. Usama Bin Ladin, et al., S(7) 98 Cr. 1023 1884, Southern District Court of New York, January 8, 2001, p. 1994.
33. United States of America v. Usama Bin Ladin, et al., S(7) 98 Cr. 1023 1505, Southern District Court of New York, February 27, 2001, p. 1624.
34. United States of America v. Usama Bin Ladin, et al., S(7) 98 Cr. 1023 1635, Southern District Court of New York, February 28, 2001, p. 1645.
35. Oriana Zill, "A Portrait of Wadih El Hage, Accused Terrorist," PBS/Frontline, <http://www.pbs.org/wgbh/pages/frontline/shows/binladen/upclose/elhage.html>.
36. United States of America v. Usama Bin Ladin, et al., S(7) 98 Cr. 1023 8369, Southern District Court of New York, February 27, 2001, p. 8395.
37. Ibid., pp. 8396-8397. Defendant Khalfan Khamis Mohamed evidence # MMK28T.
38. Peter Bergen, "The Deep Sleeper," (extract from Holy War Inc.) The Weekend Australian, November 17, 2001.
39. United States of America v. Ali Mohamed, S(7) 98 Cr. 1023 1, Southern District Court of New York, October 20, 2000, pp. 11, 12.
40. Bergen, "The Deep Sleeper."
41. United States of America v. Ali Mohamed, S(7) 98 Cr. 1023 1, Southern District Court of New York, October 20, 2000, pp. 25, 26.
42. Bergen, "The Deep Sleeper."
43. United States of America v. Ali Mohamed, S(7) 98 Cr. 1023 1, Southern District Court of New York, October 20, 2000, p. 27.

CHAPTER 39

How Can Weak Powers Win?[1]

Yang Shaohua

This paper by Yang Shaohua argues that an assymetrical strategy is essential for the weak. After a survey of the various Western theory of asymmetric war, the author focuses on Ivan Arreguín-Toft "How the Weak Win Wars," (2005), which he rejects as having some contradictions. Then, drawing on other studies (including Arreguín-Toft) he constructed a database of 36 cases from the period 1945–2007. This secondary analysis showed strong support for Arreguín-Toft's model of asymmetrically direct-indirect strategies. Indeed, Yang found an even stronger effect in the recent 1945–2007 period than Arreguín-Toft found for his long-range 1809–2003 period. Implied for both strong and weak actors is that strategies that involve deception can mitigate the asymmetric advantages of an opponent.

Mr. Yang was in Jinggangshan, China, as an assistant professor at the China Executive Leadership Academy, an official institution of the Chinese Communist Party.

Although the history of war can be characterized as the strong defeating the weak and the many defeating the few, there are examples throughout history of 'Davids defeating Goliaths'. They include the guerilla war against Napoleon's occupation of Spain, the American war of independence, Yugoslavian guerilla attacks against Nazi fascism, the Chinese Communist Party's war of resistance against the Japanese, the American war in Vietnam, and the present wars in Iraq and Afghanistan. All are examples of asymmetric conflicts, and of the tendency on the part of the weak to defeat the powerful (or to be more precise, of the inability of the powerful to defeat the weak) in such conflict. Asymmetric conflicts constitute the majority in the post-World War II period, according to available statistics. As international armed conflict and its outcomes are important phenomena in international relations, and taking into account how many instances there have been in recent years of weaker powers triumphing over stronger adversaries, why the weak are so often able to defeat the powerful is a critical question.

Current Explanations and their Limitations

To understand why it is that the weak are, on occasion, able to defeat the strong, we must first define asymmetric conflict. For the purposes of this article, 'asymmetric conflict' refers to 'armed transnational conflict in which severe disparity of material power exists between actors'. Two main factors identify asymmetric conflict: (i) whether or not there exists a disparity between belligerents of military power; and (ii) whether or not the conflict takes place across national boundaries. The belligerent with comparative advantages with respect to resources, according to the above definition of 'asymmetric' conflict, is referred to as the 'strong power' and its adversary as the 'weak power'. The power ratio between belligerents is critical to gauging whether or not there is disparity between them. For

1. Yang Shaohua, "How Can Weak Powers Win?" *Chinese Journal of International Politics,* Vol. 2, No. 3, Summer 2009, pp. 335-371.

the sake of this article, so-called 'disparity' in the relative power of the two belligerents refers to a situation in which the power of the stronger actor is at least five times that of the weaker.[1] As regards what constitutes victory or defeat, what must be considered is whether or not at the end of the conflict the political goals which led to hostilities in the first place are achieved. If the objectives are fulfilled, the outcome is victory; if not, the conflict ends in defeat. The stronger power may be unbeaten on the battlefield, but if it fails to achieve the objectives that motivated it to launch the conflict, it may be considered to have suffered defeat, according to criteria herein defined.

Contemporary research on 'asymmetric conflict' is in general limited.[2] The most outstanding studies have been conducted by Western, largely American academics.[3] To provide background for the study at hand, this article first describes and evaluates current articulations of asymmetric conflict, focusing on three elements: power, motivations, and strategy.

The Perspective of Power

Most scholars focusing on the element of power are students of realism. One of the fundamental tenets of realism is that 'relative power' is the most important difference between international actors, and that power as defined in military terms is the determinant factor in international conflict.

1 Material power here refers to the comparative values of the populations and military power of belligerents at the start of the war. There is no accurate indicator of this equation. For more on measuring material power, see John Jacob Nutter, 'Unpacking Threat: A Conceptual and Formal Analysis', in Norman A. Graham, ed., *Seeking Security and Development: the Impact of Military Spending and Arms Transfers* (Boulder: Lynne Rienner Publishers, 1994), pp. 29–49.

2 For detailed information, see Yang Shaohua, 'Bu duichen chongtu yanjiu shuping', ('A Review of Work on Asymmetric Conflict'), *Ouzhou yanjiu* (*Journal of European Studies*), No. 3 (2007), pp. 137–138.

3 Main scholars and works include: T.V. Paul, *Asymmetric Conflicts: War Initiation by Weaker Powers* (New York: Cambridge University Press, 1994); Gil Merom, *How Democracies Lose Small Wars: State, Society, and the Failures of France in Algeria, Israel in Lebanon, and the United States in Vietnam* (New York: Cambridge University Press, 2003); Ivan Arreguín-Toft, *How the Weak Win Wars: A Theory of Asymmetric Conflict* (New York: Cambridge University Press, 2005). In addition to books, a number of academic dissertations and research reports have also been published, for example: Frans Osinga, 'Asymmetric Warfare: Rediscovering the Essence of Strategy', in John Andreas Olson, ed., *Asymmetric Warfare* (Oslo: The Royal Norwegian Air Force Academy, 2002), pp. 267–318; David L. Grange, 'Asymmetric Warfare: Old Method, New Concern', *National Strategy Forum Review*, Winter 2000, http://blackboard.jfsc.ndu.edu/html/jfscPublications/assets/docs/cam_grange.pdf; Ike Skelton, 'America's Frontier Wars: Lessons for Asymmetric Conflicts', *Military Review*, Vol. 81, No. 5 (2001), p. 22; N. Seminar, 'Asymmetric Warfare: The only Thing New is the Tactics', http://www.ndu.edu/library/n2/n005605n.pdf; Kenneth F. McKenzie, Jr., 'The Revenge of the Melians: Asymmetric Threats and the Next QDR', (Washington D.C.: Institute for National Strategic Studies, National Defense University, 2000), http://www.au.af.mil/au/awc/awcgate/ndu/mcnair62.pdf.

Put more simply, actors that are relatively powerful win and those that are relatively weak suffer defeat.[4] From a realist standpoint it is difficult to accept that a weaker power can defeat a stronger power, which is why realists never ask the question, 'Why can the weak attain victory over the strong?' A realist would say that if the weak are victorious it is most certainly because they have overcome their limitations with respect to military power, and by definition, transcended their status as a 'weak' power. The question that realists most often ask is hence: How can an actor who is weak at the outset transition into one more powerful?

Realists addressing this question raise several relevant explanations. First is Robert Gilpin's power transition theory which, although not developed directly with respect to asymmetric conflict, can still be used to explain the phenomenon of the weak overcoming the strong. Gilpin argues that after reaching an historical peak hegemonic power necessarily expects decreasing marginal returns, because as it continues to expand territory under the empire's preview, the costs associated with maintaining order within it escalate and the empire goes into decline.[5] This line of logic reaches the inevitable conclusion that resistance and challenges within the empire's vast territories grow in proportion to its gradual decline and loss of control, triggering a set of circumstances that empower actors in certain regions to rise from a weak to relatively powerful status.

A second explanation by American military scholar Jeffrey Record is known as the theory of foreign assistance.[6] Record raises historical cases of asymmetric conflict, pointing out that in nearly all instances of the weak defeating the strong, the weaker actor received external assistance sufficient to reduce or eliminate the military advantage of the stronger actor on the battlefield, to the extent that the weaker power was victor.

A third type of explanation from Hans Morgenthau is known as the theory of technological expansion. It holds that the collapse of Western colonialism after World War II is attributable to loss of the technological, economic, and military advantages Western Europeans formerly had over Asians and Africans, and which had been a primary factor in Europe's subjugation and control of the rest of the world.[7]

The above realist explanations apply to the majority of wars, including many asymmetric conflicts, and their outcomes throughout history. Of all instances between the years 1800 2003 of asymmetric conflict, the victors

4 Power implies many positive outcomes for a state; it can deter invasion by other states, compel states to moderate their stance, or defeat by war its adversaries. See John Mearsheimer, *The Tragedy of Great Power Politics* (New York: W. W. Norton & Company, 2001), chapters 1–3; Kenneth Waltz, *Theory of International Politics* (New York: McGraw-Hill, 1979), pp. 188–92.

5 Robert Gilpin, *Shijie zhengzhi zhong de zhanzheng yu biange* (*War and Change in World Politics*), trans. Wu Jun et al. (Beijing: Zhongguo renmin daxue chubanshe, 1994), p. 228.

6 Jeffrey Record, 'Why the Strong Lose?' *Parameters*, Vol. 35 (2005/06), pp. 16–31.

7 Hans Morgenthau, *Politics Among Nations* (New York: Knopf, 1978), chapters 11–12.

in 71.5% of cases were stronger powers, according to research by Ivan Arreguín-Toft.[8] The general trend during this period hence conforms to realist logic. But realism does not explain the outcomes of all conflicts, especially those in the wake of World War II. Realist explanations, for example, do not explain the outcomes of the 1946 1948 oppression of Pakistan by the UK, the French war in Algeria from 1954 to 1962, the Israeli occupation of Lebanon from 1982 to 1985, and the defeats suffered by the two major world superpowers during the Cold War in Vietnam and Afghanistan. In every case the strong power maintained a military power advantage from the beginning to the end of the conflict, but not one outcome complies with the rigid realist expectation that the strong triumph over the weak.

The Perspective of Motivations

In contrast, and in addition to the scholars emphasizing material power in explanations of conflict are the many who explore the asymmetric motivations universally found in asymmetric conflict. The most representative of these scholars is Andrew Mack, who offers in his dissertation published in 1975 the most convincing argument for the role of motivations in conflict[9] In the course of exploring instances of failed military intervention by large states Mack points out: 'In these cases, the success of violent organizations does not depend on military victory on the battlefield... but relies on the continuous wearing down of the political will of the other state.'[10] He takes the argument a step further, holding that as the existence of a strong state is never threatened by a weaker power it is unlikely to achieve maximum military mobilization, because as war casualties mount and economic damage continues to rise, so inevitably do the voices of opposition within the strong state. In the weaker state, whose very existence an external power threatens, interest groups formerly in conflict unite in the common goal of opposing the enemy. They thus have stronger motivation and a greater resolve to attain victory no matter what the cost. Put simply, the logic of Mack's theory is that 'relative power' determines 'relative interests', but relative interests highlight the different 'political fragilities' of belligerents that cause strong states to suffer defeat.[11]

Other scholars hold similar views to Mack. For example, Steven Rosen points out that maintaining the ability to withstand the costs of war is vital to achieving victory, and that a weak power can overcome the military advantage of a belligerent state by demonstrating resolve to withstand

8 Ivan Arreguín-Toft, *How the Weak Win Wars*, p. 3.

9 Andrew Mack, 'Why Big Nations Lose Small Wars: The Politics of Asymmetric Conflict', *World Politics*, Vol. 27, No. 2 (1975), pp.175–200.

10 *Ibid.*, p. 175.

11 *Ibid.*, pp. 195–200.

terrible damages.[12] Gil Merom argues that countries with authoritarian leadership rarely suffer defeat (or when they do it is because power is not centralized) in asymmetric conflict, but democratic governments differ in this respect. As modern democratic institutions give society the power to intervene in national affairs, the moral values of politically empowered middle classes are often reflected in the extreme sensitivity of democratic countries to barbaric war strategies and war-inflicted damages, thus lowering their tolerance of war.[13]

Research on motivational issues helps to expand our point of view, and these explanations certainly imbue a degree of explanatory power as regards asymmetric conflict. It is important, however, not to overlook the shortcomings of such theories. First, comparisons of 'relative motivations' are difficult to observe and measure, hence often limited. Even though asymmetries in motivations are often seen as panaceas that can explain phenomena that realism cannot, if no specific standards exist through which to assess relative motivations, it is almost impossible to tell whether or not intervention by large states might be successful or whether or not these states will be forced to decamp. Second, although the motivations of the stronger power are generally weaker than those of the smaller, the former is nevertheless almost universally capable of inflicting immense damage on the latter. Any damage the larger power itself sustains is, generally speaking, much less significant. Under such situations, considering differences in motivation without taking into account costs does not produce a reliable determination of the outcome of an asymmetric conflict. Moreover, the view that democratic countries suffer defeat because they lack resolve is not fully persuasive. It does not, for example, explain why strong authoritarian states also suffer defeat in asymmetric conflict (e.g. Nazi Germany in Yugoslavia and the USSR in Afghanistan). Neither is the argument that democratic states are necessarily more fragile or more benevolent so clear throughout history (France in Algeria and the USA in Vietnam both experienced astronomical costs yet managed to continue over the long term, and the barbarism these two countries enacted in these wars far exceeded that of so-called authoritarian countries in other conflicts). A causal relationship does not, therefore, necessarily exist between state political institutions, and victory in war.[14]

12 Steven Rosen, 'War Power and the Willingness to Suffer', in John Vasquez and Marie Henehan, eds, *The Scientific Study of Peace and War* (Lanham: Lexington Books, 1999), chapter 11.

13 Gil Merom, *How Democracies Lose Small Wars*.

14 For discussions of this, see Michael C. Desch, 'Democracy and Victory: Why Regime Type Hardly Matters', *International Security*, Vol. 27, No. 2 (2002), pp. 5–47.

The Perspective of Strategy

In addition to the scholars that regard power and motivation as critical are those who focus on strategy (primarily military strategy) i.e. how military power is applied. These studies all argue that superior strategies, and strategic choices and strategic interaction between belligerents, often determine 'whether an advantage might translate into victory'.

Ivan Arreguín-Toft's studies in 2001 and 2005 of strategy in asymmetric conflicts are most representative of this perspective.[15] Toft separates strategy into two types: 'direct strategy' and 'indirect strategy', and subdivides direct strategy into conventional offensive strategy and conventional defensive strategy. The objective of both remains that of destroying the military capacity of the adversary through direct engagement. Indirect strategy consists also in two types: guerilla and barbaric strategy. Their objectives are to inflict damage sufficient and sustained enough to destroy the adversaries' resolve to continue waging war. Toft explicates the relationship between strategy choice and conflict outcome through a statistical study of asymmetric conflicts from 1800 to 2003. He finds that if the strategic interaction between the belligerents is uniform (direct direct or indirect indirect), then the stronger power prevails; if the strategic interaction is nonuniform (direct indirect, indirect direct), the weaker power is victor.[16]

Toft's primary contribution is that of considering the element of time in his investigation of strategic interaction. Toft argues that time is the most important causal mechanism in strategic interaction. When, for example, both sides select the same strategy it is impossible for the weaker power to eliminate the military advantage of the stronger, which enables the stronger power to win a rapid and decisive victory. If, however, each side selects a different strategy, the advantage of the more powerful state is reduced, and as a result the duration of the war is lengthened.[17] In this case, time has opposing effects on the belligerents: the longer the duration, the worse off the stronger power, and the better off the weaker power. When strategic interaction results in the extraordinary prolongation of the process of achieving political or military objectives in an asymmetric conflict, the stronger power tends to be defeated.[18] The primary logic of Toft's argument is that the interaction of different types of strategy results in a prolonged asymmetric conflict, which increases the probability of the weaker power triumphing over the stronger power.

There are, however, several important questions outstanding in Toft's theory of strategic interaction. First, there is a contradiction in his argument

15 Ivan Arreguín-Toft, 'How the Weak Win Wars: A Theory of Asymmetric Conflict', *International Security*, Vol. 26, No. 1 (2001) pp. 93–128; Ivan Arreguín-Toft, *How the Weak Win Wars*.

16 Ivan Arreguín-Toft, *How the Weak Win Wars*, pp. 23–48.

17 *Ibid.*, p. 34.

18 *Ibid.*, p. 35.

as regards strong powers' use of barbaric tactics to combat the guerilla tactics deployed by weaker powers which, according to Toft's model, is the strong power's sole option if it is to defeat the weaker power. But at the same time he emphasizes that barbaric strategy is in practice almost always counterproductive (for example, it increases the intensity of the weaker power's resistance, or provokes an international response).[19] He thus essentially implies that it is impossible to win against modern guerilla warfare, a conclusion not consistent with historical reality.[20] Second, Toft limits research in his theoretical model to consideration of the impact of different strategic interactions upon conflict outcome, thus purposely evading the issue of strategy choice. If it is indeed the case that there is always an optimal strategic choice, why would actors experiencing defeat in conflict not make that optimal choice? Even taking into account that the freedom of choice of weak powers, owing to limited training, technology, or power, is the narrower of the two, there is still nothing to stop a strong power from changing to a more effective strategy. Toft is aware of this problem, but nonetheless fails to consider the issue in his research, instead regarding the strategic choice of actors as something exogenous to the model.[21] Last and most important, the 'time benefits the weak but places the strong at a disadvantage' logic of Toft's explanation of the argument is not convincing.

Most instances now of weak powers defeating strong powers are indeed cases of prolonged wars in whose counter-intuitive outcomes time plays an important role. Yet, time could also help the stronger power; history has ample examples of prolonged wars ending in the stronger power's victory.[22] In the view of the author, the claim that time works to the advantage of the weak and to the disadvantage of the strong is alone insufficient; the concept of cost should be considered and also the question: Why is it that certain strategies employed by weak powers (primarily asymmetric

19 See Ivan Arreguín-Toft, 'The Futility of Barbarism: Assessing the Impact of the Systematic Harm of Non-combatants in War', http://www.ksg.harvard.edu/wappp/research/working/ivan_toft.pdf.

20 See J. Bowyer Bell, *The Myth of the Guerrilla: Revolutionary Theory and Malpractice* (Hardcover: Knopf, 1971); Richard Gott, *Lading Meizhou youjizhan yundong* (*Guerilla Movements in Latin America*), trans. Department of History, Fudan University (Shanghai: Shanghai renmin chubanshe, 1975).

21 Toft does not actually discuss why actors would select a certain strategy. At the same time he also does not explain why certain actors in the process of a conflict change their strategy but others do not. Conversely, Toft views an actor's strategic choice as fixed, consequently that when determining their cause one need only consider the impact of different strategic interactions upon outcomes. See Ivan Arreguín-Toft, *How the Weak Win Wars*, p. 36.

22 For example, the British conquered Wales by absorbing costs and continuing to hold out for victory. Later, after the years of colonial oppression from 1948 to 1957 Malays in Malaysia rose in an anti-British rebellion. See Archer Jones, *Xifang zhanzheng yishu* (*The Art of War in the Western World*), trans. Liu Kejian et al. (Beijing: Zhongguo qingnian chubanshe, 2001), pp. 449–501.

strategies) are able to reduce the will of stronger powers to continue the war, and ultimately cause them to abandon their war efforts?

The theoretical answer to this question generally follows one of two lines of thinking: One, with respect to a specific time or place, the balance of resources actually allocated to conflict by the two sides begins change in the weak power's favor, bringing it victory. Two, the concept of power is widened from that of weapons and equipment, steel production, total GDP, and other hard factors to include motivation, resolve, institutions, culture, and other soft factors. States which are weak in material resources might have the advantage as regards resolve or strategy. They consequently stand to increase their overall power sufficiently to compensate for material disadvantages and gain an advantage over their opponents, ultimately becoming the stronger power (or at least no longer a weak power). On the whole, although the focuses of these two lines of thought differ, objectively speaking both conforms to the theoretical assumption that 'strong defeat weak', and can therefore be seen as examples of realist theories.

This article argues that expanding the concept of power does not reach the core of the problem. Generally speaking, whether or not actors can achieve political objectives depends on two key factors: (i) in relative terms, whether or not one of the belligerents already has significant resources that can be either immediately mobilized or potentially deployed; and (ii) how these resources might be used to achieve political objectives.[23] In the case of asymmetric conflict, the first of these two factors is largely fixed (otherwise it would not fit the definition of asymmetric conflict earlier laid down). It is consequently the second factor that becomes the primary variable influencing the outcome of asymmetric conflict. This factor, also known as strategy, refers to 'the art and science of mobilizing all available power and resources in order to realize a fixed policy objective'.[24] As such, in considerations of asymmetric conflicts the question is not 'what resources does the weaker actor have that can be utilized', but 'how should these resources be used?' Following this line of reasoning, this article raises the theoretical assumption that it is the strategy the weaker actor selects which determines the outcome of the asymmetric conflict. To prove this assumption, the article relies primarily on empirical research methods in two specific components: (i) a statistical analysis of cases of asymmetric conflict from 1945 to 2007 to determine whether or not there is any causal relationship between dependent and independent variables; and (ii) a test of assumptions based on a case study of the US War in Iraq from 2003 to 2007. The article also makes a more detailed analysis of the causal mechanisms at work in asymmetric conflict through which the weak defeats the strong.

[23] Li Shaojun, ed., *Guoji zhanlue baogao* (*International Strategy Report*) (Beijing: Zhongguo shehui kexue chubanshe, 2005), p. 57.

[24] *Ibid.*, p. 29.

An Empirical Evaluation of Modern Asymmetric Conflict

According to the definitions set out in this article of asymmetric conflict, and criteria for what constitutes power disparity, conflicts in which there is a material power disparity ratio of less than 5:1 between belligerents is not considered asymmetric; civil wars also do not qualify as asymmetric conflicts. Proceeding from these criteria, this article considers asymmetric conflicts that occurred after the end of World War II.[25] Using these factors as selection criteria, 36 asymmetric conflicts are identified (Table 1) in which the rate of victory on the part of weak powers is almost 50% higher than in any other period of history.[26] Why is it the case that powerful belligerents have such difficulty winning these conflicts? The four classic explanations for this phenomenon most widely accepted in the field are: Andrew Mack's theory of asymmetric interests; Gil Merom's theory of democratic institutions; Jeffrey Record's theory of external aid; and Ivan Arreguín-Toft's theory of strategic interaction.[27] Symmetry of interests is thus the primary explanatory variable in Mack's theory, democratic institutions is that of Gil Merom's, Jeffrey Record's theory uses external aid, and Ivan Arreguín-Toft's the type of strategy and strategic interaction. Below is a brief statistical test of these variables.

A Statistical Test of Four Theoretical Explanations

To test these theoretical hypotheses this article uses a dataset containing 36 cases of asymmetric conflict that occurred from 1945 to 2007. Proceeding from the four theories earlier described, it uses four primary independent variables to gauge whether or not the interests of belligerents were asymmetric, whether or not the strong power was a democracy, whether or not the weak power received external aid, and whether or not the weak power employed a asymmetric strategy; the dependent variable is the outcome of the asymmetric conflict.

25 Different asymmetric conflicts are often broken down into different periods or times. The parties participating in the conflict are also not necessarily limited to two, and might change as the conflict develops. For ease of analysis, this article assumes that the actors participating in the asymmetric conflict are the two main parties to it. If the conflict should happen to have been broken down into different periods of different characteristics, this article considers only the most important of such periods.

26 According to Toft's data on the outcomes of almost 200 years of asymmetric conflicts, from 1800 to 1849, weak powers won approximately 11.82% of asymmetric conflicts; from 1850 to 1899, this ratio rose to 20.5%; from 1900 to 1949, it increased again to 34.9%; and from 1950 to 1999 it surpassed 55%. See Ivan Arreguín-Toft, *How the Weak Win Wars*, p. 4.

27 Andrew Mack, 'Why Big Nations Lose Small Wars', pp. 175–200; Gil Merom, *How Democracies Lose Small Wars*; Jeffrey Record, 'Why the Strong Lose?' pp. 16–31; Ivan Arreguin-Toft, *How the Weak Win Wars*.

Table 1 Basic Information regarding Contemporary Asymmetric Conflicts (1945–2007)

	Name of Asymmetric Conflict	Major Belligerents	Duration of Conflict	Involved Core Interest of Weak Actor	Strong Power a Democracy	Weak Power Received Substantial Foreign Assistance	Weak Power Employed Guerilla Tactics	Result of Conflict (Strong Power)
1	Indochina War	Vietnam–France	1945.9–1954.7	Yes	Yes	Yes	Yes	Defeat/Tie
2	Madagascar Uprising	Madagascar–France	1947.3–1948	Yes	Yes	No	No	Victory
3	Indonesian Uprising against the Netherlands	Indonesia–the Netherlands	1947.7–1949.11	Yes	Yes	No	Yes	Defeat/Tie
4	Malaysian Uprising Against the British	Malaysia–England	1948.6–1957.8	Yes	Yes	No	Yes	Victory
5	Tunisian War for Independence	Tunisia–France	1952.1–1956.3	Yes	Yes	No	Yes	Defeat/Tie
6	Kenyan War Against British Aggression	Kenya–England	1952.10–1963.12	Yes	Yes	No	Yes	Victory
7	Sao Tome War Against Portuguese Aggression	Sao Tome–Portugal	1953–1953	Yes	Yes	No	No	Victory
8	Moroccan Struggle Against France	Morocco–France	1954.8–1956.3	Yes	Yes	No	Yes	Defeat/Tie
9	Algerian War for Independence	Algeria–France	1954.11–1962.3	Yes	Yes	No	Yes	Defeat/Tie
10	Cyprus War for Independence	Cyprus–England	1955.4–1960	Yes	Yes	No	Yes	Defeat/Tie
11	Omani Struggle Against England	Oman–England	1955.12–1960	Yes	Yes	No	Yes	Victory
12	Cameroon War Against France	Cameroon–France	1956.7–1960	Yes	Yes	No	Yes	Defeat/Tie
13	Hungarian Revolution	Hungary–USSR	1956.10–1956.11	Yes	No	No	No	Victory
14	Congo Struggle Against-Imperialism	Congo–Belgium	1960.7–1964.11	Yes	Yes	No	Yes	Defeat/Tie
15	Angolan War of Independence	Angola–Portugal	1961.2–1975.1	Yes	Yes	No	Yes	Defeat/Tie
16	Vietnam War	Vietnam–USA	1961.5–1975.4	Yes	Yes	Yes	Yes	Defeat/Tie
17	Guinea Bissau War of Independence	Guinea-Bissau–Portugal	1963.1–1974.8	Yes	Yes	No	Yes	Defeat/Tie
18	Southern Yemeni Rebellion Against England	S. Yemen–England	1963.10–1967.11	Yes	Yes	No	Yes	Defeat/Tie
19	Mozambique War for Independence	Mozambique–Portugal	1964.9–1974.9	Yes	Yes	No	Yes	Defeat/Tie
20	Zimbabwe Rebellion Against England	Zimbabwe–England	1969.4–1979.12	Yes	Yes	No	Yes	Defeat/Tie

Name of Asymmetric Conflict	Major Belligerents	Duration of Conflict	Involved Core Interest of Weak Actor	Strong Power a Democracy	Weak Power Received Substantial Foreign Assistance	Weak Power Employed Guerilla Tactics	Result of Conflict (Strong Power)
21 Cambodian War of Anti-American Aggression	Cambodia–USA	1970.4–1975.4	Yes	Yes	Yes	Yes	Defeat/Tie
22 Turkish Invasion of Cyprus	Cyprus–Turkey	1974.7–1974.8	Yes	Yes	No	No	Victory
23 Indonesian Occupation of East Timor	East Timor–Indonesia	1975.10–1976.7	Yes	Yes	No	No	Victory
24 Western Saharan War	Minurso Front–Morocco	1975.11–1983.6	Yes	No	No	Yes	Defeat/Tie
25 Ogaden War	Somalia–Ethiopia	1977.7–1978.3	Yes	No	No	No	Victory
26 China Vietnam Border Conflict	Vietnam–China	1979.2–1979.3	No	Yes	No	No	Victory
27 Soviet War in Afghanistan	Afghanistan–USSR	1979.12–1989.2	Yes	No	Yes	Yes	Defeat/Tie
28 US Invasion of Grenada	Grenada–USA	1983.10–1983.12	Yes	Yes	No	No	Victory
29 US Invasion of Panama	Panama–USA	1989.12–1990.2	Yes	Yes	No	No	Victory
30 Iraqi Invasion of Kuwait	Kuwait–Iraq	1990.8–1990.8	Yes	No	No	No	Victory
31 Persian Gulf War	Iraq–USA	1991.1–1991.2	No	Yes	No	No	Victory
32 Somalia Peacekeeping Conflict	Somalia–US	1993.4–1993.11	Yes	Yes	No	Yes	Defeat/Tie
33 NATO Attack on Yugoslavia	Yugoslavia–USA	1999.3–1999.6	Yes	Yes	No	No	Victory
34 War in Afghanistan (Operation Enduring Freedom)[28]	Afghanistan–USA	2001.10 onwards	Yes	Yes	No	Yes	Defeat/Tie
35 Iraq War[29]	Iraq–USA	2003.3 onwards	Yes	Yes	No	Yes	Defeat/Tie
36 Second Lebanon War	Lebanon–Israel	2006.7–2006.8	Yes	Yes	No	Yes	Defeat/Tie

Data source: Melvin Small and J. David Singer, *Resort to Arms: International and Civil Wars, 1816-1980* (Beverley Hills: Sage Publications, 1983); Ivan Arreguín-Toft, *How the Weak Win Wars*, pp. 228–32; Li Jujian, *Zhanzheng yu heping: shidai zhu xuanlv de biandong* (*War and Peace: Changes in Patterns over an Era*) (Beijing: Xuelin chubanshe, 1999), pp. 390–401.

Owing to the structural characteristics of asymmetric conflicts, whether or not the conflict affected the critical interests of the weaker power is the criterion which the article primarily considers when determining if it was indeed asymmetric. Interests that are related to the survival and development of the state, according to Li Shaojun's categorization of the layers of national interests, are known as Critical Interests of State Survival, which any state will take all steps possible to protect;[30] protecting less defined state interests incurs lower costs. In the majority of modern asymmetric conflicts, characterized by powerful states engaging weaker states in armed conflict within the latter's territories, the weaker actor, objectively speaking, is incapable of threatening the central interests of the stronger actor. When considering, therefore, whether or not a conflict involved asymmetric interests, we must determine whether or not it threatened the central interests of the weak actor, or more specifically, whether or not the central interests of maintaining sovereignty and national territory hinged upon the outcome of the conflict.

In instances where the weaker actor received no external support, the variable of whether or not the weak actor obtained substantial external assistance is quite straightforward. Determining the value of the variable in cases where the weaker actor did receive support is more difficult, as the issue then becomes one of gauging the importance of such support. This article first determines support in the form of material power that is to say, rhetorical support or sympathy from the international community does not constitute significant external support; and second whether or not external support changed the original power dynamics between the belligerents. Even if support did not give the weaker actor enough power to overcome the stronger actor, it might at least have brought a measure of balance to the battlefield. For the specific purposes of this article, any open, public support of a weak state from a major power (specifically any permanent Member State of the United Nations Security Council) constitutes major external assistance. Other types of support are not considered significant external assistance.

Although academics have long been familiar with the theory of democratic peace, determining whether or not a state has democratic political institutions is nevertheless no easy matter. In one sense, any state with established

28 Since the War on terror started in Afghanistan in 2001, based on the current situation in Afghanistan, the effectiveness of the attack on terrorism, and appraisals of the effort offered by international pundits, this article considers the war in Afghanistan to have failed. For more, see http://news.xinhuanet.com/world/2008-01/31/content_7529686.htm.

29 The War in Iraq is similarly far from being over. Even though early on in the war the American military made rapid progress, by the later period it had slide into a violent quagmire, and even Bush himself has publicly stated that this war has not been won. See http://news.xinhuanet.com/world/2006-12/21/content_5514572.htm.

30 Li Shaojun, 'Lun guojia liyi,' ('On National Interests'), *Shijie jingji yu zhengzhi* (*World Economics and Politics*), No.1 (2003), pp. 4–9.

democratic political institutions constitutes a democratic state. It is important, however, to consider whether or not the state is democratic in practice. When considering the logic from the standpoint of democratic peace theory, or Merom's theory of democratic institutions, both emphasize the ability of the common people through some type of domestic political institution to influence or determine state war policy.[31] As regards the questions of occurrence, development, and conclusion of asymmetric conflict, therefore, whether or not the will of the people has strong influence over policy is the key to determining whether or not the state should be considered democratic or nondemocratic.

Type of strategy is the last independent variable. Findings of modern international politics show that when weak actors use direct strategies to engage strong powers the result is disastrous, and that such a strategy choice is, to say the least, not a rational one. A main characteristic of asymmetric conflict is that weaker actors almost universally select asymmetric strategies, that is to say, they choose what Toft refers to as guerilla strategies. Time evolution has brought about a new type of asymmetric strategy terrorism which has become a main form of conflict, particularly evident in today's global battle against terrorism. To determine this variable, therefore, we establish whether or not the weaker actor used asymmetric strategies (specifically guerilla or terrorist tactics) and whether or not there is a relationship between type of strategy and outcome of an asymmetric conflict. The relevant data for asymmetric conflicts from 1945 to 2007 is shown in Table 1.

As the data in Table 1 shows, when an asymmetric conflict involves the weak power's critical interests, the weak power is able to defeat the strong power in 63.9% of cases; when the stronger power is a democracy, the weak power is a victor in 61.1% of cases; when the weak power obtains substantial external aid, the weak power wins in 50% of cases; and finally, when the weak power opts for an asymmetric strategy, its ratio of victory reaches 91.7%. On the whole, the relationship between external aid and the outcome of the conflict appears the weakest: Even when the weaker power does not receive any assistance, the stronger power achieves victory only 43.8% of the time. The relationship between type of strategy and outcome of conflict, however, is the strongest, from which we conclude that the type of strategy the weaker power selects is critical to determining the final outcome of an asymmetric conflict.

31 Immanuel Kant, *Lishi lilun pipan wenji* (*Selected Writings from the Theory of History*), trans. He Zhaowu (Beijing, Shangwu yinshuaguan, 1991), pp. 106–7; Gil Merom, *How Democracies Lose Small Wars*, p. 15.

Test of Strategic Choice as Controlling Other Factors

To explore the causal relationship between asymmetric strategy and the outcome of the conflict, this article applies statistical controls to the three other factors of whether the conflict threatens the weaker power's critical interests, whether the stronger power is a democracy, or whether the weaker power receives external aid. That is, not to say that these factors have no role in influencing the outcome or that they are not important, only that they take a temporary 'back seat', while the article concentrates on a deeper explication of the effect of strategy choice. The values of the three other independent variables, in addition to factors related to strategy choice, are considered in turn later in the article.

As the data in Table 1 shows, most modern asymmetric conflict proceeds as follows: A relatively strong external military force (consisting either of the military of one particular state or a group of allied states) attacks a country in a weaker position, or a weaker group within a particular country's territory. In these circumstances, the attack has serious impact on the central interests of the weaker power. The assumed value of the 'asymmetric interests', therefore, is 'yes' or 1. As most of the major or strong powers participating in contemporary asymmetric conflicts are western capitalist democracies, this article also assumes that the value of the 'democratic political institutions' variable is 'yes'. Finally, instances of weak powers receiving substantial external support in asymmetric conflict are, generally speaking, limited. The article hence also assumes the value of 'external support' as 'no'. For the purposes of this article, these criteria are used to select the cases that are ultimately included in the study; the remaining 10 instances of asymmetric conflict are discarded. Cases eliminated include: The Indo-China War, the Hungarian Revolution, the Vietnam War, the Cambodian War of Anti-American Aggression, the Western Sahara War, the Ogaden War, the China Vietnam Border Conflict, the Soviet War in Afghanistan, the Iraqi Invasion of Kuwait, and the Persian Gulf War. The remaining 26 cases represent 72.2% of asymmetric conflicts that occurred during this period.

In these 26 cases, the other three variables (whether the conflict involves a core interest, whether the stronger power is a democratic country, and whether the weaker power receives major external assistance) are all constants. This enables a detailed exploration of the effects of the strategic choice variable as shown in Table 2.

As the above test results show, in all 26 cases where the values of the other three variables are identical, and where a weaker power selects an asymmetric strategy (primarily either guerilla or terror tactics) it is a victor in 88.5% of cases.

If we were to include the 10 cases of wars that do not meet the asymmetric conflict criteria earlier laid out, and instead use SPSS to conduct

Table 2 Conflict Outcome Statistics after Controlling other Variables

Weak Power Selects Asymmetric Strategy	Strong Power Victory	Weak Power Victory/Tie	Test Results (Overall Ratio)
Yes	15.8% (3)	84.2% (16)	88.5% (23)
No	100% (7)	0% (0)	

Table 3 Multivariate Regression Analysis of Modern Asymmetric Conflict

	Coefficient Value (β)	T	Level of Significance (Sig.)
Intercept (Constant)		0.158	0.876
Weak Power Uses Asymmetric Strategy (X_1)	0.822	7.357	0.000
Involves Core Interests of the Weak Power (X_2)	−0.006	−0.052	0.959
Stronger Power has Democratic Political Institutions (X_3)	−0.028	−0.269	0.790
Weak Power Receives Substantial External Assistance (X_4)	0.091	0.890	0.380

Note: The independent variables are X, and the dependent variable is Y (outcome of asymmetric conflict); $Y = 0.822X_1 - 0.006X_2 - 0.028X_3 + 0.091X_4$; Adjusted $R^2 = 0.672$.

a multivariate regression analysis of all 36 instances of asymmetric conflicts in the post-World War II period, the results would nonetheless show a similarly high correlation between weak powers using asymmetric strategies and 'the weak defeating the strong'. The results of this procedure are presented in Table 3.

The Table 3 shows that the value of the coefficient in the relationship between 'weak power uses asymmetric strategy' and 'outcome of asymmetric conflict' is 0.822; the result even at 0.000 is significant. This implies a strong level of correlation between these two variables.

This article argues, based on the results obtained after controlling the other variables, that there is a strong relationship between a weak power selecting an asymmetric strategy to wage an asymmetric war and the outcome. In other words if, under contemporary conditions, the weaker power chooses an asymmetric strategy, it is likely to emerge victorious; if, however, the weak power does not use an asymmetric strategy, the stronger power is almost certain to win the conflict.

Basic Logic of the Strategic Victory

The high correlation between type of strategy and conflict outcome established to deepen understanding of this phenomenon is more important still to explicate the mechanism driving the relationship between the two. This article draws primarily on the concept of asymmetric war and the related concept of asymmetric strategy that academics and political pundits generally use to explain this relationship.

Logic of Interaction Between Belligerents

A 'symmetric strategy' refers to a belligerent engaging an enemy using conventional military force and, through one or more decisive battles, defeating the enemy's conventional forces, gaining control of its population, territory, cities, important industries or commercial centers, and thereby winning the war. The most common form of symmetric strategy is that of the military of one state taking an offensive posture and attacking another's vital strategic targets (capital city, industrial centers, key transportation networks, main bridges, and roads). The defensive state then mobilizes a resistance, engaging the attacker in one or more conventional battles, and the war continues until one side is subdued. The direct objective of symmetric strategy is to destroy the opposition's military forces, thereby rendering powerless its physical capacity, and consequently motivation, to continue the conflict. The primary characteristic of symmetric strategy is that both the involved parties use military force, follow the same rules of war, and share the same understanding of victory and defeat.

Asymmetries are defined in reference to symmetries, and this study is no exception to that rule. Generally speaking, an asymmetric strategy refers to tactics other than those used in a symmetric strategy (or what is often referred to as a conventional war). In modern times, such tactics are generally either guerilla or terrorist in nature. Asymmetric, in contrast to symmetric strategies, do not seek to, and are not capable of destroying the other party's military capacities. Their primary objective is rather to wear down the adversary's motivation to wage the war. Actors using asymmetric strategies generally avoid full-frontal conventional warfare. They instead use indirect means of engagement that cause the enemy's war costs to rise to the extent that it loses the motivation to carry on with the conflict.[32]

Symmetric and asymmetric strategies have the common objective of using all available means to 'compel the enemy to succumb to one's will'.[33] But there the similarity ends, because there are major differences in the types of violence that the two strategies employ. Symmetric strategies conventionalize violence by concentrating its use and limiting its scope. Unlimited use

32 It should be noted that, according to Mao Zedong's thinking on guerilla war, guerilla warfare ultimately evolves into a full-scale conventional war. But it is often not possible in modern asymmetric warfare to develop asymmetric strategy into the preparatory phase for a major counterattack. See *Maozedong xuanji* dier juan (*Selected Works of Mao Zedong*, Vol. 2) (Beijing: Renmin chubanshe, 1991), p. 478.

33 Carl von Clausewitz, *Zhanzheng lun* shangjuan (*On War*, Vol. 1), trans. the Academy of Military Sciences of the PLA (Beijing: Jiefangjun chubanshe, 1964), p. 12.

of violence, however, is the main feature of asymmetric strategies. This article argues that it is unlimited violence that constitutes the main reason why weak powers are generally the victors of asymmetric conflicts. Two contradictory yet related developmental trends with respect to war that have existed throughout history are those of either limited or limitless application of violence.[34] The more that violence in war is limited by objective factors or willful acts of the people, the more civilized international society may be said to have become. But historical trends imply that war is developing towards even greater intensities of violence.[35] A fundamental reason may be that unlimited violence has never gone out of style, a logic that Clausewitz proposed and explicated in the 19th century. Clausewitz argues that war is a form of violent behavior and that the use of such violence is without limit.[36]

In Clausewitz's view, as long as the objective of war is to 'destroy the enemy' (and to avoid being destroyed by the enemy), belligerents will use all power and means possible and deploy all possible weapons to wage war.[37] In other words, as long as humanity has no peaceful means of resolving conflicts, as long as conflict and war have impact upon the core interests of one or more parties, and as long as violent resistance remains the ultimate determinant of victory or defeat, the natural decision of parties engaged in conflict is to use all means necessary to secure victory. It is precisely these 'natural tendencies stemming from the nature of war' that have led to a movement throughout the history of warfare towards broader and more intensive applications of limitless violence. But as Clausewitz understood, it is necessary often to reconsider this logic based on reality.[38] In other words, limitless use of violence in actual conflict scenarios is not inevitable; it is possible that moderation by parties to the conflict may result in conscious limitation of violence.

This being the case, which factors determine whether or not humanity opts for limitless violence or conscious restraint? It is argued here that the answer lies in recognition of an adversary's type. Mao Zedong once said: 'The foremost questions to be asked in revolution are: who is our enemy? Who are our allies?'[39] The fundamental purpose of war is to determine the nature of hostile relations between parties; the factor determining the extent of violence used within the context of war is hence none other than the

34 Li Jujian, *War and Peace*, p. 35.
35 *Ibid.*, p. 44.
36 Carl von Clausewitz, *Zhanzheng lun* xiajuan (*On War*, Vol. 2), trans. the Academy of Military Sciences of the PLA (Beijing: Jiefangjun chubanshe, 1964), pp. 15–16, 46.
37 *Ibid.*, pp. 15–16.
38 Carl von Clausewitz, *On War*, Vol. 2, pp. 16–18.
39 Mao Zedong, *The Selected Works of Mao Zedong*, Vol. 1, p. 3.

nature of hostile relations.[40] Different understandings of an actor's type result in different decisions on use of violence in war. Determination of the adversary's level of hostility is also inseparable from state interests. When battling for a critical interest, i.e. one necessary for survival, actors regard the adversary as 'absolute enemy'.[41] Under these circumstances, there is no room for moderation or subjective restraints as regards means of waging war. If, however, the conflict involves a secondary interest, i.e. one not crucial to survival, actors regard one another as 'limited enemies'. Under these circumstances, the objective is not to destroy an adversary, but to use a level of violence sufficient only to compel compliance. An actor will hence exercise subjective restraint on its use of violence in the process of deciding the means to achieving its objectives. Under modern conditions, this translates into observing widely accepted principles of international law. In other words, actors select means to achieving the ultimate objective that also avoid a vicious cycle of growingly violent reprisals and counter reprisals.

We can categorize, according to each actor's determination of the type of enemy it faces, international conflicts into three types: 'absolute enemy vs. absolute enemy', 'limited enemy vs. limited enemy', and 'limited enemy vs. absolute enemy'. The first type of conflict is commonly observed in history: For example, the Romans ploughed over and heaped salt on the lands of the Carthaginians, rendering them uncultivable. Another example is that of the emperor Domitian issuing a decree 'prohibiting the Monistrol people from existing'. Yet another is Ghengis Khan's ransacking of Khwarezmia's empire.[42] These examples demonstrate the use of all possible means to wage war. The second 'limited enemy vs. limited enemy', scenario is found more often in modern warfare involving European countries. Such wars generally have a definite beginning (declaration), and end (treaty). The countries involved in such wars often have political objectives, and the military means used to wage war are generally accepted by the international community as observing its rules and regulations.[43] Although killing on the battlefield might intensify, adversaries at least acknowledge their respective sovereign status. The continued existence of a country defeated in war is

40 Deutsch has categorized armed conflict into: 'Fight until the death' conflict and 'Mutual Survival'. See Carr Deutsch, *Guoji guanxi fenxi* (*The Analysis of International Relations*), trans. Zhou Qipeng et al. (Beijing: Shijie zhishi chubanshe, 1992), pp. 173–4.

41 The status of an enemy, of course, may change with the situation: Once the strong power has made a substantial concession (one that acknowledges and respects the core interest of the weak power), the weak power's perception of the stronger power's status as enemy changes from that of absolute enemy to limited enemy, or perhaps even ally.

42 Archer Jones, *The Art of War in the Western World*, p. 504.

43 Amstutz referred to such types of war as classic wars. See Mark R. Amstutz, *International Conflict and Cooperation: An Introduction to World Politics* (Boston: McGraw-Hill College, 1999), pp. 255–61.

basically guaranteed, and there is also hope of its acceptance once again into the international system.[44] The third type of scenario, 'limited enemy vs. absolute enemy' is that of asymmetric conflicts, the subject of this study.

Asymmetric conflict refers to cross national military conflict involving actors between whom there is a substantial disparity of material power. Parties to such conflicts also differ greatly in their choice of method of engagement; there is also asymmetry between the two actors' objectives or interests. Contemporary conflicts that end in weaker powers defeating the stronger power are generally characterized by a strong power entering the sovereign territory of a weaker power to engage in war. The objective may be to establish or maintain colonial control over the weaker power, or to maintain its interests by prolonging a puppet regime, but the overall intention is to secure interests which maintain its status as a great power. The weaker state, on the other hand, often fights the stronger power to preserve the freedom of its people, or to maintain sovereignty, its objective being fundamental survival. Under circumstances where extreme disparities exist between both means and objectives, parties to the conflict maintain vastly different views of the type of enemy they face. There is also a huge variance in belligerents' respective choice of the most appropriate form of violence to deploy against the enemy.

From the perspective of the stronger power, the interests that constitute the objective of the conflict have no impact upon its survival. It consequently fights a limited conflict because it does not regard the adversary as an absolute enemy. The strong power's determination of the enemy's status determines to a great extent its choice with respect to the type of violence to deploy in the conflict: As its objective is limited, the strong power generally hopes to achieve it through limited violence. Although the strong power typically maintains its capacity and means to increase the level of violence, doing so inevitably raises the costs of continuing the conflict. Taking into consideration the mathematics of cost and benefit analysis, increasing the level of violence is not necessarily in the stronger power's best interests. The stronger power also hopes that the target of its attack will similarly limit the use of violence, as it may face exorbitant costs if the weaker power is willing to continue fighting to the death. The solution, as Napoleon once said: 'one must fight as a guerilla in order to face guerillas in war', which then results in 'a downward spiral towards increasingly grotesque acts of terrorism and terrorist reprisals',[45] the costs of which far

44 Kaplan in his well-known 'Six Processes of the International System' described the characteristics of the 'equilibrium process', of which the most important was that which 'allows defeated states to re-enter the system'. See Morton Kaplan, *Guoji zhengzhi tixi yu guocheng* (*System and Process in International Politics*), trans. Bo Zhiyue (Beijing: Zhongguo renmin gong'an daxue chubanshe, 1989), p. 23.

45 See Carl Schmitt, *Zhengzhi de gainian* (*The Concept of the Politics*), trans. Liu Zongkun (Shanghai: Shanghai renmin chubanshe, 2003), pp. 359, 372.

outweigh the benefits in the long term. The history of modern warfare also demonstrates that wars which deploy absolute means of violence evolve into horrific long-term battles often resulting in devastating costs and damages to both sides. It is out of precisely such considerations that great powers tend towards establishing over the long term a certain set of principles governing the nature of war, to prevent limitless cycles of increasingly intense violence. In *Law of Nations*, French enlightenment thinker Voltaire says that if humanity hopes to have order and principles in the practice of armed violence, then it must establish a means of limiting occurrence of calamitous war, and always ensure that the door to peace remains open. This is to say that everything should conform to the needs of conventional war.[46] In reality, the restraint of violence is fundamentally none other than the practice of a symmetric strategy, because fighting according to such principles suits the interests of the stronger power. It thus follows that as they benefit from the conventional mode of war, strong powers maintain a positive attitude towards establishment of principles and codes of warfare.

From the perspective of a weak power, as its very survival hinges upon the outcome of an asymmetric conflict, its approach is one of absolute conflict. This perception of the enemy, from the perspective of subjective motivations at least, makes a weaker power less inclined to limit the use of violence. The tremendous gap in material power between the belligerents means that following the 'rules' when mounting a resistance to the greater power will almost certainly result in defeat. The will to survive, therefore, motivates weak powers to use whatever means of violence are at their disposal, and as their central interests are threatened, they are prepared to absorb whatever costs are necessary to maintain constant violent warfare (as was the case in the Vietnam War and the Algerian War). The main concern of weaker powers is whether or not victory is possible, in contrast to the cost benefit considerations undertaken by the stronger powers. A weak power is strongly motivated towards using a form of strategy that is within its means and that can bring victory. These strategies have throughout the history of modern asymmetric conflict been asymmetrical, more specifically the guerilla and terrorist tactics that have evolved in the post-Cold War period. The basic characteristics of such strategies is that they take no account of the 'rules of the game' imposed by the stronger powers, and hence place no limit on the duration or space in which to deploy violence. They thus transcend the traditional distinctions between times of war and peace, battle grounds and civilian areas, soldiers, and civilians. The breakdown of these distinctions magnifies the costs to both sides of the

46 J. F. C. Fuller, *Zhanzheng zhidao* (*The Conduct of War*), trans. Zhanxu (Beijing: Jiefangjun chubanshe, 1985), pp. 7–10.

conflict, to the extent that after reaching a certain level the original objective of the stronger power becomes meaningless.

The asymmetric strategy approach to deploying violence began with the Spanish guerilla resistance to Napoleon. Having been utilized by T. E. Lawrence, Mao Zedong, Nguyen Giap, and Che Guevara, it developed into a more sophisticated form of resistance with theoretical and practical frameworks.[47] This compiled body of work makes it relatively easy for contemporary weak powers to apply these strategies. Additionally, within the contemporary context of globalization, the information age, and the rise of the global economy, information, there is a free flow of people and technology. The knowledge economy is easily replicated, and the ever-quickening pace of scientific breakthroughs has considerably widened the scope available to weak powers of models and tools relative to applications of violence in asymmetric conflicts.

From the analysis above it is evident that the fundamental difference between 'symmetric strategies' and 'asymmetric strategies' is that of whether principles and codes of law with respect to the use of violence are observed or violated. An outcome in which the weak subdues the powerful is a direct result of the assessment of costs and benefits made by each of the parties to the conflict.[48] Consideration of the costs by different actors of employing different strategies thus determines the outcome of the conflict. Their behavior is essentially that of a market-based transaction: 'Only under the following conditions can a consensus be reached: the extent of the costs a victor is willing to bear in terms of the military efforts needed to secure a political goal must be the same as the costs the defeated are willing to accept as a result of complying with the demands of the stronger power.'[49] An offensive party may obtain a major military victory, but upon discovering that the enemy is not willing to accept the reality of defeat may decide that the efforts it has already expended to secure victory have reached a point where the costs of continuing outweigh any possible benefits. It is also important to point out that a weak power which maintains its will to war has the ability to continue to raise the level of violence it deploys against the stronger power, whose cost benefits calculus thus changes to the extent that the costs of continuing the war exceed any benefits. This ultimately causes the stronger power to give up the war effort out of political considerations.

47 See Adam Lowther, *Asymmetric Warfare and Military Thought* http://www.allacademic.-com//meta/p_mla_apa_research_citation/0/6/9/1/2/pages69123/p69123-2.ph.

48 Costs can also be divided into 'actual costs' and 'expected costs'. For arguments relative to this, see Robert A. Pape, 'Soft Balancing against the United States', *International Security*, Vol. 30, No. 1 (2005), pp. 36–8.

49 Archer Jones, *The Art of War in the Western World*, p. 487.

Cost Analysis of the Two Strategies

The application of strategy to military force to achieve political military objectives is central to the art of war. But cost benefit analyses of a conflict are significantly impeded by the effects on strategic actors of emotional elements and 'blind' compulsive urges. Taking into consideration long-term trends, however, and the resultant reality of practical interests, actors in conflict situations ultimately return to rational considerations. As costs and principles are two concepts central to strategic interaction in asymmetric conflict, it is only on the assumption of rational actors that it is possible to make a logical delineation of cost control within the context of the two different strategies.

With respect to symmetric strategy, the conventional military means that strong powers employ have two characteristics: (i) they are based on modern industry and information technology, and encompass a wide scope of choice as regards scale of violence; (ii) conventional war places strong limitations on actual use of violence, thus demonstrating the broad gap between objective capabilities and subjective will. This gap, however, reflects precisely the dilemma strong powers face as regards the need to control the costs of violence: Rationally speaking, if a strong power with limited objectives expects to achieve victory in an asymmetric conflict, it is critical for it to control costs incurred in engaging the enemy. This means controlling the costs of its own campaign and at the same time causing the other party's costs to rise beyond the tolerance threshold. But also critical is that in the process of achieving the latter objective the strong power does not itself incur costs to a point where this tactic becomes counterproductive. In other words, actors must avoid raising costs to the extent that they produce an outcome where neither party wins. This logic may be observed when two strong powers in an asymmetric conflict continuously raise the level of violence (for example, by violating the prevailing laws of war), while at the same time purposively establishing certain limits on the use of violence (or ostentatiously and unilaterally following particular laws to influence the behavior of the opposition). Limits on violence are regularly established by strong powers and serve their interests, but these principles at the same time reflect rational concerns relative to controlling costs.

There are, moreover, examples of cost control considerations throughout the development of modern symmetric strategy. History shows that most of the traditional wars fought in modern times employed symmetric strategy to ensure that violence remained concentrated on the battlefield, and that disputes were settled through rapid, decisive wars, thus controlling costs to within acceptable limits.[50] It soon became obvious, however, that wars

50 Herfried Munkler, *Xin zhanzheng* (*The New Wars*), trans. Zhang Lin (Beijing: Zhongyang bianyi chubanshe, 2006), pp. 100–4.

fought on the basis of symmetric strategy were impossible either to control or tolerate: The two World Wars made clear to all parties involved that no country could tolerate war on such scale again. Warfare at this stage had developed beyond the limits of what even the most patriotic of peoples were willing to endure.[51] Under such circumstances, waging war remained irrational unless a new strategy were to be devised that could significantly restrict costs of conflict.

The horrifying loss of life and immense military costs of World War I gave rise to two different symmetric strategies whose purpose was to restrict war expenditures: Maginot's defensive strategy and the offensive blitzkrieg strategy. The emphasis of Maginot's defensive theory was on banking up the defensive capabilities of a state to a pitch of impenetrability, thereby shifting military costs in favor of an asymmetric strategy. In other words, the Maginot Line minimized expenditure on defense by virtue of the deterrent effect of the intolerable cost to the enemy of mounting an offensive against it. History, however, proved that the so-called impenetrable Maginot Line, consisting of a high concentration of tanks on the German border, failed to repel the German attack.

The theory of the blitzkrieg war took the opposite approach of increasing military offensive capacity and reducing the size of the military to avoid a prolonged battle and astronomical expenditure of resources.[52] The idea was to develop offensive power to a point where wars could be brought to a rapid close. It was perhaps the simultaneous development of these two strategies to reduce expenditure that led to disaster: with both offensive and defensive capabilities raised to optimum level, neither was able to control war costs.

Although the two strategies imbued major differences, both fundamentally attempted to accomplish the same thing: Reduce military costs relative to those of the opposition, and control the overall costs of war in a way that shifted the war trajectory in a direction that could better one's interests. There are shades of Maginot's defensive theory in American attempts today to establish an anti-missile system. But it is the 'blitzkrieg' concept that has become the classic form of symmetric strategy in modern warfare. Although the Gulf War engendered new terminology such as 'information warfare', 'precision missiles' and 'land air maneuver operations', the emphasis on technological enhancements as a means to victory still does not signify any departure from the concepts central to symmetric war of maintaining control over costs and ensuring a speedy decisive victory. The English scholar Martin Shaw defines modern western war tactics as 'translating risk into war', and at the core of the 15 principles he lists in this respect are

51 *Ibid.*, p. 183.

52 William S. Lind, 'Understanding Fourth Generation War', *Military Review*, Vol. 84, No.5 (2004), p. 13.

indeed those of controlling war costs and winning a rapid and decisive victory.[53]

But western nations are nonetheless unwilling to get involved in such warfare because widespread guerilla and terrorist attacks render useless the technological advantages of the American and European militaries. Western nations now prefer to control their costs of conflict through air strikes or missile attacks. But these practices still carry the obvious problem apparent in the difficulty a stronger power experiences in achieving political objectives before actually stepping on to the battlefield.[54]

A weak power using an asymmetric strategy faces problems similar, but with main differences as regards controlling costs, to those of a strong power using a symmetric strategy. Disparities in power constitute these differences, as if the weak power fights a conventional war it is almost certain, according to hard realist logic, to lose; the greater the advantage the strong power has over the weak power, the more capability it has to launch an attack using modern technology. Weak powers consequently have strong incentive to break the rules, and as the asymmetric strategies they use break them in a way that emphasizes the characteristics of violence, such powers are often perceived as insane or unreasonable. But even if we acknowledge that 'blind' impulses exist in conflict, it is important to note that weak powers do not necessarily use such strategies on pure impulse, but rather on basis of rational calculations.

The rational logic behind guerrilla tactics, explicated in detail by many theorists, also shows that the appearance, leadership, organizational structure, logistical strategies, anticipated consequences, and external ties of terrorists all conform to the principles of utility sought by modern society. Clausewitz long ago pointed out that 'if one party uses force without restraint, and without regard to the loss of life, while the other party does not behave in the same manner, it is inevitable that the former gains an advantage.'[55]

If we admit that a weak power's choice of an asymmetric strategy is a rational outcome, how then does this strategy place the weak power in a position of advantage with respect to the balance of costs and benefits? From the perspective of raising the strong power's costs, attackers that choose an asymmetric strategy generally have two advantages: First, their targets of attack and the channels through which such attacks are made are unknown, and attackers have a wide, unrestricted range of targets from which to choose. Even a country like the USA with all its tremendous advantages with respect to technology, economy, and military power, still

53 Martin Shaw, *The New Western Way of War: Risk Transfer War and its Crisis in Iraq* (Cambridge: Polity Press, 2005), pp. 71–98.

54 Jeffrey Jarkowsky, 'Boots on the Ground–Will U.S. Landpower be Decisive in Future Conflicts?' pp. 19–20, http://www.smallwarsjournal.com/documents/jarkowsky.pdf.

55 Carl von Clausewitz, *On War* Vol. 1, p. 13.

lacks the capacity to ensure the safety of all of its people and infrastructure. 'Although the public terrorist activities happen in a small space, related though and what is unsettling is that terror and universal anxiety are not enclosed within any small space.'[56] Uncertainty as regards possible threatened targets of attack allows an asymmetric strategy to fragment a centralized defense effort. This creates even more opportunities to launch attacks, and to mobilize resources that home in on the enemy's weak points. Second, attackers often have the advantage over their opponents as regards the ability to withdraw or retreat rapidly.[57] As control of territory is not one of the objectives of asymmetric strategy, there is nothing to stop attackers from vacating immediately after a target has been attacked. The retreating party generally also has the advantage, even in cases where the adversaries use similar weapons systems. In pre-modern times, attackers would cut down trees, burn bridges, use snipers, or cross rough terrain to escape from enemy pursuit. This becomes even easier in modern guerilla or terrorist attacks, as attackers have no uniform and, after getting rid of their weapons, vanish into crowds of people and become practically indistinguishable from ordinary civilians. Modern suicide bombers do not even consider a way out. These factors all increase the prospects for success.

The development and spread of modern technology has made asymmetric strategic attacks even more effective. After Nobel invented smokeless powder, an anarchist newspaper ran an article that pointed out: 'If a person with a bomb uses it at the right time and place, he can take out an entire unit of foot soldiers... the entire process of war has already been completely revolutionized by scientific innovation.'[58] A weak power can also inflict a deadly and calamitous blow on a strong power with the even deadlier weapons of mass destruction of today. Modern communications and media also make such attacks more effective, especially in the light of the growingly globalized markets around which people's lives are organized. Assaults on these markets cause a rapid loss of control, consequently an even greater ripple effect. They give grassroots organizations with little power the chance to gain the same footing as the world's sole contemporary superpower. History shows that when weak powers consistently develop new asymmetries, the scope of war hugely expands.

Weak powers also have an advantage with respect to costs. As asymmetric conflict is their means to continued survival, cost is no object, but they do have special channels through which to reduce both costs and risks. Asymmetric strategy must also bear the additional costs of prolonging the conflict, wearing down the enemy's resources, and the need for covert

56 Carl Schmitt, *The Concept of the Politics*, p. 419.
57 Archer Jones, *The Art of War in the Western World*, pp. 449–501.
58 Michael J. Schaack, *Anarchy and Anarchists* (New York: Arno Press, 1977), pp. 87–8.

operations. These dragging-out measures are applied according to the conventional forces of the enemy's military. As the weak power is unable to achieve a rapid or decisive victory, its only course is to prolong the duration of the war, continually causing casualties and other losses that eventually wear down the opponent's political will to carry on with the conflict. Lengthening the duration of the war, however, requires the will to exchange space and resources for time, which often implies submitting one's national territory to repeated attacks by the enemy's superior firepower. But popular support for an asymmetric strategy that resists foreign conquest of national territory generally enables a protracted war. The ability to hide among or blend into local populations also helps to protect nonconventional militant forces, and can inhibit the enemy from using superior firepower to launch a disastrous attack. And as the stronger power expands the scope of its own attack, the number of civilian casualties it inflicts also inevitably rises. Consequent pressure from the international community could drive the stronger power into a self-defeating political trap.

The weaker power's status as nonstate actor is another characteristic of contemporary asymmetric conflict, one which renders untenable any moves by the stronger power to deter the enemy with an expanded scope of attack. The main difference between state and nonstate actors is that of state actors having sovereignty, state mechanisms, and wide-ranging interests. Nonstate actors, on the other hand, are in many ways equivalent to landless laborers whose lack of material possessions makes them far less easy to intimidate and deter. This is exemplified by the many state actors throughout history whose interests have forced them to accept the bitterness of defeat rather than employ asymmetric tactics (for example, Prussia in the Napoleonic Wars and Saddam Hussein during the Gulf War). Conversely, the cost benefit analysis of nonstate actors is generally confined to the position of the resistance group itself an entirely different perspective from that of a whole state, military organization, parties to a civil war, or local soldiers, whose cost benefit analyses must take a far more into account.[59] The effectiveness of the stronger power's strategy of fighting a war through aerial attacks rather than ground troops is also limited; once the enemy's infrastructure has been bombed to ruins after years of aerial attacks, there remain few available targets. Moreover, destroying these targets does not, generally speaking, put political pressure on the opposition; an actor subjected to such attacks can just sit tight until the aerial raids end. When the cost of bombs and missiles begins to exceed the value of the destroyed targets, the stronger power then faces the reality that a strategy based on aerial attacks is no longer effective.[60]

59 See Herfried Munkler, *The New Wars*, pp. 139–51.

60 This is to a great extent because after World War II, the strategic aerial bombing deployed by the USA, the UK, and other nations was never applied again.

As earlier discussed, a main difference between symmetric and asymmetric strategies is that of whether violence is restrained or the laws of war violated. Whether the weak can defeat the strong is directly dependant on assessment and ultimate balance of costs and benefits in relation to strategic interaction between the actors. From a costs perspective, an asymmetric strategy enables a weak power to hedge against a stronger power with superior military forces and resources. Such a strategy often helps the weaker power to achieve a relative advantage by virtue of the large space of possibility it gives the actor in conflicts with the stronger power, and of its minimal status in the 'game' against it. Owing to this uncertain space, the conflict rarely ends in annihilation of one or other of the parties to the conflict, but rather in belligerents moving towards a more moderate position that allows for 'mutual existence'. In modern asymmetric conflict this moderate position is often manifest in the stronger power's recognition of the sovereignty or independence of the weaker power, or in its withdrawal from the other actor's territory.[61] Strictly speaking, stronger powers typically never submit to military defeat, but the disparity in power between the parties to the conflict and the strong expectation that the strong will defeat the weak, often result in a 'draw' outcome being construed as the weaker power's defeat against the stronger.

A Case Analysis of the Iraq War

On March 20, 2003, allied forces under American command attacked Iraq with the objective of overthrowing Saddam Hussein. This is without doubt an example of a conflict in which there was tremendous disparity between the belligerents as regards relative capabilities. Despite its unparalleled military might, however, the USA has become bogged down in this conflict to the extent that it now finds itself lacking a suitable exit strategy. Even though the war has yet to conclude, the USA can be said to have lost the war in a number of senses.[62] This being the case, this article conducts

61 Even in an asymmetric conflict where it has been acknowledged that the stronger power has won, the stronger power must still make certain political concessions to eliminate the roots of the conflict. For example, the reason why the British were able to emerge victorious from the Malaysian rebellion from 1948 to 1957 was because Britain conceded before the colonial authorities that it would 'permit Malaysia to become independent in the near feature'. In doing so, it managed to separate Malaysia's two largest ethnic groups, the Chinese and Malays. See Jeffrey Parker, *Jianqiao zhanzhengshi* (*The Cambridge History of War*), trans. Fu Jingchuan (Changchun: Jilin Renmin Chubanshe, 1999), p. 567; Jay Gordon Simpson, 'Not by Bombs Alone: Lessons from Malaya', pp. 91–8, http://www.dtic.mil/doctrine/jel/jfq_pubs/1622.pdf.

62 It should be explained that this article is based on information available as of December 31, 2007. It was on December 19, 2006, that American President George W. Bush admitted for the first time in an interview with the Washington Post that the USA had not won the Iraq War, http://news.xinhuanet.com/video/2006-12/22/content_5518547.htm.

a simple analysis of the war from the standpoint of the strategies deployed by the respective belligerents.

American Strategy

On March 20, 2003, US British allied forces launched the pre-emptive attack on Iraq known as Operation Iraqi Freedom that started the war in Iraq. On May 1, 2003, US President George W. Bush made an announcement from the aircraft carrier USS Abraham Lincoln that the war had been brought to an end. As of that point in time, the US and British forces had successfully defeated the Iraqi army in just 43 days and achieved their objective of bringing down the Saddam Hussein government. In stark contrast to their impressive display in the first phase of war operations in Iraq, however, the US military was patently not able to achieve its expected results in the second, post-invasion phase. Anti-American forces began to mobilize after President Bush announced the end of major military operations in the country and repeatedly attacked US British forces. The allied forces launched a series of military operations in late 2003 to quash these attacks. But they had the effect worsening rather than stopping attacks on the allied forces, the Iraqi government and Iraqi civilians.[63] Since the end of major combat operations, the conflict between the US-led allies and Iraqi transitional government on one side and the numerous anti-government militants on the other has been both sustained and characterized by extreme violence. American ground forces have also become bogged down in a long-term, complex war of intertwining religious conflict and attempts to contain violence. Anti-American forces in Iraq are presently exhausting substantive components of the US military advantage.

American accomplishments in the first phase of the war were splendid. The USA completely annihilated the Iraqi conventional forces of over 400 000 troops in a 20-day long lightning attack. As Saddam Hussein's regime collapsed, the 'shock and awe' military doctrine of retired US military General Harlan K. Ullman seemed to scream out loud: 'For all of the western governments that participated in the military operations the fundamental question is how to effectively control political risk leading up to the election period.'[64] The USA exemplified this military principle, as all of its military plans revolved around one objective: obtaining a rapid and definitive victory. The 'shock and awe' doctrine, which emphasizes rapid dominance, gained the support of then Secretary of Defense, Donald Rumsfeld, and became the primary guide of US army strategy policy for the war in Iraq. Evidence shows that application of the doctrine indeed

63 Li Xiaogang, 'Yilake zhanzheng zongshu', ('A Full Review of the Iraq War'), *Junshi lishi* (*Military History*), No.5 (2007), p. 54.

64 Martin Shaw, *The New Western Way of War*, p. 106.

brings rapid military victory, particularly in punitive military attacks or when rapid progress is the main objective. 'Shock and awe' is essentially the equivalent of blitzkrieg in the information age. Its advantage over the German mechanized blitzkrieg deployed during World War II is that its efficiency is guaranteed through reliance on information, the speed at which it can be deployed is even shorter, and opposition to it is even less substantial. In addition, it allows far better coordination of several types of military force, which makes the ultimate military results even more brilliant. But these features represent only improvements in military and military strategy. 'Shock and awe' is actually no different from the German 'blitzkrieg', insofar as neither provides an answer to the question of what to do once a war has been so rapidly won.[65]

America's defeat in the second phase of the Iraq war is the result of many factors, but mainly that of the strategic culture that has developed over the years in the US army. Strategic culture refers to the summation of beliefs and attitudes embedded in army organization which determine the preferences with respect to how military power is deployed.[66] Several weak points of American strategic culture were apparent in the Iraq war. First of all, the US army repeatedly and consistently made destruction of the enemy's military apparatus the mission objective, failing to consider a more flexible political goal. 'They are like young people wielding hammers they think that everything is only nails.'[67] That the USA has overlooked the political element undoubtedly does not bode well for ultimate resolution of the Iraq problem. Second, the US army has always perceived itself as a conventional force, and as such has strongly resisted engaging in any form of nonconventional warfare. Despite the bitter experience of Vietnam, the army has never taken unconventional warfare seriously. The USA has relied on high technology to win conventional wars for the past 60 years. It was this use of the 'strategy of yesterday (conventional war)', that resulted in the strikingly different outcomes of the first and the second phases of the Iraq war.

In reality, it is not the case that American scholars have failed to recognize the military's lack of adaptability. The problem is rather that the Pentagon has not been receptive to the work and policy advice of these scholars. Similar to other major bureaucratic organizations, the Pentagon ultimately

65 See Harman Ullman and James Wade, *Zhengse yu weiju: xunsu zhidi zhi dao* (*Shock and Awe: Achieving Rapid Dominance*), trans. Teng Jianquan (Beijing: Xinhua chubanshe, 2003).

66 Robert M. Cassidy, 'Winning the War of the Flea', Military Review, September–October 2004, http://www.au.af.mil/au/awc/awcgate/milreview/cassidy2.pdf.

67 Joseph S. Nye, 'American Foreign Policy after Iraq', http://www.project-syndicate.org/commentary/nye43.

prefers to engage in activities it likes and at which it excels.[68] The military has been through certain types of adjustments on many occasions, but they focus almost exclusively on how advanced technology may be used to enhance defense and conventional military capacity, the assumption being that any problems that occur relate to shortfall in conventional capacities.[69] For example, then US Vice President Dick Cheney argued that as long as the US expanded the conflict and attacked what he saw as the behind-the-scenes support for rebels from Iran and Syria, victory in Iraq was 'within view'.[70] It is not fair to say that the American army lacks flexibility, but if the preferences of this organization so deeply structured around the use of conventional war strategies are to change, a fundamental revolution needs to take place in the principles it applies to organizing, arming, controlling, and training. The Iraq war is only one case in which, 'America was not nearly prepared enough, nor did it have the needed will or technology.'[71]

Strategy of Anti-American Militants

Although Iraqi anti-American militants are organizationally complex, their overall capacity is far from sufficient to engage US military forces. Organizations of this type use crudely constructed roadside bombs, simple explosive devices, submachine guns, mortars, grenade launchers and other light weaponry. Their forces are no more than 20 000 30 000, and their funding is greatly limited. These capacities in themselves present no difficulty to the US army, but the asymmetric strategies deployed by the anti-American militants do. Rumsfeld admitted that 'primary battle operations have already proved largely successful, this is most obvious. In the second phase of the war though, progress is not good enough nor is it fast enough. This is also blatantly obvious... to put it very bluntly, our country does not have any experience in doing what needs to be done: to control these violent, immoral extremists, and force them to accept our demands. They have no army, no navy, no air force. All they do is lie in the dark shadows of corners and launch their attacks.'[72] After the end of the Cold War, the entire world spoke of the unilateral transparency

68 Jeffrey Record, 'The American Way of War: Cultural Barriers to Successful Counterinsurgency', *Policy Analysis*, No. 577 (2006), pp. 9–10, www.cato.org/pubs/pas/pa577.pdf.

69 Brigadier Nigel Aylwin-Foster, 'Changing the Army for Counterinsurgency Operations', *Military Review*, Vol. 85 No. 6 (2005), p. 14.

70 Kelly Eaton and Andrew Wedeman, 'Yilake zhanzheng yu Bushi zhuyi de mingyun', ('The Iraqi War and the Fate of the Bush Doctrine'), trans. Ren Donglai, *Nanjing daxue xuebao-shekeban* (*Journal of Nanjing University-Social Science Edition*), No. 2 (2007), p. 55.

71 Jeffrey Record, 'The American Way of War', pp. 16–17.

72 Sui Lin and Wang Chenglong, 'Yilake zhanzheng Meijun zhanlue shiwu jiexi', ('An Explanation of the American Strategic Mistakes in the Iraq War') *Guofang keji* (*National Defense Technology*), No.3 (2007), p. 90.

of American military might. America has now experienced another type of unilateral transparency.

From the perspective of the American military stationed in Iraq, the country has no frontline, because the battlefield is all encompassing. Iraq has no army, but there are armed militants everywhere. Anti-American Iraqi militants are not conventional forces, and do not receive support from a major superpower, as in the war with the Vietcong. Anti-American militants nonetheless have an advantage that stems from their otherwise weak position: They are willing to use any means possible, and inflict any kind of damage regardless of the consequences. The American army, on the other hand, has ample reason for concern. It must take account of the limitations placed upon it by numerous organizations, laws, and moral standards which make impossible the unleashing of full military power. Should it begin to increase the destructiveness of its offensive, the USA would face political consequences as a result of the harm it inflicts on civilians. In sum, the hordes of anti-American militants in Iraq have no common strategic plan; but objectively speaking their actions share several commonalities: (i) They are able to expose the weakness of the occupation authorities by working to prevent Iraq from having a government, that is to say 'the more chaotic things become, the better'; (ii) by demonstrating their courage and provoking the Americans to engage in reprisals, they increase support for their movements; (iii) by continuously causing casualties among the allies, militants gradually wear down the will of the Americans to wage the war. In simple terms, the work of militants is simply to ensure that no-one gets the impression that things in Iraq are improving.[73]

At the outset, Iraqi anti-American militants also used more formal strategies of resistance. But the heavy death toll and casualties from conventional battles in Fallujah and Ramadi taught them that it was best to switch to the strategy of 'hit and run', employing a mixture of urban guerilla tactics and terror attacks. Early on in the second stage of conflict, anti-American militants engaging in ambushes or surprise attacks at short-range were killed or captured in US army counter-attacks. As time went on, anti-American guerillas gradually adopted a strategy that enabled them to attack from a distance with remote controlled bombs, mortars, and other weapons. In response to increased American defense of potential targets of attack, anti-American militants again changed their strategy to focus on 'soft targets'. These included the armies of other allied member countries, Iraq government officials and their security forces, electric power plants, and crude oil pipelines, United Nations organizations posted to Iraq, civilians in Iraq who were nationals of countries assisting the US army, and also

[73] Steven Metz and Raymond Millen, *Insurgency in Iraq and Afghanistan: Change and Continuity*, p. 8, http://www.dni.gov/nic/PDF_GIF_2020_Support/2004_05_25_papers/insurgency.pdf.

Iraqis cooperating with America. They even went so far as to launch direct attacks on overseas targets, obvious examples including the repeated bombing of sites in Saudi Arabia, and explosions in Istanbul and at the Atocha train station in Madrid.

The approach of anti-American militants is known as a 'netwar' strategy.[74] John Arquilla and David Ronfeldt first raised the concept of a netwar; they used the term to refer to 'coordination between small organizations using networks to engage in war without the presence of any clear command structure'.[75] Unlike traditional guerilla movements, anti-American forces in Iraq have no power base or clear leadership, and have made no attempts to occupy geographical locations. They lack a clear and unified ideology and, more important, have no recognizable system of rank.[76] Within this confused environment of rapid and constant change, a number of organizations share information, trade weapons, train together, and coordinate attacks. After achieving one particular goal they split up, probably never to meet or cooperate with one another again. It is these characteristics of decentralization, high levels of autonomy, and the presence of different objectives across groups that enable anti-American militant organizations to survive. At the same time, information technology, globalization, and the growing ability to cross borders provides the necessary conditions for such militants to exchange information and develop new links.[77] In support of this point, intelligence indicates that the Global Jihad and terrorist organizations provide anti-American militants in Iraq with strategic advice.[78] The netwar strategy provides a high level of adaptability that is difficult to respond to; in the words of American military scholar Bruce Hoffman, the war in Iraq might be referred to as one of 'post-modern insurgency'.[79] Looking ahead, it is likely that the USA will be unable to defeat anti-American militant forces, which implies that it will not achieve a compelling victory in Iraq. In all likelihood, the war in Iraq will continue over the long term.[80]

74 Paul Staniland, 'Defeating Transnational Insurgencies: The Best Offense is a Good Fence', *The Washington Quarterly*, No. 6 (2005), pp. 27–8.

75 John Arquilla et al., 'Networks, Netwar, Nd Information-Age Terrorism', in Ian O. Lesser et al., *Countering the New Terrorism* (Santa Monica: The RAND Corporation, MR-989-AF, 1999), p. 47.

76 Bruce Hoffman, *Insurgency and Counterinsurgency in Iraq*, 2004, p. 16, http://www.rand.org/pubs/occasional_papers/2005/RAND_OP127.pdf.

77 Steven Metz and Raymond Millen, *Insurgency in Iraq and Afghanistan*, p. 4.

78 Jim Krane, 'Iraqi Insurgents Use Tactics They Learned from Chechens, Taliban, Al-Qaida', *ArmyTimes.com*, January 5, 2004.

79 Bruce Hoffman, *Insurgency and Counterinsurgency in Iraq*, p. 18.

80 *Ibid.*, p. 20.

A Theoretical Consideration of the Iraq War

Since fighting in Iraq began in 2003, and the limited period of initial victory, America has become trapped in another Vietnam-type quagmire. The situation shows no signs of improving. In 2003, 486 Americans were killed and 2413 wounded. During the next four years these numbers rose to yearly highs of 800 900 killed and 6000 8000 wounded. The year 2007 was by far the bloodiest year since American started this conflict.[81] With the exception of President Bush and a select few other individuals who have consistently commented that 'the situation is improving, and we are achieving victory', all American scholarly articles and media reports imply either that America has lost the war or that victory is not possible.[82] Can the theory of asymmetric conflict be used as a reasonable explanation of this situation?

Andrew Mack's theory of asymmetric interests argues that 'relative power' determines 'relative interests' and 'relative interests' create different political weaknesses across pairs, which determine the outcome of a conflict.[83] From the perspective of Iraq, the conflict without doubt involves its core interests. Saddam Hussein's regime was overthrown, and he was hanged on December 30, 2006. From March 2003 when America invaded Iraq to June 2006, approximately 650 000 Iraqis were killed. The war has also made large numbers of Iraqis homeless, 1.6 million have been displaced, and another 50 000 are forced to leave their country every month, according to estimates by the Iraqi government. Nonetheless, the theory of 'asymmetric interests' cannot be used to explain America's relative interests. On the one hand, Iraq has extraordinary oil reserves, on the other, Iraq controls from its location at the center of the Arab world important sea-lanes critical to the transportation of energy throughout the Gulf region. It is thus an ideal base from which America might penetrate the Eurasian landmass. In addition, America's withdrawal from Iraq would undoubtedly benefit Islamic extremists and other terrorist cells, as American allies (especially Gulf States) would then question the credibility of American protection. In the face of strong pressures to pull out of the country, therefore, American policy makers maintain a strong desire to press on, which reflects the magnitude of American interests in this area. But this resoluteness cannot bring about an American victory.

81 http://icasualties.org/

82 This article considers conflicts through to the end of 2007 only. At the end of this period, America, on the whole, had failed in its war on Iraq. The Iraq war is, of course, still being fought, and there are occasional report newspaper reports regarding positive changes taking place on the ground, or worsening conditions in Iraq. It is impossible to determine the final outcome of this war from similar contemporary wars, but it is undoubtedly the case that if it persists for a long period of time America will pay an ever-higher monetary price.

83 Andrew Mack, 'Why Big Nations Lose Small Wars', pp. 195–200.

Gil Merom's theory of democratic institutions is used especially within the context of explaining why democratic countries lose in small-scale wars. He argues that the obstructions created by citizens with the ability to engage in political participation result in strong powers suffering defeat in war.[84] But this theory, so far at least, fails to explain the Iraq war. At any time a major adjustment in diplomatic approach becomes necessary, especially one calling for a move towards admitting defeat, states are forced to engage in a long, difficult process.

Even more important, the issue of the Iraq war has become the dominant diplomatic artifact of the Bush administration; defeat or victory in this war will undoubtedly determine Bush's place in history. Consequently Bush will never accept the reality that 'America has lost the Iraq War'. As far as the Democratic Party is concerned, even though it could constitutionally use the congressional 'power of the purse' to end the war, the 2008 American presidential election has shown that the Democratic Party must avoid being portrayed as cowardly or giving in to terrorists. Also, as many members of the Democratic Party previously supported the war, and as they are not able to propose a credible strategy for getting out of Iraq, objectively speaking it is impossible for them effectively to oppose Bush's war policy. Consequently, there was in 2007 another increase in the number of American troops sent to Iraq, setting a new record of a total 165 000.

Jeffrey Record's 'theory of foreign assistance' is similarly unable to provide a reasonable explanation of the Iraq war. When considering the financial resources of anti-American militants, they come from three sources: (i) funds remaining from the Hussein period; (ii) external support; (iii) proceeds from kidnappings or other criminal activities. Practically speaking, the equipment used by these militants is inexpensive, and the cost of maintaining such forces is much lower than that required by conventional forces.[85] As far as arms are concerned, Iraq itself has around 10 000 publicly known weapon stockpiles, and those of gunpowder are as high as 1.2 million tons. Before the Iraq war, the Iraqi military stored large amounts of weaponry in schools, mosques, and village temples that might be used in a guerilla war, and distributed 7 million guns to civilians.[86] American claims that 'foreign assistance' (including organizations in Iraq, Syria, and Iran with connections to the foreign Jihad) is to blame for the current situation are, therefore, correct to an extent, but the impact of such external forces, compared with internal opposition, is limited. 'In comparison to the number of

84 Gil Merom, *How Democracies Lose Small Wars*, p. 15.

85 Robert Pape, 'Coercive Air Power in the Vietnam War', *International Security*, Vol. 15, No. 2 (1990), pp. 126–9.

86 'Guimo pangda chaoguo Meijun Yilake you yiwan ge junhuoku', ('Greater in Size than the US Military: Iraq has over 10,000 Weapons Caches'), *Huanqiu shibao* (*Global Times*), October 29, 2004, p. 10.

weapons stolen from Sadam Hussein's weapons storage facilities, the caches from Syria and Iran are secondary.'[87] In addition, violent activities among Iraqis have already exceeded attacks by external terror organizations, according to an assessment report released by the oversight office of the US CIA on February 2, 2007, based on information compiled from 16 American intelligence agencies.[88]

Relatively speaking, Toft's 'theory of strategic interaction' best matches the reality of the war in Iraq. During the first part of the war that focused on bringing down Sadam's regime, the American army deployed a lightning attack using conventional forces which defeated Iraqi troops in just 43 days. During the anti-guerilla phase of the war, however, the American military, accustomed to the use of conventional tactics, was ultimately unable to contain attacks by anti-American militant groups. Even though scandals regarding the murder of civilians and the abuse of prisoners have been continuously exposed, strategically speaking, terrified as it is of the ubiquitous international media and great pressure of international public opinion, America has not shifted towards use of 'barbaric tactics'.[89] It can be said then that America's defeat is almost certainly related to the assumptions found in Toft's theory of strategic interaction. Although Toft does not explain in great depth the logical mechanisms that drive his theory, instead tending towards pure description of objective phenomena, his observation is useful that: It is 'very difficult to defeat radical militants.... Not only is it highly unlikely that dependence on pure and largely unstructured military power can win a war, it is almost certain that it cannot win the peace.'[90]

A detailed consideration of the Iraq war makes plain that the fundamental perils America faces relate to its inability effectively to check the costs of war. On the one hand, anti-American militants are not fighting a conventional war. Unless absolutely necessary, they will not engage the US army in a traditional battle, but conversely, will use roadside bombs, car bombs, mortars, assault rifles, secret attacks, kidnappings, and even suicide bombings to maintain their attack on the American military and its allies. Beyond these tactics, anti-American militants also continue relentlessly to destroy basic infrastructure, to the extent that efforts to reconstruct water plants, oil pipelines, and other infrastructure cannot keep pace with attacks. As a result of kidnappings and executions, progress in

87 Kelly Eaton and Andrew Wedeman, 'The Iraqi War and the Fate of the Bush Doctrine', p. 51.

88 http://news.nen.com.cn/72341298102665216/20070322/2182654.shtml.

89 In reality, America has also admitted that 'barbarism' could not be a successful strategy because it is impossible to implement all the way. This is also because the concept of autonomous ethnic rule has impassioned popular resistance. See Ivan Arreguín-Toft, *How the Weak Win Wars*, p. 227.

90 *Ibid.*, pp. 226–7.

rebuilding political and economic institutions is also impeded. As Bruce Hoffman notes, 'key to the use of violence is that it deprives civilians of a sense of security'. Radicals attempt through violence and bloodshed to create an atmosphere of terror that makes apparent the government's weakness and lack of capacity with respect to maintaining security.[91]

Another important point to keep in mind is that the limited ranks of the American army are ultimately insufficient to impose effective control over Iraqi territory. From past experience, the ratio of army troops to guerilla forces needed to control the guerilla resistance alone is approximately 20:1; these figures do not take into account the forces needed to defeat guerilla forces. 'Considering that America has already taken control of Iraq, this means that it needs two million troops there to control the situation.'[92] It is virtually impossible for the American army, highly dependent as it is on advanced technology, to achieve control. In addition, keeping in mind that the goal of the use of force is to establish long-term control and effective development, it must be acknowledged that the military alone is insufficient to accomplish this goal, and in certain circumstances could even be counterproductive. The extreme reliance of the American army on the use of force and the deficient efforts to rebuild Iraq render the USA incapable of winning the recognition of the Iraqi people.[93] This general focus constricts the space within which the US forces can work, and places it in an even more difficult predicament. In the end, Kissinger's words perhaps best describe the predicament America faces in Iraq: 'As long as guerilla forces do not lose they are victorious, as long as traditional superpowers do not win they will suffer defeat.'[94]

Conclusion

This article considers contemporary asymmetric conflict and the trend towards the 'weak defeating the strong' from the standpoint of strategy. In modern asymmetric conflict, on the one hand, as the weaker power in an asymmetric conflict normally faces a threat to a core interest, it is not likely to maintain reservations with respect to the use of violent force. On the other, taking into account its growingly decentralized nature, the characteristics of its networks and the context of contemporary globalization and developments in science and technology, the weaker power has a possibility of achieving victory in the conflict through use of an

91 Bruce Hoffman, *Insurgency and Counterinsurgency in Iraq*, p. 15.

92 Henry C. K. Liu, 'Yilake zhanzheng yu chaoji daguo shidai de zhongjie', ('The Iraq War and the End of the Age of Superpower'), trans. Lin Yinjian, *Guowai lilun dongtai (Foreign Theoretical Trends)*, No.1 (2007), p. 65.

93 Hannah Allam, 'Outlook is Bleak for Success in Iraq', *Philadelphia Inquirer*, May 9, 2004, p. 1.

94 Henry Kissinger, 'The Vietnam Negotiations', *Foreign Affairs*, Vol. 48, No.2 (1969), p. 214.

asymmetric strategy. In a conflict of limitless violence and counter-violence, the more powerful state, unable to see the light of victory at the end of the tunnel, ultimately gives up the war as a result of the exorbitant cost of carrying on. On the whole, as the use of asymmetric strategies in modern asymmetric wars becomes more sophisticated and prevalent, and as the weaker powers in such wars have a definite advantage when compared with stronger powers as regards controlling the costs of war, 'the weak defeating the powerful' in contemporary asymmetric conflict is becoming a more frequent outcome. It is for this reason this article argues that asymmetric strategies and the use of unlimited violence are the source of a weak power's ability to defeat a stronger power.

This article argues, based on earlier data and case analysis, that the term 'asymmetric strategy' may be used to describe the phenomenon whereby a weak power defeats a strong power. The weak power's use of an asymmetric strategy which causes an obvious change in the stronger power's cost benefit ratio enables it to weaken the stronger power's will to wage war, and win an asymmetric conflict. Theoretically speaking, this conclusion goes a long way as regards refuting the realist 'power determination theory'. Power is undoubtedly a useful theoretical concept; in contemporary asymmetric conflict, however, the weak can employ 'asymmetric strategies' to reverse what might at first seem to be an unfavorable situation.

It must be noted, however, that this research offers considerations only from the perspective of military strategy, and does not deny the explanatory value of other factors. In reality, every asymmetric war is different, and in some instances the successful experiences of one conflict do not necessarily apply to others. Neither is it the case that the rise of asymmetric strategies implies powerful states are doomed to defeat in asymmetric conflict. This research shows that, within the contemporary context, as the weaker power has an advantage with respect to adaptability, the greater power faces considerable hardship when it comes to controlling the costs of war.

As this article illuminates the commonalities and trends to be found in contemporary asymmetric war, it has certain practical implications for the strategies and policies of a number of countries. It could help weak states to increase their confidence when waging war, and guide the direction of their efforts; it could help strong states make a more practical assessment of the issues they may face in advance of such conflicts, and act as forewarning against entering into war without serious consideration of the consequences.

Section IX: Ambush and Counterambush

Introduction

Ambush

Except for sniping and counter-sniping, the ambush would generally be the smallest-scale, organizationally and logistically simplest, and least complex type of military deception operation. Still, as we'll see in the following cases, the same sophisticated psychological principles that we find in all deception operations are at work from the most organizationally complex to the simplest. The lesson is that deception is truly a mind game.

CHAPTER 40

How to Ambush a NATO Fortress[1]

Commander Jan Berglund, Norwegian Navy

Commander Jan Berglund is a commanding officer in the Norwegian Special Forces. He graduated with his Masters of Science in Defense Analysis from the Naval Postgraduate School in Monterey, California.

This case highlights the point that the greater the sense of security, the greater the opportunity for surprise.

Main Circumstances

In 1996, as the commanding officer of the Norwegian Naval Special Forces, I got the assignment to test the security and the close defense of one of our National Defense Headquarters. The headquarters (HQ) was a hard target where all the operational functions and communications were built deep inside of a mountain. The facility was designed to resist a nuclear attack and was self-sufficient, with water, electricity and food for a certain number of weeks.

The close defense was organized in depth, with the outer perimeter's first gate approximately one mile from the mountain entrance. At this checkpoint, cars and buses were searched and individuals were checked for identification. The second gate was 200 meters inside the tunnel entrance to the facility itself. Here, an electronic identification card would be inserted and matched with a personal pin code. Some hundred more meters inside and behind the massive doors protecting the facilities from a potential nuclear, biological, and chemical (NBC) attack was an additional post. This post functioned as a surveillance center and a rallying point for military police and sentries on duty.

The defending force during the exercise was a reinforced company-sized security force. That force could be further reinforced by the home guard in case of a crisis or by mobilized regulars in case of high tension or war. The target seemed quite impenetrable, and therefore, would require a well-developed deception plan to infiltrate even a single commando, much less a whole team. Forced entry was not considered a viable option. However, even if a clandestine infiltration succeeded, another problem remained: how could they smuggle in enough equipment, weapons, and explosives to inflict substantial material damage?

After a considerable planning period, we developed three courses of actions (COAs) that seemed feasible. In order of priority, they were:

1. Commander Jan Berglund, Norwegian Navy, "How to Ambush a NATO Fortress," research paper, Department of Defense Analysis, Naval Postgraduate School, 1996.

- Blow up one or more of the four shuttle buses that brought people to work every morning (classical ambush).
- Get one or more men inside and inflict critical damage to either the computer/communication section or assassinate the Commander and the command group (deception entry).
- Attack the power supply, antennas, and water supply cables leading into the mountain (classical commando raid).

To give the opposing force the most training value, all three COAs were executed. However, I would have realistically preferred the first or second alternative. In the following account I will focus on the first COA, the ambush.

Prior to the ambush, a team of four men conducted close reconnaissance for several days. The reconnaissance was done both clandestinely and overtly. Clandestinely by patrolling the terrain in order to find the best spot to attack the buses along a 20 km route from the city. A 24 hour observation post (OP) was also established from a secure distance to monitor the bus and car arrivals, including their regularity and punctuality. Overtly, the reconnaissance team dressed like civilians and drove or jogged by the facility to get detailed information about the first checkpoint.

The ambush was to take place 300 meters in front of the checkpoint. At this particular point, the buses had to slow down to 5–10 mph because of a narrow bridge followed by a 90-degree curve. In addition, our reconnaissance observed that the buses often queued at this point since they tended to arrive almost at the same time. Our main objective was to take out at least one of the buses. Normally, about half of the employees, officers and civilians, would arrive at the HQ by bus. The other half arrived in their own cars. The assumption was that if we took out 50 to 100 people in one blow, some key persons would be included, which in turn would hamper ongoing operations inside the HQ. Many of these individuals were high-ranking officers or highly specialized technicians. Consequently, the loss would be considerable and not easily replaceable. We also identified four flag officer vehicles that arrived predictably every morning with the Commander and his deputies from the Army, Navy, and Air Force. Although they all seemed like tempting and easy targets, we decided to prioritize the buses and take the flag officers with snipers if we had the opportunity. Of the four buses, two were more important than the others because they arrived from downtown where we knew that most of the key personnel would embark. These buses were also more crowded.

In addition to the deception supporting the reconnaissance, we had to place three dummy 12-kilogram claymore mines and hide them until the ambush. It was considered a risky operation in itself because it was only 300 meters from the checkpoint, and security forces frequently patrolled the area. The team, dressed in camouflage uniforms placed the mines the night before.

The mines were effectively camouflaged close to the road. Close passing security patrols never detected them during the morning hours. Attached to the dummies were a small pyrotechnic devices that would simulate detonation and provide some smoke. The detonation device was remotely controlled by radio and set off from the main OP, which had a perfect view overlooking the target. The remote control device provided for perfect timing, flexibility, and security for the ambushers.

In the morning, the ambushers (now in the main OP) got the word by radio from a forward OP of the sequence in which the buses would arrive. When the target bus passed the claymores at low speed, the first and second claymore went off as planned. The third was detonated immediately after, probably inflicting damage to the trailering bus. The four flag officers were not targeted because they had already entered the facility and an attempt on them would have compromised the main attack.

In my view the ambush was successful because:
The security forces never anticipated the chosen COA. Hence, they were taken by surprise. They had limited their responsibility to the immediate fence perimeter and physical facilities. Consequently, they forgot that their "humanware" can be just as valuable as their hardware as demonstrated by the loss of 50–100 people in one single, simple blow. Tactical deception worked. The group successfully masked their reconnaissance by traditional means of camouflage and stealthy movement under the cover of darkness.

Additionally, the reconnaissance force mimicked passing civilians without drawing attention. The mine emplacement was also masked successfully, blending the three 12 kg claymore mines into the terrain. The security forces had the illusion of control. The mines were placed within their security perimeters, right in front of their noses, just a few meters from the checkpoint. To them, the situation seemed normal until the blast.

What could the ambushers have done better?
They could have synchronized the operations better and taken out multiple targets, such as the flag officer vehicles and more buses. Other simulation techniques like decoying could have been tried to create a traffic jam for the arriving buses and cars.

What could the ambushed victims have done better?

- They could have planned better by wargaming against clandestine attacks.
- They could have avoided falling prey to routines and driving patterns
- They could have expanded their security responsibility to the time the buses leave their garages
- Time bombs could easily have been planted on the buses even before any passengers embarked. They could have checked for the buses daily prior to use.
- They could have looked for the unnatural—civilians could have been randomly stopped for a closer check.
- The terrain could have been examined more closely.
- They could have familiarized themselves with their own environmental conditions. Security forces often restrict themselves to roads and paths, especially in the dark.
- They could have added more sensors (IR, thermic ray cameras, movement sensors, etc.) to supplement the human sensors.
- They could have established barriers (such as obstacles, minefields, or booby traps) to canalize any assaulting enemy force.

What lessons were learned?
The important lesson is that there often is an alternate indirect approach to a problem if one spends enough time looking for one. In this case, a direct attack or a single small group penetration into the HQ would have been very risky, maybe suicidal. In addition,

the result or effect of this COA is difficult to estimate because it is likely the attack would be uncovered before completion.

Another important lesson is the importance of stealth in order to achieve surprise. Camouflage, environment blending, and slow movements are crucial. Leaving the impression that everything is normal until the last crucial moment is one of the key factors for success. However, once surprise is lost it must be compensated for by speed.

CHAPTER 41

High Desert Ambush: Hard Lessons Learned the Hard Way[1]

LTC (Ret) Lester W. Grau

"Les" Grau, a specialist in Russian military strategy and tactics, retired from the U.S. Army with the rank of lieutenant colonel sometime shortly after 1992 while assigned to the staff of the Foreign Military Studies Office at Ft. Leavenworth, Kansas. There he remained through the publication of this article.

For the perspective of the Mujahideen & Taliban as ambusherers of Soviet forces see Ali Ahmed Jalali & Lester W. Grau, "The Other Side of the Mountain: Mujahideen Tactics in the Soviet-Afghan War," Quantico, VA: U.S. Marine Corps, Studies and Analysis Division, 1999.

Afghanistan is not Europe, yet the Soviet Army that occupied Afghanistan in late December 1979 was trained to fight NATO on the northern European plain. Consequently, the Soviet Army had to reequip, reform and retrain on-site to fight the insurgent mujahideen [holy warrior] guerrillas. The Soviets were forced to revise their tactics and tactical methodologies in order to meet the demands of this very different war. One of the tactical areas which the Soviets thoroughly revised was the conduct of ambushes. The Soviets planned to use ambushes in the European theater, but they were primarily ambushes against attacking or withdrawing NATO armored columns. The Soviets constructed most of their ambushes around tanks and tank units. They planned to employ concealed individual tanks, tank platoons and tank companies along high-speed avenues of approach or withdrawal to engage the enemy from the flank and then to depart. Such ambushes were part of security zone defensive planning as well as planning for the deep battle and pursuit. The Soviets also trained their squad and platoon-sized reconnaissance elements to conduct dismounted ambushes to capture prisoners and documents. They employed a command element, a snatch group and a fire support group in these small-scale ambushes.

In Afghanistan, the Mujahideen seldom used armored vehicles and seldom advanced along high-speed avenues of approach. Instead, they infiltrated light-infantry forces through some of the most inhospitable terrain on the planet to mass for an attack or ambush. The Soviets soon discovered that they had difficulty maintaining control of the limited road network which constituted the Soviet lines of communication. The guerrillas constantly cut the roads and ambushed convoys carrying material from the Soviet Union to the base camps and cities in Afghanistan. The Soviet ability to maintain

1. Lieutenant Colonel (Ret) Lester W. Grau, "High Desert Ambush: Hard Lessons Learned the Hard Way," *Red Thrust Star,* 3 Parts: July, 1995; October, 1995; and October, 1996. [Excerpts of introduction and conclusion].

its presence in the country depended on its ability to keep the roads open and much of the Soviet combat was a fight for control of the road network. During the war, the guerrillas destroyed over 11,000 Soviet trucks (and reportedly even more Afghan trucks) through ambush. The Soviets learned from Mujahideen ambushes and used the ambush to interdict the guerrilla supplies coming from Pakistan and Iran. The Soviets conducted ambushes mainly with reconnaissance and other special troops (airborne, air assault, spetsnaz and elements from the two separate motorized rifle brigades which were designed as counter-guerrilla forces). The composition and employment of ambush forces differed with the units involved and the part of Afghanistan in which they were employed.

Contemporary Issues

Since the war, the Russian army has studied their ambush experience in Afghanistan.

As Russian forces continue to serve on the Tadjikistan/Afghanistan border, in Chechnya and other mountainous areas, the high-desert ground ambush remains a valid tactic for Russian ground forces. Recently, the Russian military press began debating the proper size and composition of an ambush force. The debate questions the exclusive use of reconnaissance and special troops to conduct dismounted ambushes. According to some Russian professionals, ambush forces should be constituted not only from reconnaissance troops, but also from motorized rifle, tank, antitank, artillery, flamethrower, helicopter aviation, fixed-wing aviation, antiaircraft, chemical, electronic, engineer and, in the future, combat robot troops. Ambushes can be antitank, anti-personnel, anti-air, anti-landing, reconnaissance, phoney and special (electronic, chemical, engineer and combat robots).

The Russians have taken the following lessons from their Afghanistan experience:

First, the ambushing force sometimes encounters a much larger force than anticipated. To avoid disaster, the ambushing force must be able to disengage quickly and withdraw. All ambush sites require local security to prevent the ambushing force from being surprised.

Second, weapons systems present in platoon and company ambushes are usually restricted to the organic equipment of that unit. Supplemental tanks, IFVs, self-propelled artillery and self-propelled air-defense guns can usually be incorporated into these ambush sites and hidden from enemy reconnaissance. The effect of the surprise fires of these weapons systems at close range can be devastating on an enemy force in the ambush kill zone.

Third, aviation and artillery planning and close coordination are essential for successful ambush execution. Artillery forward observers and forward air controllers should be included in ambush sites where they can bring increased destruction on the enemy and provide necessary cover during the disengagement and withdrawal of the ambush force. High-precision munitions should be used in support of an ambush.

Fourth, ambush forces need to be able to extricate themselves from encounters with larger enemy forces and to avoid counter-ambush while they withdraw. Bronegruppa (armored groups) can be incorporated on the flanks of the ambush where their weapons systems can contribute to the deadliness of the ambush. Further, bronegruppa can be located forward of the ambush in a security zone. From these forward positions, bronegruppa can engage the enemy and then withdraw, pulling the enemy force into the ambush kill zone. The principal function of the bronegruppa in the ambush, however, is to provide for the quick, safe withdrawal of the ambush force.

Fifth, the Russians studied the mujahideen ambushes (Figure 1). They characterized mujahideen ambushes as having three components in addition to the command group — a diversionary group, a combat group and a snatch group. The diversionary group sealed the exit of the ambush with barricades, rockslides and minefields. The diversionary group would open fire when the Soviet force attempted to clear the obstacle. The Soviet attention would be focused on this diversionary group and the traffic would stack up in the kill zone behind the obstacle. Then, the combat group would open up on the stationary force in the kill zone. The snatch group would move forward to grab prisoners, documents, weapons and munitions. The ambush force would then withdraw over secured paths and roads.

Soviet soldiers caught in an ambush zone would try to drive out of the kill zone. If they could not move, soldiers would usually dismount the personnel carriers and shelter behind the carriers. A favorite mujahideen technique was to plant directional mines (claymore-type) on the opposite side of the road facing into the ambush. When the soldiers would shelter behind the carriers, the mujahideen would detonate the directional mines and kill the sheltering soldiers.

The Russians see a place for new technology in the ambush. Infrared sights and viewer, combat robots, remote-controlled weapons and sensors can be used. Scatterable mines, helicopter mobile obstacle construction detachments and controlled minefields

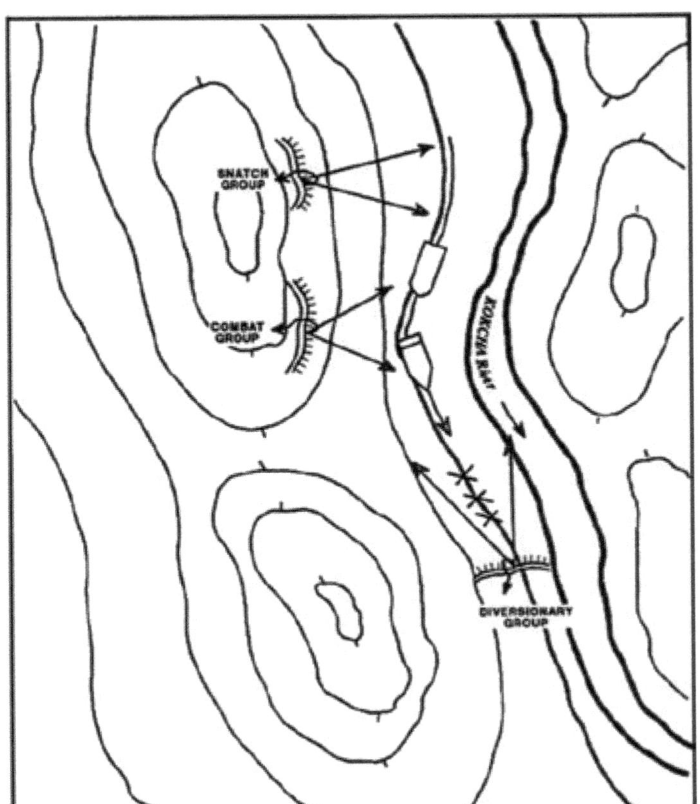

Figure 1 The Mujahideen ambushes

can fix the enemy in place while artillery, close air support and remotely piloted vehicles destroy the enemy.

For the more immediate future, the Russians have been studying the United States Army experience in Vietnam and, based on their studies, have proposed the creation and training of combat strike groups [udarno-boevaya gruppa-UBG]. These are three or four man teams armed with automatic weapons with silencers, portable anti-tank grenade launchers, mines and radios. Up to 18 UBGs could be trained in a company and up to 54 UBGs could be trained in a motorized rifle battalion. UBGs would operate at night on enemy-occupied territory—such as the security zone—once the enemy has crossed it. UBGs would set up and change ambush sites two or three times a night and destroy enemy personnel and equipment. In order to increase their survivability, a network of bunkers with stores of ammunition, explosives, food, communications equipment, medicine could be set up in advance. The UBGs could also use these bunkers to rest in during the day. This is a significant departure from Afghanistan where the Soviets only deployed large ambush forces and only against hard intelligence.

In conclusion, the Russians are reviewing their Afghanistan ambush experience and trying to adapt it to their present situation and future wars. Many of their ideas seem relevant to other modern armies who may fight in high desert or in a counter-guerrilla role.

CHAPTER 42

A Modern Day "Trojan Horse"—Operation JAQUE and the Use of Stratagem in a Hostage Rescue Operation[1]

Colonel Greg Wilson, US Army

The U.S. Military has been training, advising, and equipping the Colombian Armed Forces for decades. Interestingly, while the fruits of these long-term investments were clearly on display during operations designed to pressure the FARC hostage-holding network, nothing about these efforts helped to prepare the Colombians to develop the type of stratagem used in Operation JAQUE. This is a case where the student clearly became the teacher.

Colonel Greg R. Wilson graduated from the United States Military Academy in 1985. He completed Special Forces qualification and training in 1989 and since then has served in variety of Special Operations assignments from the tactical to strategic levels. COL Wilson served as the SOC Forward, Colombia, Commander and senior U.S. military advisor to the U.S. Ambassador and Colombian Special Operations forces during the planning and execution Operation JAQUE.

On 2 July 2008 Colombian Commandos surprised the world by executing one of the most complex and audacious hostage rescue missions in thirty years—Operation *JAQUE*, or "Check." Without firing a shot, twelve unarmed Colombian Commandos rescued three Americans along with Ingrid Betancourt[2] and eleven Colombian hostages, held prisoner by the FARC for over five and a half years. However, this was not a stereotypical hostage rescue operation built around the surgical application of force. Instead, it relied on a cunning deception plan that tricked the FARC into handing over the hostages to the Colombian government by creating a modern-day Trojan Horse. The operation delivered a devastating blow to the FARC by removing its high-value political prisoners, or "crown jewels," representing one of its few political bargaining chips remaining in their 40-year struggle against the Colombian Government. It also represents a successful model for US security assistance as well as a tangible return on long-term U.S. investment in Colombia and its defense forces. AMB William Brownfield made this point when he said, "Colombia represents the most successful U.S. nation building effort in the past 40 years."

1. Colonel Gregory Wilson, US Army, Special Forces, "Operation JAQUE and the Use of Stratagem in a Hostage Rescue Operation," (Naval Postgraduate School, 2013).
2. Ingrid Betancourt Pulecio is a Colombian politician, former senator and anti-corruption activist. Betancourt was kidnapped by the Revolutionary Armed Forces of Colombia on 23 February 2002.

Figure 1 U.S. Hostages after rescue by Colombian Commandos.

This chapter provides an insider's view into the campaign planning and execution of Operation Willing Spirit (OWS). It includes both U.S. advisory efforts and planning efforts designed to pressure the FARC hostage holding network and the Colombian rescue operation named Operation JAQUE. However, before we examine the details leading up to the events of 2 July 2008, it is important to note that this successful military deception and rescue operation cannot be viewed as a discrete event. Countless individuals and organizations in Colombia and the U.S. devoted time and resources to what became both a labor of love and a source of frustration for many—the safe return of Tom Howe, Marc Gonzales, and Keith Stansell.

The U.S. hostage crisis began on 13 Feb 2003 when three U.S. government contractors were captured while flying a counternarcotic spray mission over the remote jungles of Colombia, and disappeared into the FARC hostage holding network. Their light, single engine aircraft experienced a mechanical problem and crash landed in the remote jungles of Southern Colombia. Initial efforts to locate the hostages proved unsuccessful as the dense and isolated Colombia jungle seemed to swallow the plane and its crew. The survivors of the crash had entered the sophisticated FARC hostage-holding network[3].

From its inception, hostage taking has been part of the FARC's strategy. Most have been low-level kidnap-for-ransom operations designed to generate money for its activities. However, taking political hostages seemed to offer additional leverage as demonstrated by the kidnapping of Ingrid Betancourt in 2002 and the U.S. hostages in 2003. To protect these valuable assets, the FARC developed a sophisticated hostage-holding network. This network consists of a series of camps geographically dispersed throughout the remote jungles of Southern Colombia, hidden beneath triple canopy foliage making it very difficult to identify. The network is quite vast and once Colombian defense forces apply pressure near a hostage holding camp, the FARC simply picks up the hostages and moves them to another remote camp. Security at camps is extensive

3. Colombians aboard the aircraft were executed at the crash site.

and often consists of multiple rings of security around the site. Hostages are typically chained together and held in the center of the camp, housed in pens, with armed guards instructed to execute the hostages if there is a rescue attempt. The inability to locate the U.S. hostages and disrupt the FARC hostage holding network resulted in this being one of the longest hostage crises in U.S. history.

Operation Willing Spirit

The U.S. government was initially slow to react to the unfolding hostage crisis with the military lacking proper authorities to act until the inception of Operation Willing Spirit in May of 2005. Operation Willing Spirit (OWS) represented the U.S. Government's effort to safely recover the hostages. Department of Defense (DoD) efforts in support of OWS fell under the responsibility of U.S. Southern Command (SOUTHCOM) who designated Special Operation Command South (SOCSOUTH), a sub-unified command, to lead all DoD planning and hostage recovery efforts. Operation Willing Spirit played a key role in building the foundation for the rescue mission, called operation JAQUE. The mission centered around three lines of effort: building Colombian military capacity, campaign planning, and Colombian and U.S. Special Operation Forces (USSOF) efforts designed to stress the FARC hostage holding network. Operation Willing Spirit also resulted in the deepening of relationships between U.S. and Colombian Special Operations Forces. These relationships intensified overtime and resulted in a level of transparency, integration and partnership seldom achieved between foreign military forces[4].

Partnership—U.S. and Colombian Special Operational Forces

Colombia and the United States share a strategic partnership that is very strong and dates back to the Korean War. Colombian forces served alongside the U.S. forces in Korea and won a Presidential Unit Citation as part of the UN effort to repel the communist Chinese.[5] USSOF military capacity building efforts in Colombia go back more than 30 years. General Yarborough, then Commander of the Special Warfare Center and School, led an assessment team to Bogota in 1959 to advise and assist the Colombians in developing a counterinsurgency strategy and a military training plan.[6] Early efforts included building Colombian conventional capabilities to thwart an ongoing insurgency. USSOF played an important role in helping to professionalize the Colombian defense forces by establishing the Lancero Course, a small unit leader's course modeled after the U.S. Army Ranger School.[7] Additional USSOF efforts included assistance in training Colombian Mobile Brigades and Counter-guerrilla units that have proven to be highly effective in operations against the FARC.

"Plan Colombia", implemented in 1999, authorized additional funding and resources for Colombia. Mobility was a critical shortfall and this new plan provided U.S. helicopters to the Colombian air force. These assets allow Colombian forces to reach the FARC in the remote jungle areas. SOF advisory teams helped these forces plan operations and perfect helicopter insertion, extraction and resupply techniques. These

4. Author's first-hand account.
5. Barbula and Old Baldy, March 1953, Charles H. Briscoe Veritas, ARSOF in Colombia, pg 15.
6. Plan Lazo, Evaluation and Execution, Charles H. Briscoe Veritas, ARSOF in Colombia, pg 38–39.
7. Colombian Lancero School Roots, Charles H. Briscoe Veritas, ARSOF in Colombia, pg 30.

techniques coupled with President Uribe's counterinsurgency policy allowed Colombian forces to regain control of much of the countryside previously controlled by the FARC.

In 2001, USSOF capacity building efforts shifted focus to the development of Colombian Special Operations Forces (COLSOF). With USSOF assistance, the Colombian defense forces developed their own special operations forces with expertise in jungle warfare and direct action missions. During 2002, the CCOPE[8] was established as the Colombian national mission force and quickly became their premier unit for going after the FARC leadership and conducting hostage rescue operations. These long-term investments were crucial to the success of this rescue operation.

Operation Willing Spirit Campaign Planning

In February 2003, U.S. priority began shifting to the recovery of the hostages. Efforts include establishing an aggressive psychological warfare program designed to pressure the FARC by encouraging defections through the rewards for justice program and accumulating actionable intelligence on the hostage-holding network. In May 2005, USSOF were finally given the official go-ahead to support hostage recovery efforts in support of Operation Willing Spirit.

Gaining knowledge about the FARC hostage-holding network was deemed critical to the planning and execution of a hostage rescue operation. FARC hostage holding camps are typically located in remote jungle areas under triple canopy foliage, so gaining detailed information is very difficult. Site reconnaissance and exploitation assisted in acquiring the needed intelligence by allowing specially trained USSOF teams to visit abandoned FARC hostage holding camps to gather evidence that the hostages were alive and to develop a better understanding of the FARC's hostage holding network.

Additionally, developing the ability to conduct combined U.S. and Colombian special operations greatly expanded military options available to the planners. Training, equipping and advisory efforts, under OWS authorities, greatly enhanced the CCOPE's ability to conduct long-range reconnaissance and sustained operations designed to stress the FARC hostage holding network. Combined rehearsals during training events allowed SOCSOUTH and CCOPE staffs to develop integrated staff procedures in the event a combined operation was deemed appropriate. At the tactical level a combined reconnaissance capability was developed to facilitate U.S. and Colombian reconnaissance on suspected hostage holding camps.

Combined planning and operations required detailed coordination by U.S. and Colombian SOF as well as the interagency within the respective governments. The OWS Campaign Plan and Playbook were developed to codify procedures and provide a menu of U.S. capabilities to support a variety of hostage rescue option. These options included a unilateral U.S. military operation, a combined US/Colombian operation and a unilateral Colombian option. The OWS Playbook allowed SOCSOUTH to synchronize efforts and make possible the quick and efficient flow of U.S. support assets into Colombia. It also provided key decision makers in the Interagency, U.S. and GOC a common framework which served to facilitate decision making as new intelligence altered the operational picture.

8. COMANDO COJUNTO DE OPERACIONES ESPECIALES or Joint Special Operations Command.

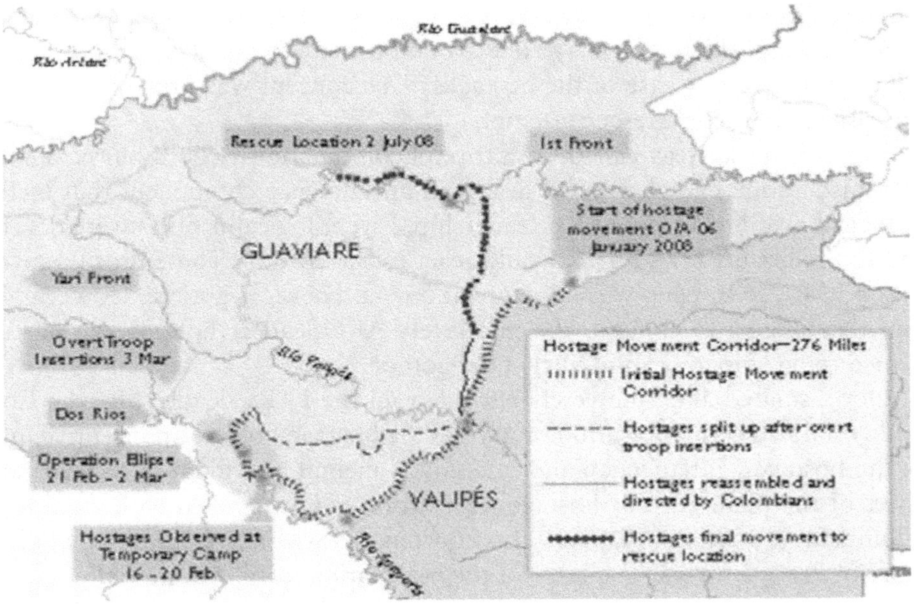

Figure 2 Map of Hostage Movement Corridor.

As Operation Willing Spirit planning and operational capacity was improving, Colombian Defense Force's were achieving unprecedented success against the FARC. President's Uribe's democratic security strategy was starting to take hold through successful targeting of FARC leadership and counterinsurgency operations to reclaim large portions of the countryside. In late 2007, the Colombian military was finally ready to mount a campaign to pressure the hostage holding network. As a result of this campaign, it became clear that the FARC 1st Front was holding the high value hostages. By January 2008, the sustained pressure against the 1st Front forced the need to transfer the hostages to the Yari Front for safe keeping.

The Reconnaissance Plan and Operation Ellipse

All told, the hostages would move 276 miles from January to July 2008 on remote rivers and through dense jungle trails before arriving at the location where they would ultimately be rescued. The movement of the hostages opened the door to increased surveillance and reconnaissance and created a significant vulnerability for the FARC. The hostages were being moved on foot and along a series of rivers in small indigenous boats. They remained chained and under tarps to minimize detection. But slowly, information about the hostages began to emerge and Colombian SOF planners and U.S. advisors developed a reconnaissance plan focused on the Apaporis River Valley. This plan included the emplacement of motion detection sensors along a series of rapids. Planners determined the hostages would have to disembark their small boats to negotiate the rapids at one particular location, one of the only places on the river where observing the hostages could be possible.

With reconnaissance efforts underway, planners shifted focus to developing rescue options. Minimizing the risk to hostages was key. Accordingly, a jungle rescue operation against a mobile target was deemed too dangerous. This forced planners to develop a different approach which came to be known as the Cordon Strategy and was named

Operation Ellipse. Once the location of the hostages was identified, Operation Ellipse called for establishing a security zone around the site to facilitate negotiations with the FARC for the safe release of the hostages. This concept was preferable to mounting a high-risk, direct action rescue attempt.

The cordon strategy called for a careful combination of both physical and psychological isolation. The plan relied heavy on an influence component that included psychological profiling of the FARC leadership. Psychological profiles were developed and used to predict how the FARC would react to the cordon. These profiles revealed that the top FARC leadership was hardcore and would never negotiate. However, if isolated from their higher command, the mid-level FARC leaders holding the hostages were deemed to be vulnerable and likely to negotiate.

Once isolated and completely cut-off from the FARC leadership, the Colombians would establish communications and offer the hostage holders an incentive in exchange for the hostages. Interagency and intergovernmental coordination was critical to this aspect of the plan. An FBI hostage negotiation team travelled to Bogota to train the Colombian negotiators. Planners also coordinated with the Colombian High Commissioner of Peace who was in charge of the negotiation team. Additionally, a communications system was put in place to facilitate decision making for Colombia's leadership.

While planning continued, the reconnaissance teams were patrolling the Apaporis river valley positioning themselves at key locations along the riverbanks. They had limited fields of view and while observing boat traffic was easy, identifying the hostages chained and under tarp in a moving boat would be difficult. Fortunately, persistence paid off when a Colombian reconnaissance team found the "needle in the haystack" when in mid-February they actually spotted the hostages bathing on the Apaporis River. This was the first time they had been observed since their capture.

Now Operation Ellipse swung into full motion with Colombian SOF accompanied by U.S. advisors moving rapidly by helicopter to positions on the northern and southern sides of the hostage site. The forces landed 15 kilometers from the hostage site to minimize the risk of detection. The cordon force's movement was very slow due to the difficult jungle terrain. Unfortunately, the FARC and their precious cargo slipped out of the noose just hours prior to the final positioning of the cordon forces.

Operation Ellipse transitioned into a pursuit operation with cordon forces moving great distances on foot up the Apaporis River valley attempting to encircle the hostages at two other locations. As they approached Dos Rios, it became apparent that the hostages could slip undetected into the Yari Front and it would be very difficult to re-acquire them. To counter this, the decision was made to conduct a series of overt Colombian troop insertions in and around the Dos Rios area to discourage further movement west. The idea was to flood the zone with helicopters and troops to turn the FARC back.

With Colombian forces losing the element of surprise, the FARC altered their standard procedures by dividing the hostages into three separate groups and melted into the jungle. The trail went cold in March but the Colombian Special Forces continued operations to pressure the FARC and attempt to reacquire the hostage trail.

Problems From Within

With the hostage trail going cold, pressure on the FARC leadership network was increasing. On 1 March, the Colombian's killed Raul Reyes in a cross border raid into

Ecuador. Raul Reyes was a member of the FARC Secretariat and in charge of the FARC's International front. Mono JoyJoy, the FARC's top military commander, was also under intense military pressure and out of communications with the FARC's leadership. In fact, Colombian intelligence had been able to establish that the guerrilla leader guarding the hostages, Cesar, had not been in direct contact with the high command of Mono Jojoy for over three years. There also was a power struggle within the ranks of the FARC since the death of Manuel *tirofijo* Marulanda, the FARC top leader. It had been said that the military leader, Mono Jojoy, was unhappy when political strategist Alfonso Cano was chosen to succeed Marulanda.

Cracking the FARC Communication Codes

As pressure on the FARC leadership network intensifed, Colombian intelligence efforts to infiltrate the FARC communication network moved into high gear. Collecting intelligence for this operation began long before it was actually carried out. Defectors and former guerrillas provided pieces of the puzzle. These defectors came out of the jungle periodically from 2003-08. Former FARC hostage (1998-2007), Policeman Frank Pinchao, escaped in May 2007, 14 months prior to the July 2008 mission. Pinchao provided excellent information on the FARC routine. Frank had lived with Ingrid Betancourt and the three U.S. captives.[9] A critical development were laptops, seized in the cross border raid into Ecuador to kill Raul Reyes, that revealed some of the FARC's darkest secrets and information vitial to understanding their operations and communications networks. Armed with this intelligence, some innovative Colombian intelligence professionals hatched a scheme to insert themselves into the FARC communication network. This was the key to JAQUE and was only possible after 20 years of studying FARC communications procedures and vulnerabilities.

During a scheduled change of frequencies and encryption codes, the Colombians successfully inserted themselves as the radio operators for two senior FARC leaders, Mono Jojoy and Cesar. This required identifying individuals with the same voice tones and accents to impersonate the FARC radio operators. Remotes jungle sites were also constructed to replicate the conditions under which the messages would be transmitted to duplicate background ambient noise. The imposter operators entered the FARC communications network and slowly tested their ability to deceive the FARC leadership. Up to this point, the government of Colombia had to rely on traditional intelligence sources like communications intercepts and jungle reconnaissance to glean information on the hostages. This new capability allowed the GOC to manipulate the FARC chain of command and actually issue orders about what to do with the hostages. This was a total game changer.

Operation JAQUE—"Check"

On June 26, the Colombians presented the deception and rescue concept to the U.S. Government. During the meeting, a bold and decisive deception operation was outlined that would essentially trick the FARC into unwittingly handing over the hostages to the Colombian government. In it simplest form, the Colombians were preparing a force to

9. Interview with Kevin Higgins, U.S. advisor to the GOC from 1999-2011.

act as a fictitious International Organization, that would go deep into FARC controlled territory, unarmed and pick up the hostages using civilian helicopters, and transport them to freedom. But the devil was in the details.

The plan called for the Colombians to leverage their capability to infiltrate the FARC's communications network and send radio messages directing Cesar, the FARC 1st Front Commander, to prepare the hostages for transport to another FARC-controlled area that was currently not under Colombian military pressure. The Colombians would assume full responsibility for the operation and the President of Colombia had approved the plan the day prior. They also acknowledged the risk—worst case the Colombians would lose a helicopter and 13 additional hostages to the FARC. The Colombian's emphasized the need to act quickly as delay would increase the chance that the FARC would discover the deception plan.

The initial U.S. reaction to this plan was skeptical at best. But as time passed and the details on the plan became known, it was clear that it had tremendous potential. The rescue team was carefully selected from volunteers who knew this was potentially a one-way trip. Team members included various ethnicities to replicate an international organization and everyone received acting classes in preparation for their roles. The deception included the construction of a web site to backstop the international organization and detailed preparation of the government MI-17 helicopters to look authentically civilian.

Deception operations rely heavily on the enemy's perception of the current environment and for this operation anything out of the ordinary would risk compromise. To reduce the possibility of compromising the operation, the U.S. support signature in Colombia had to be minimized. Surprise was deemed critical and any spike in U.S. presence could tip off the FARC and spoil the rescue attempt. Accordingly, the decision was made to primarily leverage the existing USSOF advisory effort in country to support the mission.

The stratagem called for the FARC rebels to hand over the hostages to Colombian soldiers posing as members of a fictitious nongovernment organization that supposedly would fly the captives to another FARC camp to meet rebel leader Alfonso Cano. Soldiers would impersonate cameramen, a journalist from a Latin American TV station, two would pose as guerilla fighters and four troops would dress as aid workers. Several aspects of the mission were designed to mimic a previous Venezuelan hostage transfer, including the actual composition of the group, type of helicopters used and a location near the previous transfer spot. In case of mission compromise or failure, Colombian and U.S. planners had prepared an armada of over 20 helicopters to ferry approximately 400 troops and U.S. advisors, in less than 15 minutes, to cordon off the rescue location and force the FARC to negotiate a hostage release or suffer the consequences.

The GOC selected July 2, 2008 for the rescue. Despite all the careful preparation, the operation did not begin well. Throughout the morning a thunderstorm between Bogota and the pick-up location prevented transmission of the final instructions to Cesar to assemble the hostages and move them to the pick-up location. Finally, around noon, the message went thru, and the rescue operation sprung into full motion with the helicopters taking off to the selected pick-up location.

Around 1 PM, one Mi-17 helicopter landed at the FARC pick-up location in the Guaviare district of Colombia. Once on scene, Colombian agents wearing Che Guevara T-shirts linked up with FARC security and arranged to transfer the hostages. The hostages were flex cuffed and loaded aboard. Cesar, previously invited to come, escorted the hostages in anticipation of meeting with the new FARC Leader, Alfonso Cano, at the transfer destination. Cesar along with an additional rebel were persuaded to hand over their pistols and were quickly subdued after the helicopters took off by Colombian forces. The hostagess were told: *"Somos el Ejército Nacional. ¡Ustedes están en libertad!" We are the national army. You are free!*

After 22 minutes the radio transmission verified that helicopters had taken off and the hostages were secure. Twenty minutes later the MI-17 touched down in a secure landing zone. The unbelievable had just happen—the hostages were free and the FARC 1st Front Leadership was captured, all without firing a shot. The Colombians had planned and executed one of the most complex and audacious special operations mission in recent time.

Conclusions

Deception is often enabled by a technical component or capability. The Colombian's carefully developed this capability through studying FARC communications for over 20 years and exploiting its vulnerabilities through breaking codes and identifying communications schedules and routinely used frequencies. With the know how to infiltrate the FARC communications network, the Colombians identified people with the right voices and tones to impersonate select FARC radio operators. This included constructing radio transmission sites in the jungle to establish the proper background ambient noise.

With the capability to imitate FARC leaders' communications, the Colombians proceeded to craft a strategy to recover the hostages. Instead of relying on ground reconnaissance in the dense jungle to locate the hostages, the Colombians replicated the FARC chain of command and manipulated subordinate FARC commanders into unwittingly handing over the hostages to the Colombian Government.

Sustained Colombian military pressure against the FARC hostage holding network and against the FARC military leadership ultimately set the conditions for the rescue and significantly contributed to both the friction and fog of war experienced by the FARC. It was military pressure that set in motion the hostage transfer between FARC fronts by increasing the network's vulnerability to detection and interdiction during movement. These vulnerabilities allowed Colombian SOF to increase pressure on the hostage holding network and place the FARC 1st Front leadership under sustained duress during Operation Ellipse from January to June 2008. Mono JoyJoy was also under sustained military pressure, constantly on the move and often out of communications with the hostage holders. This pressure prevented the FARC from successfully transferring the hostages to the Yari Front increasing the FARC's vulnerability to the deception by allowing for the introduction of an alternative solution to the problem of moving the hostages out of the FARC 1st Front area.

Furthermore, creating the alternate reality exploited a signature familiar to the FARC—Venezuelan President Hugo Chavez's periodic interventions to arrange Red Cross helicopters to pick up hostages from the FARC and using a location near a site previously used for the hostage pick-up made the ruse even more believable.

A "Trojan Horse" has come to mean any trick or stratagem that causes a target to invite a foe into a securely protected bastion. This was clearly the case when Cesar, the FARC 1st Front Commander, essentially invited the Colombian commandos, disguised as an international organization, to fly in and pick up the hostages for transfer to what they thought was another FARC controlled area. The FARC actually picked the transfer site, readied the hostages for pick-up and unwittingly assisted in loading them on the rescue helicopter. The Colombian's went a step further and cleverly invited Cesar to escort the hostages, which he readily accepted as an honor.

Operation JAQUE represents one of the finest examples of military deception in modern history. Deception is the deliberate misrepresentation of reality to gain a competitive advantage. The Colombian government cleverly concealed their capability to infiltrate the FARC's communications network and carefully tested their ability to manipulate the FARC hostage-holding network ultimately choosing to exploit their capability at precisely the right time to rescue high-value political hostages and thus achieving a strategic effect.

Section X: Deception Planners and Planning

Introduction

Deception Planners and Planning

Successful deception necessitates overcoming numerous challenges—some obvious and expected, others not so obvious and frustrating. The first challenge of successful deception is the enemy. An enemy who knows more about his would-be deceiver than the deceiver knows about his target is apt to mitigate the effects of stratagem. In theory, the competence of the enemy should be the single greatest challenge for the deception planner. Unfortunately and most frustrating, policy and strategic culture, inexperienced planners and a military force that is less skillful than required undermine deception, perhaps more than the enemy.[1] Deception challenges can be divided into two categories—preconditions and process.

Preconditions

Preconditions include—but are not limited to—knowing the enemy; possessing a strategic, ulture that prizes deception; having policies that cultivate, reward, and institutionalize deception as an instrument of statecraft; and developing skilled practitioners. To reiterate, a deceiver must know more about the adversary than the adversary knows about the deceiver. This requires somewhat of an outward focus designed to understand the culture and institutions of the enemy sufficient to identify and exploit vulnerabilities. All successful deception requires a knowledge advantage.

This knowledge advantage needs to be supported by an institutional structure that systematically recognizes deception as a legitimate tool. Policies should provide direction, establish goals and constraints, and consider risk. An institutional structure and sound policy are a clear reflection of commitment.[2]

Another precondition for deception success is the practitioners. Deception is more of a military art than a science. It can and should be taught in military schools but teaching art is not a foolproof way to nurture the artist. While it is possible that classroom instruction or self-study can produce competent deception practitioners, Bart Whaley suggests that stratagem, being an art, is best passed on from teacher to student. Implicit in Whaley's suggestion is that the teacher can identify the promising student.[3]

1. Colin Gray makes a similar claim referring to strategists.
2. Walter Jajko, "Deception: Appeal for Acceptance; Discourse on Doctrine; Preface to Planning," *Comparative Strategy*, Vol. 21, 2002, p. 355.
3. Bart Whaley, *Stratagem: Deception and Surprise in War*, Norwood, MA: Artech House, 2007, pp. 3–6.

Process

The process of deception involves planning, evaluating, coordinating, deconflicting, integrating, gaining approval, executing, measuring effectiveness, adjusting, terminating, and recording results.[4] This process, with minor variations, reflects standard military planning practice and is derived from numerous conceptual models and theories about deception.[5] All of these models and theories have similar objectives. These objectives include:

1. To provide a commander with freedom of action to carry out his mission, by deluding the enemy as to his intentions and by diverting the enemy's attention away from the action being taken.
2. To mislead the enemy and persuade him to adopt a course of action that is to his disadvantage and that can be exploited.
3. To gain surprise.
4. To save the lives of one's own troops.[6]

The process is simple and logical and involves the following:

1. The purpose of the deception—along with specific objectives—must be established.
2. An appraisal must be made to determine whether or not deception is the preferred tool. Valuable assets needed for deception should not be put at risk if sufficient knowledge of the enemy is lacking or if achieving the objectives listed above do not require stratagem.
3. A thorough assessment of the enemy is required to determine his strengths, weaknesses, vulnerabilities, intelligence capabilities, and the decision making process used by the decision maker. This assessment calls for the deceiver to view himself through his enemy's eyes. Mirror imaging will likely lead to a faulty assessment.
4. The deception story or mental picture that will be revealed to the enemy must be put together. This story must be integrated with the real plan so that they are mutually supporting.
5. Develop the details of the plan. The details include, as a minimum: identifying or developing channels for transmitting information, physical displays, troop movements, staged compromises of information, diplomatic exchanges and establishing clear indicators, and a feedback mechanism. These are needed in order to verify whether or not the stratagem is working.

Finally, every deception operation, whether it succeeds or fails, should be formally terminated. Termination may require a separate micro-operation designed to protect agents, sources, and communication channels for future use.

The comments above are not intended to address every element necessary for the successful use of stratagem. Still, the two categories discussed—preconditions and process—must be taken into account. Properly used, deception is an effective and economical tool for both the offense and the defense. However, it is not a substitute for an integrated strategy that applies all instruments of statecraft towards achieving a realistic political outcome. Stratagem can enable good strategy but it cannot rescue bad strategy.

4. Jajko, 351–353.
5. For a concise summary, see Michael Bennett and Edward Waltz, *Counterdeception: Principles and Applications for National Defense,* Norwood, MA: Artech House, 2007, pp. 20–31. Also see Bart Whaley's article in this section.
6. Jon Latimer, *Deception in War,* New York: Overlook Press, 2001, p. 62.

CHAPTER 43

A Commander Improvises His Own Deception Planning Team

Fleet Admiral William F. Halsey

Admiral William F. Halsey was a U.S. Naval officer who commanded the South Pacific Area during the early stages of the Pacific War against Japan. After Pearl Harbor, Halsey led a series of raids against the Japanese, and then carried out the Doolittle Raid on April 18, 1942. Halsey's phrase, "Hit hard, hit fast, hit often" soon became a slogan for the Navy.

Admiral William F. "Bull" Halsey, Commander of the Third Fleet and (until 15 June 1944) the South Pacific Area, was involved in strategic deception at least as early as October–November 1943 in the Bougainville campaign.[1] But, as that operation was conducted jointly with General MacArthur, it is difficult to separate Halsey's contribution, if any, to the deception plans.

In reminiscing about a later period (November 1944), Halsey passed the following intriguing remark:[2]

> After the movie, I sat in on the nightly meeting of my Dirty Trick Department — Mick Carney, Ham Dow, Doug Moulton, Harold Stassen, and Johnny Lawrence — and listened to them concoct new methods of bedeviling our gullible enemy. (The Navy prefers me to drop this topic right here.)

When Halsey published this tantalizing vignette in 1947, it is surprising that the Navy censor passed even that brief but provocative mention of deception. At any event, this five-man "Dirty Trick Department" consisted of:

- Rear Adm. Robert B. "Mick" Carney, Chief of Staff, SOPAC and Third Fleet.
- Lt. Cdr. Leonard J. "Ham" Dow, Halsey's Communications Officer.
- Capt. H. Douglas "Doug" Moulton, Halsey's Air Operations Officer.
- Lt. Cdr. Harold E. Stassen, Halsey's Flag Secretary.
- Lt. Cdr. John E. "Johnny" Lawrence, an Air Combat Information Officer.

Here we glimpse a Commander working intimately (and apparently effectively) with his deception planners. If naval deception was new to Halsey, intrigue was not. In setting the scene for the victory at Midway on 2–3 June 1942, Halsey had connived with Nimitz to circumvent the direct orders of Admiral King in Washington.

1. For deception in the Bougainville operation see Whaley (1969/2007), Case A39.
2. Fleet Admiral William F. Halsey and Lieutenant Commander J. Bryan III, *Admiral Halsey's Story*, New York: Whittlesey House, 1947, p. 235.

CHAPTER 44

Deception Planning in 145 Different Disciplines: Lessons from Behind Other Hills[1]

Barton Whaley

How They Deceive

The prime task of the military commander and intelligence analyst was defined in the Duke of Wellington's famous remark about always seeking to estimate what the opponent is up to by peering at "the other side of the hill." This is the task of all other types of "detectives," other analysts. To fulfill this task, all of these analysts rely on several theories and methods. This chapter sets forth seven of these approaches that have proven highly effective for various nonmilitary planners and analysts, particularly when deception is suspected. While some political-military planners and intelligencers use some of these ideas and procedures, few have used them as systematically or even thought them through as much as in those disciplines where they are routinely practiced.

This chapter draws on my continuing research during the past four decades on 145 different types of detectives. They ranged from political and military intelligence analysts to arms control inspectors, cryptanalysts, police detectives, forensic scientists, psychologists, con artists, practical jokers, and joke writers; to such *spies of nature* as biologists and physicists. Let's now look at some theories and methods used to deceive that are usually overlooked by most military deception planners and are seldom spelled out in official doctrine. Eight of these deserve special notice here.

The Essence of Deception: Hiding the Real and Showing the False

- Philosopher of science: Sir Francis Bacon (1597);
- Forensic document examiner: Albert Osborn (1910);
- Artist: Solomon J. Solomon (1920);
- Physicist: Dr. R. V. Jones (1942 and 1957).

The notion of hiding-and-showing as a linked pair or dichotomy was already old in 1597, when English philosopher, Francis Bacon, raised this everyman's notion to the level of philosophy in his essay, "Of Simulation and Dissimulation." The reader will appreciate that this seemingly simple concept of *simultaneously* hiding one thing and

1. Barton Whaley, "Lessons from Behind Other Hills: Planning Deceptions in 145 Different Disciplines," Original version submitted under contract to Dr. Jack Dziak on January 15, 2007.

showing something else in its place is the basis for both the theory of deception and all attempts to analyze deceptions.

It's no accident that we owe the refinement of this concept to Bacon rather than some other Elizabethan humanist. Francis Bacon was, after all, a great admirer of Machiavelli. Moreover, like his Italian model, Bacon was an historian, writer, diplomat, and an ambitious politician whose career ups and greater downs came at the whims of tyrants. Additionally, Bacon had personal experience since age 16 as a secret agent in the cut-throat politics of the English royal court.

Albert Osborn, that pioneer American questioned-document examiner, understood the simulation-dissimulation linkage as found in the forgery of signatures. He stressed that even any half-successful forgery must be a *double process,* one where the forger tries simultaneously to *simulate* the characteristics of the hand being imitated while disguising his own. Even then, neither the simulation nor the dissimulation is apt to be perfect.[2] Obviously, Osborn's injunction can be extended to all types of forgery and, I insist, all types of deception.

Solomon J. Solomon was a London-born, British-Jewish portrait painter whose book *Strategic Camouflage* (1920) was a study of Allied and German camouflage on the Western Front in the Great War.[3] His landmark book offered the first primitive theory of camouflage and introduced to military theory the old, two-part simulation-versus-dissimulation notion using those two words. It also first made explicit the concept of strategic-versus-tactical camouflage. Unfortunately, instead of building on the wisdom of Solomon, most of the later camouflage field manuals have degraded theory while concentrating on updating specific methods.

Indeed, this concept of deception as a two-fold hiding-plus-showing process is not as clearly understood in any other part of military, particularly Western military doctrine or practice as it is in such other professional doctrines as conjuring, con artistry, psychic fraud, counterfeiting, and forgery. Even the leading theorist and practitioner of military deception, Dr. R. V. Jones, was uncertain on this point. On the one hand, he often seemingly considered deception as *combining* dissimulation and simulation. On the other hand, he sometimes clearly defined a two-part typology of *The Principles of Security* which are "in a sense negative" in contrast to *The Principles of Deception* which are "positive".[4] And other British and American writings on the subject similarly introduce, but inconsistently use, the mutually exclusive dichotomies "camouflage and deception" (C&D), "cover and deception" (C&D), and "denial and deception" (D&D). However, I prefer the prior definition of deception not merely because it is elegant, (Occam-elegant), but because it emphasizes the psychological fact that its two components, dissimulation and simulation, always occur simultaneously.[4] To give a military example: If an insurgent team camouflages (hides or dissimulates) an RPG nest overlooking a road junction, they simultaneously imply (show or simulate) that there is no ambush there. Indeed, to enhance the notion that they pose no threat, they might even set up a visible dummy (simulated) ambush nest somewhere out of range of the junction.

A conclusion in deception is fundamentally simple in theory. Even in practice, this simplicity can be maintained. Tangled webs and other such complexities are sure

2. Albert S. Osborn, Questioned Documents, Rochester: The Lawyers' Co-operative Publishing Co., 1910, p. 12.
3. Solomon J. Solomon, *Strategic Camouflage,* London: Murray, 1920.
4. R.V. Jones, "Intelligence and Deception," in Robert Pfaltzgraff, Uri Ra-anan, and Warren Milberg, eds. Intelligence Policy and National Security, London: Macmillan, 1981, pp. 17–18.

symptoms of incompetent deception planning because they unnecessarily multiply the blatant clues for the deception analyst to work with. Conversely, a sure sign of a competent deception analyst is one who systematically applies incongruity analysis.

What They Hide and Show: The 5 Ws

- Journalists: Rudyard Kipling (1902), etc.

Because news reporters are a numerous and important type of detective, it is natural that they developed, or at least adopted, a few standard investigative procedures of their own, such as the data bank that they call their in-house library or, more colloquially, morgue. However, the most important of their adaptations has become a popular memory aid, as seen in the following poem:

I keep six honest serving men
They taught me all I knew;
Their names are What & Why & When
And How & Where & Who.

Just so, rhymed English ex-newspaperman, Rudyard Kipling, back in 1902. I agree with Prof. Stuart Kind, an outstandingly successful English forensic scientist, that this bit of journalistic doggerel also "encapsulated much of the equipment of the good crime investigator." Kipling's is the earliest citation I find to the mnemonic traditionally used by media journalists in reporting any and every news story. Usually they call it "The Five Ws"—Who, What, Where, When, Why. Sometimes they call it "The Five Ws Plus H" (for the "How") or just "The 5 Ws and the 1 H". These formulations were commonplace in the journalism trade as early as 1932.

The Five Ws is every journalist's checklist of the essential elements that make up a news story—any news story, every news story. Once he's collected these five (or six) tidbits of information the story is technically complete, ready to report—ideally summarized in the lead paragraph. Sociologists and political scientists—and a few military intelligence analysts who analyze human communications interactions adopted this model—call it the Lasswellian Paradigm. They were copied by a few military intelligence analysts beginning with myself in 1969 when I first proposed adapting Harold Lasswell's five-category model to the analysis of military deception operations. Eventually (by 1982) I expanded this checklist to nine categories.[5] The otherwise admirable Alexander George, a Stanford professor of political science who surely knew of the Lasswellian Paradigm, didn't credit anyone in 1979 with his retrograde proposal that the major "dimensions" of surprise were "whether, what, where, when, why." A decade later, Israeli political scientist Ariel Levite would explicitly copy George's "five dimensions" and Israeli strategist Ephraim Kam would compile (without credit) an inferior 4-point list of "aspects" he called "whether, when, where, how."

However, the 5-Ws analytical model and its variants precedes even Kipling—indeed all journalists, all social scientists. It goes back to an anonymous Roman jurist who gave us the maxim:

5. Barton Whaley, *Stratagem*, Cambridge, MA.: CIS, MIT, 1969, pp. 210–215; Whaley & Bell, 1982, pp. 428–431.

"Quis, quid, ubi, quibus auxiliis, cur, quomodo, quando?" [Who, what, where, with what, why, how, when?]

Various similar 7-, 6-, and 5-mode versions were also adopted by pioneering American criminologists like Edward Heinrich, British forensic scientists like Professor John Glaister, and that fine Hungarian-American mathematician, George Polya, who advised:

> *Your five best friends are What, Why, Where, When, and How. You ask What, you ask Why, you ask Where, When and How—and ask nobody else when you need advice.*
> —G. Polya, *How to Solve It* (1945), 197

Polya didn't need "Who" because math problems don't normally involve a second human actor.

The Five Ws and all their prior and derivative models are extraordinarily powerful tools. If they seem obvious or simplistic, they aren't. Research psychiatrist, Professor Daniel X. Freedman, notes the ease with which medical and social science fraud enters undetected into the scholarly journals. He believes that this is because most researchers and editors conventionally omit information about when, where, or who collected the raw data and which of several researchers authored what parts of the conclusions. Dr. Freedman urges that, in order to make detection easier, "The *who, what, when, where, how,* and *why* of an enquiry must be reported."[6] Particularly serious, even potentially deadly, examples occured when political-military analysts like Cruickshank (1979), George (1979), Levite (1987), or Kam (1988) omitted any of these dimensions.

More generally, many media reporters, social scientists, and detectives regularly overlook one or more of these crucial elements, thereby missing one or more parts of the whole story. Witness the accounts in almost every morning newspaper and most evening TV news shows. The media journalists, having early on adopted this powerful but simple aid to memory, have almost forgotten it in their zealous dumbing-down preoccupation with maximizing entertainment at the expense of information. I was not surprised to notice on the morning of the capture of Saddam Hussein that only one American TV news anchor cleaved to the spirit of the Five Ws. He even named them. This was CBS's Dan Rather, the oldest and arguably the only thoroughly professional journalist left on American TV until 2004 when even he self-destructed over his inexcusable gullibility when handed a clumsy forgery.

I'll leave this model by just noting in passing that there are several derivative, although unacknowledged, versions posing under various acronyms. Those nameless committees in the U.S. military who edit field manuals and books on joint doctrine are among the main offenders. The earliest example I can find is STASM, which originated in World War II with the Propaganda Branch of the U.S. War Department General Staff. This acronym stands for the five elements that are involved in the conduct of psychological warfare: Source (including media), Time, Audience, Subject, and Mission. The most recent (by 1987) example is SALUTE, which stands for Size, Activity, Location, Unit, Time, and Equipment.

My personal list, which gradually expanded from Lasswell's list, has nine categories. I see perception and misperception in general and deception in particular applying to only

6. Daniel X. Freedman in Michael Hersen, Richard M. Eisler, *Progress in Behavior Modification*, ed. Peter M. Miller (NY: Sycamore Publishing, 1994), 194.

nine types of things. These are the categories that a deceiver tries to dissimulate or simulate, hide or show. They are the things about which we can be deceived or surprised. Let me count the ways:[7]

The nine categories of perception and misperception

1. *Pattern:* What is the structure and process of the event?
2. *Players:* Who are the actors?
3. *Intention:* What does the main actor hope to accomplish?
4. *Payoff:* With what potential costs and benefits?
5. *Place:* Where will the event occur?
6. *Time:* When is it scheduled for?
7. *Strength:* How much effort can or will the main actor bring to bear?
8. *Style:* What, if any, new methods or technologies might appear?
9. *Channel:* Through which media will the main actor communicate?

While this is the most comprehensive list, I use it only when doing systematic paper-and-pencil analyses. The shorter 5 Ws or 5Ws+H lists are usually sufficient, quick-and-dirty tools, particularly when pressed for time.

In conclusion, a few basic check-lists, such as The 5 Ws, simplify the work of both deception planners and deception analysts by efficiently organizing their planning and analyses.

Bluff and Double Bluff

- Magicians: Thayer (1914), etc.;
- Card sharps: Scarne (1945);
- Mathematician: Von Neumann (1947);
- Economists: Morgenstern (1947) and Schelling (1960);
- Counterintelligence officers: Pinto (1952), etc.;
- Sociologist: Goffman (1969).

The first theory of Poker was proposed in *The Theory of Games and Economic Behavior,* published in 1944 by two immigrant Americans, Hungarian mathematician John von Neumann and German economist Oskar Morgenstern.[8] Their ground-breaking book included a mathematical analysis of poker that proved the rightness of expert players' intuitive belief that successful poker depends not merely on randomly intermixing bluffs with straight play, but that it is essential to try to bluff weak hands, even though these are likely to be called, in order to make the high hands pay. Indeed, during

7. Based on Bart Whaley, "A Typology of Misperception; or The Ways We Can Be Wrong," draft March 1980, p. 111. Some of this material was summarized in Whaley & Bell (1982), p. 428–431, where I added the category of "pattern."
8. John von Neumann and Oskar Morgenstern, *The Theory of Games and Economic Behavior,* Princeton: Princeton University Press, 2nd edition, 1947, pp. 186–189. *The Theory of Poker* and specifically the *Theory of Bluffing* were primarily the work of Von Neumann, during the period 1926–1928. See also Lawrence Friedman, "Optimal Bluffing Strategies in Poker," *Management Science,* Vol. 17, No. 12, August 1971, pp. B764–B771.

the mid-1800s poker was also called "Bluff." The best situations for bluffing in Poker occur in the following circumstances:

1. Any time the ratio of the price of being called is high compared to the value of the pot.
2. Whenever all the loose, that is, reckless players (particularly novices), have folded and all but one or two tight, that is conservative, players remain in the game.
3. In high stake games, because these are more likely a) to yield situation number 1 above and b) to discourage loose players from calling just to keep the other player honest.
4. Whenever you have evidence that the other players still in the pot perceive you as playing tight.
5. If, holding a strong hand early in the game, you check and then re-raise.
6. When you are sitting last.

This theory of bluffing and this list, although applied specifically to one particular type of card game, is generally true of all forms of bluffing. Harvard Professor Thomas C. Schelling, a political scientist and Nobel Prize winner, applied it to military deterrence.[9] How often do we hear someone described as good at war strategy, diplomatic negotiation, or everyday business dealings because he is a good poker player. Of course there are exceptions. Thus, von Neumann himself was usually a loser in those many poker games with his fellow A-Bomb scientists at Los Alamos. This was attested to by his fellow player, Polish mathematician Stanislaw Ulam, who offered the old losers' excuse that they (Ulam, von Neumann, and Edward Teller) would "think about completely unrelated subjects during the bidding or betting."

> *The double bluff is a trick of the kind so prettily described by conjurors as a "Sucker."*
> —Charles Waller, *Magic from Below* (1929), 163

Magicians call it a *sucker effect* or sucker gag. This refers to their method of leading the audience to believe they've detected a trick's method and then pulling a double bluff to surprise them even more. It is the magicians' equivalent of the gambling sharps' the cross from the popular expression double cross. Magicians use it as insurance to cover the secret of all their tricks, from the smallest tricks to the largest stage illusions. They are one of the two most nearly detection-proof effects in magic, the other being the One-Ahead Gag, which is discussed later ("Time Out of Joint").

Of course, the old double bluff does exist outside magic and con games. However, seldom on a routine basis and then only in the spy versus counterspy games where spies are conventionally given a legend (a cover story) and often even a reserve story (a second-level story-within-a-story) as backup in case the initial phony story was blown. Political-military deception planners should at least be alert to opportunities for incorporating sucker gags into their plans.

In conclusion, bluff and double bluff can profitably be employed by military (or political) planners in substantially more situations than at present. This seems to be currently routine only in the Chinese Communist military doctrine. At the very least, it increases uncertainty in the minds of opposing analysts.

9. Thomas C. Schelling, *The Strategy of Conflict*, Cambridge, MA: Harvard University, 1960, pp. 3–20.

The Theory of Outs

- Magicians: Outs;
- Con artists: The Blowoff;
- Military: Plan B.

If the performer knows enough outs, he will never fail to bring a trick to successful conclusion.
—Jerry Mentzer, *Magician Nitely* (1974), 23

There are only four circumstances in which deception will fail. It fails if the target:

1. Takes no notice of the intended effect;
2. Notices but judges it irrelevant;
3. Misunderstands its intended meaning;
4. Detects its method. In most cases, the deception game ends at that point.

However, both the successful detective and the persistent deceiver can—if they have prepared—continue the game one stage further. The successful detective of deception can switch from passive to active counterdeception. However, this will fail if the deceiver takes certain steps to rescue success from impending failure by using outs.

Interestingly, among all types of deceivers, only confidence tricksters and magicians have developed any standard operating procedures for recovering from the unfortunate effects of the discovery of their deception. All others leave themselves only the option of hasty flight, if even that.

In Con Games

Never give a sucker an even break or smarten up a chump.
—W. C. Fields, *Poppy* (1923, serio-comic musical play), an improvised line, probably copying gambler Milton Mizner.

Confidence tricksters and other hustlers have developed a standard technique to buy enough time for a safe escape just before the victim sees the operation for the scam it is. This technique is called *cooling out* or the *blow-off*. It is built into the final stage of the operation and is itself a deception. Professional liars, like con artists, don't have to improvise when questioned about discrepancies in their story. They always prepare in advance by, as lie expert Paul Ekman explained, "memorizing a credible reply for each and every question they think their quarry might raise."[10]

Cooling the mark out is usually only a matter of convincing the disappointed and/or irate sucker that his loss involves neither loss-of-face nor was the responsibility of his

10. Paul Ekman, *Telling Lies: Clues to Deceit in the Market Place, Politics and Marriage,* New York: Norton, 1985, pp. 14–15.

friend, the hustler. If played carefully, the hustler may even appear so blameless that he can hit the mark again later.[11]

In Magic

Magicians share with con artists the concept of the blow-off. In fact, they borrowed this bit of jargon from the confidence tricksters. The blow-off in magic is simply any climax of a trick that sends the spectators away puzzled because of a final twist in logic. Where the con artist seeks only to buy time for his or her own safe get-away, the magician is concerned to ensure that the trick's method remains secret.

Magicians have gone the con artist one better. In addition to the blow-off they also have their own concept of the out. It applies to all magic tricks. It *could* be systematically applied equally to all military deception operations but, so far, has not. First conceived in 1785, the magicians' *Theory of Outs* is simply a much earlier statement of the military deception planners' *Theory of Alternative Objectives*. The out is any preplanned alternative deceptive method in case the original plan fails.

French magician Henri Decremps foreshadowed *The Theory of Outs* in 1785 as Rule 2 in his list of principles of magic, where he wrote: "Whenever possible have several ways of doing the same trick."[12] This rule first appeared in English translation in an anonymous book, plagiarized also as Rule 2 but elaborated with a reason as:[13] "Endeavour, as much as possible, to acquire various methods of performing the same feat, in order that if you should be likely to fail in one, or have reason to believe that your operations are suspected, you may be prepared with another."[13] Magicians to this day observe Decremps' Second Rule, quoting or elaborating on it, although—having little regard for their own history—never crediting its originator.[13]

Robert-Houdin expanded on Decremps in 1868 when he wrote: "However skilful the performer may be, and however complete his preparations for a given trick, it is still possible that some unforseen accident may cause a failure. The only way to get out of such a difficulty is to finish the trick in some other manner. But to be able to do this, the performer must have strictly complied with this important rule: *never announce beforehand the nature of the effect which you intend to produce.*" He went on to explain: "However awkward the position in which you may be placed by a breakdown, never for one moment dream of admitting yourself beaten; on the contrary, make up for the failure by coolness, animation, and 'dash.' Invent expedients, display redoubled dexterity, and the spectators, misled by your self-possession, will probably imagine that the trick was intended to end as it has done."[16] Robert-Houdin need only have added that the invention of expedients should be done in advance in anticipation of the more likely types of failure.

11. David W. Maurer, *The Big Con*, Bobbs-Merrill, 1940, pp. 48, 289; Erving Goffman, "On Cooling the Mark Out," *Psychiatry*, Vol. 15, 1952, pp. 451–463; Robert C. Prus and C.R.D. Sharper, *Road Hustler: The Career Contingencies of Professional Card and Dice Hustlers*.
12. Robert-Houdin, *Les secrets de la prestidigitation* (1868) as translated into English as *The Secrets of Conjuring and Magic*, London: Routledge, 1878, pp. 32–33.
13. See, for example, John Henry Anderson, *The Fashionable Science of Parlour Magic, 6th edition*, Self-published in London, 1844; *The Magician's Own Book*, New York: Dick & Fitzgerald, 1857, 79, taken verbatim from Anderson; Ellis Stanyon, *Conjuring with Cards*, London: Gill, 1898, p. 3.

In General

The Essence of Deception mentions the two modes of deception—dissimulation (hiding) and simulation (showing). This simple two-part model can be elaborated to give deception's basic structure:

The Structure of Deception

Dissimulation	*Simulation*
Masking	Mimicking
Repackaging	Inventing
Dazzling	Decoying

Earlier, we saw that in practice, as in theory, all three ways of hiding the real can accompany the three ways of showing the false in any of their possible combinations. Here, the three ways to dissimulate and the three to simulate, have been arranged in descending order of assumed effectiveness. J. Bowyer Bell's hypothesis was that the most effective way to dissimulate would be by *masking*. Therefore, if masking fails to confer enough invisibility to the real object or event and it is detected, the deceiver can then resort to *repackaging* to disguise it. *If* repackaging fails and the thing is recognized for what it is, then dazzling can be used as a last-ditch measure to at least confuse the target about some characteristics. Similarly, Bell assumed that the most effective way to simulate would be by *mimicking*. Therefore, should *mimicking* fail to show the false by a convincing imitation, the deceiver can resort to *inventing* to create an alternative reality. If inventing also fails, *decoying* can still be used at least to attract the otherwise undivided attention of the target away from it. Bell suggested that deception planners not only also have the possibility of an out but indeed, a series of priorities for them—the poor, better, best in each of the two columns of the above chart.

Consequently, I thought it likely that the most effective deceptions will dissimulate by *masking* and simultaneously simulate by *mimicking*, while the least effective deceptions would be those that combine *dazzling* with *decoying* to achieve only a mere *razzle-dazzle* effect. Therefore, while dazzling-decoying deceptions might get invented, few—if any—would survive frequent operational experience and be soon dropped from the repertory.

To test Bell's hypothesis and my subsidiary one, I chose to use the tricks of magicians, because among all professional deceivers, they have by far the most frequent experience of both failed and successful deceptions. Accordingly, 60 magic tricks that had been collected to illustrate a previous paper were fitted into a 3×3 matrix as follows:

Simulating

Dissimulating	*Mimicking*	*Inventing*	*Decoying*
Masking	19	10	10
Repackaging	4	3	1
Dazzling	10	2	1

Source: Whaley (1981, table, based on an opportunity sample of 60 tricks).

As can be seen in practice, or at least in magicians' practice, both hypotheses are substantially verified. Masking and mimicking are not only overwhelmingly the most common methods used for dissimulation and simulation respectively; they also are the two used most often in combination. Conversely, only one example of dazzling-decoying was discovered, despite an even wider search among magic tricks. Thus, this is an apparently useful procedural guide for designing the potentially most effective types of deception and avoiding the lesser and least effective ones.

Ruses of the dazzle-decoy type are also quite rare outside of conjuring. For example, of the dozens of carnival *flat store* (cheating) games of pretended chance or skill, only one is of this type. Carnies call it, appropriately, Razzle-Dazzle—but only among themselves; the suckers know it by such disguised names as Bolero, Double Up, or Ten Points. This "game" uses a method of scoring too complicated for the sucker to follow in the short time permitted so that the "flat-joint" operator need only false-count to make the sucker lose.[14]

This model of hiding and showing can also be turned around to serve as a guide for the analyst in assigning priorities in his search for and analysis of deceptions.

In War

Some academic military deception specialists take the sterile view that the risk of a failed deception is so great that it's better to have no deception at all. Even one Cold War U.S. Army manual on camouflage directed that no subordinate army units initiate any camouflage operation without prior approval by senior command. This is overly pessimistic advice. Fortunately for their careers, few military commanders listen to this advice from the so-called experts. In researching *Stratagem* (1969), I found no cases of backfire[15] and that even unsuccessful deception operations often served to at least confuse enemy intelligence.

Neither soldiers' manuals nor generals' war plans tell them what to do when their deception operations fail. The usual attitude is "Well, we tried." Because deception seldom costs much, most commanders just write it off as a small loss and either abort the real operation or, more likely, charge on in without the advantage of surprise. In either case, the logic is that a failed deception is no worse than not having tried at all. I agree with the logic but would hope for a more creative solution. I know of only two cases where military officers overcame their self-crippling sense of helplessness and redesigned their deception plans to take positive advantage of the new situation.

The first example takes place around mid-1942. Major Oliver Thynne was a novice planner with Colonel Dudley Clarke's "A" Force, the Cairo-based British deception team. Thynne had just been told by Intel that the Germans had learned to distinguish the dummy British aircraft from the real ones because the flimsy dummies were supported by struts under their wings. When Major Thynne reported this to his boss, Brigadier Clarke, the "master of deception", fired back:

"Well, what have you done about it?"

"Done about it, Dudley? What could I do about it?"

"Tell them to put struts under the wings of all the real ones, of course!"[16]

Of course? Hardly. A commander with a straightforward mind, having recognized a telltale flaw in the dummies, would have ordered the camouflage department to correct

14. John Scarne, *Scarne's Complete Guide to Gambling*, New York: Simon and Schuster, 1961, pp. 478–489.
15. Since then, I have found a single example, a 1944 U.S. example.
16. David Mure, *Master of Deception: Tangled Webs in London and the Middle East*, London: Kimber, 1980, p. 98.

it. However, Clarke's wonderfully devious mind immediately saw a way to capitalize on the flaw. By putting dummy struts on the real planes while grounded, enemy pilots would avoid them as targets for strafing and bombing. Moreover, it would cause the German photo-interpreters to underestimate the number of real RAF planes based locally and falsely conclude that many of the real ones must be based elsewhere.

Clarke had the wit to see the out as a general principle. One of his favorite maxims that he taught his staff was: "When you put over a deception, always leave yourself an escape route."[17]

The second example of a necessary military out that worked was an audacious and risky one, a last minute improvisation forced upon the deception planners by sheer urgency.[18] Operation OLIVE was General Alexander's decisive offensive to break the German line across central Italy in the summer of 1944. In early June, Alexander planned his main attack with his American and French divisions to go through the mountainous center, with a feint by his British divisions along the eastern flank on the Adriatic. On July 5, Alexander learned that London and Washington had decided to strip him of many of his American and all his French divisions, sending them away for the amphibious invasion of South France. Consequently, on August 4, he reluctantly decided to make the real attack up the Adriatic coast with his British troops, leaving the denuded American force to conduct the feint at the center—in other words—the very strategy that the current deception operations were communicating successfully to the Germans.

This reversal of strategy now required a plausibly readjusted deception. Moreover, with OLIVE D-day set for August 25, Alexander's deception planners had only three weeks to reverse the enemy's perceptions—a feat that military history shows to be very rare indeed.

The essence of the new deception plan was to work a double bluff by having the Germans now believe that the old evidence fed to them had been and still was the Allied deception operation. This was carried out by a new and elaborate program of radio deception. As far as I am aware, this is the only case where a military deception plan was turned back on itself—a stratagem to attempt to discredit earlier disinformation by exposing it for the deception it had been. In the event, the new deception was successful enough to gain surprise of place and strength and of timing as well—two German divisions were uselessly tied up in reserve for a *possible* attack at the center, one division at the point of real attack was caught being relieved, and both the field commander and a key divisional commander were still off on leave.

The out works in military deception because of the "principle of security of options."[19] Deception, by its very nature, provides its own best security. Although the term *security of options* was my original coinage, the concept was only a generalized extension of Liddell Hart's principle of *alternative objectives,* first formulated by him in 1929.[20] Alternative objectives means simply that every football play, most military offensives, and so forth, can have two (or more) goals. In most cases, commanders should

17. Mure, *Master of Deception,* 1980, p. 221.
18. Whaley, *Stratagem* (1969), A414-A418; and Sir David Hunt, *A Don at War,* London: Kimber, 1966, pp. 267–270.
19. First stated in Whaley, *Stratagem,* (1969), pp, 225–226.
20. The concept of "alternative goals" or "alternative objectives" was first stated by B.H. Liddell Hart in his *The Decisive Wars of History: A Study in Strategy,* London: Bell, 1929, pp. 141–158. See also his *The British Way in Warfare*, London: Faber & Faber, 1932, p. 302; *Strategy: The Indirect Approach*, New York: Praeger, 1954, pp. 161–164, 333–372; and *Memoirs, Vol. I,* London: Cassell, 1965, pp. 166–168. See discussion of the evolution of this concept in Whaley, *Stratagem* (1969), pp. 128–139.

be prepared to accept a second-best goal rather than battering away at their primary target.

If the deception operation succeeds in anticipating the preconceptions of the victim and playing upon them, deception security is absolute. In that case the victim becomes the unwitting agent of his own surprise, and no amount of warning (such as security leaks) will suffice to reverse his fatally false expectations. Even if the deception plan runs counter to or fails to play upon the victim's preconceptions, the very fact that it threatens alternative objectives will usually assure enough uncertainty to delay, defuse, or otherwise blunt the victim's response.

The worst possible case would occur if the deception plan itself were prematurely disclosed to the victim. However, such potentially disastrous disclosure is rare in general and so far unknown in war.[21] The closest to this was during the Battle of Midway in 1942 when faulty Japanese security permitted the U.S. Navy to see *around* the primitive Japanese deception operation and set an ambush.

Even if the deception plan was compromised, all would not necessarily be lost. First, the disclosure itself would have to be believed. Second, if the deceiver knows or even suspects disclosure, he can actually capitalize on this by switching to one of the alternative courses of action or simply adopting a new deception plan to reverse appearances, as General Alexander did in breaching the German Gothic Line in Italy in 1944 by successfully reversing an already successful deception operation by deliberately disclosing the first one. Even if the direction or objective of the attack has been compromised, the planner can still manipulate the victim's perception of the timing or strength or even the intent or style of the attack.

In conclusion, *The Theory of Outs* is sadly neglected by political and military operations planners. Most military manuals mention the need for anticipating alternatives when old plans crash; but commanders and their planners tend to place that activity last—usually when time has run out. Magicians and con artists teach us that these Outs (alternatives) should be considered at every planning stage.

Time Out of Joint

- Magicians: The One-Ahead Gag (1693), The One-Behind Gag;
- Psychic frauds: Billet Reading (1862), It's Later Than You Think;
- Practical jokers: It's Earlier Than You Think;
- Con artists: Past Posting; or, It's Later Than You Think;
- Military: The Not Yet Ruse (1944).

Time itself is fair game for four types of deceivers, namely magicians, pseudo-psychics, practical jokers, and confidence artists. Although most military planners and intelligence analysts are aware of some of the possibilities for playing deception games

21. See Whaley, *Stratagem* (1969), A286-287. A recent example from strategic international politics was the disclosure in 1986 of the CIA's disinformation campaign against Libyan leader Gadhafi. At less exalted levels we see this in "tip-offs" of con games and other scams and commonly in "poison-pen" letters that reveal marital or other infidelities.

with time, few of them make it routine practice. Consequently all planners and analysts can profitably learn from magicians, practical jokers, con artists, and soldiers.

Magicians—The One-Ahead Method

> *The old 'one ahead' I contend is the strongest move in magic.*
> —Larry Jennings, lecturing, 1985

The *One-Ahead Principle* underlies the magicians' method of keeping a step ahead of the spectators while the trick or series of tricks unfolds. American magician Bruce Elliott has given a concise explanation of the one-ahead:

"In a series of seemingly repetitious movements or visible actions, one action unseen by the audience precedes the visible one. This unseen action, in advance of the audience's knowledge, is what allows the trick to work."

A one-ahead mentalist card trick (involving sighting the next card while naming and showing the previous one) was used and exposed by magicians as early as 1693 (by Ozanam; then in 1716 by Neve; in 1740 by Guyot; in 1763 by Dean; in 1859 by *Secret Out*). It was later also applied to Billet Reading by spiritualist mediums, first by Charles H. Foster by 1862. Thus, while some clever spiritualist may have independently re-invented this method, they didn't invent it, as is widely assumed.

The one-ahead method is surprisingly rare inside magic and even rarer outside. Among the few real-world examples are a few mentioned next.

Practical Jokers: It's Earlier Than You Think

The clock gag is the type of practical joke where clocks are set forward, as when the college roommate, having studied long into the night for a 9 AM exam, awakens to find his alarm clock and wristwatch both showing nine and rushes madly off to a class that does not start for another two hours. The clock gag was a common practical joke in old-time live radio. There the dupe, sabotaged by the other performers and staff, would enter the broadcast studio to find a clock set ten or fifteen minutes ahead and thinking he was (already) on the air, be sabotaged by the other performers and staff.

Con Artists: It's Later Than You Think

The Sting became the highest-grossing movie of 1974 and won seven Oscars. It was also one of the best movies ever done on con artistry. Out for vengeance, young Robert Redford recruits retired con artist Paul Newman to mastermind a big con to bilk a crime boss out of a fortune. Although filmed from an original screenplay by David S. Ward, the scam that formed the film's gimmick was based accurately on a particularly clever con game called a *Wire Store*. Invented in 1898 by American con artist Joseph "Paper Collar Joe" Kratalsky, it had been continued with distinction by the Gondorf brothers from 1906 to 1915. The Wire Store is a large office fitted out to look like a real illegal horse betting parlor, filled with hired actors dressed to look like well-heeled gamblers. The operators tell their rich but greedy sucker that they have a confederate at the track who telegraphs or telephones the key race results to them a minute or so before the official results are posted and then transmitted to the betting parlors. They then tip the sucker to enough real winners that he bets big—at which point they tip him to a loser.

Ward moved that sting forward to the 1930s and then added, now anachronistically, Scott Joplin's ragtime music that had been composed in the 1910-1915 period.[22]

A fresh wrinkle was introduced in this grand old scam in 1949 when the Whitey Mob neatly fleeced a Chicago bookie by interrupting the bookie's phone-line, tape-recording the incoming horse-race results, betting the winners, and releasing the delayed information a few minutes later to the unsuspecting bookie.[23]

By creating an unperceived time delay, The Wire con notionally put time out of joint to produce an it's-later-than-you-think effect or, what I call, the Time Distortion Effect.

Soldiers

The weather was still very stormy that week, unfavorably so for the enemy to mount its expected amphibious assault. So Red Force was greatly surprised that morning when Blue had begun launching its main diversionary attack across one of Red's Left Flank Beaches. Although Blue's diversion appeared strong enough to secure the beachhead and move slowly beyond, Red decided that this threat could be contained by the local garrison without further reinforcement. Consequently, Red's main force—including all its heavy armor—stayed in place behind the Center Beach, where Blue's main invasion force was expected. Red maintained this policy for more than two weeks before they realized that the diversion was, in fact, Blue's main and only assault. It was later than they'd thought. By then it was too late. That earlier morning's landing had indeed been Blue's true D-Day. That beachhead expanded steadily after June 6, 1944 until Hitler's Fortress Europe was reconquered.[24] This two-stage time-place deception had been initially suggested by Field-Marshal Montgomery's wiley Intelligence Officer-turned Chief of Staff, Maj. Gen. Francis de Guingand and perfected by Lt.-Col. Roger Hesketh at Eisenhower's deception-planning unit. It proved spectacularly successful, possibly even decisive, and has been widely studied. Nevertheless, this type of play upon an opponent's sense of timing is seldom exploited, far less often than it could be.

Imagine some general of the future, directing battle from his immaculate H-bomb-proof command center deep within the Earth—computers and remote sensors his only link to the terrible reality of the "electronic battlefield" above.[25] He must *believe* that his instruments are giving him the truth, the whole truth, and nothing but the truth. But how can he know they are not Shakespeare's treacherous "instruments of darkness"?

In conclusion, the ruses that can be employed to play games with a victim's perception of time and timing range from the ridiculously simple to the sublimely clever and difficult. Try it as a brainstorming exercise.

The Theory of Indirect Approach

- Military: Sun Tzu (ca. 350 BC);

22. Jay Robert Nash, *Hustlers and Con Men*, New York: Evans, 1976, pp. 20, 256–265; David W. Maurer, *The Big Con*, Bobbs-Merrill, 1940.
23. John Scarne, *Scarne's Complete Guide to Gambling*, New York: Simon and Schuster, 1961, p. 72.
24. John P. Campbell, "Roger Hesketh and the de Guingand Letter," *Intelligence and National Security*, Vol.15, No.4, Winter 2000, pp. 131–142.
25. For the recent and still influential military concept of the electronic battlefield see, for example, Colonel William V. Kennedy, editor, *Intelligence Warfare*, New York: Crescent Books, 1983, pp. 76–95.

- Military: Liddell Hart (1929).

This theory is the only one given in this paper that originated with soldiers. It is the only one that is widely known to soldiers, particularly in present-day China and, I presume, North Korea. However it deserves to be closely reexamined by all deception planners and deception analysts because it is a key part of deception theory and practice. Lip service is not enough.

Direction, Misdirection, and Indirection

> *Perhaps the most elementary principle of all deception is to attract the enemy's attention to what you wish him to see and distract his attention from what you do not wish him to see. It is by these methods that the skilful conjurer obtains his results.*
> —Field-Marshal Sir Archibald Wavell, *Ruses and Stratagems of War* (1942)

The magician's key concept of *misdirection* works hand-in-glove with direction. For any trick to succeed, the spectator's attention must be directed toward the intended effect and misdirected away from the secret method. The magician directs and misdirects by both words (patter) and body language.

The magician may, for example, patter about the hand being quicker than the eye to induce his audience to look for sudden movement while, with slow stealth, he makes his move right under their eyes. Friar Roger Bacon, the 13th Century English philosopher, stands first in a long line of uninformed observers to perpetuate this most common of all the false magical theories when he wrote of "persons who create illusions by the rapidity of the movements of their hands."

The magician may also use body language to both direct and misdirect. This is neatly illustrated by the oft-quoted maxim of the Scottish master of sleight-of-hand, John Ramsay:

> If you want somebody to look at something, look at it yourself.
> If you want somebody to look at you, look at them.

Ramsay's first sentence is a non-verbal technique for direction, one also used by the practical joker who stands and stares up at nothing until a crowd gathers, peering to spot the mysterious thing while the trickster vanishes. His second sentence is a non-verbal method for misdirection, one that even few magicians can resist. As Ramsay joked, "I work alone, but I have an invisible assistant—Miss Direction."

Misdirection sets the scene for deception. If the misdirection is detected, the deception will fail. As applied to military theory I believe this key concept should be credited to English military theorist Captain Sir B. H. Liddell Hart.[26] Admittedly inspired by Sun Tzu, he first proposed his notion of an indirect approach strategy in 1929 in his *The Decisive Wars of History* and repeatedly enlarged on it until 1954 in his *Strategy: The Indirect Approach*. Unfortunately, Liddell Hart's critics seized on his somewhat ambiguous word indirection to grossly over-simplify his original insight and dismiss his theory as mere circular reasoning by twisting his argument to mean that only successful

26. As I first argued in Whaley, *Stratagem* (1969), 116-121.

military operations get labeled indirect. Even his two greatest champions, Col. T. E. Lawrence and Gen. Archibald Wavell complained privately to him about this circularity—a remarkable fact given their own strong stratagemic bent.[27] Liddell Hart's theory might have fared better had he stressed the words direction and misdirection and deception and surprise and spelled out the logical and causative connections within each pair.

In his more or less systematic survey of 280 campaigns in 30 European wars during the period from 490 B.C. to 1913 A.D., Liddell Hart found that the strategically "direct approach" was the usual one, while a purposefully indirect approach involving deception was the rare exception. Moreover, he discovered a direct relationship between the degree of military success and the degree of strategic indirectness. Thus of the 32 campaigns that produced decisive victories, only 6 succeeded by a direct approach while the 26 others (81 percent) were credited by him to use of an indirect strategy.[28]

My own close study of 216 major battles in 21 wars from 1914 through 1973 confirmed and extends Liddell Hart. I found 33 victories that greatly exceeded their planners' expectations. Of these, only 5 involved a direct approach while the 28 others (85 percent) involved at least some calculated effort to deceive. Furthermore, success correlates with surprise through deception. That is, the more sophisticated the stratagems, the better the chance for major surprise (as perceived by the victim) and, consequently, the greater the degree of success (as measured by the deceiver's expectations).[29]

In conclusion, *The Theory of Indirect Approach* applies to both deception and counterdeception in all disciplines studied.

27. Wavell and Lawrence to Liddell Hart, 15 March 1934, *Liddell Hart Papers*, States House, Medmenham, England.
28. B.H. Liddell Hart, *Strategy: The Indirect Approach*, London: Faber and Faber, 1954, pp. 161–162. Originally B.H. Liddell Hart, *The Decisive Wars of History: A Study in Strategy*, London: Bell, 1929, pp. 141–143, studied only 240 campaigns in 27 wars during that same period, crediting 26 as decisive.
29. Whaley, *Stratagem* (1969), pp. 133–139, 179–192, plus additional cases. My analytical categories, while not entirely matched to those of Liddell Hart, are on inspection of the specific cases generally comparable.

Section XI:
Naval Warfare

CHAPTER 45

How to Cheat on Naval Tonnage Treaty Limits 1919–1939[1]

Donald McLachlan

Donald McLachlan was born in Islington, Scotland, in 1898. He was a journalist, author and the founding editor of "The Sunday Telegraph." McLachlan was educated at Magdalen College in Oxford. During the Second World War, he initially served in the British Army's Intelligence Corps, but was transferred to the Naval Intelligence Division by Admiral John Henry Godfrey. In 1941, he became head of the Naval Propaganda sub-section NID 17Z, which focused on propaganda efforts against the Kriegsmarine. He served in this capacity for the remainder of the war and reached the rank of commander. Among his colleagues was Ian Fleming, who would go on to create James Bond.

Many men conspired, if not always deliberately, and many events linked up to induce the Admiralty to be content with the 14-in. gun *King George V* in face of the 15-in. gun *Bismarck* with its greater size, endurance and protection. The error, as it turned out, was not fatal—but it might have been, as the official account of the chase of the *Bismarck* in 1941 shows. Wishful thinking, that ever lurking temptation for politicians dealing with military affairs - and for serving officers involved in politics—is even more conspicuous in this episode than in that of the U-boat sinkings.

The story begins with the Anglo-German Naval Treaty of June 1936 which released the Germans from the limits set to their navy by the Treaty of Versailles. As this had been repudiated by Hitler in March 1935, it was thought wise in London to start afresh with the new regime in the hope that the German Navy could be kept to an agreed fixed percentage of the British. So the Admiralty argued. Considering the attitude towards Britain then uppermost in Hitler's mind, the calculation was not unreasonable; for it is now known that he was not planning an early or direct challenge to our naval predominance in Europe and hoped to avoid war with Britain.

The weakness of the arrangement was that Britain had already been restricted in its naval construction by an agreement to which the Germans were not parties when the *Bismarck* was laid down—the Washington Naval Agreement and the later London conference which had limited battleships to 35,000 tons. If the Germans were to build bigger ships, so that two or three or more British major units would be needed to pursue and destroy one of theirs, then the principle of the treaty—which was that German strength should never exceed thirty-five per cent of the British—would lose its value. It was doubtless on that calculation that Admiral Raeder ordered that the *Bismarck* and *Tirpitz* should both be of 45,000 tons and deliberately cheated on his Government's undertakings.

1. Donald McLachlan, "How to Cheat on Naval Tonnage Treaty Limits, 1919–1939," Room 39, Atheneum: New York, 1968, pp. 134–142.

How precisely was the game played and at what points were the British outwitted? On 1 July 1936 the German Embassy informed the Foreign Office in confidence that the particulars of battleship F would be 35,000 tons, beam 36 metres (118 ft), calibre of largest gun 38 cm. (14.8 in.), mean draft 7.9 metres (26 ft). The Director of Naval Construction is on record as commenting (5 September) some weeks later that the large beam (15 in. more than that planned for the *King George V)* was presumably dictated by the comparatively shallow draft of the ship, which in turn was made necessary by the shallow water in the Kiel Canal and the Baltic. The same month Plans Division in London, which had been closely involved in negotiating—and therefore believed in—the various treaties limiting naval armaments, made their comment. Captain Tom Phillips, later to be Churchill's first Vice-Chief of Naval Staff, remarked: 'The present design of German capital ships appears to show that Germany is looking towards the Baltic with its shallow approaches more than in the past.' In other words, they felt, these battleships might be regarded as aimed more at the Russians than at ourselves. Yet Phillips was by nature a sceptical, aggressive man.

In the German section of NID, however, there were suspicions of the German figures, partly inspired by secret reports that they were inaccurate, partly by a reasonable assumption that Nazis might lie. But at a meeting to discuss the *Bismarck* in early 1937 a split in opinion was revealed between on the one side the technicians, represented by the Director of Naval Construction and the NID technical section, and on the other side the non-technical officers. The Director of Naval Intelligence, with his advisers divided, was in no position to be tenacious for scepticism against optimisim; and later that year on 5 August DNC wrote:

> The figures given are not in themselves sufficient to warrant the conclusion that the standard displacement of 35,000 tons is being purposely exceeded . . . From our experience with Gorizia it is probable that in these countries the designers are not pressed to be meticulously accurate. To put this into more precise terms, taking into account displacement variations of completed British ships before displacement by treaty was laid down, the designers of the 35,000-ton ship would not be blamed if the displacement on completion was 36,000 tons, and such a figure would be more in conformity with the length and beam reported than 35,000 tons.

This reference to the *Gorizia* should have alerted the Board to the risks of optimism about *Bismarck*. For physical examination of the Italian cruiser while in dock at Gibraltar had shown that she had an armoured belt and exceeded the permissible treaty 10,000 tons by between two and four thousand tons. There was no reason to believe that Hitler's capital ship policy would be notably more honest than Mussolini's. Yet the Director of Plans allowed himself to write: 'Our principal safeguard against such an infraction of treaty obligations lies in the good faith of the signatories'.

Who is to blame Phillips for this attitude when the Whitehall line, set by the politicians and by what they believed to be the mood of the voters, was to hope for the best and give the dictators a little more rope to hang themselves? Plans Division's job was not to argue against the treaties, but to make the best plans they could within their limits.

Is the Director of Naval Construction's department entitled to that particular defence? It appears that the mystery of the *Bismarck*'s shallow draft was not investigated as thoroughly as it should have been by those who were designing and building the equivalent British battleships. On the other hand they were hard pressed with the new programme and deeper probing of the German design would almost certainly not have led them to a change in our own designs which would mean breaking treaty limitations. When, in 1941, information from the survivors of the sunk battleship and from the Russian Admiralty confirmed NID's suspicions that *Bismarck*'s tonnage was nearer 45,000 than 35,000 tons, it became clear, even to DNC, that the 'mystery' of her shallow draught was a mystery no more. The *Bismarck*'s draught was in fact not very different from that of the British capital ships and its shallowness was simply a device to give colour to the gross understatement of her displacement.

What irked Godfrey at the time was the readiness of his colleague at Bath to believe immediately from the Russians something that he would not accept from NID.

From the German naval documents we know what was going on behind the scenes in Berlin while the Admiralty hoped and guessed. On 11 February 1937 the Plans Division of the German Admiralty wrote to Admiral Raeder:

> In view of the difficult conditions in our harbours, a displacement of 42,000 tons should be the limit of ships to be built, unless we widen the harbours and channels and also the curves of the Kiel Canal. . . . Politically it must be decided if the displacement should be still further increased above the treaty limits.
>
> The Constructional Office considers that no further increase should take place as long as we are tied to the 35,000-ton limit which we have already exceeded by 7,000 tons, so that a further increase could hardly be concealed . . . The Constructional Office would like to build an extra ship rather than exceed the 35,000-ton limit.
>
> Even if other Navies renounce the limits of the London Agreement or go in for calibres above 38 cm . . . we should be reluctant to increase them owing to the state of our channels and our Treaty obligations . . . a decision cannot be reached till the various aspects including the military factors and the clauses of the Treaty have been considered.

Here, clearly, someone was trying to be honest; but the argument which counted in his mind was the technical rather than the moral one. Indeed, the morals of the matter were not his business.

For German Plans Division it was a different matter. They stated in a memorandum to the Chief of Staff dated 18 February 1938 that the *Bismarck*'s true displacement was twenty per cent greater than that announced to the British; but they argued that it would be wrong to indicate a greater tonnage than had already been announced to Britain, Russia and Japan because 'we shall be accused of starting an armament race'.

What, then, was the British Naval Attache doing? Was he too misled and if so, by what methods? Were the suspicions of NID due to anything reported by him? On 23 December 1936 the yearly report of Captain Troubridge, Naval Attache in Berlin, to the Foreign Office (copies to DNI and C-in-C Home Fleet) said:

> The Anglo-German naval agreement was one of the master strokes of policy which have characterized Germany's dealings with her ex-enemies since the war. When the time is ripe, as history shows, it will unquestionably go the same way as other agreements: but the time is not yet.

The italic passage was omitted from the printed version of the Ambassador's annual report (Sir E. Phipps to Mr. Eden dated 12 January 1937 C357/357/18) and the Director of Plans already mentioned in this chapter, Captain Phillips, wrote a minute on 1 January 1937 on the NID docket sent to him with Troubridge's unexpurgated report:

'It would be of particular interest to know the basis for the very definite statement on page 2 of the memorandum.'

This general expression of scepticism seems to be as far as Troubridge got and there is no doubt that the intelligence coming to the Admiralty from Germany was thin and poor. Why that should have been so is not clear; the Nazis after 1934 never tried to conceal their successful rearmament and were always anxious to show what they could do. For example, the French Air Attache, Stehlin, was given the fullest opportunity to see what the Luftwaffe would be able to do to France (see Chapter 15). It seems likely that the Secret Service was not specially pressed to find out the state of German naval building just because of the official optimism about the working of the treaty.

Ten years later, Troubridge was asked by Godfrey if he could explain how we had been so misled and he wrote as follows:

- I fancy that generally speaking the reason that we were 'fooled' by the Germans, both as regards submarines and the battleship tonnage, was that we were inclined at the time (after the Anglo- German agreement) to trust them in these matters.
- I confess that Raeder's earnestness and apparent sincerity, assuring me that Germany meant to adhere strictly to the agreement, may well have influenced DNI when my reports of these interviews came through. I started my time pretty well open-minded, though having studied Great Britain and the German Navy by Woodward, I was pretty well on my guard.
- As evidence of my state of mind in 1936 I mentioned in the NA's annual report, which is annexed to the annual report of the Ambassador, something to the effect that the Anglo-German naval agreement remains the cornerstone of German naval policy. It will unquestionably be torn up when the moment seems propitious, but the time is not yet.

The boys in the Chancery refused to put this paragraph in, but it will remain on the record none the less in the copy sent to DNI and Commander in Chief Home Fleet.

Doubt about the tonnage of the *Bismarck* came shortly after the launching, when information from Williams, the vice-consul in Hamburg, indicated that she was drawing a good deal more than she should.

That Raeder lied to me and also to my predecessor is undeniable. Muirhead-Gould [Naval Attache Berlin before Troubridge] always said he had, but I fear I was lulled to a certain extent by his apparent sincerity. Moreover I found it hard to understand why he found it necessary. But then to understand is to understand the tortuous workings of the Teutonic mind. Sometimes they lie just for the sake of lying.

It will be interesting to see, when the records are opened, whether historians can find out the reason for this significant excision by Sir Eric Phipps from his Naval Attache's despatch. The fairest presumption is that he thought his subordinate should stick to naval facts and abstain from political generalization.

The lies, the realities and the contrast between them are most easily grasped in tabular form on p. 141.

The mistakes and omissions here recorded were made for the most part in the prewar NID which faced difficulties of getting information and staff which have been described elsewhere. But there is this to be said in their defence. The Naval Staff in peacetime were strongly disinclined to accept from their intelligence advisers any views or facts that might be 'awkward', that is to say, in sharp conflict with current strategic doctrine or political appreciation. They probably felt that if Germany was in fact cheating and could be proved treacherous, no one in the Foreign Office or Downing Street would handle the accusation effectively; why then make it?

But there seems too to have been reluctance among our own technical experts to believe that the *Bismarck* could combine the high speed, long endurance and good protection with the mounting of 15-inch guns; just as there was reluctance to believe that U-boats could dive as deep as they did dive, or that underwater speed could be increased to the degree that it was by 1944. Perhaps the British underrated the tremendous advantages of starting again, as the Germans did, from scratch and were inhibited by too many decisions already taken and too many economies rigidly enforced. Be that as it may, they were disinclined to listen to uncomfortable truths of intelligence. Godfrey himself looks back on this episode as illustrating three lessons for the craft:

1. The unwillingness of authority to believe information that has awkward political implications.
2. The tendency of naval officers and others who have taken part in negotiations to become advocates of the integrity of the persons with whom they secured agreement, and to lose the scepticism which is part of vigilance.
3. Our technicians may not be the best judges of enemy intentions and achievement. They find it hard sometimes to believe that what they cannot do or have not thought of doing has been done by the other side.

Facts about the Bismarck

Authority	*German Embassy Official Statement*	*Bismarck survivors and notebook*	*Russian Admiralty*	*German Records*	*King George V*
	Apr. 39	Oct. 41	Oct. 42	Oct. 44	
Displacement	35,000*	41,150*	45,000	53,000† 45,000*	35,000
Length	792	792	792	792	745
Beam	118	118	118	118	103
Draft	26	33 ¾	34	34	34
H.P.	80,000	150,000	150,000	150,000	111,000
Armament	8-15"	8-15"	8-15"	8-15"	10-14"
Thickest armour	9"	13"	13"	12½	15"
Full speed	27	30¼	30¼	30¼	28¼
Endurance	?	?	17,500 miles at 18kn	17,500 miles at 18kn	14,500 miles at 10kn
Fuel stowage	?	?	8,500 tons	8,500 tons	3,860 tons

*Standard displacement.

† Full load.

Definitions:

Full load displacement with all oil stores, ammunitions, etc., on board vessel fully equipped and ready for sea.

Standard displacement—as for full load, but excluding all fuel and reserve feed water.

Normal displacement—figures may differ in different countries but generally cover the condition of vessel fully equipped and ready for sea but with a portion only of stores, fuel and reserve feed water.

CHAPTER 46

Harwood and the First Deception of WWII, 1939[1]

A Book Review by Frank Stech

Eric Grove's superbly researched and written "The Price of Disobedience" recounts the first major naval battle of World War Two. This British victory hinged on the first major deception of the war.

The German Panzerschiff (pocket battleships) Deutschland and Graf Spee were at sea when war broke out in September 1939. Deutschland sank two ships in the North Atlantic, then was ordered to return to German waters. Graf Spee headed to Cape Horn and began sinking British shipping in both the South Pacific and South Atlantic. By December 1939, Graf Spee had sunk nine British ships.

Captain Harry Harwood, commanding the British cruisers Exeter, Ajax, and Achilles, used some remarkable deduction and found Graf Spee off Uruguay. His ships engaged on 13 December 1939. Exeter, severely damaged, was sent back to the Falklands, two of Ajax's four turrets were disabled, and the magazines of Harwood's squadron were down to a seventh of their ammunition.

Graf Spee, hit over twenty times, was seriously damaged and her captain, Hans Langsdorff, ran her up the neutral River Plate to Montevideo, where Langsdorff hoped to make repairs. Harwood summoned the cruiser Cumberland to join the squadron at the mouth of the Plate, while the Admiralty ordered the battleship Renown, the carrier Ark Royal, and an escort of cruisers and destroyers to sortie from Cape Town to reinforce Harwood.

Churchill, as First Lord, orchestrated the British and neutral press coverage of the engagement to deceive Lansdorff and the Kriegsmarine into believing Graf Spee was bottled up, by not just Harwood's hors de combat squadron, but the Renown battle group, soon to be joined by the French battleship, Dunkerque. The New York Times' front-page headlines claimed a "reinforced allied fleet" awaited the Graf Spee. Rumors were circulated and printed that Renown and Ark Royal had refueled in Rio de Janeiro (when the British ships were 2,500 miles away). In German-leaning Argentina, the two largest newspapers quoted "reliable sources" that "more than five cruisers were waiting" for Graf Spee.

Uruguay ("England's butcher") heavily favored British interests and, refusing Langsdorff's pleas for repair and resupply, Uruguay ordered the Graf Spee to sail under the neutrality rules, i.e., in forty-eight hours. The British, however, wished Graf Spee to remain in Montevideo until the Renown battle group had reinforced Harwood. The British naval attaché ordered a British merchantman to depart Montevideo. Under international law, Graf Spee could not follow an adversary's departure for 24 hours. The British attaché stretched out British sailings over several days (until the Uruguayans

1. Frank Stech, "Harwood and the First Deception of WWII, 1939," Amazon review March 18, 2004. http://www.amazon.com/review/R35UPIBTP5430Q. Review of Eric Grove, *The Price of Disobedience,* Annapolis: US Naval Institute Press, 2001.

caught onto his ruse) and used clever diplomacy to trap Langsdorff and the Graf Spee in Montevideo.

Meanwhile, Harwood skillfully maneuvered in the approaches to the Plate and signaled to Cumberland as if she were the Renown battle group, while the BBC broadcast an ongoing account of the mythical fleet waiting the trapped German Panzerschiff. Masthead spotters on Graf Spee reported the Cumberland's rigging as Renown, which the German attaché in Buenos Aires "confirmed," having fallen for the deceptive refueling story. The BBC vividly reported Ark Royal (still far distant) joining Harwood, and the German spotters on Graf Spee, expecting to see the British carrier, did. The BBC reported Dunkerque and a second British battleship would soon join Ark Royal and Renown.

Langsdorff believed the Graf Spee was doomed. Hitler ordered Langsdorff not to allow the ship to be interned, and to breakout or scuttle. Langsdorff scuttled her. When he and the German crew arrived in Buenos Aires, Langsdorff learned the Ark Royal and Renown had just reached Rio, over 1,000 miles from where Graf Spee lay scuttled. Knowing Hitler's likely reactions to his obedience, Langsdorff wrapped himself in the Kriegsmarine's ensign and shot himself. Pictures of the burning scuttled Graf Spee circled the world. So did news of Langsdorff's suicide. The Battle of the River Plate, the first great naval victory for the Allies, resulted from highly skilled and coordinated naval deception.

Section XII:
Air War

CHAPTER 47

MiG Sweep—Operation BOLO[1]

Colonel Walter J. Boyne

Colonel Walter J. Boyne received his wings as a Second Lieutenant in the United States Air Force in 1952. Boyne flew the B-50 Superfortress as a member of the 330th Bomb Squadron of the 93rd Bomb Wing and was then trained to fly B-47 Stratojet. In 1957, Boyne returned to school earning a Master's degree from the University of Pittsburgh. Boyne then returned to flying as a test pilot with the 4925th Nuclear Test Group. Upon his retirement in 1974, he had a prolific writing career, publishing over 50 books and 1,000 magazine articles.

1. Colonel Walter J. Boyne, "MiG Sweep – Operation BOLO," *Air Force Magazine,* Vol. 81, No. 11, November 1998, pp. 46–51. Reprinted by permission of the Air Force Association.

The North Vietnamese thought they were attacking bomb-laden F-105s. What they ran into was Robin Olds and the Wolfpack, flying Phantom F-4s.

MiG Sweep

On Jan. 2, 1967, with aircraft losses in Southeast Asia on the rise, the United States Air Force resorted to an elaborate combat sting. The mission, called Operation Bolo, constituted an electronic Trojan Horse concealing the hard-hitting F-4 Phantoms of USAF's 8th Tactical Fighter Wing within a radiated image that simulated bomb-laden F-105 Thunderchiefs.

Despite adverse weather and a few surprises, the "MiG Sweep" did what it was designed to do: trick the increasingly elusive MiG-21s of North Vietnam into engaging F-4s rigged for aerial combat.

Until the latter part of 1966, MiG aircraft had not been as great a threat to USAF strike forces as the Surface-to-Air Missiles and anti-aircraft fire. Ironically enough, the introduction of the QRC-160 (ALQ-71) electronics countermeasures pod on the F-105s changed this. The QRC-160 was effective in neutralizing the radar controlling the SAMs and flak, and the resilient North Vietnamese responded by increasing their use of MiG fighters to prey on vulnerable F-105s configured for bombing.

Operating under ground control, and making maximum use of both cloud cover and the almost benevolent American rules of engagement, the enemy aircraft were adroitly employed. The MiGs, especially the later model MiG-21s armed with heat-seeking missiles, sought to attack the strike flights and make them jettison their bomb loads prior to reaching the target areas. Their mission was fulfilled if the Thuds were forced to drop their bombs prematurely, but they tried to score kills wherever possible.

The air war in Southeast Asia, while unique in many respects, harkened back to earlier conflicts in terms of the relative missions, forces, and equipment. As in World War II and Korea, the mission of US forces was to obtain air superiority, destroy the enemy air forces, and conduct long-range bombing operations. The mission of the enemy forces was to defend their most important targets by choosing to engage the American bombers on a selective basis.

Thud, Phantom, Thud

There were other parallels. To achieve the air superiority mission, the American fighters had to have a long-range capability and still be able to defeat the enemy fighters over their own territory. What the Mustangs and Sabres did in their wars, the F-4 Phantom II was required to do in Southeast Asia. Flights of F-4s, carrying a mixed ordnance load of bombs and missiles, would be sandwiched in between Thud flights at four- or five-minute intervals. If the F-105s in front or behind were attacked, the F-4s would drop their bombs and try to engage. If they were not, the F-4s would drop bombs right along with the Thuds.

A final, tragic parallel is the price paid to execute the missions that were often laid on for statistical rather than tactical reasons. Flying Phantoms or Thuds was dangerous work. As a single example, by late 1967, more than 325 F-105s had been lost over North Vietnam, most to SAMs and anti-aircraft fire.

The North Vietnamese air force consisted of slow but heavily armed and maneuverable MiG-17s and a handful of modern delta-wing MiG-

Col. Robin Olds came up with the strategy of luring North Vietnam's MiG-21s into battle with F-4s that were masquerading as the more vulnerable F-105s. In the painting "MiG Sweep," at right, aviation artist Keith Ferris portrayed a successful encounter in Operation Bolo, as Olds and his backseater rolled out of the vertical and acquired a MiG-21 above the cloud deck.

By Walter J. Boyne

FP 680

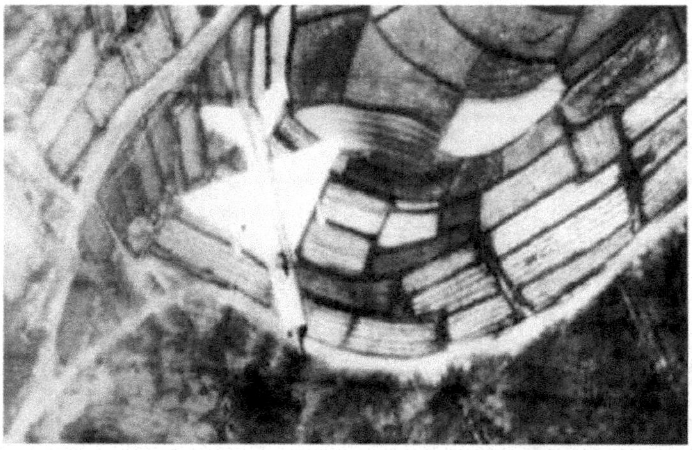

The MiG-21 was about half the size of the F-4 and a formidable opponent. Only 16 were estimated to be in the theater, and Operation Bolo aimed to either destroy as many of them as possible or run them out of fuel.

21s. The MiG-17s were semiobsolete but still effective in their defensive role. (The MiG-19 did not enter service with the North Vietnamese air force until February 1969.)

The MiG-21 Fishbed was roughly half the size of the Phantom and was designed as a high-speed, limited all-weather interceptor. It could carry two cannons and two Atoll infrared homing air-to-air missiles which had been developed from the US AIM-9B Sidewinder. At altitude, the MiG-21 could outfly the F-4 in almost all flight regimes. It had spectacular acceleration and turning capability. At lower altitudes, the F-4s used their colossal energy in vertical maneuvers that offset the MiGs' turning capability, for they lost energy quickly in turns at low altitudes. The MiG-21s were operated under tight ground control. They typically sought to stalk American formations from the rear, firing a missile and then disengaging. If engaged, however, its small size and tight turning ability made the MiG-21 a formidable opponent in a dogfight.

The Phantom had been intended originally to be a fleet defense aircraft, but it proved to be versatile in many roles, including reconnaissance, Fast Forward Air Control, Wild Weasel, bombing, and air superiority. The F-4Cs were armed only with missiles, although gun pods could be fitted.

The air war in Southeast Asia had grown progressively intense, and Dec. 2, 1966, became known as "Black Friday" when the Air Force lost five aircraft and the Navy three to SAMs or anti-aircraft fire. Air Force losses included three F-4Cs, one RF-4C, and an F-105. The Navy lost one F-4B and two Douglas A-4C Skyhawks.

These ground-fire losses were accompanied by the marked increase in MiG activity during the last quarter of 1966. Because the rules of engagement prohibited airfield attacks, the men of the 8th Tactical Fighter Wing were determined to blunt the enemy's efforts by luring the MiGs into air-to-air combat and then destroying them.

The reluctance of the MiG-21s to engage did not mean that the North Vietnamese pilots were lacking in either courage or skill. At the time, the US estimated that there were only 16 MiG-21s in the theater, and the enemy had to employ them selectively to maximize their utility.

The New Boy

Brig. Gen. Robin Olds, USAF (Ret.), recalls himself as the proverbial "new boy on the block" with the 8th TFW, as yet unproven in the jet air war of Southeast Asia. When he arrived at Ubon RTAB, Thailand, as a colonel, to assume command of the Wolfpack on Sept. 30, 1966, Olds, who was 44 and stood six feet two, struck some as more the Hollywood concept of a combat commander than an Air Force regular officer. This was, at least in part, because he was married not to the girl next door but to film star Ella Raines.

Son of Maj. Gen. Robert Olds, one of the most influential generals in the Army Air Corps, the new commander of the 8th was a World War II ace. Olds would later remark that he never flew one mission over Germany that was as tough as any mission over Hanoi.

Olds' war-ace status was marred somewhat by a reputation for being a maverick. Olds had often argued forcefully against contemporary Air Force training. He was an outspoken advocate of intensive training in the arts of war he learned in Europe. Unable to wangle his way into the Korean conflict, he had continued to press for training in strafing, dive-bombing, and other conventional warfare techniques at a time when US fighters were being adapted to carry nuclear weapons and fight a nuclear war. His advice, though not well received, was a realistic forecast of what would be required for war in Southeast Asia.

Olds knew he would have to prove himself to the combat-hardened veterans of the 8th as a leader in their war. He wished to use his past beliefs in a plan that would confirm his present status. He had first presented his idea for a MiG ambush to Gen. Hunter Harris Jr., Pacific Air Forces commander. Harris ignored him. Olds next went to the commander of 7th Air Force, Gen. William W. "Spike" Momyer. It was in early December 1966, at a cocktail party in the Philippines, that Olds edged next to Momyer. After a few polite remarks, Olds said, "Sir, the MiGs are getting pesky" and went on to describe ways to bring them to battle. Momyer's expression of deep disinterest didn't change. He moved away, leaving Olds with the uncomfortable impression that he had blown a good opportunity.

He Listened

However, Momyer had listened after all, and a week after their conversation, Olds was called to Saigon to discuss the concept of tricking the MiGs into combat. Momyer told Olds to develop a plan, one that specifically excluded attacks on North Vietnamese airfields for political reasons.

By Dec. 13, Olds was working closely with four top veterans of the 8th, striving to develop his idea. In brief, the concept called for F-4s to simulate F-105s, and Olds gave his planners specific guidelines to work by. Central to the concept was that, while no North Vietnamese air fields could be attacked, the MiGs would be prevented from landing; flights of Phantoms would orbit above the airfields, cutting off MiG escape routes to China. Olds hoped either to engage the MiGs in combat and destroy them or to simply run them out of fuel by denying them access to their airfields.

The planning group included Capt. John B. Stone, Lt. Joe Hicks, Lt. Ralph F. Wetterhahn, and Maj. James D. Covington, a wing staff officer. They worked under the tightest security; those aircrews that would fly the missions were themselves not briefed until Dec. 30.

It was a perfect combination—Olds providing the overview and the major decision elements, and the younger officers, more experienced in the theater, breathing life into a concept. The team worked long hours to develop key details on force structure, refueling points, and altitudes, ingress and egress routes, radio communications, flak suppression, electronic countermeasures, and all the other details the mission required.

The planners determined that, if the MiGs engaged in combat, their endurance from takeoff to landing would extend only for about 55 minutes. F-4 flight arrival times were set five minutes apart to ensure maximum opportunities for engagement. The group planned for a concerted strike by a "west force" of seven flights of F-4Cs from the 8th at Ubon and an "east force" made up of five flights of F-4Cs from the 366th TFW at Da Nang AB, South Vietnam.

Everything hinged on getting the MiGs airborne, where they could be destroyed. Luring the MiGs into battle would not be easy, for the communists often declined to attack if they thought the weather would seriously impair the bombing accuracy of US attacking aircraft. The North Vietnamese had many advantages. All of the targets were in the midst of the most heavily integrated air defense system then in existence. Their geography and the onerous rules of engagement under which American forces operated had severely reduced the F-105s' options in Rolling Thunder missions. The number of approach routes was limited, as were the targets permitted to be attacked.

The Pod Deception

Olds took these factors into account and called for a plan that depended upon a basic deception. The strike force would imitate the route, speed, and radio chatter of a normal F-105 mission. However, the force would comprise not bomb-laden Thuds but rather F-4Cs, each armed with four AIM-7E Sparrows and four AIM-9B Sidewinders. Maj. Gen. Donavon F. Smith, chief of the Air Force Advisory Group in Vietnam, suggested the Phantoms carry the QRC-160 electronic countermeasures pod that the Thuds had been carrying.

Simply acquiring the necessary QRC-160 pods was a logistic effort that extended all over Southeast Asia and all the way back to the United States. It was the first of a series of events that engaged many disparate elements of the Air Force.

Also at play was another factor, one that Olds hoped would be the key factor in success. The first three flights entering the combat area would have "missile free" firing options. For a few precious minutes, the Americans would know exactly where all friendly aircraft were. Any other aircraft could be assumed to be hostile and be fired upon without visual identification. This gave many advantages, including surprise, isolation from counterfire, and, most of all, time to let the missile do what it was designed to do under the most favorable conditions, without excessive g forces to trouble the missile systems.

On Dec. 22, Olds briefed Momyer in Saigon. The commanding general accepted the plan without a change. Execution was set for Jan. 2, 1967. The force would contain 96 fighters—56 F-4Cs, 24 F-105s, and 16 F-104s. The force also would include KC-135 tankers, EB-66s electronic countermeasure–support aircraft, EC-121 Big Eye surveillance aircraft, and rescue forces.

Eight days after briefing Momyer, Olds canceled all leaves at the 8th TFW and postponed the New Year's Eve party. Then, bad weather moved in, and it was obvious that the mission would not be flown on Jan. 1. Most thought it probably would not occur on Jan. 2, either. The party was reinstated for the evening of Jan. 1—a mistake, for soon the mission was reset for the morning of Jan. 2. Olds agreed to go forward, despite the probability of bad weather, because the QRC-160 pods were "on loan" to him for only seven days.

Normally, the computers at 7th Air Force developed the code words assigned to flights, targets, and routes. Because timing was so critical, however, code terms for Operation Bolo were carefully picked. The Wolfpack flights were given the names of cars, with mission commander Olds leading Olds Flight. (Olds was dismayed by this; he felt that the flights should

MiGs armed with heat-seeking missiles had been attacking strike flights of bomb-laden F-105s and also forcing the Thuds to jettison bombs prematurely. Operation Bolo F-4s mimicked the route, speed, and radio chatter of an F-105 mission.

have been given names similar to those used by the F-105 flights. In his pre-mission briefing he told his pilots to use first names for their radio calls.) MiG base locations were identified by the names of US cities. Phuc Yen, northwest of Hanoi, was called "Frisco," while Gia Lam, south of Phuc Yen, was "Los Angeles."

Distillation

It had required a massive Air Force-wide effort to bring Bolo into being. The entire 8th TFW's energy was thrown into overcoming last minute problems, with the support troops working all night long. (A typical glitch involved the sway braces on the F-4C. They were located differently than on the F-105, and the shell of the QRC-160 pod had to be reinforced in order to fit well.) However, as the aircraft rolled for takeoff, the long days of nonstop planning, the assembly of resources, the intense training of munitions crews, crew chiefs, pilots, and backseaters now began to condense into a 13-minute dogfight. The historic battle would be fought in a slice of sky that ranged from 10,000 to 18,000 feet in altitude and within a 15-mile radius of Phuc Yen airfield.

Olds carefully emulated the F-105 flight profile, flying a fluid-four formation at 480 knots until reaching the Red River. At that point, he accelerated to 540 knots and assumed the QRC-160 pod formation. This was similar to the standard fluid four but with a separation of about 1,500 feet. The aircraft would weave up and down, and the combined effect of the pods was to jam the enemy acquisition radar.

The force maintained this Thud feint for a full three minutes after the Olds Flight arrived at its target. By that time, Olds expected the North Vietnamese to have realized what they were dealing with. Olds arrived over Phuc Yen at 1400 Zulu, exactly on schedule, but he was disconcerted to find that the MiGs were not airborne. There was a complete undercast, with tops at about 7,000 feet, and the communist ground controllers had delayed the MiG takeoffs by about 15 minutes. Olds had no way of knowing this and had to contemplate calling the mission off for the inbound flights.

He passed over Phuc Yen airfield to the southeast and then made a 180-degree turn to the northwest. The first sign of enemy activity proved sterile as Olds 3 picked up and then lost a bogie moving swiftly in the opposite direction. Knowing that Ford Flight, led by his longtime friend Col. Daniel "Chappie" James Jr., was due over the target, Olds now canceled the missile-free option and made another 180-degree turn.

Ford Flight burst into the battle area exactly on time and simultaneously with the first appearance of MiG-21s popping up out of the undercast. Ford 1 called out a MiG-21 closing on Olds Flight. Olds turned to throw off the MiG's aim and attacked another MiG that appeared in his 11 o'clock position, low and a little over a mile away.

This version of the F-4 participated in Operation Bolo. Note its inboard pylon ECM pod. The F-4's initial lack of an internal gun put it at a disadvantage in a close-in fight, but this one's two red stars indicate success over enemy MiGs.

First Trip

It was Olds' first trip to the Hanoi area, and his first engagement with a MiG. With his backseater, Lt. Charles Clifton, he set up for a Sparrow attack as he closed to get positive identification. When he saw the silver delta shape of the MiG he fired two Sparrows and a Sidewinder—but none of them guided. Olds sighted another MiG—they were appearing everywhere now—and used the Phantom's power and energy to vector roll behind it. This time he fired two Sidewinders and the first one made impact, blowing the MiG-21's right wing off and scoring the first of the MiG kills. The pilot did not eject.

Wetterhahn, one of the key planners, had been disappointed to be flying as Olds 2, but in the course of Olds' attack he was able to slide behind a MiG-21. Working with his GIB (the Guy In Back), 1st Lt. Jerry K. Sharp, he salvoed two Sparrows. They lost sight of the first one, but the second Sparrow caught the MiG just forward of its stabilizer and blew it up. Two down.

Olds 4, flown by Capt. Walter S. Radeker III, with 1st Lt. James E. Murray III in the back, saw a MiG-21 tracking Olds 3. Radeker experienced some difficulty getting a solid tone on his Sidewinder before firing, yet the missile guided perfectly, striking just forward of the MiG's tail and sending it spinning into the undercast. Three down.

The next MiG fell to Capt. Everett T. Raspberry and 1st Lt. Robert W. Western in Ford 2. Two MiGs had closed on Ford 3 and 4, overshot, then pressed an attack on Chappie James in Ford 1, overshooting him as well. The MiG broke into a hard left turn, and Raspberry rolled to wind up at the MiG's six o'clock position. He fired a Sidewinder that guided up the MiG's tailpipe, blowing it up. Four down.

Rambler Flight had arrived exactly on time, to find itself in the midst of the MiG melee. One of the most important of the planners, Stone, was the Wolfpack's tactics officer. He was flying with Lt. Clifton P. Dunnegan Jr., as the backseater. Over Phuc Yen, Stone picked up two MiGs, 4,000 feet below and two miles away. Uncertain of his lock-on, Stone fired three Sparrows. The second missile struck the MiG's wing root, and the pilot ejected. Five down.

Two young first lieutenants, Lawrence J. Glynn Jr. and Lawrence E. Cary, in Rambler 2 had been on Rambler lead's wing all through its combat maneuvers. Just after Rambler 1 scored, Glynn locked on to a MiG-21 and fired two Sparrows. The second missile hit the MiG in its wing root, the debris damaging Rambler 2 slightly. The enemy pilot ejected and Glynn saw his parachute open. Six down.

Maj. Phil Combies in Rambler 4 was flying with Lt. Lee Dutton in the backseat. After Dutton had locked on to a MiG-21, Combies tracked a fighter carefully, pulling no more than 4g's, and fired two Sparrows. He didn't observe the first missile at all but was able to track the second from launch to impact. It struck in the tail section. So swiftly did the parachute appear that Combies later speculated that the pilot must have ejected when he saw the missile coming.

That made seven MiG-21s down. It was the final confirmed victory of the day.

Combies and Dutton had latched on to a second MiG and had fired four Sidewinders. They saw the first two detonate just below the enemy's tailpipe, with the last two tracking well, but then they had to break hard right when they heard "F-4C, I don't know your call sign, but break right." The message was intended for Stone, but the break caused Rambler 4 to claim only a probable. (Maj. Herman L. Knapp in Rambler 3 also claimed a probable.)

The F-4 carried four close-range AIM-9 Sidewinder heat-seeking missiles and four radar-guided AIM-7 Sparrows like those shown here. In the opening moments of Operation Bolo combat, Olds scored his first MiG-21 kill with an AIM-9.

It's Over

Suddenly, the MiGs were gone, and the four remaining Wolfpack flights (Lincoln, Tempest, Plymouth, and Vespa) arrived to find the action was over. The 366th, out of Da Nang, had flown up the coast to a point off Haiphong, evaluated the weather, and elected not to participate in the western part of the mission. Operation Bolo was over.

Seventh Air Force was elated with the Wolfpack's results. Twelve F-4Cs had engaged 14 MiGs and shot down seven, with no losses. It is worth noting that of the 14 crew members who scored victories, only one, Glynn, had ever seen a MiG in air combat before. (Olds had seen MiGs at a distance.) The Phantom crews, despite their relative inexperience in combat and their lack of dissimilar aircraft combat training, used vertical maneuvers to put themselves in firing position.

For dogfighting, the F-4C proved clearly superior to the MiG-21, and the AIM-7E Sparrow and AIM-9B Sidewinder proved to be highly effective weapons. Only 10 Phantoms had fired their missiles. Eighteen Sparrows had been launched; of these, only nine guided, but these nailed four MiGs. Twelve Sidewinders were launched, seven guided correctly, and they destroyed three MiGs.

The QRC-160 ECM pods had apparently worked very well, although the presence of MiGs in the combat area undoubtedly inhibited both missile and anti-aircraft fire. Only five SAMs were spotted and a light burst of 85 mm anti-aircraft fire seemed to be aimed at random.

The battle proved beyond doubt the importance of the largely unsung GIB, the backseaters, who locked the radar on the target and who, despite the continuously changing g forces, kept their heads on a swivel watching out for enemy aircraft and SAMs.

Finally, the battle proved Olds to his men. He made sure that all who participated in Operation Bolo, whether in the air or on the ground, were given full credit for their contributions. The general effect of Bolo on Air Force morale was positive, in Southeast Asia and the US.

There was a postscript. The MiG force had retaliated by attacking an Air Force RF-4 reconnaissance airplane, and this inspired 7th Air Force planners to use another deception.

Two F-4Cs, fully armed, were to fly in close formation so that they would appear as a single blip. They flew a mission as a reconnaissance aircraft would on Jan. 5, without any enemy reaction. They did it again on Jan. 6 and were rewarded by being bounced by four MiGs. The F-4Cs shot down two of the North Vietnamese aircraft, meaning that nine of the 16 MiG-21s had been shot down. The MiG-21s went through a three-month stand-down, during which both sides studied the lessons of the battle. ▪

Walter J. Boyne, former director of the National Air and Space Museum in Washington, is a retired Air Force colonel and author. He has written more than 400 articles about aviation topics and 29 books, one of which is Beyond the Wild Blue: A History of the United States Air Force, 1947–1997. *His most recent article for* Air Force Magazine, *"The Easter Halt," appeared in the September 1998 issue.*

CHAPTER 48

Aerial Deception since the Second World War[1]

Jon Latimer

Jon Latimer served for sixteen years with the Royal Monmouthshire Royal Engineers, the Royal Welch Fusiliers, and the Royal Regiment of Wales, as platoon commander, assault troop leader, and battalion intelligence officer. He has been published widely in military journals, and is also the author of "Operation Compass 1940" and, with Jim Laurier, "Tobruk 1941". He lives in Wales.

Tactical deception and developments in electronic warfare made a considerable contribution to Allied success in the Second World War and saved countless aircrew lives. Perhaps more significantly, air units were formed for the first time and equipped specifically to carry out this role. As is so often the case, these lessons were quickly forgotten in the immediate aftermath of the war, but were then revived and developed, and since the war there have been phenomenal developments in both aircraft and sophisticated avionics systems. Aircraft themselves are faster and more powerful, ranging from fast agile fighters through versatile multi-role fighter-bombers to the colossal B-52 bomber, which dwarfs the Lancaster and Super Fortress of the Second World War. More significantly, aerial warfare no longer involves getting as close as possible to one's target in order to shoot directly at it with cannon or machine-guns, or to lob free-fall bombs. Instead, it relies on sophisticated missile technology, based on radar or infra-red systems. Similar systems now equip ground-based air defense units and tactics have been transformed, leading in turn to defensive flares and chaff (bundles of WINDOW) to act as decoys and draw off attacking missiles. Flying at low altitude and high speeds may put aircraft under effective radar cover, but also makes them vulnerable to ground fire.

The development of 'Wild Weasel' electronic counter-measure aircraft led to specially equipped fighter-bomber units, designed to seek out and destroy enemy radar-controlled gun and missile systems in a role called Suppression of Enemy Air Defense (SEAD). If an enemy radar is activated, it can be located and attacked with anti-radiation missiles that home in on radar emissions. Since the Vietnam War such units have formed a crucial element in securing control of the air. More recently, so-called 'stealth' technology has been designed to make aircraft radar 'invisible', using radar-absorbent materials, carefully designed shapes and 'cancellation' technology (avionics that predict an aircraft's reflective signal at a given frequency and angle in order to transmit a signal which will cancel the signature). 'Stealth' technology also includes techniques to mask

1. Jon Latimer, *Deception in War: The Art of the Bluff, The Value of Deceit, and the Most Thrilling Episodes of Cunning in Military History, from the Trojan Horse to the Gulf War,* UK: Overlook, 2003, pp. 202–204. Reprinted by permission of Overlook Press.

sound and infra-red sources such as engine exhausts. Although the primary threat was seen for a long time as being from radar-guided weapons, a US Department of Defense study showed that in the ten years to 1985 ninety per cent of tactical aircraft destroyed were victims of infra-red systems. Such technology is increasingly being applied to land and naval units while towed decoys, in use at sea since the 1940s as a method of countering acoustic torpedoes, have been adopted since the 1980s in the form of a small radar jammer towed behind an aircraft. However, counter-'stealth' techniques were soon identified, since it seems likely to continue for many years to come.

CHAPTER 49

Detection of the B-2 Bomber and a Brief History on 'Stealth'[1]

Tao Yue

This article is a handy unclassified introduction to stealth and its real and potential anti-stealth countermeasures, including bi-static radar which was undergoing development since 1997 by the British firm of Roke Manor Research.

Tao Yue, then a staff writer for the MIT student weekly periodical, took degrees in both Electrical Engineering and Computer Science and Mathematics in 2004 and Masters degrees from the University of Washington and Princeton University. He has done work in various fields of computing, including multi-tier applications, graphics, and computational biology. In 2001, he investigated the feasibility of counter-stealth technologies for The Tech, the MIT student newspaper.

Scouting for Surveillance

Detection of the B-2 Stealth Bomber and a Brief History on "Stealth"

"Cell phones uncover stealth bombers."

In early June, the news was filled with such headlines. Newspapers put them at the top of the front page, magazines printed colorful diagrams, and television networks ran the story as the lead on their evening news broadcasts.

And why not? The story was irresistible. Stealth technology is the most potent symbol of America's military supremacy in the post-Cold War world. Though other nations have worked on similar technology, thus far none have been as successful as the United States. That something as commonplace as cellular telephones could bring down this symbol of America's military-industrial complex was simply too ironic for the media to resist. In news stories, the technology was described as new and revolutionary, with numerous analogies to David and Goliath.

Within a week, though, the story had practically disappeared from the media. The U.S. military did not launch a crash program to counter this threat. No cellphone-based stealth detection systems were sold.

We are left wondering: "What happened?"

Overview of stealth technology

Stealth technology was originally developed at Lockheed Martin's legendary Skunk Works research facility. The group had designed a number of pioneering military aircraft:

1. Tao Yue, "Scouting for Surveillance: Detection of the B-2 Bomber and a Brief History on "Stealth", *The Tech*, November 30, 2001.

the P-80, America's first jet fighter; the U-2, a high-altitude reconnaissance aircraft that photographed Soviet nuclear missile installations in Cuba in 1962; the SR-71, which remains the fastest operational jet aircraft ever built; and of course, the F-117 Nighthawk, the Stealth Fighter.

Long before the Stealth Fighter's existence was acknowledged publicly, rumors about its capabilities had been circulating in the aerospace and defense community. Tom Clancy featured the Stealth Fighter prominently in his novel Red Storm Rising, a political-military thriller describing a conventional war between the Warsaw Pact and NATO. Testors, maker of accurate scale models of cars, ships, and aircraft, felt confident enough of the Stealth Fighter's existence to sell a model, based upon alleged sightings. Experts agreed that the fighter would be known as the F-19, the next available number in the aircraft designation system.

When the F-117 was finally revealed, observers were surprised by more than just its unexpected designation. The plane simply didn't look like a modern jet fighter. Instead of a sleek, aerodynamic profile optimized for supersonic performance, the F-117 was blocky and featured many flat surfaces. Its wing was swept back so sharply that the plane had difficulty developing enough lift to take off.

There was a reason for this strange shape. Stealth technology had begun with coatings that reflect less radar than the aluminum commonly used on airplanes. Indeed, the now thirty-year-old SR-71 reconnaissance aircraft had already made use of radar-absorbent coatings to reduce the risk of detection. But there is no material that would absorb radar perfectly. Skunk Works took the concept of stealth one step further, designing a shape for the F-117 that would reflect a radar beam in a different direction rather than back towards the transmitter.

Due to the limited computing power available in the 1970s, the plane was designed with flat surfaces to reduce the number of calculations required. Since each additional surface would add an extra direction in which radar could be reflected, the number of surfaces used was also kept to a minimum. The strange shape made the plane aerodynamically unstable about all three axes, so fly-by-wire capability was required to allow the pilot to control the airplane. Stealth had to be built into all aspects of the airplane's design: enclosed bomb bays, special pilot canopies, special seals at all joints, and special cooling vents for the engines.

The F-117 had a radar signature about 1/100 that of conventional airplanes, making it appear little larger than a bird on radar scopes. The B-2 Stealth Bomber, the next stealth airplane in America's arsenal, benefited from the greater computing power available by using a complex curved shape that further reduced its radar signature. The newest stealth fighter on the drawing board, the F-22, uses a still more advanced shape.

Potential vulnerabilities

Stealth required years of research and massive computing power to develop. Defeating stealth was a correspondingly daunting task. F-117 Stealth Fighters flew over 1300 sorties in the Gulf War without a single one being shot down. No stealth aircraft was lost in combat until 1999, when Yugoslav forces shot one down in Kosovo. This feat was, however, not repeated.

From the beginning, though, it has been recognized that stealth does not make an airplane invulnerable to detection. An airplane may be detected not only by radar, but also by other means. Stealth aircraft typically do not use radar or send any radio communications while in combat, in order to avoid detection by passive radio receivers. Stealth aircraft can be picked up visually in the daytime, making them usable only at night. Engine exhaust, though cooled to minimize the aircraft's infrared signature, is nevertheless hotter than the ambient air. It was this vulnerability that permitted Russian-made SA-3 infrared air-to-air missiles to lock onto the F-117 shot down over Yugoslavia.

Those problems can be solved operationally, though, by limiting the use of stealth warplanes to favorable military situations. A more serious problem is the inherent imperfection of the airplane's surfaces. No matter how precisely they are manufactured, they will degrade naturally during flight as a consequence of atmospheric friction. Dust will collect on the aircraft and begin to reflect radar. Rain will damage the radar-absorbent material. Care must be taken to repair nicks and scratches, and to seal joints where one manufactured part is attached to another. However, these tasks must be performed by maintenance crews working under a great deal of time pressure to get each plane out for another attack run. All of these factors contribute to the fact that a stealth plane will always reflect some amount of radar.

The Roke Manor system

The stealth-detection system announced over the summer was developed at Roke Manor Research, a British defense firm based in Romsey, Hampshire in the United Kingdom. It does not try to detect emissions from careless stealth aircraft, which would be an easily-countered move. Instead, the Roke Manor system attacks the foundations of stealth technology by detecting the radar waves that inevitably will reflect off it.

John Hansman, a professor of Aeronautics and Astronautics at MIT, explains, "Some stealth aircraft, like the F-117, are specifically designed to have a low radar cross section to monostatic, or conventional, radars. They are not stealthy to some bi-static configurations."

Conventional monostatic radar places the transmitter and receiver in the same location. This makes it a trivial problem to calculate an aircraft's position from the time and angle of the return signal. Bi-static or multi-static radar would position the receiver at a different position from the transmitter. Since stealth aircraft do reflect some radar signal and intentionally direct it away from the transmitter, bi-static radar could conceivably receive the reflection and detect the stealth aircraft.

The problem then becomes one of scale and coordination. The stealth aircraft will be visible only if an ideal alignment exists so that the transmitter bounces a signal off the stealth aircraft to the receiver. Stealth aircraft, however, are vulnerable from a very small subset of angles. It would be prohibitively expensive to build an enormous number of radar sites to obtain adequate angular coverage of stealth aircraft.

The Roke Manor system solves this problem with computing power and some creative thinking. Radar is simply a specialized application of radio, and we now live in a wireless age in which radio waves are ever-present. Notably, industrialized nations now have cell phone towers every few miles, sometimes every hundred feet in particularly populous areas. Telephone companies know where the towers are located and also have telephone lines hooked up to them.

In effect, the Roke Manor researchers have envisioned the use of cell phone towers as an extremely dense network of radar transmitters and receivers, interconnected via communications links. The sheer quantity of cell phone towers makes detection much more likely than with dedicated but less-numerous radar sites.

"A lot of stealth technology deals with redirecting radar waves," said Greg Duckworth, a Principal Scientist at BBN working on underwater acoustics in an area very much analogous to radar." It's very effective against monostatic radars. However, if you have bistatic radars, in particular a very large number of sources, so that you excite the target from a wide range of angles, and you have a multiplicity of receivers in many locations, you essentially will get around the stealth target's redirection capabilities. It is highly likely that an incident wave from a cell tower will be redirected towards one or more receivers."

Having invalidated the assumptions behind the stealth aircraft's shape, the multistatic radar system then pieces together all the data from the cell phone towers. Until recently, this problem would have been insurmountable. However, increased computational power and advanced signal processing techniques have made it possible to sort through all the signals and form a coherent radar picture. Ironically, the further development of the same computing technology that originally made stealth possible has now made it possible to detect stealth aircraft.

A natural technological progression

Given a dense cell phone network, massively parallel computers, and the Roke Manor software, how much can one determine about a plane? Quite a bit, as it turns out.

"If you can get a radar return, you can get all kinds of information from the return signal if you can process it sufficiently," Hansman said. "For example, if you look at the Doppler shift of the returned signal, you can get aircraft velocity. If you are sensitive enough, you can see frequency effects, such as engine rotation or structural vibration. If you have several receivers or different imaging angles, you can begin to reconstruct an image of the target."

These data further reduce the effectiveness of stealth technology. While stealth has always returned a small signal, even to monostatic radars, that signal is so small that it is usually filtered out either by the radar scope or by the operator. The intermittent nature of the monostatic radar contact also makes it difficult to calculate velocity. However, ubiquitous radio coverage makes it likely that stealth airplanes can be tracked, complete with velocity and shape information. This makes it considerably easier to distinguish stealth aircraft from birds in the sky.

Ernie Rockwood, a researcher for Sensis Corporation, a company that specializes in air traffic and air defense, said that he was "not surprised" by this development. "Some of my co-workers and I worked on novel bistatic battlefield radar techniques to improve survivability. We also submitted a proposal to Rome Labs for an operational concept using multistatic techniques."

Defense researchers generally agree that the technology is sound. Some believe this to be a natural development in radar technology.

"Underwater, they've already gone to multistatic systems because the reflectivity of targets is such that they don't naturally bounce stuff back," said Greg Duckworth. "Not because they tried to, as was the case with stealth technology, but because the physics makes them do that naturally."

Duckworth also drew an analogy between cell phone towers and television transmissions.

"Televisions have improved quite a bit, and comb filters have gotten better," said Duckworth. "On older TV sets, though, when an airplane goes over your house, a reflective wave from the aircraft ends up interfering at your antenna, and you see lines and artifacts on your screen. To the extent that a stealth aircraft does not absorb the wave, the remnants of it still interact with the airplane and result in detectable interference patterns."

The television analogy is particularly apt, as Lockheed has been working on a project that operates on the same principles as Roke Manor's anti-stealth system. In this project, called Silent Sentry, FM radio stations and VHF television broadcasts are used to provide the dense network of radio waves that interacts with stealth aircraft. While there are fewer FM and VHF transmission towers than cell phone towers, each individual station transmits with greater power. The smaller number of stations would also reduce the computational requirements of the system.

Military consequences

How far-reaching are the implication of this anti-stealth technology? As with all military technologies, it depends on the particular application.

Owen Cote, Associate Director and Principal Research Scientist of MIT's Security Studies Program, explained, "Even if this system works, it wouldn't be useful if you couldn't shoot the aircraft down. You'd have to find some way of guiding a missile very close to the target before an infrared or illuminating radar could achieve a lock on the aircraft."

"This is not very mobile technology," he continued. "Your cell phone towers are in fixed locations. While it would be close to impossible to destroy them all, they are susceptible to jamming just like conventional radar. Stealth might very well be a technology with a very short half-life. However, against foes such as Serbia or Iraq whose technology is not yet competitive with ours, I see stealth as having a much longer life. As a proof of concept, this bistatic technology sounds right. The actual implementation, though, is another matter."

Still, Dr. Cote saw some long-term effects of a successful system. "No offensive advantage lasts," he said. "Often there is a relatively cheap defense counter to match new offensive technology. We may find ourselves moving further away from manned delivery platforms and focusing more on cruise missiles, tactical ballistic missiles, and short-range missiles with incredible accuracy."

The technology is widely acknowledged to be feasible, and Roke Manor claims to have working prototypes. However, bistatic radar is neither a miracle nor a disaster that renders worthless decades of stealth research. It is yet another battle in the ongoing war between armaments and armor.

Section XIII:
Ground War: Theory and Practice

CHAPTER 50

Night Action by Gideon at Moreh—Perhaps 1249 B.C.[1]

Colonel A.P. Wavell

General Archibald Wavell was born in Colchester in 1883, and gained his commission in 1901. He fought at the Battle of Ypres, where he lost his left eye. After the injury he remained active, and went on to create the Middle East Command, charged with protecting the Suez Canal and the nearby oil reserves. Wavell also served as Commander-in-Chief of British troops in India. After Pearl Harbor, the primary task was to protect British territories against Japanese attack. In July of 1943, Wavell was made the 1st Earl of Cyrenaica and retired in 1947.

> *Except that I had to substitute science for divine inspiration, I worked somewhat along the same lines as Gideon.*
> —General Duncan in The Cavalry Went Through

All that we know of Gideon, son of Joash, warrior and statesman, is contained in three chapters of the Bible—Judges vi to viii. But the record of his character and actions given in those chapters is sufficient to rank him very high: with the possible exception of Joshua, he was the best general and shrewdest head of State that Israel ever brought forth.[2] And he was no mere rough soldier. His reply to the men of Ephraim (Judges viii. 2) could not have been bettered by the sleekest and readiest diplomat that ever purred French; while his dealings with the "non-cooperators" of Succoth and Penuel show that his diplomacy masked no weakness when he had to do with insolence and disaffection amongst his own people or tributaries. Wiser than Saul, he refused the crown of a hereditary Kingdom which the Israelites offered him. Of his talents as a soldier we can judge from the account of how he carried out two of the most testing operations of war—a night attack and a sustained pursuit. His brilliant execution of these prove that his skill and determination were those of a really great captain.

And now to paint the picture of the opening situation (as umpires and directors of tactical exercises say) for his night attack. Gideon's enemies, the Midianites, nomad Arabs of the desert, probably differed very slightly indeed from their descendants of today. Sheik Faisal ed Dowlish, whose activities on the Iraq border our Air Force has lately been engaged in curbing, is direct in tradition from Gideon's opponents, Zebah and Zalmunna, gallant caterans who met their end unflinching. And the story of Israel's

1. This is the date given by the learned Dr. Angus, who is just as likely to be wrong as any of the many others who have tried their hand at Old Testament chronology; Colonel A.P. Wavell, "Night Action of Moreh—circa 1249 BC." Wavell retells the story of Gideon's deception of the Midianites.
2. After which judgment, it is a little disconcerting to turn to St. Paul's catalogue of famous and faithful worthies of old (*see* Hebrews, chapter xi). He gives that frivolous but resourceful lady, Rahab, a verse to herself, while Gideon comes only in the "amongst-others-present-we-noticed" class. Paul's catalogue need not be taken too seriously, however. It omits Joshua altogether. Paul was probably writing in haste to catch the last post to the Hebrews.]

subjection to the Midianites has been common form round the borders of the Arabian desert for centuries. A raid or two by the nomads into the cultivated lands finds the settled peoples weak and divided, the gendarmerie and frontier guards inefficient and timidly handled. The raiders grow bolder and more numerous (some failure of pasture in the desert may urge them on) until at last the various tribes sink their differences, unite for once, and come up in the springtime "like grass-hoppers for multitude," with their tents and their cattle and their camels (just as described in Judges vi. 5) to eat up the pasture and the crops in the settled land.

The camp of the Midianites was pitched in the eastern part of the great plain of Esdraelon (not far from the modern El Afule, where Allenby's horsemen cut the Turkish communications). We must conceive of it as a laager rather than as a fortified camp, of the Midianites as armed guards to their cattle and camels rather than as an organized force, of their object as to pasture themselves and their beasts on the unaccustomed plenty rather than as to fight. It was, in fact, more of a gigantic annual picnic, than an invasion. But it was a serious matter to the Israelites. Gideon's men occupied the western end of Mount Gilboa on the hills to the south. The battlefield is almost the same as that on which Saul was routed and slain by the Philistines some two hundred years later. In all fighting in Palestine, water plays a chief part.

And it was so here. The only water available for the Israelites lay at the foot of the hills on which they had taken up their position, and thus close to their enemy. This explains the fitness of the test by which Gideon chose his three hundred (the same number, it may be noted, as made history under Leonidas at Thermopylae). The majority of his men, parched by the heat on the bare, rocky hills, flung themselves down full length by the stream when their opportunity came, and drank heedless and careless. Only the seasoned warrior, with experience of snipers and ambushes, kept his weapon in one hand and his eyes towards his foes, while he dipped the other hand in the water and lapped from it, ready for action at the slightest sign of danger. Gideon must have sorely felt the need of these trained warriors. The motley levy that had flocked to his standard[3] can hardly have inspired him with much confidence. Remember that for seven years the Israelites had hidden in the hills when the Midianites invaded the lowlands they had sown; that more than two-thirds of his original gathering had already shown that they had no stomach for the fight (Judges vii. 3); and that Gideon was an unproved commander and a man of no particular weight or influence (see Judges vi. 15). He had soon realized that his Army was no fit instrument for a pitched battle, and that he must depend on stratagem and guile for success rather than on numbers. It was comforting to find even three hundred seasoned men whose discipline and steadiness he could trust.

But the mortar that bound together the various clans of his enemy was also loose and weak. The Arab of the desert is a guerilla, an irregular, accustomed to fight in small bodies with plenty of manoeuvre room. The cramped conditions of battle within the great Midianitish host must have given him a feeling of unease akin to claustrophobia. Close-order was not his style of fighting at all. Nor did religious fanaticism impel him to rush fiercely on his foes; he had still to wait some two thousand years for his revelation and his prophet, for his Paradise and his darkeyed houris. So that when tales began to float down to the Midianite camp (as Gideon took good care that they should) of the

3. Any one now remember the speech of a well-known member of the Government some years before the War, who asked "what need there could be of compulsory service for Home Defence? On the day that a German Army sets foot in England, said he, the people of this country will "flock to arms" without any compulsion.

marvellous signs and portents that had marked the rise of the Israelites' new national leader, they fell on nervously receptive minds. Manured by rumour, panic once sown would spring like mushrooms from such soil.

To return to Gideon. His whole plan was based on the possibility of creating such panic in his enemy's ranks. He prepared his attack, as all night attacks must be prepared, with the greatest care and attention to detail. First he organized his force. He divided the three hundred into three companies, which were doubtless subdivided into platoons. The remainder he sent off—the Book says "Every man to his tent"—but, as subsequent events showed, he must at least have arranged for a part to hold the fords of Jordan and to complete the discomfiture of the enemy, once the three hundred had got him on the run. Gideon's next care was the issue of equipment—to every man a trumpet, a pitcher and a torch, strange weapons to cause so complete a rout. Then he made his personal reconnaissance, taking with him his batman Phurar (who probably acted also as battalion runner). This reconnaissance showed him that the hostile patrols were inactive, while the nervous talk he overheard amongst the outposts proved that the Midianites were ripe for panic. He now returned, issued his final orders, and gave out the famous password, "The sword of the Lord and Gideon." Each man lit his torch and carefully concealed it within the pitcher, so that no light could give away the movement until the critical moment; slung his trumpet ready to hand, and grasped his sword or spear. So the companies moved off, on an accurately worked out time-table, to their appointed stations round the enemy's camp. Gideon's signal was to be given about midnight, at which time, it had been ascertained, the Midianites were in the habit of changing sentries ("they had but newly set the watch"—Judges vii. 19).

The success of the stratagem was complete. The startled Arabs suddenly found themselves, as they imagined, beset by a host on every side. Panic spread; the loose cohesion of the undisciplined horde broke up; and tribe fought tribe in the darkness, with those shattering trumpets, those waving torches and that exultant battle-cry all around them. There was no need for the three hundred to strike a blow; they stood where they were while their enemies fought each other and fled.[4]

Slaughter overtook the demoralized Arabs in their flight, for Israel had seven long years of oppression to avenge, and the whole countryside rose now that their enemies were on the run. And when the remnant with their leaders had, as they thought, outdistanced the pursuit and were secure, Gideon, relentless and untiring, fell on them again and completed their disaster. So that "the day of Midian" became a proverb in Israel for completeness of victory.[5]

The principles on which Gideon acted—the value of training and discipline in work at night, the need for attention to detail, the importance of personal reconnaissance, the moral effect of a night surprise—are all still valid to-day.

A close parallel to Gideon's plan can be traced from our own naval history — in Drake's scattering of the Armada with fireships at Calais, or in Cochrane's similar though less-known exploit against the French Fleet in the Basque roads, 11th of April, 1809.

4. One wonders whether memories of Joshua's trumpets at Jericho had any part in suggesting to Gideon the ruse he employed.
5. See Isaiah IX 4.

CHAPTER 51

BCTP: Be Unpredictable, Take Risks—Or Lose[1]

Major R. Powl Smith, Jr.

This article is an account of training exercise PRAIRIE WARRIOR 96, an annual training exercise at Ft. Leavenworth, Kansas, that confronted student officers of the Command and General Staff College (CGSC) playing BLUFOR against the Battle Command Training Program (BCTP)'s famous world-class opposing force (WCOPFOR). These computerized exercises generally—and in this case specifically—simulate corps-level combat.

PRAIRIE WARRIOR 96, played in May 1996, was the first time [since BCTPs had begun in the early 1980s] that a student BLUFOR class not only won, but definitively defeated the WCOPFOR. They did so by abandoning the conventional conservative low-risk U.S. strategy of "two up, one back" that the opposing WCOPFOR had been conditioned by previous exercises to expect, substituting a high-risk unorthodox strategy that paid off, contrary to the urging of many mentors, senior observers, and instructors.

Smith, a 1982 West Point graduate and 1996 graduate of the Command and General Staff College (CGSC), then a major, simulated BLUFOR's II Corps Chief of Staff for PRAIRIE WARRIOR 96 as co-author of its operations and deception plans. Leaving the Army with the rank of lieutenant colonel after 26 years of service, he became a vice president of Sabot 6, an American management consulting firm.

1. Major R. Powl Smith, Jr., "BCTP: Be Unpredictable, Take Risks—or Lose," *Field Artillery*, March–April, 1997, pp. 16-21. Reprinted by permission of the Fires Bulletin.

BCTP: Be Unpredictable, Take Risks—or Lose

by Major R. Powl Smith, Jr.

For the first time in history, the 1996 Command and General Staff College (CGSC) students defeated the Battle Command Training Program's (BCTP's) vaunted world-class opposing force (WCOPFOR) during the annual Prairie Warrior exercise at Fort Leavenworth, Kansas. And they beat the WCOPFOR quite soundly.

Senior BCTP Observer General (Retired) James Lindsay said it was the best performance of a student corps and one of the best ever performances of a blue force (BLUFOR).[1] How did the II (Student) Corps defeat the WCOPFOR? II Corps exercised one of the most overused buzz phrases in the Army: "think 'outside the box.'"[3] In short, II Corps studied the WCOPFOR's application of the principles of war of *security* through his masterful use of deception, *mass* and *surprise* and turned the tables on him.

BLUEFOR commanders and their staffs usually are predictable—they opt for the safe, conventional, "two up, one back" answer to tactical problems. In part, this is due to their adherence to another principle of war: simplicity. In war, even the simplest tasks are difficult—no one will deny that. However, slavish application of "simplicity" can preclude the application of other, no less important principles—such as security, mass and surprise.

Simple Plans Are Not Always Good Plans

In most Warfighter exercises, the WCOPFOR is successful for two reasons. First, he emphasizes security; he prevents the BLUFOR from knowing his plan, usually through a well executed deception effort. Second, he quickly and easily deciphers the predictable BLUFOR plan.

He has been deciphering the plans of American military leaders for years and is very good at it. Therefore, if it's a simple plan, the WCOPFOR probably can figure it out very quickly—which puts him inside the BLUFOR decision cycle, giving him the initiative.

Having lost the initiative, the BLUFOR must rely on its technological edge.[2] It must use its better intelligence sensors and better attack helicopters, tanks and artillery to win—a matter of brute force instead of tactical finesse. This tends to result in the BLUFORs conducting a series of almost Napoleonic frontal assaults.

But the WCOPFOR is ready for the high-tech bludgeoning. Time and again he has countered brute force attacks by applying the principles of security, mass and surprise. He maintains security, masses his artillery on "strike sectors" and then masses his maneuver forces in a surprise envelopment. The WCOPFOR fights to win while the BLUFOR fights to gain a draw or a pyrrhic victory.

II Corps adopted the WCOPFOR's own "recipe" to beat him: one defeats a high-quality opponent by getting inside his plan and denying him yours. At Prairie Warrior 96, the II Corps Commander and staff chose security over simplicity. They developed and fully resourced a deception plan and executed a complex operational plan, both of which exploited the WCOPFOR's own predictability. II Corps took risks to achieve mass and surprise.

Know Thy Mission

In mission and task analysis, II Corps made two assumptions. First, to win, the corps had to take and keep the initiative. More experienced at fighting in the corps battle simulation (CBS), the WCOPFOR had advanced pucking (computer) skills, battle drills and tactical command and control (C²) procedures that could be devastating if he gained the initiative as he usually did. Therefore, II Corps had to control the fight from the start.

Secondly (and no surprise to anyone experienced at BCTP), II Corps had to beat him decisively in the deep fight. The corps had to desynchronize his plan, coax and corral him and defeat his center of gravity, his massive artillery force. II Corps could not let the WCOPFOR close unscathed on friendly units or mass his considerable fires. It had to neutralize or destroy his artillery, especially the many long-range multiple rocket launchers (MRLs).

The corps mission was difficult: rapidly penetrate an infantry army defending behind a series of major rivers, then fix and rapidly defeat the advancing army group reserve. The WCOPFOR defense consisted of two weakened infantry divisions augmented by more than 350 army-level guns and MRLs. Although he had to cover a sector 60 to 70 kilometers wide, he had the advantage of defending from behind a series of rivers, all of which required engineer support to cross. While II Corps had a 3:1 combat power advantage, the artillery was 1:1. That left II Corps conducting opposed river crossings under heavy artillery fire, a huge tactical challenge.

After crossing the rivers and penetrating the WCOPFOR's defense, II Corps faced a meeting engagement with a force of comparable or greater size and combat power: the 39th Mechanized Corps with more than 500 T-80B tanks, 1,000 BMP-2 infantry fighting vehicles, 400 artillery tubes and 100 MRLs plus additional artillery and maneuver units attached from the army level. II Corps was tasked to meet and defeat this powerful corps.

The ground II Corps would fight on was a tactical nightmare (see Figure 1). The corps area of operations was bisected and dominated by the Harz Mountains with major rivers bordering both ends. The high ground of the key terrain on the eastern side of the mountain mass dominated both the Salle River crossing and "The Turntable," a road nexus that controlled lateral north-south movement in the corps area of operations. The corps had to decide whether to send the main effort north or south around the Harz Mountains (through the Harz was even considered).

The mission was force-oriented—fix and defeat the 39th Corps—so II Corps had to go where the 39th went. II Corps couldn't afford to guess wrong about which way the 39th Corps was going and let him get into its rear. So, II Corps had to decipher the WCOPFOR's plan.

Know Thy Enemy

The WCOPFOR's doctrine states he prefers the meeting battle, but not an unconstrained free-for-all Americans often envision. His maneuver of choice is a find-fix-set-envelop battle drill tailored for the meeting battle (Figure 2 on Page 18).

Further, the WCOPFOR likes to isolate tactical fights, singling out and massing fires and maneuver against one battalion, one brigade at a time. If forced to fight in more than one direction, he can't mass effectively.

One of his greatest weapons is his operational and tactical security, allowing him to execute deception plans that enable him to achieve mass and surprise at the decisive point and time. In the January-February 1996 Center for Army Lessons Learned (CALL) Newsletter, "News from the Front," the article, "Deception and the WCOPFOR," by the (then) WCOPFOR Commander, Lieutenant Colonel Richard T. Lambert, explains the WCOPFOR's most effective tool—deception.

The WCOPFOR's unwritten motto is "Trust in 'Blue,'" meaning BLUEFOR leaders, American military leaders, are imminently predictable. This predictability starts with the over-application

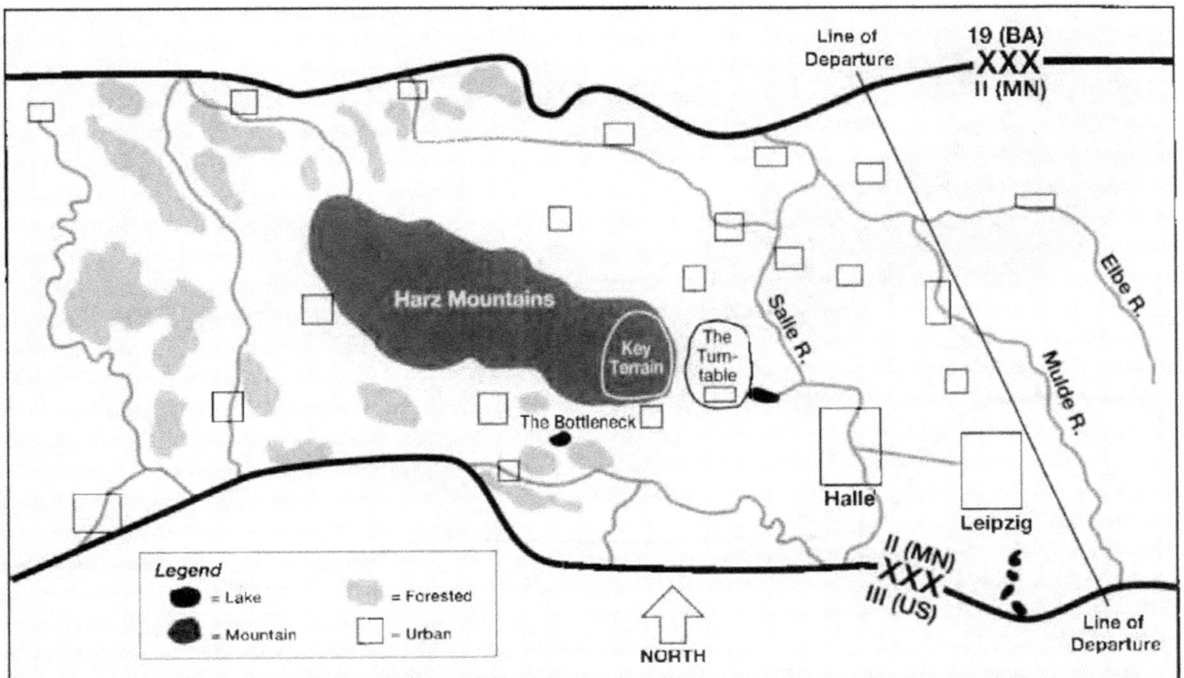

Figure 1: II Corps Area of Operations. II (Multinational) Corps' mission called for it to attack from east to west across a series of rivers.

of simplicity, but it goes much further. As noted American Army historian and frequent consultant to the Chief of Staff of the Army, Dr. Roger Spiller, recently wrote, Army staffs tend to produce consensus products that are "the safest, [and therefore] most nearly mediocre answer."[3] It is a stinging indictment of our staff analysis process that staffs tend to gravitate to the safest courses of action: i.e., the main effort on the best avenue of approach/terrain and using the least risky form of maneuver and task organization. In other words, they choose a defensive "two up, one back" stance every time. Too often, Americans optimize to obtain a draw rather than risk losing (or winning).

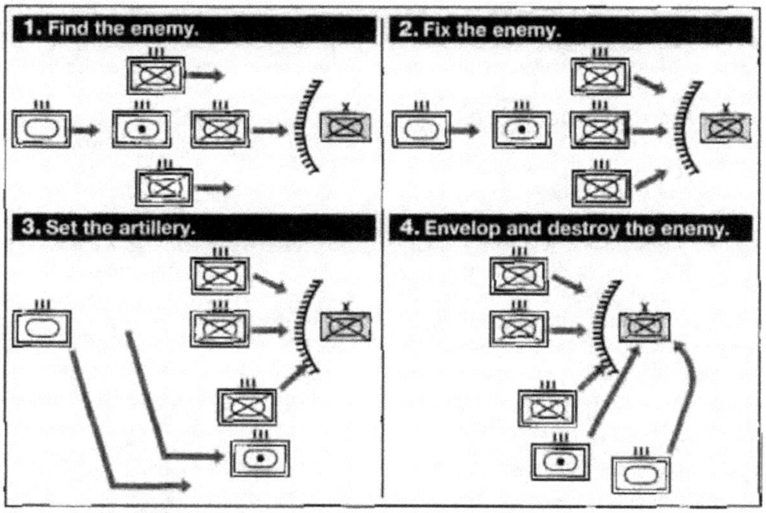

Figure 2: WCOPFOR Battle Drill: Find-Fix-Set-Envelop

The WCOPFOR bases his course of action on this military predictability. The WCOPFOR "Trusts in Blue" to select the least risky, most favorable (and most obvious) course of action.

Therefore, II Corps' planning premise was that the WCOPFOR expected—

(1) II Corps to execute the main attack on the optimal avenue (the northern avenue shown in Figure 1).

(2) II Corps to weight the main attack with a preponderance of heavy maneuver forces, preferably US-pure forces for ease of command and control in the main effort. In this case, a heavy US division covered by an armored cavalry regiment (ACR) or heavy brigade were the odds-on favorites.

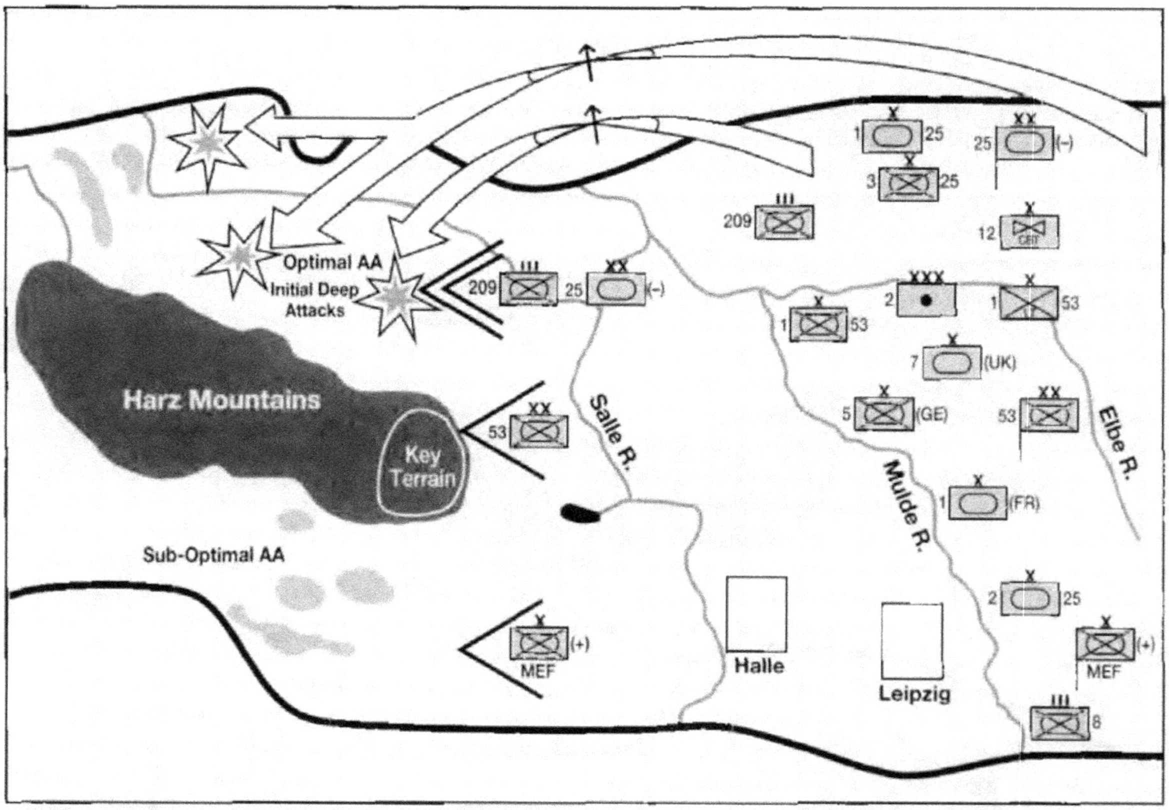

Figure 3: II Corps Deception Plan. To meet the WCOPFOR's expectations, II Corps initially positioned the 25th Armored Division with the 209th Armored Cavalry Regiment in the north as if to attack the optimal avenue of approach (AA) with the unwieldy 53d Multinational Division in the center and the Marine Expeditionary Force south as supporting efforts. The deception plan was reinforced with initial deep attacks on the Optimal AA by the artillery, Army aviation and USAF.

(3) II Corps to weight the main attack with the preponderance of general support (GS) artillery. If the WCOPFOR knows where the multiple-launch rocket system (MLRS) battalions are on the battlefield, he usually has found the reinforcing and GS artillery aligned to follow and support the main effort.

(4) II Corps to try to execute a poorly resourced deception plan, depicting the main attack along a suboptimal avenue of approach (the southern avenue).

(5) II Corps to use the II Marine Expeditionary Force (Forward) (MEF)—roughly the equivalent of a motorized infantry brigade with some armor and a large fixed-wing air component in a semi-independent economy-of-force role. This is in lieu of integrating this self-contained, task-organized, air-land MEF structure into the main fight with dissimilarly organized US Army units—contrary to the principle of simplicity. Americans tend to see the MEF as an excellent force for a supporting effort in a heavy corps. The WCOPFOR counts on the presence of the MEF to confirm the location of a secondary attack/effort.

(6) II Corps not to use allied brigades or the multinational division as the main effort. II Corps' 53d Infantry Division was the most powerful division in terms of raw combat power; but its multinational structure of two US brigades (one light and one heavy) and two allied brigades was a command and control challenge. Using this division in the main effort, again, would be contrary to the principle of simplicity.

(7) II Corps, after penetrating the initial defense, to rush forward to the "key terrain" and "set the defense" for the fight with the WCOPFOR heavy corps. The WCOPFOR knows most US commanders avoid a meeting engagement (very risky) and don't like to attack (risky and complicated); they prefer to defend (safe and easy).

(8) II Corps to keep its divisions and brigades in their own "lanes" once across the line of departure. Moving divisions and brigades laterally across the corps zone is too complex. If the WCOPFOR knows where the major BLUFOR units are at the line of departure, essentially, he knows how they'll come at him for the rest of the fight. The WCOPFOR's excellent intelligence capabilities tells him where the BLUEFOR units are at H-Hour. (For purposes of BCTP, the "excellent intel capabilities" translate into the WCOPFOR knowing 80 percent of BLUEFOR unit locations at the start of the exercise.)

(9) II Corps to use its attack helicopters at first dark and every night thereafter to find and attack the WCOPFOR artillery. The WCOPFOR moves his artillery at dark, spreads it out, does not expose it too early, deceives as to its location and protects it with air defense artillery (ADA) concentrations. In fact, he lays ADA ambushes, especially at deception locations.

Deceive Thy Enemy

The WCOPFOR briefs that in his deception plans, he simply shows the BLUFOR commander indicators of what the commander already believes. The WCOPFOR doesn't try to change the commander's mind, he reinforces the commander's perceptions.

It sounds pretty simple. So II Corps decided to try the deception strategy too. II Corps planned to deceive the WCOPFOR based on his expectation that II Corps would employ conventional, safe mediocre tactics. This became the basis for II Corps' scheme of maneuver. Unlike typical US planning, the deception plan was not an afterthought.

II Corps reinforced the WCOPFOR's expectation of where the main attack would occur by positioning the 25th Armored Division (less one brigade) and 209th ACR (a total of 10 ground maneuver battalions) to attack on the optimal northern avenue of approach. (See Figure 3.) The GS artillery and the corps aviation brigade were positioned in the center/north to reinforce the appearance of weighting the main effort.

To reinforce the WCOPFOR's expectation of a supporting effort on the southern avenue, the MEF (with a brigade from the 25th Armored Division under its operational control for a total of seven ground maneuver battalions) was positioned east of Leipzig. In the center, facing the Harz Mountains, II Corps positioned the multinational 53d Division led by the 5th German Mechanized Brigade. This huge but unwieldy division (17 ground maneuver battalions, 13 of which were allied) painted the perfect image of a force for the supporting effort, a force that would move ponderously forward, secure the key terrain and protect the flank of the main attack to the north. To further ensure the friendly deception plan was convincing, the initial II Corps Air Force and Army aviation deep attacks focused on the WCOPFOR's artillery in the "optimal" northern avenue.

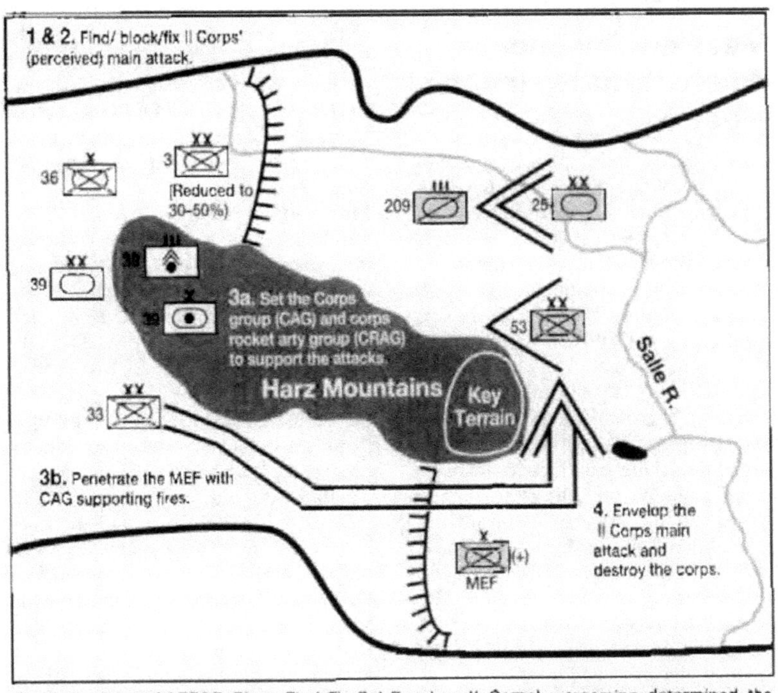

Figure 4: The WCOPFOR Plan—Find-Fix-Set-Envelop. II Corps' wargaming determined the WCOPFOR's plan would look like this. The WCOPFOR Commander would take advantage of II Corps' main effort to the north by launching his main attack in the south.

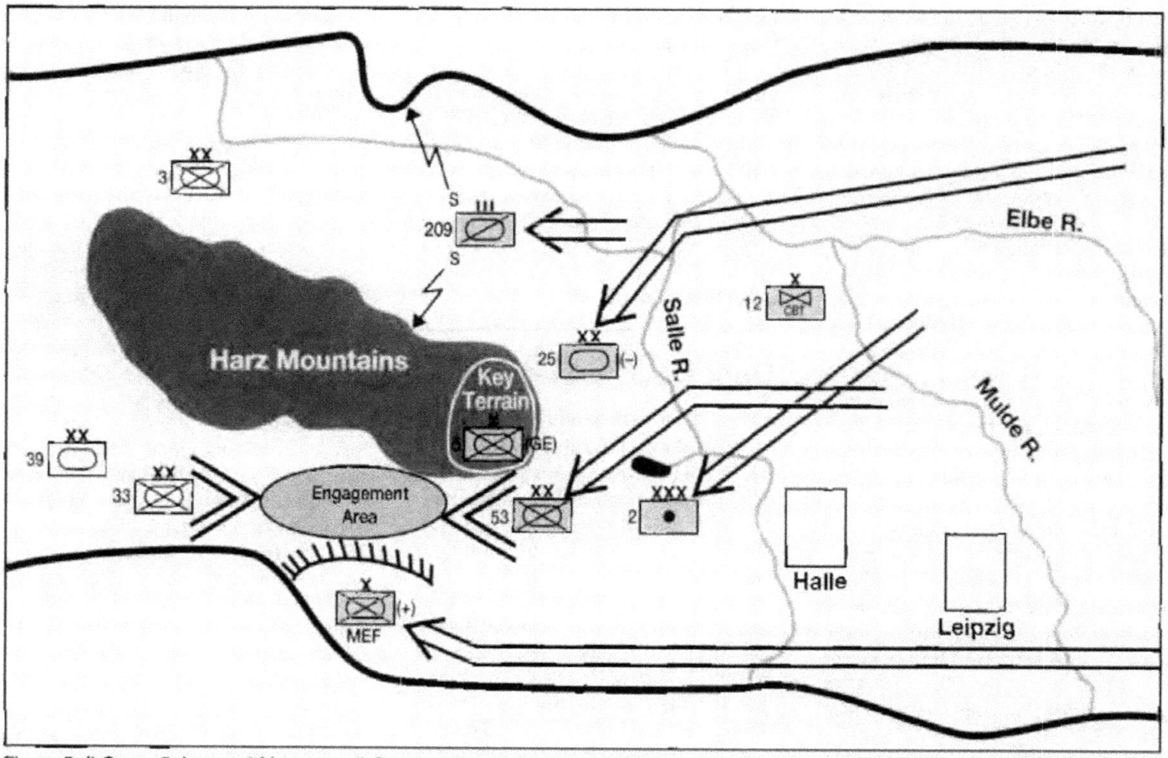

Figure 5: II Corps Scheme of Maneuver. II Corps had to make an awkward, unconventional and risky lateral move, shifting its combat power from the north and center axes to the southern one in order to defeat the WCOPFOR. II Corps' advantages were surprise and mass.

Mass on Thy Enemy Where it Hurts Him the Most

II Corps' deception plan was to reinforce the WCOPFOR Commander's expectation that II Corps's main attack would be to the north, so he would launch his main attack along the southern avenue in an attempt to envelope II Corps (Figure 4, Page 19). Therefore, to defeat the WCOPFOR's main effort, II Corps had to mass its combat power against the WCOPFOR in the southern corridor. The challenge was to move enough forces fast enough from their initial deceiving positions in the north to defeat the WCOPFOR in the south. II Corps had to conduct an awkward, unconventional and risky lateral move, shifting its combat power from the north and center axes to the south—the 25th Armored Division had to make a 50-kilometer move that would take eight hours under good conditions.

The corps' scheme of maneuver to accomplish this was complex. This was the first and greatest risk the student corps accepted. Many mentors, senior observers and instructors strongly advised against such a complex plan, especially when executed against the BCTP's World-Class OPFOR. Yet II Corps sacrificed simplicity to gain mass and surprise at the critical place and time.

In the first phase of the battle, II Corps attacked across a broad front with its initial deep attacks focused on reducing the WCOPFOR heavy artillery by 60 percent while the maneuver forces destroyed the two defending WCOPFOR infantry divisions. The ACR followed by the 25th Armored Division crossed the Elbe River and made a show of a major push due west into the northern avenue, the "main effort" (Figure 5).

In the center, the 5th German Brigade crossed the Salle River and then attacked west to secure the key terrain of the eastern Harz. The multinational 53d Division followed, clearing the zone for future operations. In the south, the Marines used their unique organization to launch a lightning-quick attack, blasting holes in the WCOPFOR defense with their air power, seizing river crossings with their air assault and amphibious capabilities and launching their light armored units followed by the Army's armored brigade to race ahead and secure the bottleneck in the southern avenue around the Harz.

By this time, the WCOPFOR's 39th Corps was committed to attacking through the southern corridor of the Harz. The narrow southern corridor gave II Corps some advantages. First, the WCOPFOR regiments and divisions would have difficulty using their preferred battle drill: find-fix-set-envelop. They would, however, still be able to mass artillery fires against II Corps in "strike sectors," allowing WCOPFOR to overwhelm the defending II Corps units. Therefore, II Corps' success rested equally on its ability to defeat the WCOPFOR artillery with deep operations and counterfire.

As the WCOPFOR piecemealed his regiments into the southern avenue, his artillery (and unavoidably his maneuver forces too) were subjected to II Corps' unrelenting deep attacks by USAF and Marine aircraft in air interdiction (AI) and "deep" close air support (CAS), Army aviation and the Army tactical missile system (ATACMS).

In the southern corridor, the enemy expected to be able to mass one of his heavy divisions and his corps artillery group (CAG) and corps rocket artillery group (CRAG) against the defending, isolated MEF at 3:1 or better odds. Instead, his lead division was canalized by the restrictive terrain and caught in a two-direction attack from the MEF and the 53d Infantry Division while II Corps deep attacks destroyed his center of gravity, the supporting CAG and CRAG.

II Corps took a lot of risks. Would the WCOPFOR buy the deception plan and go south? (If II Corps guessed wrong and the 39th Corps went north, the 209th ACR stretched across 40 kilometers could not hold off the WCOPFOR until the rest of II Corps turned around and redeployed to the north. The deception plan had to work, or II Corps was finished.)

Could the ponderous, multi-lingual, multi-doctrinal 53d Division shift directional axis and then get there in time to spring the ambush? Could the corps deep attacks neutralize enough of the WCOPFOR artillery? Could the 25th Armored Division move laterally from the north to south axis in time to be ready to follow and assume the attack?

So, even when the deception worked, II Corps was hardly assured of victory. Successful deep operations and rapid, large-scale movement to bring maneuver mass to bear were the keys to victory. With the WCOPFOR in the narrow southern trap, the next requirement was to strip him of his best means of extricating himself—his artillery.

Win Deep or All is Lost

The II Corps Commander from the outset emphasized the importance of winning deep. II Corps created a deep operations coordination cell (DOCC) based on a combination of the V and III Corps' DOCCs and standing operation procedures (SOPs) and tailored its structure for Prairie Warrior 96.

II Corps' deep operations won not only the deep battle, but the entire fight as well. Deep operations accounted for an astounding amount of damage to the WCOPFOR when compared to damage inflicted in the close fight; ten times more ADA systems, cavalry and infantry fighting vehicles, anti-tank systems, artillery pieces and tanks were destroyed in the deep and counterfire fights than in the close fight. In the first 12 hours alone, II Corps' deep operations and counterfire destroyed more than 60 percent of the WCOPFOR's army artillery group (AAG) and army group of rocket artillery (AGRA) assets—more than 200 pieces.

The success of II Corps' deep operations was due both to excellent G2 analysis and to an unswerving focus on finding and destroying the WCOPFOR's air defense and artillery. Only one deep aviation attack was targeted specifically at his maneuver forces.

II Corps Artillery set a new record for the number of WCOPFOR artillery systems acquired and attacked—in fact exceeded the WCOPFOR's own usually enviable performance. By the end of the exercise, deep operations and counterfire had destroyed or damaged more than 1,300 enemy artillery systems. The WCOPFOR tactical center of gravity had been neutralized within hours of the start of the exercise and, thereafter, was hounded to destruction.

II Corps' deep operations were so successful that on the second night, the BCTP staff declared that all AH-64 attack helicopters in the corps were grounded due to "bad fuel" and that half the ATACMS in theater were "blown up by saboteurs." The next day, fixed-wing aircraft were declared grounded because of "bad weather" over the corps zone. II Corps' deep operations were too successful and had to be artificially turned off—otherwise, a close battle might not occur in Prairie Warrior 96. Unfortunately for the 39th Corps, it was already too late.

Get Close To Thy Enemy and Polish Him Off

The 39th Corps reacted as planned, advancing rapidly toward the southern corridor. Fooled by the deception plan, the lead division steered directly for the Marines and the ambush, only to be brought up short by a corps deep attack. Before the division ever got into the corridor, it was nearly destroyed by AH-64s.

The WCOPFOR's trail division fared little better, finding itself hounded by fixed-wing air in daylight and attack helicopters at night. By the time the division reached the ambush, the MEF needed no help in polishing it off.

The plan had worked well, but the planners hadn't counted on the enemy units that kept appearing. As II Corps shattered the first team, the WCOPFOR introduced more units "to level the playing field." By the time "change of mission" was declared, the WCOPFOR included four more divisions: an airborne division, armored division, an entire heavy corps and several new artillery brigades. The fight concluded with II Corps in a "greenish/amber" status but ready to continue the attack.

The II Corps CGSC students won, not because of computer anomalies or overworked WCOPFOR personnel but because they "thought outside the box." They avoided predictability, took real risks and got inside the opponent's plan. II Corps believed it was possible and made it happen.

Major R. Powl Smith, Jr., is the S3 of the 3d Battalion, 13th Field Artillery, part of the 214 Field Artillery Brigade, III Corps Artillery, Fort Sill, Oklahoma. A 1996 graduate of the Command and General Staff College (CGSC), Fort Leavenworth, Kansas, Major Smith was the II Corps Chief of Staff for Prairie Warrior 96, where he co-authored the operations and deception plans and established and supervised the corps' Deep Operations Coordination Cell (DOCC). Before attending CGSC, he was a Fire Support Observer/Controller for the Battle Command Training Program (BCTP) at Fort Leavenworth, observing 10 corps/division Warfighter rotations. He was the Deep Battle Coordinator for V Corps Artillery in Germany, and commanded A Battery, 2d Battalion, 32d Field Artillery (Multiple-Launch Rocket System), 41st FA Brigade, part of V Corps.

Notes:

1. Remarks made by General (Retired) James Lindsay during the Prairie Warrior 96 post-exercise After-Action Review, 15 May 1996.

2. The BLUFOR technological edge is typified by corps deep assets. The corps has access to intelligence from a combination of high-tech intelligence collection systems such as the joint surveillance and target attack radar system (JSTARS), Rivet Joint, Guard Rail Common Sensor, unmanned aerial vehicles (UAVs), etc. The information is then fused by the all-source analysis system (ASAS) and exploited by high-tech deep killers: AH-64s Apache attack helicopters and the Army tactical missile system (ATACMS).

3. Roger J. Spiller, "In the Shadow of the Dragon: Doctrine and the US Army After Vietnam," *From Past to Future: The Australian Experience of Land/Air Operations*, Army History Conference, 29 September 1995, Jeffrey Grey and Peter Dennis, ed., (Australian Defence Force Academy: 1995), 20-21.

CHAPTER 52

The Monkey's Paw[1]

Major James R. Koch

This article represents the most concise recognition of the highest levels of Soviet deception plans and operations against Japan (Manchuria) in 1945 and compares with British-American sophisticated deceptions for D-Day 1944. It also describes the Soviet plans as too tightly held and too centralized when compared with the looser and more diffused Allied system.

Koch, a 1974 graduate of St. Lawrence University, had retired by 1997 at the rank of Lieutenant Colonel.

1. Major James R. Koch, "The Monkey's Paw," *Military Intelligence*, Vol. 14, No. 4, October 1988, pp. 26–28.

The Monkey's Paw

by Maj. James R. Koch

The United States, for all of its power and technological might, has only recently come to fully recognize the importance of a potentially great combat multiplier. This area is deception, both at the tactical and operational levels. The lessons that we learned under the tutelage of the British during World War II are only now being reexamined and treated with the respect that they deserve.

The Soviets, on the other hand, have not forgotten. Their military science is based, in large part, on their experiences in the "Great Patriotic War." The importance of deception, or to use their more encompassing term *maskirovka*, has not been forgotten. Their military science is aimed at the operational level of war, with victory at that level. Losses at the tactical level, while painful, are acceptable if they contribute to the overall operational-level victory.

It is this dichotomy between their system of *maskirovka* and our deception doctrine that will be examined. We can still learn from the Soviet experience, and our doctrine can profit from it. Our efforts in revitalizing deception, while laudable, are aimed at the tactical level. Until we give proper credit to the operational level, we shall remain at a distinct disadvantage in the area.

Maskirovka[1]

Deception is one of the eight methods listed by Soviet Maj. Gen. S. P. Solov'ev for achieving operational surprise. It is considered so important that it is a critical segment of all plans, from the strategic to the tactical level. Soviet history is filled with examples of its use in warfare and its contribution to victory. *Maskirovka* is more than just deception; it is an all-encompassing discipline utilizing passive techniques, such as concealment and camouflage, as well as such active techniques as simulation, disinformation and maneuver.

Maskirovka is highly touted in Soviet literature, and its use is up to the imagination of the commander. At the lower levels, due to lack of time and resources, *maskirovka* may consist only of camouflage, concealment and the use of decoys. At the strategic level, diplomatic assurances that nothing is going to happen would probably precede hostilities. Some manufactured event, such as an uprising in one of the Warsaw Pact countries, could be used to screen the mobilization of forces. In a limited objective war, this could catch the Supreme Allied Commander, Europe off guard just enough so that the ability to respond might be compromised.

C.J. Dick lists eight basic tasks of *maskirovka* at the operational level. They are:

- Masking an increase in or redeployment of forces or weapons which the enemy has spotted.
- Blocking the enemy's perception or identification of new weapons.
- Distracting the enemy's attention from other activities.
- Overloading the enemy's intelligence analysis capability.
- Focusing the enemy's attention in an area where there is no threat by portraying a false threat there.
- Presenting strength as a weakness, thereby creating a secure feeling.
- Conditioning the enemy with patterns to hide preparations.
- Confusing the enemy's expectations, leading him to misunderstand Soviet actions, thereby selecting an improper response.

Given this list, we could expect to see any number of these tasks worked into a Soviet operational plan. While we may see all of it, we will not know that it is a deception effort until the fighting begins. Even then, we might not recognize it! The reason for this is the Soviets' never ending quest for inventiveness in their deception efforts. Basically, their goal is to never let us see the same thing twice, especially at the operational level. This lesson was learned by the United States during World War II and is still valid today.

Dick enumerates other principles, some of which closely parallel our own:

- Have a clearly defined objective to support the operational scheme.
- Exercise strict control at the highest possible level.
- Preparation, execution and supervision must be thorough.
- The plan must be a credible, plausible alternative to the real one.
- The enemy must be made to work for its intelligence; it can't be given too easily.
- Feed as many intelligence collection sources as possible.
- Remain flexible to react to enemy actions.
- Security is vital to the success of the plan.

Given these tasks and principles, a basic understanding of the Soviet system and an appreciation for the historic roots of *maskirovka*, we can see that they have clearly thought out their deception doctrine and are prepared to implement it. They have created units whose sole purpose is deception execution. Yet, the one thing which may cause them problems is their almost fanatical emphasis on security, resulting in a close-hold plan. While I advocate having only a limited number of people aware of a plan, those people must be carefully selected. Innocent innovations, though tactically sound, on the part of a subordinate commander, could ruin the plan. The Soviets' overcentralized, top-down approach to deception could be their Achilles' heel.

U.S. Deception Doctrine[2]

U.S. Army doctrine has deception cells at division and corps level, as well as those planned for the operational level. This decentralized approach could profit by having its center of power raised to a higher level. The basis for our current doctrine is found in 10 maxims elaborated upon in Field Manual 90-2, *Battlefield Deception*. I consider four of these maxims (Magruder's principles, a choice among types of deception, the importance of feedback and "The Monkey's Paw") critical when comparing our system to that of the Soviets.

Magruder's principles espouse the theory that it is easier to make the enemy maintain a pre-existing belief than to present notional evidence to him to make him change that belief. When looking at the Soviet system and their reliance on "norms" to calculate battle plans, you will find that a knowledge of those norms will tell the deception planner what the Soviet reaction will be, given his view of the battlefield. A plan to modify or exaggerate that view could result in more predictable outcomes of our deception effort. Therefore, it is extremely important that we get to know the Soviet operational system as well as his forces.

Along with Magruder's principles is the maxim of a choice among deceptions. Whenever possible, our deception plan should be geared toward reducing uncertainty in the mind of the target. The plan should reduce the level of uncertainty and force acceptance of the false picture. The Soviet response should therefore be more predictable, though initiative at his operational level may counter some of these efforts.

Next, the importance of feedback is critical to determine if the bait has been taken. This maxim has its roots in World War II when we used ULTRA intercepts to read the German High Command's mind. We can never again expect to have a system this good (although it can be argued that ULTRA caused us to be tactically surprised at the Battle of the Bulge when German plans didn't show up on Ultra intercepts). We must use our intelligence system to verify enemy intentions, and the commander and the deception planners must review these intentions in light of our known deceptive actions. The key here is to rely on every source of feedback available, no matter how trivial it may seem at the time.

The last maxim to be discussed is "The Monkey's Paw." This states that deception may produce subtle and unwanted side effects. The more deception plans are included in operational schemes, the more likely "The Monkey's Paw" will become a factor. We should therefore have a more centralized deception planning framework.

Finally, there is one maxim that I believe should be on the list but is not — plausibility. The deception story must be plausible and within your capability to carry out. Otherwise, the enemy will most likely ignore the intelligence feeding him the story as an aberration within the system. The only case that could be made for ignoring plausibility is when you intentionally structure your effort to make it unbelievable, causing the enemy to focus his attention elsewhere at your real deception effort. This is a classic case of having a plan within a plan.

Using these maxims, with particular attention focused on the ones highlighted, we can begin to fill in the five components of a deception operation: objective(s), target, story, plan and events. While these are discussed in some detail in Field Manual 90-2, one point needs clarification. Many believe that the target of the plan is the enemy intelligence analyst who will receive and process the information. The argument is made that if the analyst doesn't believe the story, it won't reach the commander. This has some validity, but since most analysts "comment" on their reports, there is a chance that poorly conceived deception information will still reach the decision maker. In either case, the target is the decision maker. The analyst is an important middle man in the process and has the power to influence what the decision maker sees and thinks, but he does not have the power to make the decision. Therefore, when considering what you want the target to *do*, remember to consider what you want the analyst to *think*.

Historic Examples

The most recent historical evidence showing the command and control relationships used in deception at both strategic and command levels comes from experiences during World War II. At the strategic level are the examples of Soviet operations on the Manchurian Front and the Anglo-American operation known as BODYGUARD.

The movement of Soviet forces from the Western Front to the Eastern Front involved traveling between 9,000 and 12,000 kilometers and utilizing some 136,000 railway cars. *STAVKA*, the headquarters of the Supreme High Command, recognized that a movement of this size could not go unnoticed. The scale, timing and location of the redeployment had to be masked. Many separate, distinct efforts were undertaken, all directed by *STAVKA*. High ranking commanders transferred disguised as subordinates; the Trans-Siberian Railroad was camouflaged to conceal movement; movements were made at night; and assembly areas were located far from attack positions. The effect of these operations was to make the Japanese believe that the Soviets could not open an offensive earlier than late 1945 and probably not until early 1946. By July 25, 1945, all forces had been transferred. After the end of the war with Germany in April, over 1,000,000 men and their equipment were moved without drawing attention to the mission.[3] This effort is probably the greatest strategic military deception in the history of warfare, and it led to virtually complete operational surprise. Note that the vast majority of the planning and implementation was conducted at *STAVKA* level, with only a handful of planners aware of the magnitude of the actual preparations.

On the Anglo-American side are the experiences of the London Controlling Section (LCS), which developed deception plans in conjunction with the European invasion planning conducted by the Supreme Allied Command. The LCS kept the Germans guessing as to the actual location of the attack by creating a myth that the attack could come at anytime, anywhere, from the Balkans to Norway. Recognizing that this would not keep the Germans guessing for long, it was decided that when it was obvious that the main attack would come cross-channel, the attack would be covered with a double-bluff deception. The deception would indicate that the main attack at Normandy was a feint and depict a notional unit poised in Southeastern England ready to strike across the Pas de Calais. The extent of the planning for this deception was handled almost exclusively by the LCS. However, since they did not control the assets needed to make it appear real, LCS tasked various units to fulfill pieces of what was obviously a deception effort. Some high-level commanders were not particularly cooperative in this effort and could have caused its failure. However, due to the general weakness of the German intelligence capability and the Allied use of double agents and ULTRA, the effort was successful.[4]

The Soviets maintained their characteristic tight control on the deception effort, masking it even among their own forces. The English and Americans were more open in their efforts with the staffs, involving many subordinate commanders.

At the operational level are the U.S. effort at crossing the Rhine (Operation *VIERSEN*) and the Soviet effort to cross the Dnepr River north of Kiev in October/November 1943.

In March 1945, First Army had just completed crossing the Rhine at Remagen, and Ninth Army was set to force a second major crossing. The deception objective was to deceive the enemy as to the location and timing of the second crossing. The XVI Corps, chosen to make the crossing, was the weakest of the three Corps in the Ninth but was situated on the best crossing sites. To make up for the deficiency in strength, a division was chopped to the XVI from each of the other two Corps. *VIERSEN* was to simulate a notional river crossing staging area in another Corps section, while the actual force was assembling in XVI sector. The portrayal involved 23rd Special Troops, a dedicated deception unit and a number of subordinate units. Demonstrations were conducted in the notional unit's zone, while operations security covered the assembly of the crossing force. The fact that this unit portrayed the false as real and the real as false contributed significantly to the success of *VIERSEN*.[5] Note that this deception was planned at Army level and executed at Corps level and below.

In a similar situation the Soviets, in October 1943, had forced numerous crossings of the Dnepr River around Kiev and had decided that the main thrust was to occur at the Bukrin bridgehead. The Germans repulsed every effort made in that region, but success was being enjoyed further to the north. In view of that fact, *front* requested from *STAVKA* permission to shift the main effort to that success and that they be given an additional tank army to ensure the success. *STAVKA* permitted the shift of effort but denied the additional troops. Instead, they insisted on secretly transferring the 3rd Guards Tank Army to the effort. The *maskirovka* plan involved the regrouping of numerous units within a six-day period, to make the attack a success. In the Bukrin area it was business as usual, with aggressive patrolling, strengthening of defenses and normal patterns of behavior. The 3rd Guards Tank Army was disengaged from combat, moved north over 200 kilometers, conducted three river crossings and got into position to conduct the attack within the allotted time. *Maskirovka* efforts, including the use of simulations, were complete in Bukrin. The secrecy of the move of the 3rd Guards Tank Army was also carefully monitored. All of this led to complete success of the plan.[6]

The major difference between this move and *VIERSEN* is that the latter was planned at Army level and conducted at Corps and below, while the Kiev operation was planned at *STAVKA* and executed at that level, with very few people aware of the plan.

Conclusions

It has been argued that the old Field Manual 90-2 gave the impression that deception was something for the strategic or operational level, that few examples were included for the tactical level.[7] The new version corrects that perception with relevant tactical discussions but fails to clearly show the link between the strategic, operational and tactical deception plans. "The Monkey's Paw" will be a reality if this link is not effectively realized.

In only one place is the approval of lower deception plans by higher headquarters discussed. This review and approval is necessary to avoid "The Monkey's Paw." Further, too little attention is given to planning at the operational level. Only now are we beginning to recognize the need for a deception cell at echelons above corps, the level at which we talk about operational war fighting. If we don't win the operational level of war, our tactical successes will be for nothing.

The deception cells at division must concentrate on tactical deception, ensuring that every effort that they take is properly coordinated with corps. Corps-level cells should also concentrate on the tactical level, though it is conceivable that, in some instances, they could be used at the operational level. In those cases where they may have both a tactical and an operational mission, the 12-man section may be too small to properly accomplish the task.

At the operational level, the 10-man staff is probably adequate for wartime planning, but in peacetime it is too large. It will end up being used by the G3 for other tasks not associated with deception, thereby detracting from its usefulness as a planning tool. Recognizing that each theater's requirements will differ, it would be more effective to use a three or four-man peacetime cell used solely for operation planning and contingency planning. This cell could be expanded in wartime to the size required by drawing on corps assets. These assets would, in turn, draw on subordinate divisions for augmentation. This expansion could be accomplished by pulling the planners up to the next higher level, while leaving those charged with the execution of the deception mission in place. With centralized planning there will be less of a need for the planning function at the tactical level and more at the higher, operational level. Planning could still be carried out by the deception executors at division and corps levels, when necessary, thus allowing us to centralize with experienced personnel during wartime operations while not depriving the corps and division commanders of all their expertise.

In peacetime, we should practice deception at lower levels in our command post exercises and field training exercises in order to gain experience. In times of crisis, however, that expertise should be drawn to higher levels for wartime tasks. We are too decentralized in our current level of effort. Decentralization may lead to too much innovation and initiative which, in the end, could be counterproductive. While the close-hold, commander only or *STAVKA* system of the Soviets is too centralized for our use, it has worked. It has probably worked more efficiently than anything we've accomplished at the operational level. We should learn from this and take deception as a tool used by the few for the benefit of the many. Given this and an understanding on the part of the commander that deception planning is a time consuming effort requiring detailed planning and utilizing precious assets, we may be able to use this combat multiplier to win the operational-level battles of the next war. ★

Footnotes

1. C. J. Dick, "Catching NATO Unawares: Soviet Army Surprise and Deception Techniques," ***International Defense Review***, No. 1, 1986, pp. 24-26.

2. Field Manual 90-2, *Battlefield Deception*, (Final Draft), pp. 1-5, 1-32.

3. David M. Glantz, "August Storm: The Soviet 1945 Strategic Offensive in Manchuria," *Leavenworth Papers*, No. 7, 1983, pp. 1-4.

4. Charles G. Cruickshank, **Deception in World War II** (New York: Oxford Univ. Press, 1980), pp. 87-93.

5. Thomas A. Savoie, "Tactical Deception: A Lost Art?", *SAMS Monograph*, 1985, pp. 15-17.

6. David M. Glantz, "38th Guards Tank Army's Deception as Part of the Kiev Operations, Nov. 1943, **Soviet Deception in Transition, 2d Period of the Great Patriotic War, November 1942 to December 1943** (unpublished), pp. 197-210.

7. Savoie, p. 22.

Maj. James R. Koch received a bachelor's degree from St. Lawrence Univ. and a master's degree from the Univ. of Oklahoma. He is a distinguished graduate of the Field Artillery Officers Advance Course. He is also a graduate of the Tactical All Source Intelligence Officer Course and the Command and General Staff College. Koch currently is serving as AFSCOORD and FA Intelligence Officer, 3rd Armored Division Artillery in Hanau, Germany.

CHAPTER 53

1st Cav in Desert Storm—Deception, Firepower and Movement[1]

Brig Gen Tommy R. Franks and Patrecia S. Hollis

Brigadier General Tommy L. Franks served as Assistant Division Commander (Maneuver), 1st Cavalry Division, during the 1991 attack into southern Iraq. A graduate of the Armed Forces Staff College and Army War College, Franks was the U.S. general leading the 2001 invasion of Afghanistan. Patrecia S. Hollis was the managing editor of "Field Artillery" magazine at the time of the interview. In the interview, Franks quotes from Sun Tzu, citing him as having inspired the successful opening deception plan.

1. BrigGen Tommy R. Franks and Patrecia S. Hollis "1st Cav in Desert Storm—Deception, Firepower and Movement," *Field Artillery,* June 1991, pp. 31–34.

INTERVIEW

Brigadier General Tommy R. Franks, Former Assistant Division Commander for Manuever, 1st Cavalry Division

Sam Orr

1st Cav in Desert Storm

Deception, Firepower and Movement

Interview by Patrecia S. Hollis, Managing Editor

Brigadier General Tommy R. Franks was Assistant Division Commander for Maneuver with the 1st Cavalry Division, when the "First Team" deployed to Saudi Arabia in September 1990 for Operations Desert Shield and Storm. By mid-October, the division, then attached to XVIII Airborne Corps, was closed in Saudi Arabia and assembling in the desert about 160 kilometers west of Dharhan. By early January 1991, the First Team had repositioned almost 500 kilometers to the northwest, in the vicinity of King Khalid Military City (about 50 kilometers south of the Iraqi border). When the air campaign started, the division was in defensive positions along the Wadi Al Batin, a historic avenue of approach that runs from Kuwait and Iraq into Saudi Arabia. The next 40 days were exciting for the First Team and key to the success of Operation Desert Storm.

What were the 1st Cavalry Division's missions during Operations Desert Shield and Storm?

While attached to the XVIII Airborne Corps, our mission was to defend Saudi Arabia and deter attack by the Iraqis. After repositioning, the division was attached to VII Corps, and our mission was to defend the historic approaches into Saudi Arabia. It was during these defensive operations that we started planning the counteroffensive. At this time, the division was "on a string" to CENTCOM [Central Command] as the theater reserve.

During January, we moved north through a series of defensive positions to the tri-border area on the Wadi Al Batin. By tri-border, I mean the juncture of the Kuwait, Iraq and Saudi borders. We focused on security operations, still defensive in nature, but initiated planning for offensive operations.

In the first week of February, soon after the air war started, the division received the mission to be a part of the deception force. Our purpose was to deceive Saddam Hussein into believing that Coalition Forces would attack up the Wadi Al Batin into Kuwait. We wanted him to mass his forces in the area across from the 1st Cavalry Division. So we set up operations along the border and conducted feints and raids prior to G-Day [Ground-Day] on the 24th of February.

From our positions on the border, we conducted massive rocket raids. I recall one occasion when our division artillery, augmented by the fires of an additional MLRS [multiple launch rocket system] battalion, fired more than 400 rockets deep into Iraq in less than 90 seconds. We were after Iraqi artillery, command and control targets of the front-line divisions and his air defenses. We found out later the raids had been very effective, and the forces immediately across the border

An S-60 57-mm Iraqi anti-aircraft gun destroyed in Desert Storm.

INTERVIEW

from us (several infantry divisions) had been fixed in place and effectively blinded to the repositioning of the VII and XVIII Corps.

We continued these raids up until G-Day and, in fact, conducted mounted operations 30 to 40 kilometers into Iraq on several occasions. Our purpose was twofold. One, to continue to gain and develop intelligence on the enemy. And second, to reinforce Saddam's belief that the Coalition would attack Kuwait by coming up the Wadi Al Batin.

During deception operations, several things happened. The division took almost 2,000 prisoners along the border in the two weeks just prior to G-Day. And it was through the interrogation of these prisoners that we confirmed the enemy's tactical template—the locations of his front-line divisions.

Then, on G-Day, we launched a limited attack into the major trench and obstacle works of the divisions immediately to our front. This was to continue to persuade the Iraqis that we were the spearhead of the main effort.

Meanwhile, the campaign plan unfolded, and both US Corps were brought to bear well out to the west to envelop Iraqi front-line divisions and destroy the Republican Guard's divisions further to the north—the flanking operation General Schwarzkopf called the "Hail Mary."

Sun Tzu said 2500 years ago, "Make your way by unexpected routes and attack unguarded spots." That's what the campaign plan called for, and that's exactly what we did . . . it was a masterstroke.

Once G-Day kicked off, the deception was complete, and the division was pulled off line, brought to the west 40 to 50 kilometers and moved up through the breach created by the 1st Infantry Division. The 1st Cav broke contact, moved to the north almost 300 kilometers, rearmed, refueled and linked up with the 1st Armored Division in less than 24 hours. We were prepared to continue operations into Al Basra when the ceasefire was announced.

I could not be more pleased with the incredible performance of our soldiers, leaders and equipment.

In the 1st Cav's fast-paced operations, how did the artillery do keeping up with the maneuver forces?

The artillery did a super job of keeping up because Colonel Jim Gass, our Div Arty Commander, and his guys planned, prepared and rehearsed the actions necessary to keep up. The division first moved 160 kilometers to the west of Dhahran and established itself in an assembly area in the desert. Next we moved almost 500 kilometers up to the northwest to the vicinity of King Khalid Military City, and then we moved through a series of defensive positions up to the Iraqi border. You might say, we had a lot of practice before G-Day. Finally, we moved another 300 kilometers into Iraq.

Colonel Gass had worked our division artillery very hard in practicing desert movement. It was a challenge to move command and control and cannons along with maneuver forces in a movement-to-contact. But our success proved the old adage, "You do well in war what you practice in peace."

As far as specific vehicles were concerned, we confirmed that the old M548 cargo carrier is not able to provide artillery ammunition resupply during fast-paced, mobile operations. I say we "confirmed" because we've known the M548's shortcomings for some time. We need to buy more FAASVs [Field Artillery ammunition support vehicles] now, and of course we will continue to work the development of the FARV-A [future armored resupply vehicle-ammunition] for the out years.

We also confirmed that the modern generation of wheeled vehicles, such as the HMMWV [high-mobility multipurpose wheeled vehicle] and HEMTT [heavy expanded-mobility tactical truck], do well moving across country in a desert environment, while our older equipment isn't as capable.

We learned that the FIST-V [fire support vehicle] isn't a mobile enough companion for the M1 tank and the Bradley fighting vehicle. In fact, we modified some Bradleys to accomplish the FIST-V mission. By mounting GVLLDs [ground-vehicular laser locator designators] on the Bradleys, we were able to keep our fire supporters in the fight.

In the final analysis, I give a great deal of credit to Major General John Tilelli, the First Team's Commander, and our brigade commanders. They recognized the mobility strengths and weaknesses of our equipment, and movement tables were planned and executed accordingly.

A combat-ineffective Iraqi UAZ-469B Republican Guard jeep.

Iraqi T-72 tanks, casualties of the Storm.

INTERVIEW

Should we consider modifying Bradleys to use as FIST-Vs Army-wide?

Yes, I think so—we need a fire support vehicle that looks like the vehicles with which it fights. But we also need to continue our efforts to find an appropriate command and control vehicle for both the artillery and maneuver forces. Mobility and command and control will remain key ingredients of fire support into the 21st century.

Was range a problem for the artillery?

When we discuss range—more is always better. The MLRS provided range out to 30 kilometers, and on a variety of occasions, we had the Army TACMS [tactical missile system] with even greater range. Couple that with close air support [CAS] and, in some cases, BAI [battlefield air interdiction] by the Air Force, and we got the job done.

The issue isn't simply longer range, but the volume and responsiveness of fires at longer range. We need longer range cannons to provide greater volumes of fire at the direct call of the maneuver commander.

Approximately what percent of DPICM [dual-purpose improved conventional munitions] did the 1st Cav fire and how effective was it?

The preponderance of ammunition we fired was DPICM—perhaps 70 percent. It was very effective. We saw the devastating results of top-attack of armored and soft-skinned vehicles as well as trenchworks and troop targets. I think ICM proved to be a winner.

But I will also tell you, there's "room at the inn" for HE [high explosive munitions]—volumes of high explosive rounds did a great job on bunkers and built-up logistical and C^2 [command and control] sites. We need to retain a mix of munitions in response to the lessons of Desert Storm.

At this point, let me digress and talk a minute about the value of training and training devices. When we deployed to Saudi Arabia, our soldiers had fired no live DPICM and very few MLRS rockets and Copperhead rounds. But in combat, they fired tens of thousands of rounds of DPICM, thousands of rockets and about 30 Copperheads—they were on target in virtually every case. The fact is that soldiers had used training devices to practice these skills. Sergeants had trained their soldiers to standard, and our training devices worked. There's no substitute for smart, realistic training. It's a testament to the quality of our training, our training devices, and quality of our soldiers that they were able to fire these munitions so successfully when they had never fired them before.

Our suites of munitions worked well, and I must say the 1st Cav used them all. We were among the first to fire high volumes of DPICM in battle. The same with high volumes of MLRS fires—the first to fire Copperhead.

Feedback from Desert Storm indicates cannon artillery typically fired battalion or larger fire-for-effect missions. Did the 1st Cav confirm that?

Yes. The joint munitions effectiveness manuals (JMEMs) tell us we get the best results when we mass fires. That's one of the principles of indirect fire support. We fired a lot of rounds at a few targets, rather than a few rounds at a lot of targets, and I think the results speak for themselves.

We had missions where we fired several battalions of 155-mm munitions and MLRS rockets into the same target area, multiple volleys of 155—up to 10 volleys per battalion. The results were devastating.

We need to continue to reinforce this lesson from Desert Storm at our home stations and training centers.

Brigadier General Franks surveys an Iraqi T-72 tank damaged in Desert Storm.

INTERVIEW

An Iraqi BMP, the last kill of the 1st Cav in the 100-Hour War.

In your Desert Storm experience, you learned a great deal about logistical operations in the desert. AirLand Operations (the evolving AirLand Battle-Future) foresees a battle over a large area such as the desert. What recommendations for logistics would you offer to support AirLand Operations?

We need to increase the focus on combat service support [CSS] for the Army—not just for the Field Artillery or the fire support community. We referred to CSS operations in Saudi Arabia as "the long pole in the tent." The Army needs to continue to increase the mobility and materiel-handling capabilities of our logistical force—across the board.

When we mass artillery fires, the volume of ammunition required is tremendous. As you know, the Field Artillery School is working diligently to develop the materiel-handling and automated capabilities we need to supply enough ammunition for AirLand Operations.

Additionally, we have to take a look at CSS for our FA brigades—I think we'll have to give each brigade something like a forward support battalion to get the dedicated support we'll need in the future.

How did the Firefinder radars do in Operation Desert Storm?

Our Firefinder radars were extremely successful. The Iraqis were not very sophisticated in using indirect fire, at least from what I saw. When we probed their defenses, they simply fired "final protective fires"—in most cases. The Firefinders were able to pick up these firing units, and we deliberately and diligently destroyed them.

I will say that from my experience, the Q37 needs increased mobility. And we need to downsize the Firefinder over time to increase its deployability. These fixes are in the works.

I'd also like to see the range of the Q36 increased. It's a mobile and capable system now. . . .with increased range, we'll have a real winner.

It appears joint doctrine for fire support coordination and targeting may need some refinements. What were your impressions and what do you recommend?

I think "refinement" is the right word. We have sound doctrine. The doctrine establishes fire support coordination measures and fire support coordination lines; our joint appreciation of that concept needs to be developed, and we need to practice what we teach.

The relationship between the Army and the Air Force, with respect to who attacks what part of the battlefield, when, will have to evolve over time because of the new technology we're developing for command and control, target acquisition and attack systems. But for now, we simply need to train to execute the doctrine we have.

What was one of the 1st Cav's greatest training carryovers from the NTC [National Training Center, Fort Irwin, California]?

I'd say the efficiency of the fire support system (the "system of systems") as a whole—the fire support process "paired" the combat power of individual cannons and rockets with targets to be fired. The linkage between and among the fire supporters, our maneuver forces and our cannon and rocket units was exceptional. For anyone who has spent much time at the NTC, that's what it takes to win.

Our training centers focus great energy on bringing fires to bear at points critical to the maneuver scheme. That's what we took from the NTC into the desert with us. Decide, detect, deliver. . .

What message would you send Redlegs worldwide?

We have the best-trained, best-equipped and best-led Army in the world. Fire support is a key ingredient of that Army. But we can't afford to rest on our laurels. We must continue to recognize that the world is a dangerous, dynamic place. And as we evolve the fire support mission area, we must press the technology "envelope" while we train soldiers and leaders to respond to the dictates of future battlefields.

The business of fire support is not an amateur sport. We have to continue to provide fires for the maneuver commander—when and where he needs them. Napoleon once said, "God fights on the side with the best artillery." We have no choice but to continue to train and field the best artillery. The alternative is unacceptable.

Brigadier General Tommy R. Franks was the Assistant Division Commander for Maneuver of the 1st Cavalry Division in Southwest Asia during Operations Desert Shield and Storm. Currently, he's the Assistant Commandant of the Field Artillery School, Fort Sill, Oklahoma. He has served as Chief of Staff of the 1st Cavalry Division, and he commanded the 1st Cavalry Division Artillery, both at Fort Hood, Texas. He also commanded the 2d Battalion, 78th Field Artillery, 1st Armored Division, Germany; an artillery battery at Fort Sill; and a howitzer battery as well as the 84th Armored Engineer Company in the 2d Armored Cavalry Regiment in Germany. In addition to other assignments, Brigadier General Franks served in several positions with the 9th Infantry Division in Vietnam.

Ground War: U.S. Training for Conventional Field Combat

Introduction

Conventional military forces have experienced the most unconstrained, realistic combat training at the National Training Center (NTC) in California. The NTC is a dynamic field environment where commanders through privates receive immediate feedback from a competent OPFOR as a result of the decisions they make. Accordingly, the opportunities and penalties associated with the practice of deception are easily assessed. An unpublished Rand Corporation study on deception at the NTC is revealing.[1] The study sought to determine the occurrence of deception, the deception techniques used, and the impact attributed to deception efforts. An analysis of 104 individual battles fought by units undergoing training provided the necessary data.

Roughly 38 of the 104 battles in the Rand Corporation study exhibited deception designed to purposefully elicit behavior from the OPFOR for exploitation by the unit undergoing training. Deception techniques were observed in 21 additional battles, but were so sufficiently uncoordinated and disconnected from purposeful behavior that they could hardly be classified as deception operations. The remaining 45 battles provided little evidence of deception.[2]

Several deception techniques emerged from the 59 battles where some form of stratagem was evident. The table below lists the deceptive techniques observed and the frequency of occurrence.[3]

Technique	*Occurrence*
Smoke	30
Night maneuver	19
False battle position	12
Indirect approach	11
Deceptive counter-reconnaissance	7
Feint or diversion	5
Camouflage & concealment	1
Demonstration	1
OPSEC	1
Deceptive reconnaissance	1
False retreat	1
Controlled movement	1
False intelligence	1
Multiple simultaneous attacks	1

1. F. Feer, "Tactical Deception at the National Training Center," The Rand Corporation, working draft, 1989.
2. Ibid., 5.
3. Ibid., 6.

The impact of deception operations on battle outcomes is astonishing. To begin, successful deception operations resulted in significantly lower losses for the training unit and higher losses for the OPFOR. Attempted deception resulted in comparable losses on both sides. However, training unit losses greatly exceeded OPFOR losses in battles where deception was not attempted. More importantly, when deception was not attempted, success in battle occurred only 14% of the time. However, successful deception operations resulted in battlefield success 75% of the time.[4] Commanders have an obligation to consider deception when they conduct operations. Ignoring deception will almost certainly increase the cost of victory and may result in defeat.

4. Ibid., 8.

CHAPTER 54

Deception Operations in REFORGER 88[1]

Captain Paul A. Haveles

Captain Paul A. Haveles was a battlefield deception plans officer who served with 2nd Battalion, 47th Infantry Division.

This article offers a description, analysis, and criticism of Operation Hero, the deception plan employed in the corps-versus-corp training exercise REFORGER 88 held in West Germany in September 1988. Although this was "the largest peacetime deception operation conducted by the U.S. Army" [and by NATO] since WWII, it was only a "limited success." The author attributes the immediate cause to inadequate coordination between the written deception plan and the actual unit simulations and the underlying cause to the fact that the deception planners were "new to deception."

Significantly, this was the last large U.S. and NATO field exercise in Europe. Because of the high costs ($12 billion in 1990) and environmental and property damage, the REFORGER exercises—which had been an annual event since 1969—were abolished, being replaced with far less realistic command and control simulations.

Given the poor state of deception awareness and training in the U.S. Army in 1988 at corps level as revealed by this article, and that at battalion levels at the National Training Center (NTC) [as revealed by Fred Feer's RAND study] it is fortunate that the Soviet Union collapsed when it did.

1. Captain Paul A. Haveles, "Deception Operations in REFORGER 88," *Military Review*, Vol. 70, No. 8, August 1990, pp. 35–41.

Deception Operations in REFORGER 88

Captain (P) Paul A. Haveles, US Army

The importance of battlefield deception operations was again realized in a corps–level training exercise during **REFORGER 88**. *The author describes the contributions of the V Corps' newly assigned deception cell and the efforts of numerous corps and division units in a very large and complex deception operation. He offers several valuable lessons gleaned from their experiences and the observations of their VII Corps opponents in the exercise.*

THIS ARTICLE will describe the evolution of what may well be the largest peacetime deception operation conducted by the US Army in the last 20 years. It chronicles the development, coordination, composition, execution of and lessons learned from the V Corps battlefield deception plan for *CERTAIN CHALLENGE*, the field training exercise phase of *REFORGER 88*. The purpose of this article is to help provide an understanding of the complexity of deception planning and the capabilities and limitations of deception operations.

Recently, the Army decided to revitalize its use of battlefield deception in all levels of planning: strategic, operational and tactical. A result of this revitalization was the fielding of battlefield deception elements (BAT–DEs) at corps and division levels. One of the Army's first BAT–DEs fielded was at V Corps. An initial goal set for the BAT–DE was to plan, coordinate and execute a corps–level deception plan for *CERTAIN CHALLENGE*. The BAT–DE was fielded at V Corps in October 1987, and prior to *CERTAIN CHALLENGE*, there had been only a few opportunities to test deception operations. However, these previous exercises were limited in their size (division against division) and by the lack of a real enemy target. In most of these exercises, the enemies had been notional.

CERTAIN CHALLENGE was a corps–against–corps (VII Corps versus V Corps), free–play exercise, that pitted two equal strength forces against each other. Each corps fielded approximately 25 maneuver battalions. Each commander had similar forces, and each possessed equal intelligence–gathering systems. Therefore, victory would only go to the commander who most effectively read the battlefield and massed his combat power at the decisive time and place. Deception would be employed prior to and during the exercise in an attempt to cause the VII Corps commander to mass his combat power at a time or place that would give V Corps a decisive advantage.

The BAT–DE at V Corps normally works in concert with operational planners. The overall tactical planning cell for *REFORGER* included a deception planner who assisted the tactical planner in the development of plans for *CERTAIN CHALLENGE*. This ensured complete integration of the deception plan with the rest of the tactical plan. As possible courses of action for the corps were developed, the deception courses of action were also considered. The

To accomplish [its] mission, V Corps needed to convey the deception story that the 8ID(M) was moving to an attack position in 3AD's sector from which they would attack the flank of VII Corps from the northeast. The desired result would be a weakened VII Corps west flank, thus a more favorable force ratio at the point of V Corps' main effort.

deception plan that was ultimately chosen was an initial course of action presented to the corps commander. Therefore, the deception plan would be one that could have been implemented as a course of action and thus more easily believed.

The exercise scenario for *CERTAIN CHALLENGE* pitted the forces of Northland (V Corps) against Southland (VII Corps). VII Corps was, initially, the aggressor and would be attacking to link up with a notional airborne insertion to the rear of V Corps. V Corps would defend initially, then counterattack to gain the initiative. V Corps' initial array of forces included: 3d Armored Division (3AD) defending in the east flank; 197th Separate Infantry Brigade defending in the center; 10th Panzer Division (10PZ) on the west flank and 8th Infantry Division (M) (8ID[M]) as corps reserve. 8ID(M) was initially positioned to the rear of 10PZ, with contingency plans to attack along one of two axes from the west side. The preferred course of action was for 8ID(M) to attack deep from the west around 10PZ (fig. 1).

The deception objective was to influence the VII Corps commander to commit a brigade-size force from his follow-on division to counter a perceived main attack from the 3AD sector (northeast) and also to prevent him from further committing any major forces in the west. To accomplish this mission, V Corps needed to convey the deception story that the 8ID(M) was moving to an attack position in 3AD's sector from which they would attack the flank of VII Corps from the northeast. The desired result would be a weakened VII Corps west flank, thus a more favorable force ratio at the point of V Corps' main effort, the 8ID(M) counterattack. The deception plan was called Operation Hero.

To support the plan, the BAT-DE felt that the corps had to show a significant amount of 8ID(M) assets moving from the western side of the corps sector to the eastern side. The result would be a significant threat that the VII Corps commander could not ignore (fig. 2).

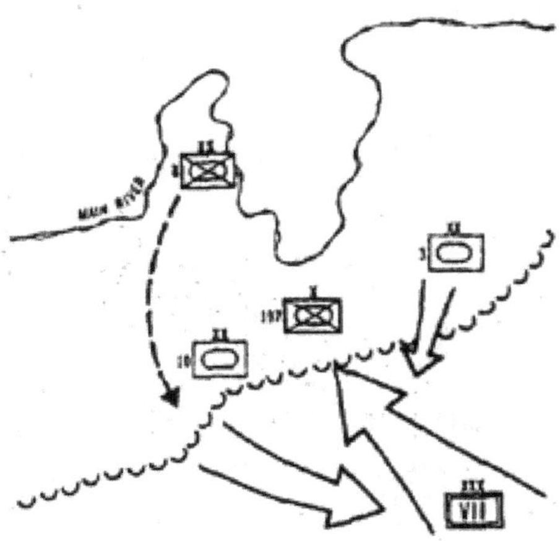

Figure 1

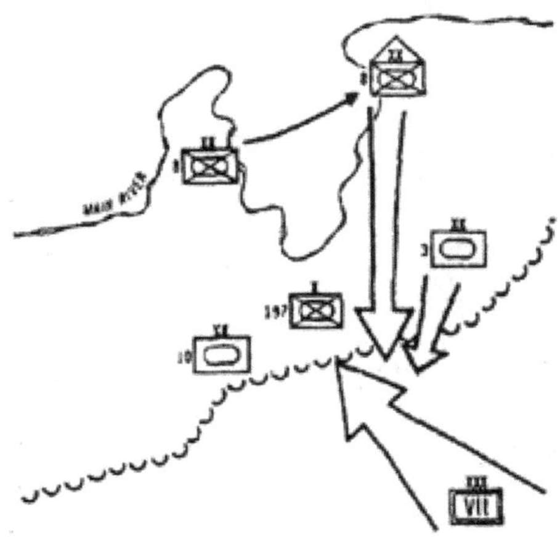

Figure 2

Deception activities were divided into two phases: those that would take place prior to the start of the exercise (STARTEX) and those that would take place on the first day of the exercise. Prior to STARTEX, deception planning began in February with the development of the actual operation plans. As was noted earlier, the deception effort depicted an actual course of action. The deception plan was always briefed as a course of action without any mention of its being a deception effort. Only a select group of individuals actually knew of the deception plan. Thus, the BAT–DE was able to begin spreading the story that V Corps would attack from the east with 8ID(M) and 3AD.

Several Class 60 routes were required to support the deception story. Therefore, written requests were processed for Class 60 routes that would support a major movement from east to west through V Corps' movement control center (MCC). Since the maneuver box for *REFORGER* was located in the VII Corps peacetime sector, the MCC had to process a similar request through the VII Corps MCC to get route information. Host nation permission was needed to conduct smoke operations in support of the deception plan; therefore, written requests had to be processed through V Corps G5 (civil affairs) to obtain approval. Again, since the maneuver box was in the VII Corps sector, the requests had to go through VII Corps' G5 section. The assumption was that this information would eventually find its way to the VII Corps G2 (intelligence) section.

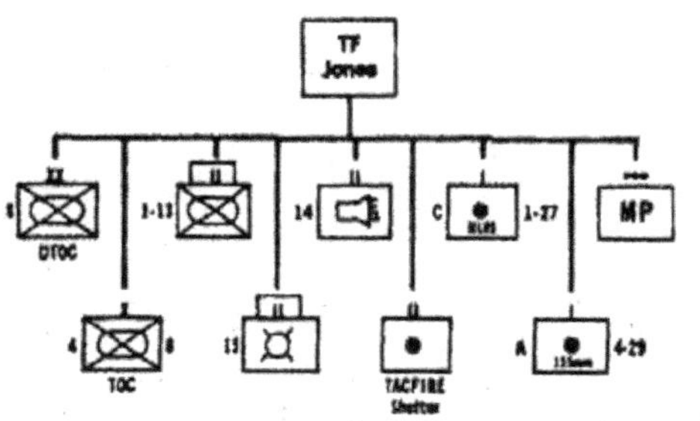

Figure 3

The final action conducted during the first phase took place over the last few days prior to STARTEX. These actions primarily involved 8ID(M). The 8ID(M) DTOC (division tactical operations center) initially occupied a site to the rear of 3AD called TAA (tactical assembly area) Fremont. Then the maneuver units that were occupying positions on the other side of the corps sector sent quartering parties to TAA Fremont to conduct route and area reconnaissance. Quartering parties were also sent from 15th Ordnance Battalion (OD Bn) and 18th Military Police (MP) Brigade in support of this phase of the plan. The intent was to get VII Corps to focus its intelligence–gathering efforts in the vicinity of TAA Fremont.

The 14th PSYOP Bn broadcasted the audio signature of numerous tracked vehicles along the routes. These operations took place during the hours of darkness and were designed to augment the audio signature of the 15th OD Bn. They also spent time crossing the float bridges playing recorded sounds of heavy tracked vehicles (M60s, M88s and M113s).

The second phase of the plan took place on the first day of maneuver. To portray the physical signature of the deception story, a task force (TF) (TF *Jones*) was assembled to replicate the 8ID(M). The TF included elements of both the 8ID(M) and the 3d Corps Support Command (COSCOM). Additional support for the operation was provided by corps combat support, service support and deception assets (fig. 3). This formidable force contained 114 tracked vehicles, 180 wheeled vehicles and more than 1,000 personnel. TF *Jones* represented 25 percent of the 8ID(M) total deployed strength, potentially painting an unmistakable physical signature. The size of the TF, combined with the extended

TF* Jones *represented 25 percent of the 8ID(M) total deployed strength, potentially painting an unmistakable physical signature. [Its size] combined with the extended duration of the operation, was the essential element of the plan. Each element of TF* Jones *was designed to replicate key elements of the 8ID(M). The DTOC and the 4th Brigade tactical operations center were used to replicate the necessary [C^2] headquarters of the division. The battalion TF, 1–13 Infantry Battalion, was used to replicate the maneuver units of the entire 8ID(M). The battalion–size logistics TF would be used to replicate the division support command.

duration of the operation, was the essential element of the plan.

Each element of TF *Jones* was designed to replicate key elements of the 8ID(M). The DTOC and the 4th Brigade tactical operations center (TOC) were used to replicate the necessary command and control headquarters of the division. The battalion TF, 1–13 Infantry Battalion, was used to replicate the maneuver units of the entire 8ID(M). The battalion–size logistics TF would be used to replicate the DISCOM (division support command). This force, although task organized under the 15th OD Bn, consisted of vehicles from various units of 3d COSCOM. The objective was to gather various types of vehicles normally found in a DISCOM. The signal center (SIGCENTER) was used to replicate the UHF/multichannel communications link between division and corps. The TACFIRE (tactical fire direction system) shelter was used to replicate the movement of the DIVARTY (division artillery) command and control. The artillery batteries also supported this portion of the story.

TF *Jones* conducted a prolonged movement from the 8ID(M)'s initial assembly area to TAA Fremont. Movement took place on the first day of the maneuver phase of the exercise and began at the same time that mounted maneuver of the rest of the corps began. The TF moved during daylight. The objective of the plan was to convince VII Corps, by 1700 that day, that the 8ID(M) was moving to the east side of the V Corps sector and would follow the 3AD as the corps' main attack. If VII Corps could be fed enough indicators that 8ID(M) was on the move, they would react to the threat.

The other units that participated, 130th Engineers, 18th MP and 14th PSYOP Bn (a Reserve Component psychological operations battalion from California), were corps assets performing other tasks to supplement our story. The 18th MP Brigade was employed to operate traffic control points (TCPs) along all routes used by TF *Jones*. The 130th Engineer Brigade was employed to establish float bridges across the Main River. These bridges were located on the east edge of the Main River loop, in areas that did not have sufficient existing Class 60 bridges that would support a major east-to-west movement. The 14th PSYOP Bn broadcasted the audio signature of numerous tracked vehicles along the routes. These operations took place during the hours of darkness and were designed to augment the audio signature of the 15th OD Bn. They also spent time crossing the float bridges playing recorded sounds of heavy tracked vehicles (M60s, M88s and M113s). All of these operations were intended to increase and lengthen the audio and physical signatures of the movement of TF *Jones* from the 8ID(M)'s sector to TAA Fremont.

Additionally, two types of electronic deception devices were employed: a communications deception system (CDS) and a prototype radar deception device. They were used to augment the electronic signature of TF *Jones*. These devices, when properly employed, can replicate the communications and radar signatures of units. These devices were deployed to replicate a communications net and the various radar signatures associated with a division.

The activities of the deception plan began with the 18th MP Brigade establishing the TCPs at 0700. TF *Jones* began its movement with TF 1–13 at 0700. Movement was spaced out so that the last convoy departed by 1400 and closed on TAA Fremont by 1600. The next unit to move was the first convoys of the 15th OD Bn. Their movement was also spread out so that they started movement at 1400, and their last unit departed the initial assembly area by 2300. In addition, the audio signature of 15th OD Bn's movement was augmented through the employment of loudspeaker teams from the 14th PYSOP Bn. Several other activities such as the movement of artillery assets into the 3AD sector, the PSYOP's broadcasts, initiation of the division's communications signature and the activation of the communications deception devices were spread out over the day.

At 0800, the bridges that would support the story were placed into the Main River. Float bridging was used so it could be camouflaged during the day and used only at night. At 1300, the artillery units of TF *Jones* moved to positions that would support the deception story. This included the TACFIRE shelter, the 155mm and MLRS (multiple launch rocket system) batteries.

By keeping the 15th OD movement spread out over a 12-hour period, the deception story portrayed a division on the move. By intermixing loudspeaker teams and broadcasting the sounds of tracked vehicles among the wheeled convoys of the 15th OD Bn, it appeared that more tracks were en route to the TAA. Additionally, by moving at night the chances of vehicle bumper numbers being identified were reduced. This also served to further confuse any long-range surveillance teams in the area. Once movement to TAA Fremont was complete, the operation began to phase out.

The first priority of the phase-out was to get the maneuver forces back to the 8ID(M). TF 1–13 began movement back to 8ID(M) at 2100 that evening. This was based upon 8ID(M)'s attack, which was scheduled for daylight the next morning. The SIGCENTER, 15th OD Bn, 14th PSYOP Bn and the CDS devices maintained the signature of the occupation of TAA Fremont through the evening. Vehicle movement on the roads in the TAA and loudspeaker teams broadcasting the audio signature of tanks and armored personnel carriers in the forested areas attempted to give the impression that 8ID(M) was still in the area. The CDS devices replicated the electronic signature of a brigade command net, which was phased out during the early morning hours of Tuesday. Phase-out was accompanied by peaking the frequency of transmissions, which is common just prior to a major movement, and then reducing the transmissions to a finish. The SIGCENTER was still in the TAA, and by keeping it on the air, the division's communication signature was maintained.

At first light on Tuesday, 8ID(M) began the attack on the flank of VII Corps. Once elements of the division made contact with VII Corps, the deception operation was completely phased out. Although 8ID(M)'s attack met significant opposition, there was definite surprise at the company and battalion levels. VII Corps did send maneuver forces to the area intended by the deception; however, it was not determined whether this was part of VII Corps' initial plan or a reaction to the deception plan. In after–action discussions with VII Corps, we did learn what they saw and how we could possibly conduct operations more efficiently in the future.

VII Corps stated that for several hours on Monday it was unaware of 8ID(M)'s actual location and that for a time was preparing to accept the deception. However, VII Corps saw some things that kept it from fully believing the deception. The first indicator of the deception was the sudden appearance of 8ID(M). V Corps made an assumption that VII Corps would be able to identify all of the initial positions of its units.

VII Corps was not able to identify the initial assembly areas of the 8ID(M) . . . Then suddenly, a very clear signal. . . showed up in the collection efforts. When VII Corps so easily found 8ID(M) on Monday, its first impression was to doubt the story; but it did not dismiss the deception immediately. Thus, for a 4–hour period, VII Corps was deceived as to the actual location of 8ID(M).

The maneuver box was in the VII Corps sector, and personnel had a week prior to STARTEX to collect intelligence on the disposition of V Corps forces. However, VII Corps was not able to identify the initial assembly areas of the 8ID(M) forces and at STARTEX, it had no idea where they were. Then suddenly, a very clear signal of 8ID(M) showed up in the collection efforts.

When VII Corps so easily found 8ID(M) on Monday, its first impression was to doubt the story; but it did not dismiss the deception immediately. Thus, for a 4–hour period, VII Corps was deceived as to the actual location of 8ID(M). 8ID(M)'s excellent OPSEC (operations security) prior to STARTEX, combined with the obvious signature presented in the deception, worked against the full success of the operation. In the eyes of VII Corps, the blatant signature of the TF *Jones* activity was not typical of a unit able to previously execute outstanding OPSEC.

A second lesson learned was that the ratio of tactical to wheeled vehicles in TF *Jones* was not indicative of a mechanized infantry division. VII Corps personnel indicated that when they counted vehicles and their type, they felt that it was an incorrect tactical–to–wheeled vehicle ratio. However, 8ID(M) did move one quarter of its tactical vehicles, and its wheeled vehicle movement was delayed. The problem here may have been that VII Corps did not see all of 8ID(M)'s tactical convoys. This does illustrate that without actual tactical vehicles in the deception area, a complete and believable story cannot be told. The general assumption used was that an opposing force analyst would round up the number of vehicles identified. During CERTAIN CHALLENGE, this was not the case. When conducting deception operations, making assumptions should be approached with caution.

The next lesson learned concerned the complexity of deception operations. Attention to detail is extremely critical to the success of any deception; any hole in the deception can lead to the enemy's reluctance to accept it. To be effective, the deception plan requires planning, coordination and a significant degree of centralization. It must include all aspects of the story that is being told: command and control, logistics, communications and tactical forces. The result is that the deception plan becomes a complete operations plan that can stand by itself. To effectively meet these requirements, it was necessary to conduct thorough coordination and tap the expertise of the entire corps staff. Having a deception planner work with the operational planner ensured that all of the staff areas that were integrated into the operational plan were included in the deception plan as well.

Security is an essential part of deception operations. The more people aware of the deception story, the greater the chance for compromise. The newness of the deception cell resulted in a greater need to seek assistance from various members of the corps staff. Since those responsible for the deception plan were new to deception operations, as well as to corps–level operations, a significant amount of coordination had to be accomplished. As a result, there was an increased risk of compromise of the plan. We felt that this risk was justified due to several factors: the inexperience of the deception personnel; the need for detailed planning; a need to get the corps staff to routinely integrate deception operations; and the need for a significant contribution from deception operations as a combat multiplier during *REFORGER 88*.

Another lesson learned was the effect of critical signatures. Each unit, regardless of its size, has certain critical signatures on which intelligence analysts key. In our situation, VII Corps keyed in on the military intelligence battalion and DIVARTY of 8ID(M). These elements, during the period of the deception operation, were tasked to support the 197th Separate Infantry Brigade and the 10th PZ. A critical error was made in not including these signatures in the deception. VII Corps quickly determined that these elements were not where they would have been if the deception story were true. Some artillery units and some command and control had been included at the battalion level. However, no attempt was made to portray the DIVARTY or the field artillery brigade headquarters that normally supports the division. This reflects the need to pay attention to detail and to portray every significant signature of the unit in order to increase the chances for success in a deception operation.

Each unit, regardless of its size, has certain critical signatures on which intelligence analysts key. In our situation, VII Corps keyed in on the military intelligence battalion and DIVARTY of 8ID(M). These elements, during the period of the deception operation, were tasked to support the 197th Separate Infantry Brigade and the 10th PZ. A critical error was made in not including these signatures in the deception.

Operation *Hero* can be called a limited success in that it did delay VII Corps from acting for a few hours. It was not fully successful in that it did not divert a brigade from the reserve division. The true success in Operation *Hero* was in its training value to V Corps. The operation caused a cross section of the entire corps to be involved in deception operations. The corps staff began to "think deception," and there are indications that such thinking also serves operational planning considerations across the corps staff. The commanders of the corps now realize the costs involved in deception operations and perhaps are now more aware of deception and its potential.

The deception element of the corps and the divisional elements also received an introduction to the complexity of deception operations and are now several steps closer to being prepared to conduct wartime deception operations. Operation *Hero* was a limited tactical success during *REFORGER 88*. But considering the growth in deception understanding, planning and cross–corps thinking, Operation *Hero* was a true training success and contributed to the objectives of *REFORGER 88*. ***MR***

Captain (P) Paul A. Haveles is a corps battlefield deception plans officer, Headquarters V Corps, Frankfurt, West Germany. He holds a B.A. and an M.A. He served with 2d Battalion, 47th Infantry, 9th Infantry Division, Fort Lewis, Washington, as a rifle platoon leader, heavy mortar platoon leader, battalion maintenance officer and company commander.

CHAPTER 55

Voices in the Sand: Deception Operations at the NTC[1]

Captain George L. Reed

This article summarizes the extensive but highly cost-effective use of deception by the OPFOR (then the 63rd Regiment of 1st Infantry Division) at the Fort Irwin (California) National Training Center (NTC). With its smooth blend of theory and practice, this article is far more practical than any of the relevant US doctrinal manuals. It is combat deception in a six-page nutshell.

1. Captain George L. Reed, "Voices in the Sand: Deception Operations at the NTC," *Armor*, Vol. 97. No. 5, September–October 1988, pp. 26–31.

Voices In The Sand:

Deception Operations at the NTC

by Captain George L. Reed

"The way to overcome superior intelligence capability is to provide the gatherer with large quantities of believable information - while the real plan is executed elsewhere."

The U.S. Army has gathered a significant treasury of lessons learned from the many rotations at the National Training Center (NTC). With experience and study, these lessons have developed into refined skills – in gunnery, command and control or, in more recent times, intelligence gathering.

The employment of new equipment and internalization of proven doctrine has greatly increased the amount and detail of intelligence data available to maneuver commanders. The ability to monitor numerous enemy radio nets and to see far behind enemy lines has given task force commanders a rather devastating weapon to add to their arsenal of combat multipliers. But, as is the case with many lethal weapons, this sword cuts both ways.

The ability to acquire, transmit, and use intelligence data stopped being original after the battle for the Marathon Plains. Since then, we have evolved the system, but the tenets remain the same: the way to overcome superior intelligence capability is to provide the gatherer with large quantities of believable information – while the real plan is executed elsewhere. Hence comes deception, "the art of the possible."

The stated doctrine for diversionary brigades, PSYOP battalions, and deception platoons, as discussed in the 100-series FMs and many other sources, is that the opposing forces (OPFOR) at the NTC focus deception activities on a Blue unit's failure to adhere to a basic precept of intelligence gathering. That grain of sand, not yet polished into a diamond, is that *you must always confirm your intelligence.*

A discussion of deception operations at the NTC, and their value and applicability to U.S. operations, can be conducted in a manner as simple as the "illusions" themselves, by explaining what, how, and why. The "what," of course, is the use of minimal internal assets, and/or deception sections from division or combined arms army-level, to attrit the enemy's combat-effectiveness by accomplishing three major tasks: confuse, deceive, and weaken his forces.

Confuse the enemy by:

- **Drawing simultaneous attention to several areas.** An effective deception will show several major actions occuring in several locations. This will spread out the enemy's intelligence assets and make confirmation more difficult. The enemy will now have less room for error, and his intelligence personnel will be under greater pressure to correctly analyze the situation.

- **Make the enemy attend to details he thought had been completed.** This is done through radio deception discussing the clearing of enemy obstacles, securing of terrain features, and other such actions that will lead the enemy to waste time rechecking details he has already spent precious time completing. The benefit of this is that time spent reconfirming known data is not spent detecting the rest of the deception plan.

• **Make OPFOR available forces appear larger than they are.** The enemy will probably have visual confirmation or reports from higher intelligence sources that tell him where the actual OPFOR elements are located. Visual and radio deceptions can make it appear that other elements of comparable size are also located in the sector. This action leads the enemy to be overcautious in his planning, thinking he is up against a much larger force than really exists.

• **Make enemy intelligence sources conflict with each other.** This is easily done by providing his electronic warfare elements with large quantities of false radio traffic. The next step is to hinder other, less "foolable" assets, such as scouts and ground surveillance radars, by screening the sector with smoke and jamming radio capabilities. The end result is a confusing barrage of intelligence that does not make sense.

Deceive the enemy by:

• **Causing him to anticipate attacks and/or counterattacks from fictional locations and keeping his attention devoted to areas other than where we want to operate.** As the enemy watches and listens with EW assets to preparations for false attacks and counterattacks, he will begin to "lean forward" by moving forces to positions of advantage against the movement he anticipates. If this is accomplished, the real attack may find little resistance in its path on the side of the sector opposite that of the deception plan.

• **Causing him to believe the OPFOR has made massive repositioning maneuvers since his last confirmed intelligence.** Enemy artillery barrages are a significant threat to OPFOR maneuver commanders, often accounting for up to 50 percent of OPFOR losses. Enemy barrages will not be on target, however, if the enemy thinks he is shooting at where the OPFOR has "moved."

• **Causing him to believe he is under threat of imminent destruction by ambush, penetration, attack, or counterattack.** The adage that "haste makes waste" is applicable when an enemy commander hastily sends part of his combat power to guard a flank that his intelligence indicates the enemy will attack.

Weaken enemy actions by:

• **Causing early deployment of reserves and/or counterattack forces.** Command and control of a task force or brigade is a difficult thing to maintain in the heat of battle. Once a reserve or counterattack force is deployed, it is difficult to get it to change course and redirect its offensive into another area. Early deployment, at the very least, deprives the maneuver commander of a timely deployment when and where he really needs it.

• **Causing unnecessary repositioning.** If the enemy is repositioning against imaginary threats, he is not rehearsing in his chosen battle position, and he is not executing sleep plans. This will help to fatigue him, as well as attrit his preparations for the coming battle.

• **Diverting attacking forces from the main effort.** Such diversions are the benefit of making the enemy believe he is about to be ambushed. Combat ratios in the actual fight change to the OPFOR's favor when the enemy commander sends teams off to guard against non-existent counterattacks. Thus, the OPFOR can defeat the enemy in detail at even or better odds.

• **Causing the waste of combat multipliers such as artillery, air defense, close air support and FASCAM fires.** A common response to unexpected "attacks" is to fire conventional artillery or FASCAM to slow the assault and attrit the attackers. This, if the "attack" was real, would allow the enemy time to ready himself for the "surprise attack." The end result is that the enemy wastes critical combat multipliers against empty ground.

• **Causing command and control elements to question the intelligence and spend needless time analyzing simple data and courses of action.** As the deception plan comes together, the enemy will spend long hours trying to figure out what is going on, and wondering why the "attack" that higher intelligence has reported as imminent never comes. Once the maneuver unit begins to doubt its intelligence sources, the value of the intelligence gatherers is compromised.

It is the general goal of deception operations to accomplish the above tasks in most phases of combat, including attacks, meeting engagements, defenses, reconnaissance activities, and night operations.

With the above description of the goal of deception planning, we can move to the methodology of how to plan and execute these operations. The key is *to depict a scenario that the enemy is willing, if not anxious,*

"Show the enemy what he expects to see; it will make him think he is in control of the situation...."

to accept as real. The success of the deception is directly proportional to the number of reliable intelligence sources that can be contaminated with the various forms of false data within the context of the deception plan. Put simply: What is seen and heard, again and again, becomes believable. (Incidentally, this is the favorite tactic of dictators; constant repetition creates belief.) Show the enemy what he expects to see; it will make him think he is in control of the situation. Few overconfident commanders take detailed precautions.

The preparation of the deception must be made at the tactical planning level, in the maneuver unit's operations order, thereby making the false operation an integral part of the real maneuver plan. This brings the deception plan into phase with all of the other combat multipliers at a maneuver commander's disposal, thereby ensuring that the illusory operation enhances the mission without hindering the commander's intent. It is unlikely that this would be accomplished if the deception was planned in the war room of a military intelligence battalion.

In common with U.S. doctrine, the OPFOR uses METT-T analysis as the core of deception planning. At the NTC, the OPFOR has developed different styles of deception for offensive and defensive operations. In offensive operations, mainly division and regimental attacks or meeting engagements, assets internal to the motorized rifle division and combined arms army are used, but parceled out to the controlling headquarters. The main emphasis of deception in the offense is to deceive the enemy with false avenues of attack and false times when the attack will occur. This allows the regiment or division to retain surprise and momentum.

Many media provide the desired picture:

- **Sound:** Tapes of a MRR's vehicles in a combat line will provide both the straightforward audio presence of a "regiment" in the pre-dawn hours, as well as an excellent background for battle traffic on the radio. This can be accomplished with $400 worth of used car stereo equipment and a few hours to fit the system to the BRDM-II.

- **Visual:** Test fires, MILES checks, heat signatures, covering smoke, and dust signatures will replicate a regimental-size march to the attack. As the enemy begins to see the "attack," he will react with security upgrades, indirect fires, and repositioning. After all, seeing is believing.

- **Artillery:** This can provide false illumination and false preparatory fires. Areas the "regiment" appears to be interested in will shortly be of interest to the enemy.

- **Radio:** Enemy intercepts of a "busy" command and control net are often the pivotal intelligence source. If the enemy hears "us" talking about doing what his people are seeing and hearing, he will believe it.

- **Flanking:** During meeting engagements especially, the enemy is concerned about flanking actions, ambushes, and surprise attacks. If the deception (including FASCAM fires) can convince him to send elements off to the various flanks, the real combat ratio is significantly improved.

- **Misfires:** Misplaced or poorly-timed FASCAM and ICM fires can greatly assist the real maneuver element. As the other elements of the deception are completed and misfired enemy artillery occurs, the enemy maneuver elements on the ground will believe their headquarters is shooting at the real attack.

Putting It All Together

These techniques can combine to provide an excellent battlefield effect, as occurred in a recent MRR night attack against an M1/M2-equipped task force in the central corridor of the NTC maneuver area. The OPFOR conducted the attack from east to west with a line of departure at Hill 720 (See Map 1.)

Phase One: Radio traffic concerning rehearsals in the north end of the sector, as well as the clearing of defiles along the march route, cluttered the air all afternoon. As a result,

- The enemy expended about 4,800 rounds of enemy artillery in vain attempts to destroy the "forces" in the north.

- The enemy was convinced that two MRBs would attack in the north, with LD no later than 2100 hours.

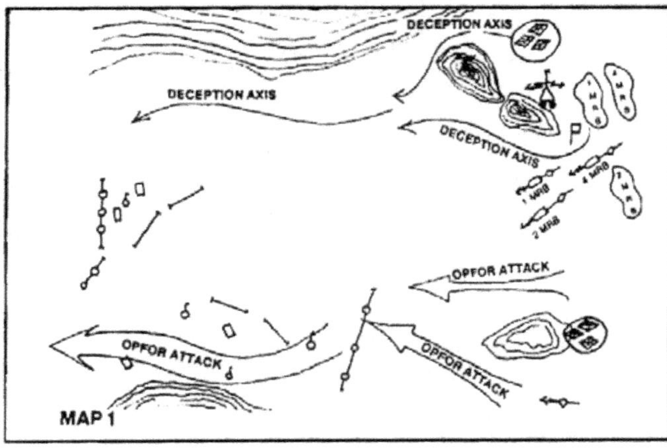

MAP 1

• The enemy was out of artillery ammunition, pending resupply, when he tried to fire indirect at the OPFOR dismounted attack at 1900 hours.

Phase Two: At 2030, the "regiment" passed the LD *en masse* with smoke, sound, illumination, dust, chemlites, and prolific "battle traffic" on the radio. As a result,

• More wasted enemy artillery ammunition.

• The brigade informed the enemy task force to expect the main attack within 30 minutes.

• The task force cancelled sleep plans, bringing the task force to 100-percent security.

Phase Three: At 2300, the "regiment" conducted a second attack, with sound, smoke, illumination, chemlites, dust, heat signatures, fake battle damage, and heavy battle traffic on the radio. The results were that:

• Ineffective enemy FASCAM fires.

• The enemy task force came to 100-percent security again.

• The enemy brigade told task force that two MRBs were confirmed to attack in the north.

Phase Four: When the real regiment attacked in the south at 0100, the deception regiment attacked north. The end result was that enemy forces repositioned to the south too late, and an MRB(+) secured the objective. The task force was combat ineffective. The benefits of the deception were that the OPFOR retained surprise, initiative, and momentum, while depriving the defender of critical artillery assets and execution time.

Deception In the Defense

At company, battalion, and regimental levels, deception has a key role in the success of the defensive mission. The primary goal of deceiving the attacking enemy is to seize the initiative, thereby draining his attack's lifeblood. Also gained by a successful defense deception is protection from air and artillery barrages, as well as mounted and dismounted infiltrations. Such combat multipliers, if the attacker uses them effectively, can unseat a defender, however well prepared. The tasks involved with defense deception planning are to deceive the enemy about where the defensive positions, avenues, and routes for counterattacks and repositioning are, and where the reserve forces are. This can be done as low as motorized company level, but must be coordinated with higher command levels, as with any combat multiplier.

The following media can be useful in developing effective illusions in the defense:

• **Sound:** Tapes of repositioning and/or counterattacking forces, as well as vehicles idling in false positions, provide a deceptive radio picture. The same inexpensive sound system on the BRDM-II will suffice here.

• **Visual:** False barriers, vehicle and infantry positions, and artificial heat signatures, not to mention simulated battle damage (supposedly destroyed vehicles in flames, etc.) will paint an extremely believable tale of a main defensive belt forward. Flashbulb trip flares will add greatly to the reality of the ruse, and are relatively cheap and easy to make.

• **Artillery:** Smoke and guiding illumination, or special signal flares, will give a physical signature to a deception plan involving repositioning or counterattacking elements.

• **Radio:** Lengthy discussions of vehicle locations, rehearsals for counterattacks, and loose talk such as, "Are routes Green and Red clear for my Charlie-Alpha?" paint a picture of an overconfident defender with poor OPSEC.

Combinations of these techniques can keep the enemy guessing for extended periods of time, as well as having the following effects:

• Misfired artillery preparation.

- Enemy wastes time probing and breaching empty positions, making himself a vulnerable artillery target.

- The enemy attack is hindered, forcing night attacks to take place in daylight.

- After the enemy has "overwhelmed" the bogus position, he ends up doing his reconsolidation in the main fire sack.

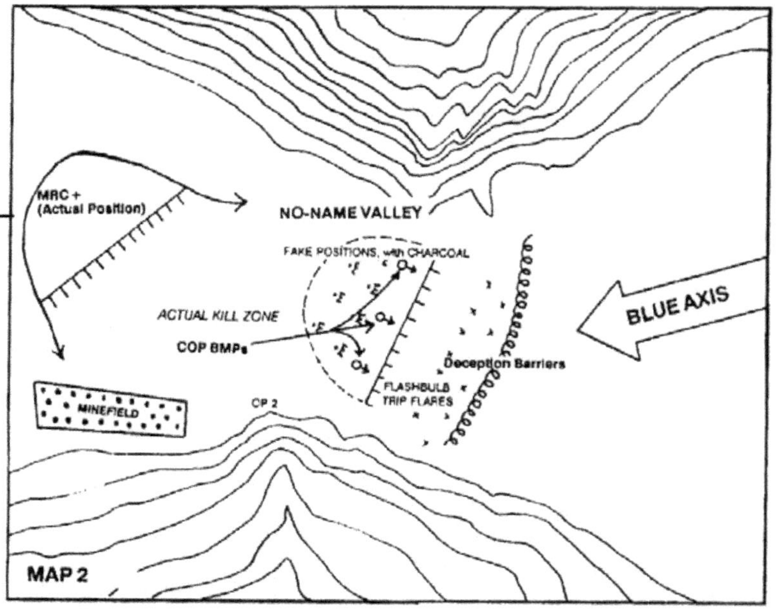

MAP 2

During a recent MRC defense, deception operations effectively helped defend the area known as "No-Name Valley" in the NTC maneuver area. An MRC(+) of six T-72s and 14 BMPs defended that terrain against a night attack from an M1-heavy task force with 41 tanks and 20 Bradleys (See Map 2).

Phase One: The MRB commander directed his MRC(+) to dig in at the western mouth of the valley, leaving the eastern end open, except for a COP consisting of three BMPs. False positions, obstacles, and vehicles went in three kilometers forward of the main defense belt. Initial laager of the MRC(+) was well forward of the defensive position. Heavy engineer activity was evident in the deception area.

Result: Enemy templated an MRC(+) forward, in the eastern end of the valley. Pre-planned artillery barrages were prepared against the "known" positions. Enemy reconnaissance went out in the late afternoon to probe the templated defense.

Phase Two: The MRB commander placed 100 flashbulb trip flares in the deception battle positions and along the perimeter of the false barriers. COP BMPs were placed in three of the fake fighting positions, while the remaining positions were prepared with charcoal fires, iron-gratings, and simulated vehicle antennas. False OPSEC violations on the radio discussed feeding Class I to the 13 vehicles east of OP-2.

Result: The COP destroyed seven enemy scouts when the Bradleys began probing the initial barrier line. The scout's dying report was that they had been destroyed in the main defensive belt. The COP withdrew to the rear of the deception position until after the enemy artillery barrage that evening. As the barrage ended, the COP reoccupied and provided harrassing fires against the enemy main body as it moved to seize the battle position. The enemy main body set off numerous flashbulb trip flares and fired on the flare pits with 25-mm cannon, thinking they were tank signatures.

Withdrawing OPFOR reconnaissance vehicles, in conjunction with the COP, destroyed the seven Bradleys near the first line of false positions.

Phase Three: Enemy tanks continued attacking the fake positions throughout the night, taking occasional casualties from the withdrawing COP.

End Result: The sun rose at 0545 as the task force was reconsolidating in the main OPFOR engagement area. The MRC(+) executed volley fire over the next two hours, destroying the confused vehicles. At change of mission, the enemy task force had lost all 41 M1s and 17 Bradleys. OPFOR casualties were one BMP by 25-mm direct fire and one T-72, an artillery fire casualty.

Throughout the course of the battle, the defender used deception to take initiative and momentum away from an enemy whose equipment was built for, and relied upon, speed and violent momentum.

As noted before, the assets required to execute effective deception operations at all levels are mini-

"...There is no reason why deception tactics cannot be used within tactical operations at the U.S. division and brigade levels. By so doing, the enemy would be forced to refine his own intelligence processes, confirming what he gathers."

mal. They include little more than the following.:

Organization & Equipment of OPFOR Deception Team

- One BRDM-II with tape deck and loudspeakers.
- One HMMWV command and control vehicle.
- Two RPG-7 launchers.
- About a hundred deception trip flares and 500 chemlites.
- Ten smoke pots and 10 false heat signatures (charcoal).

For every good plan there are drawbacks, and deception operations are no exception. The equipment listed above does not include protection against direct or indirect fires. Survival and vulnerability are a thorn in the side of a deception plan. If engaged, the deception team will be destroyed. It is not a true fighting force..

The good news, however, is that a successful deception operation has the enemy looking for at least an MRB-size element, not a lonely BRDM. During 15 deception missions conducted in recent months, NTC deception teams have sustained no casualties.

Recently, an M1 platoon drove within 15 meters of the OPFOR deception team's BRDM-II, but failed to engage it. They were looking for a regiment and disregarded the single vehicle. Perhaps the following quotation applies:

"When a man is searching for the source of the voices, he pays small heed to the sand in his eyes."

— **Lawrence of Arabia**

OPFOR troopers prepare to dismount after a successful NTC exercise.

Deception planning, within doctrinal framework, is an integral part of OPFOR mission execution at NTC. The operations are kept simple and require little manpower and few assets. More important, the operations work. In the offense, they allow the OPFOR to achieve surprise; in the defense, they disarm the attack and transfer initiative and momentum to the defender.

These are the benefits of good deception and the cost of failing to confirm intelligence, a failure that is not uncommon, although often difficult to avoid. When confirmation and effective evaluation are internalized into the intelligence process, the maneuver commander need not fear deception. He need only detect it, and then disregard the erroneous data.

On the other hand, there is no reason why deception tactics cannot be used within tactical operations at the U.S. division and brigade levels. By so doing, the enemy would be forced to refine his own intelligence processes, confirming what he gathers. The simple deception techniques illustrated above can be and, indeed, have been used against the OPFOR. After all, sand can obscure the vision of any man whose eyes are unprotected — regardless of his uniform.

Captain George L. Reed was commissioned in Armor in 1984 from the USMA. He has served in the 1-63 Armor, the OPFOR at the NTC, as mortar platoon leader, tank platoon leader, tank company XO, battalion liaison officer, and regimental fire support officer. A graduate of the AOB, IMPOC, and JOTC, he is currently assigned as the 1-63 Armor S4. His duties include OIC of OPFOR tactical deception operations.

CHAPTER 56

OPFOR Counterreconnaissance at the National Training Center

Captain Richard Randazzo

This article reports the highly cost-effective counter-reconnaissance efforts of the Red Team ("OPFOR") regiment at Ft. Irwin and the generally poor showing of the Blue Team ("BLUFOR") brigades. Incidentally, these Blues performed no better than the Blues of a decade earlier, as reported by Colonel John Rosenberger.

Richard A. "Rick" Randazzo (1993 West Point graduate), a U.S. Army (Armor) officer, served 34 rotations with the OPFOR at the NTC and subsequently as an Assistant Professor of Military Science at Southern Illinois University where he took an MBA in 1999.

FROM THE NTC:

OPFOR Counterreconnaissance At the National Training Center

by Captain Richard Randazzo

The success or failure of most National Training Center battles is determined long before the main combat forces leave the line of departure. An attacking force with good intelligence can effectively plan and maintain the initiative, while a poor intelligence effort often leads to haphazard planning and a blind, ineffective attack. With this in mind, the OPFOR places a fundamental emphasis on detecting and destroying the attacker's reconnaissance effort. This article will describe how the OPFOR conducts effective counterreconnaissance screens in order to provide ideas for BLUFOR commanders to refine the execution of their own counterreconnaissance missions, and to provide BLUFOR commanders with some techniques to defeat the OPFOR screen line when attacking. This article addresses each element of the Battlefield Operating Systems (BOS) and explains their synchronization within the OPFOR Motorized Rifle Battalion (MRB) counterreconnaissance effort.

Situation

Once a mission is received, the MRB commander, the MRB executive officer (XO), and the Motorized Rifle Regiment (MRR) scout platoon leader, will conduct a combined map reconnaissance to determine possible scout avenues of approach, possible support-by-fire positions, probable kill sacks, and positions for OPFOR vehicles. If time permits, the MRB commander, executive officer, and scout platoon leader will conduct a joint area reconnaissance to confirm or deny their initial map assessment. Once this initial planning stage is complete, and his intent for the counterreconnaissance battle is fully understood, the MRB commander will focus on preparing the main defense, and the MRB XO will assume responsibility for the counterrecon fight.

During the subsequent operations order, the MRB XO will issue the concept and intent of the counterreconnaissance operation, including proposed locations, requirements for closing lanes in obstacle belts, and guidance for patrolling obstacle belts. The scout platoon leader will then brief his vehicle locations, which positioned themselves after the initial planning session, providing eight-digit grid locations, as well as task and purpose for each scout vehicle. He will also provide the scout platoon radio frequencies, engagement and disengagement criteria, and any newly gathered intelligence.

The MRB XO will generally position the Counterreconnaissance Patrol (CRP), consisting of three BMP2s and two BRDMs, in ambush positions along probable mounted infiltration routes 800 to 1,000 meters in front of the MRB tactical obstacle belt. The XO will also incorporate AT-5 and air defense systems into the MRB counterreconnaissance plan, placing them on key terrain near the MRB reserve/quick reactionary force. In addition to the MRB assets positioned by the MRB XO, each MRC will position its own individual screen line consisting of one T-80 tank and two BMPs 100-200 meters behind the tactical obstacle belt in the main defensive area.

The MRB creates a substantial four-belt counterreconnaissance screen with scouts, CRP vehicles, MRC screens, and AT-5s in position. Although each of these counterreconnaissance forces must also prepare defensive positions for the main battle, the OPFOR commander understands that denying enemy reconnaissance is the key to victory, and therefore, that is where he weights his defensive effort.

Command and Control

The MRB XO commands and controls the counterreconnaissance fight from his BRDM. This gives him the flexibility to quickly reposition himself and provides better communications than a tracked vehicle. The enhanced communication proves vital to the XO as he must monitor both the scout's intelligence net and the MRB command net. As scouts send reports on the intelligence net, the MRB XO coordinates between the four counterreconnaissance belts on the command net, ensuring positive hand-off of enemy forces. If necessary, the XO will instruct the scout platoon to coordinate directly with a killer team, but he will continue to monitor and control the entire effort. All vehicles send spot reports on the command net to ensure attachments monitor the proceedings, but all engagements occur on internal frequencies.

To further facilitate command and control, the XO issues a specific task and purpose to each belt of the counterreconnaissance effort. Scouts are the forward eyes and identify and report approaching enemy vehicles, engaging the enemy only in self defense. Scouts will maintain visual contact with enemy vehicles until positive hand off occurs with the CRP or MRC screening forces. The CRP assumes responsibility of the enemy forces from the scouts and if it is capable, destroys them. If the force is too strong for the CRP, it will pass the enemy back to the tanks in the MRC screen line. The MRC screen line will then engage to destroy the enemy force before it locates the defense's main obstacle belt.

Positive hand-off between each reconnaissance belt is essential to help eliminate fratricide. Therefore, the OPFOR will conduct detailed counterreconnaissance rehearsals at the MRB, MRC, and MRP levels. These rehearsals stress the initial identification of the enemy vehicle and the tracking of that vehicle until it is destroyed.

Fire Support

During the planning process, the XO and the scout platoon leader plan illumination and HE targets. As the scouts and

CRP deploy following the initial planning process, they will confirm or adjust each target. They will also place VS-17 panels with chemical lights at the grid of each artillery target to further facilitate effective calls for fire.

Once the counterreconnaissance battle begins, the scouts will utilize illumination rounds to provide the CRP easy visual identification of enemy forces. If enemy forces stop, the scouts will destroy them with indirect fires, adjusting from either VS17 panels or chemical light TRPs.

Intelligence

The scout platoon, which consists of vismod BRDMs, BMPs, RKHs, GSRs and ERPs, deploys throughout the depth of sector. It usually sends two BRDMs and two RKHs to attempt identification of the enemy's line of departure prior to the attack. It positions the remaining forces along key terrain covering mounted and dismounted avenues of approach. To maximize the reconnaissance effort and ensure redundancy, each vehicle will also position a two-man dismounted observation post. Scout BMPs generally block mounted avenues of approach or are incorporated into the MRBs counterreconnaissance fight.

Air Defense

Dismounted SA-14 teams are employed on the high ground along the flanks of the main defensive area. Although they position themselves along the probable air mobility corridors, the scout platoon often identifies enemy air assets first; therefore, the ADA SA-14 teams must monitor the intelligence net to ensure they have a common view of the battle with the scouts. The air defense BRDMs will clear possible enemy landing zones and are quickly assimilated into the quick-reaction force/reserve.

Mobility/Countermobility

Both scouts and CRP vehicles emplace protective obstacles to aid in the destruction and detection of the enemy. Wire and mines are usually employed on the OPFOR side of an intervisibility line (IV) or after a turn on a single vehicle trail. Like the main obstacle belt, locations of the protective minefields must be reported higher and incorporated into the MRB's obstacle plan. Additionally, boulders and tank ditches are sometimes used to block the small avenues of approach along the flanks. When obstacles are utilized, either a scout BMP or a CRP BMP will overwatch the obstacle.

Maneuver

As defensive preparations begin, the MRB commander allocates one third of each MRC's combat power into the MRB's third counterreconnaissance line. Although substantial forces are already forward (scouts, CRP) his maneuver forces must still dig in.

During daylight, one MRP from each MRC will conduct the counterreconnaissance screen while the other MRPs prepare their defensive positions. The screening MRP will identify the CRP vehicles to its front and any friendly scout platoon positions along its flanks. The BMPs dismount their crews and conduct dismounted patrols of high ground that can observe their battle positions.

During limited visibility, the positioning of the MRP screening force becomes critical as battle hand-off between the CRP becomes more difficult. Each MRC commander positions his screening force where it can still observe main avenues of approach, but during limited visibility, he also ensures it possesses a clear hand-off from the CRP vehicles. If his sector includes a flank, MRC commanders will reinforce that area, placing two vehicles in a "backstop" position behind the existing CRP vehicle. The third MRP vehicle is responsible for the remainder of the sector and ties in with the flank MRP. All three MRCs will array in this format.

Backstopping the MRP screen lines are the AT-5 assets and the MRB reserve. Usually the OPFOR will place one or two AT-5s on a key piece of terrain to help identify any penetrating enemy vehicles with their thermal sights. The MRB reserve, made up of the MRB commander's tank and BMPs and BRDMs from the MRB, is a flexible force which reacts quickly to any penetrations of the screen line and to any threats in the rear or flanks.

Recommendations

The OPFOR is successful during the counterreconnaissance fight primarily because they echelon their counterreconnaissance forces. Unlike the BLUFOR, the OPFOR involves the entire battalion in the counterreconnaissance fight, therefore increasing the probability of detecting infiltrating forces. The four-echelon structure allows the OPFOR the flexibility to reinforce high-speed avenues of approach without risking other areas.

A second key to the OPFOR's success is centralized command and control. Having one commander who is responsible for the entire counterreconnaissance fight ensures that, not only is the mission planned, rehearsed, and executed, but it is also synchronized at the MRB level and includes all BOS elements.

The OPFOR will usually position the majority of its vehicles to cover the flanks, as they are the most likely infiltration routes leading into a sector. Therefore, "a way" the BLUFOR commander may penetrate an OPFOR screen line would be to infiltrate his scouts down the center of the defender's sector while a tank company attacks one flank. Although the terrain in the center is often open, and may not be conducive to unobserved movement, the BLUFOR will gain the element of surprise where the OPFOR is the weakest. The tank company should provide enough distraction to allow the scouts to penetrate the screen line quickly before they are detected. To improve chances of success in the center sector, scouts should also attempt to infiltrate dismounted.

A second technique to penetrate an OPFOR screenline would be to hold the LD times of infiltrating scouts until after 0200 hours. Although this would not guarantee the OPFOR is asleep, it would increase the chances of being successful.

Conclusion

The OPFOR allocates over 30 vehicles to detect and destroy enemy reconnaissance attempting to penetrate their defenses. Their four-echelon defense is very successful in destroying the enemy's reconnaissance effort. By preventing the BLUFOR attackers from acquiring the knowledge required to achieve success, the OPFOR consistently achieves decisive defensive victories.

CPT Richard Randazzo graduated from the U.S. Military Academy in 1993. He has been an MRC commander, MRB executive officer, and the division and regimental scout platoon leader at the National Training Center for 34 rotations. His next assignment will be as an assistant professor of military science at Southern Illinois University at Edwardsville.

The author would like to thank CPT Ross Brown, CPT Rob Kaderavick, and 1LT Geoff Smaltz for their help and insights with his article. Lastly, he would like to thank CPT Steve Mandes for his leadership and training.

Section XIV: Economic Warfare

CHAPTER 57

The Farewell Dossier: Duping the Soviets[1]

Gus W. Weiss

The large-scale and largely successful Soviet industrial espionage efforts against the West since 1970 are well-documented and widely known. Less known is a cunning American countermeasure. Dr. Waldo Weiss (1966 NYU Ph.D. in Business Administration) served as a senior advisor to U.S. Presidents Nixon, Ford, Carter, and Reagan and the NSC. Here, he reveals fascinating details of his starring role as a planner in the planting of false technology, including software rigged with a "Trojan Horse," on the KGB beginning in 1982. The French and Americans had been alerted to Soviet espionage efforts by French agent KGB Col. Vladimir I. Vetrov from 1981 until his discovery and execution in 1983.

The Farewell Dossier

> *We communists have to string along with the capitalists for a while. We need their credits, their agriculture, and their technology. But we are going to continue massive military programs and by the middle 1980s we will be in a position to return to a much more aggressive foreign policy designed to gain the upper hand in our relationship with the West.*
> —Leonid Brezhnev, remarks in 1971 to the Politburo at the beginning of détente.

During the Cold War, and especially in the 1970s, Soviet intelligence carried out a substantial and successful clandestine effort to obtain technical and scientific knowledge from the West. This effort was suspected by a few U.S. Government officials but not documented until 1981, when French intelligence obtained the services of Col. Vladimir I. Vetrov, "Farewell," who photographed and supplied 4,000 KGB documents on the program. In the summer of 1981, President Mitterrand told President Reagan of the source, and, when the material was supplied, it led to a potent counterintelligence response by CIA and the NATO intelligence services.

President Nixon and Secretary of State Kissinger conceived of détente as the search for ways of easing chronic strains in US-Soviet relations. They sought to engage the USSR in arrangements that would move the superpowers from confrontation to negotiation. Arms control, trade, and investment were the main substantive topics. The Soviets viewed détente as "peaceful coexistence" and as an avenue to improve their inefficient, if not beleaguered economy using improved political relations to obtain grain, foreign credits, and technology.[1] In pure science, the Soviets deserved their impressive reputation, and their space program demonstrated originality and accomplishment in rocket engineering—but they lacked production know-how necessary for long-term

1. Gus W. Weiss, "Duping the Soviets: The Farewell Dossier," *Studies in Intelligence,* Vol. 39, No. 5, 1996, pp. 121–126.

competition with the United States. Soviet managers had difficulty in translating laboratory results to products, quality control was poor, and plants were badly organized. Cost accounting, even in the defense sector, was hopelessly inadequate. In computers and microelectronics, the Soviets trailed Western standards by more than a decade.

Soviet S&T Espionage

The leadership recognized these shortcomings. To address the lag in technology, Soviet authorities in 1970 reconstituted and invigorated the USSR's intelligence collection for science and technology. The Council of Ministers and the Central Committee established a new unit, Directorate T of the KGB's First Chief Directorate, to plumb the R&D programs of Western economies. The State Committee on Science and Technology and the Military-Industrial Commission were to provide Directorate T and its operating arm, called Line X, with collection requirements. Military Intelligence (GRU), the Soviet Academy of Sciences, and the State Committee for External Relations completed the list of participants. The bulk of collection was to be done by the KGB and the GRU, with extensive support from the East European intelligence services. A formidable apparatus was set up for scientific espionage; the scale of this structure testified to its importance. The coming of détente provided access for Line X and opened new avenues for exploitation. Soviet intelligence took full advantage.

In the early 1970s, the Nixon administration had no comprehensive policy for economic relations with the USSR. The sale of strategic goods to Communist countries was governed by the Coordinating Committee of NATO (COCOM), which administered an Alliance-agreed list of products and data embargoed for sale. Nixon's policy worked within this system, and, for the export of products exceeding the approved list, special exceptions were necessary. And, in a new set of commercial and scientific arrangements, the United States and the USSR set up joint technical commissions to assess prospects for cooperation. Topics included agriculture, nuclear energy, computers, and the environment. As Kissinger noted:

> Over time, trade and investment may leaven the autarkic tendencies of the Soviet system, invite gradual association of the Soviet economy with the world economy, and foster a degree of interdependence that adds an element of stability to the political relationship.[2]

Beginning in 1972, delegations of Soviet specialists came to the United States to visit firms and laboratories associated with their commissions. Line X, ever alert, populated these delegations with its own people: in an agricultural delegation of 100 about one-third were known or suspected intelligence officers. On a visit to Boeing, a Soviet guest applied adhesive to his shoes to obtain metal samples. In another episode, the ranking scientists and managers of the Soviet computer and electronics industry obtained a visa for the specific purpose of visiting the Uranus Liquid Crystal Watch Company of Mineola, Long Island (a firm not among the *Fortune 500*). Three days before the delegation's arrival, they requested an expansion of the itinerary to include nearly all U.S. computer and semiconductor firms. This maneuver was done to observe (that is, collect) the latest technology and it was executed at the last minute so that the Defense Department would not have time to object. It was legal—Line X had studied our regulations and turned them to its advantage.

To acquire the latest aircraft technology, the Soviets in 1973 proposed purchasing 50 Lockheed transports if the firm, then in financial difficulty, would build and equip a modern "aircraft city" in the USSR. A similar proposition was put to Boeing (it besieges the imagination to ponder Brezhnev appearing from the cabin of an Aeroflot 747). Line X practiced the venerable capitalist technique of playing off competitors, and, from this bidding, the Soviets sought to gain technical data for use at home. On a less lofty technical plane, in 1972 the Soviets surreptitiously bought 25 percent of the US grain harvest, using phone intercepts of the grain dealers' network to listen to both sides of the market. The purchase led to higher grain prices for consumers, and taxpayers provided for a 25 percent a bushel export subsidy. Those of us observing these arabesques began to question the USSR's total commitment to the spirit of détente.

US Computer Export Policy

In late 1973, President Nixon asked his Council on International Economic Policy to determine which computers and associated production technology might be prudently sold to Communist countries. This study was necessary because détente implied the expansion of commercial opportunities with Eastern Europe and the USSR; a new and more liberal set of COCOM rules was required to fit these prospects, however illusory they may have been. Data processing was the most important product requiring review. I was put in charge of the project, and I was also made responsible for the broader problem of technology transfer. The computer study was the first review of technology policy within détente; it sought to assess the economic gain to the United States from computer sales set against the national security risk from those sales.

Not surprisingly, the study concluded that the USSR was short of computers and the means to pay for substantial computer imports. Our analysis presumed that the Soviets intended to use their foreign exchange to best advantage by purchasing the most powerful computers, those that also held the most national security risk (large computers were used for nuclear weapons calculations and cryptography). The report concluded that the export potential for American data processing to the USSR was small and the risk great if the more powerful computers were allowed for sale. The study recommended raising moderately the power of machines allowed for COCOM release, while at the same time restricting the sale of technology. Export of the largest computers was to be prohibited. In National Security Decision Memorandum (NSDM) 247, 14 March 1974, *U.S. Policy on the Export of Computers to Communist Countries*, President Nixon approved these recommendations, and they became the new export guidelines. As a result, the Soviets were excluded from importing significantly powerful Western computers, détente notwithstanding.

If the Soviets were to reach comparability with the United States in computers, their engineers would on their own now have to create designs and produce equipment. Line X would have to use its espionage resources to supplement what could be developed at home. NSDM 247 eliminated the West as an open source available to the Soviets, but Western intelligence was unaware of the collection apparatus the Soviets had deployed to obtain the technology.

Strong Suspicions and Skepticism

In the early 1970s, there were no U.S. intelligence collection requirements for technology transfer and scientific espionage, and few, if any, reporting sources. But, by observing the behavior of Soviet delegations visiting U.S. plants and by keeping in mind the clever 1972 grain purchase, a few government officials began to suspect that a master plan was in place to obtain our know-how. Direct evidence was nonexistent—only anecdotal clues were at hand. In their intelligence history, the Soviets could point to the success of the atom bomb spies, and they also had to their credit collection against industrial technology in Germany during the 1920s. After World War II, the Soviets copied the American B-29 and the Rolls-Royce Nene jet engine (the copy powered the MiG-15). Two former members of the Rosenberg network had set up the modern Soviet microelectronics industry. Soviet intelligence was professional at ferreting out science and technology and had the results to prove it. The Soviets were adept at copying foreign designs. In the style of Sherlock Holmes, the clues could almost speak for themselves: the USSR was behind in important technologies, their intelligence was accomplished at collection, and détente had opened a path.

Those suspicious of a Great Game in technology espionage found that the U.S. Government was not 221 B Baker Street—we could make little headway in persuading officials in charge of intelligence requirements that the United States was facing a significant threat. We received discouraging responses to our pleas for help: "No evidence" of a grand design; "not usual Soviet practice;" "no requirements and no interest;" "no sources." It seemed to have escaped these authorities that having no evidence does not mean it is not true. The system defied movement.

A few alert colleagues were dispersed among the executive departments. In one episode, the Department of Commerce discovered a Line X effort to obtain an embargoed computer through a dummy corporation set up for this one transaction; officials intercepted the shipping container and substituted sandbags. (A note was enclosed, but it would not be politically correct to quote it.) In 1975, the Apollo-Soyuz spacecraft docking was used to gain intelligence access to the U.S. space program. This project was conceived by the Nixon administration as part of détente, and President Ford had no choice but to continue the effort. To the consternation of NASA, a few weeks before the launch counterintelligence suspected that one of the Cosmonauts was a KGB officer who had been collecting away over the course of the project.

Presidential Interest

President Carter was the first chief executive to take an interest in technology loss. During his administration, CIA had begun to report the diversion of computers from the West into the Soviet defense complex, and he wanted details. In response, the Agency assigned staff to this endeavor and produced a more complete picture of technology loss than had been available since the start of Directorate T. Carter also ordered the first comprehensive study of technology transfer, Presidential Review Memorandum 31, a document that only distantly addressed the threat from clandestine collection. It was largely a missed opportunity, but Carter responded to the Soviet invasion of Afghanistan by instituting sanctions, canceling several computer sales, and stopping equipment destined for the Kama River truck plant.

President Reagan came to office intent on reversing what he saw as the "window of vulnerability" favoring the Soviets in strategic weapons. He also believed that the USSR's economy did not work and that the Soviet system was on the way to collapse. His intuition led him to believe the Cold War could be won. Joining Reagan's NSC staff were those of us who thought similarly and entertained the idea that economic pressure would have some effect. The NSC staff sought to fashion policies to take advantage of the USSR's low productivity, its lag in technology, oppressive defense burden, and inefficient economic structure. Reagan was the first president for whom this line of thought would have been even remotely acceptable.

A Defector in Place

Into the receptive climate of the Reagan administration came President Mitterrand, bearing news of Farewell—that is, Colonel Vetrov. In a private meeting associated with the July 1981 Ottawa economic summit, he told Reagan of the source and offered the intelligence to the United States. It was passed through Vice President Bush and then to CIA. The door had opened into Line X.

Vetrov was a 53-year-old engineer assigned to evaluate the intelligence collected by Directorate T, an ideal position for a defector in place. He had volunteered his services for ideological reasons. He supplied a list of Soviet organizations in scientific collection and summary reports from Directorate T on the goals, achievements, and unfilled objectives of the program. Farewell revealed the names of more than 200 Line X officers stationed in 10 KGB *rezidents* in the West, along with more than 100 leads to Line X recruitments.[3]

Upon receipt of the documents (the Farewell Dossier, as labeled by French Intelligence) CIA arranged for my access. Reading the material caused my worst nightmares to come true. Since 1970, Line X had obtained thousands of documents and sample products, in such quantity that it appeared that the Soviet military and civil sectors were in large measure running their research on that of the West, particularly the United States. Our science was supporting their national defense. Losses were in radar, computers, machine tools, and semiconductors. Line X had fulfilled two-thirds to three-fourths of its collection requirements—an impressive performance.

Interest in Technology Transfer

Overnight, technology transfer became a top priority, rising from the basement of Intelligence Community interest. CIA set up a Technology Transfer Intelligence Center, and the Pentagon created groups to assess damage and find ways to tighten technology controls. But careful study of Farewell's material suggested that more than just a few committees could come out of this wealth of intelligence. With the Farewell reporting, CIA had the Line X shopping list for still-needed technology, and with the list American intelligence might be able to control for its purposes at least part of Line X's collection, that is, turn the tables on the KGB and conduct economic warfare of our own.

I met with Director of Central Intelligence William Casey on an afternoon in January 1982. I proposed using the Farewell material to feed or play back the products sought by Line X, but these would come from our own sources and would have been "improved," that is, designed so that on arrival in the Soviet Union they would appear genuine but would later fail. U.S. intelligence would match Line X requirements supplied through Vetrov with our version of those items, ones that would hardly meet the expectations of that vast Soviet apparatus deployed to collect them.

If some double agent told the KGB the Americans were alert to Line X and were interfering with their collection by subverting, if not sabotaging, the effort, I believed the United States still could not lose. The Soviets, being a suspicious lot, would be likely to question and reject everything Line X collected. If so, this would be a rarity in the world of espionage, an operation that would succeed even if compromised. Casey liked the proposal.

A Deception Operation

As was later reported in *Aviation Week and Space Technology*, CIA and the Defense Department, in partnership with the FBI, set up a program to do just what we had discussed: modified products were devised and "made available" to Line X collection channels. The CIA project leader and his associates studied the Farewell material, examined export license applications and other intelligence, and contrived to introduce altered products into KGB collection. American industry helped in the preparation of items to be "marketed" to Line X. Contrived computer chips found their way into Soviet military equipment, flawed turbines were installed on a gas pipeline, and defective plans disrupted the output of chemical plants and a tractor factory. The Pentagon introduced misleading information pertinent to stealth aircraft, space defense, and tactical aircraft.[4] The Soviet Space Shuttle was a rejected NASA design.[5] When Casey told President Reagan of the undertaking, the latter was enthusiastic. In time, the project proved to be a model of interagency cooperation, with the FBI handling domestic requirements and CIA responsible for overseas operations. The program had great success, and it was never detected.

In a further use of the Farewell product, Casey sent the Deputy Director of Central Intelligence to Europe to tell NATO governments and intelligence services of the Line X threat. These meetings led to the expulsion or compromise of about 200 Soviet intelligence officers and their sources, causing the collapse of Line X operations in Europe. Although some military intelligence officers avoided compromise, the heart of Soviet technology collection crumbled and would not recover. This mortal blow came just at the beginning of Reagan's defense buildup, his Strategic Defense Initiative (SDI), and the introduction of stealth aircraft into U.S. forces.

National Security Directive

On 17 January 1983, to define his policy for political, military, and economic relations with the USSR, Reagan approved National Security Decision Directive (NSDD) 75, *U.S. Relations with the USSR,* a document spelling out purposes, themes, and strategy for competing in the Cold War. It specified three policy elements: containment and reversal of Soviet expansionism, promotion of change in the internal system to reduce the power of the ruling elite, and engagement in negotiations and agreements that would enhance US interests. In economic policy, NSDD 75 highlighted the need to control technology; Farewell's reports had moved those writing the Directive to put emphasis on preventing technology loss, and the President had agreed (so a KGB defector working for a foreign intelligence service put his stamp on a part of presidential policy). Later in 1983, Reagan proposed the SDI, which Gorbachev and the Soviet military took far more seriously than American commentators. SDI would, if deployed, place unacceptable economic and technical demands on the Soviet system. Even Reagan's 1983 "evil empire" speech had its economic effect, for immediately thereafter the Soviet military asked for a budget increase, this on top of already-bloated defense expenditures.

Two events beyond presidential control dovetailed with NSDD 75. The Federal Reserve's restrictive monetary policy of the early 1980s led to a fall in gold and primary product prices, sources of Soviet foreign exchange. And the discovery of Alaskan North Shore oil contributed to the 1986 fall in petroleum prices, cutting the revenues not only of OPEC but also of the USSR. Coincident events and deliberate government policy had the twin effects of adding to the burden on the Soviet system and of shifting the superpower competition to advanced technology, where the United States held a clear advantage.

Good-by to Farewell

About the time I met with Casey, Vetrov fell into a tragic episode with a woman and a fellow KGB officer in a Moscow park. In circumstances that are not clear, he stabbed and killed the officer and then stabbed but did not kill the woman. He was arrested, and, in the ensuing investigation, his espionage activities were discovered; he was executed in 1983. CIA had enough intelligence to institute protective countermeasures.

In 1985, the case took a bizarre turn when information on the Farewell Dossier surfaced in France. Mitterrand came to suspect that Vetrov had all along been a CIA plant set up to test him to see if the material would be handed over to the Americans or kept by the French. Acting on this mistaken belief, Mitterrand fired the chief of the French service, Yves Bonnet.[6]

An Important Contribution

In 1994, Gorbachev's science adviser, Roald Sagdeev, wrote that in computers and microelectronics—the keys to modern civil and military technology—the Soviets trailed Western standards by 15 years and that the most striking indication of their backwardness was the absence of a domestically made supercomputer.[7] The Soviets considered a supercomputer a "strategic attribute," the lack of which was inexcusable for a superpower. Line X did not acquire designs for such a machine, nor could Soviet computer scientists build one on their own—and NSDM 247 had stopped Western help. As for Farewell, his contribution led to the collapse of a crucial collection program at just the time the Soviet military needed it, and it resulted in a forceful and effective NATO effort to protect its technology. Along with the U.S. defense buildup and an already floundering Soviet economy, the USSR could no longer compete, a conclusion reached by the Politburo in 1987.

When historians sort out the reasons for the end of the Cold War, perhaps Farewell will receive a footnote. It would be deserved.

Notes

1. Kissinger, Henry A., *White House Years*. Boston: Little, Brown and Company, 1979, pp. 1, 142.

2. Kissinger on détente. Thomas G. Paterson and Dennis Merrill (Ed.), Major Problems in American Foreign Relations, Volume II, 1995, p. 600.
3. For a primary source from a former KGB officer, see Oleg Gordievsky and Christopher Andrew, KGB: The Inside Story. New York, Harper Collins, 1991.
4. Schweizer, Peter. Victory: The Reagan Administration's Secret Strategy that Hastened the Collapse of the Soviet Union. New York: The Atlantic Monthly Press, 1995, pp. 187–90.
5. Conversation with James Fletcher, Administrator, NASA.
6. Porch, Douglas. The French Secret Services. New York: Farrar, Straus and Giroux, 1995, p. 448.
7. Sagdeev, Roald Z. The Making of a Soviet Scientist. New York: Jolui Wiley & Sons, 1994, pp. 298–301.

Section XV: Cost Effectiveness

CHAPTER 58

The One Percent Solution: Costs and Benefits of Military Deception[1]

Barton Whaley

> A shilling's worth of radio countermeasures will mess up a pound's worth of radio equipment.
>
> Dr. Robert Cockburn, Director of the British Telecommunications Research Establishment in World War II

> An ounce of deception is worth a 240-pound tackle.
>
> Jake McCandless, Princeton football coach, 1970s

In recent years, deception has come to be recognized almost universally as a force multiplier. Deception almost always yields substantially greater benefits than expected, and the cost of deception is usually far less than we have generally been led to believe. Even at its most costly, deception need not divert more than a very small fraction of the available combat force from the real target. This holds true for the whole range of operations from grand strategy to minor tactics, for conventional and unconventional operations, and for all terrains. Deception has at least an 80 percent chance of yielding surprise, and the payoffs for surprise are impressively high. Surprise multiplies the chances for quick and decisive military success, whether measured in terms of explicitly sought goals, ground taken, or casualty ratios. Deception is seemingly never contraindicated or dysfunctional in the economics of war.

When the costs of being deceived are high, the benefits of detecting deception are correspondingly high. The costs for both deceivers and detectors may be economic – in material or in the time or lives of personnel – or they may be psychological, social, political, or ethical. In this chapter, I explore the costs to the deceiver in material and personnel. Generally, as we shall see, they are quite low.

Assessing the costs and benefits of deception

The economic cost of any deception is best measured by the proportion of the deceiver's resources (money, time, personnel, or apparatus) diverted from the main event. In war, this is the proportion of personnel and material diverted from the usual field of battle. The benefits that may accrue to the deceiver are

1. Bart Whaley, "The 1% Solution: Costs & Benefits of Military Deception," *Information Strategy and Warfare,* John Arquilla and Douglas Borer, editors, New York: Routledge, 2007, pp. 127–159.

potentially manifold: quick success at less than the normal cost in personnel and material to oneself and higher cost to an enemy.

In principle, the method of assessment is simple: it is nothing more than old-fashioned operational research, as its British inventors first called it. The problem is getting relevant and more-or-less reliable data to analyze. A deception cost-benefit analysis requires access to good data on both the deceiver and the enemy – the overall combat resources or both sides, the combat resources diverted by both sides as a result of the deception, and the losses to each side as a result of the deception.

Although there is no shortage of case studies that illustrate the general effectiveness of deception (Liddell Hart 1929, 1954; Whaley 1969, 1979), few studies have specifically measured these benefits together with the costs involved. Until we directly cross-check costs and benefits, any conclusions we draw from our case studies, however glowing in their portrayal of the overall success of deception, remain open to question. Moreover, the few existing cost-benefit studies are narrowly focused, concentrating, for example, on deception measures and countermeasures only in aerial combat or during a relative brief time span, such as World War II. This chapter is a preliminary exploration of the many dimensions of the cost-benefits of deception operations.

The general benefits of deception

The theory of deception (presented elsewhere, e.g. Whaley 1969, 1982) predicts several consequences of deception, three of which are covered here: (1) The more intense the deception (i.e. the more of types of ruses used), the greater the surprise – but only up to a point; (2) the greater the surprise, the smaller the force ratio required; (3) the greater the surprise, the fewer the casualties.

More deception yields more surprise – up to a point

The general trend can be seen in Table 6.1, which glosses the minor trends and perturbations in data from World War I through 1968.

Using Whaley's expanded but unpublished DECEPTR database of 232 military engagements from 1914 to 1973, Ronald Sherwin focused on the ninety-three strategic cases (Sherwin and Whaley 1982, 188). Table 6.2 summarizes the results.

Two interesting findings emerge from the table. First, when deception is absent, surprise is rare; conversely, when even a deception employs a moderate number of types of ruses, surprise is common. This finding was expected. The second finding was not: Sherwin found that the "optimal threshold for achieving surprise is between two and three [types of] ruses [out of a possible ten] – probably closer to two [types], depending on which two [types], which target, and in which context." This finding may seem counterintuitive; after all, it might seem reasonable to expect that the use of more deception ruses would bring more surprise. However, one of the more entrancing recent findings of cognitive psychology is that, contrary to popular belief, few decision makers (such as intelligence

Table 6.1 Relationship between intensity of deception and intensity of surprise, 1914–68

Intensity of surprise	*Intensity of deception*					
	0		*1–4*		*5–8*	
	N	%	N	%	N	%
0	53*	72.6	5	6.4	1	6.3
1	6	8.2	9	11.6	1	6.3
2	7	9.6	32	41.0	3	18.7
3	4	5.5	23	29.4	2	12.5
4–5	3	4.1	9	11.6	9	56.2
Total	73	100.0	78	100.0	16	100.0

Note

* This cell represents all 53 type C (non-surprise) examples in the database. While it is not possible to say how representative the figure may be, the bias in selecting type C was to err on the low side. Intensity of deception refers to the relative level of investment made in terms of material resources, manpower, and even actions in the field (e.g. feints). The most intense surprise occurs when the opponent is unaware that an attack is coming and is deceived even as to the likely places where an attack might arise. The least intense surprise is a situation in which an attack is expected, its location anticipated, but the specific timing is not. Middle levels of surprise generally comprise instances where an attack is expected, but not the location, or vice versa, and instances where the size of the attack may be misestimated.

Table 6.2 Intensity of deception (number of types of ruses used) and whether surprise was achieved in strategic battles, 1914–73

Result	*Number of ruses used*							*Totals*
	0	*1*	*2*	*3*	*4*	*5*	*6*	
No surprise achieved	0	0	6	0	0	0	0	6
Surprise achieved	0	3	46	20	12	7	0	87

analysts) wait until all the information is in. Instead, most people reach decisions on the basis of only around three data sets.

Lower force ratios follow from surprise

One of the legacies of the machine-gun- and artillery-dominated Western Front was a "magic" number for success: commanders came to adopt the notion that a 3-to-1 local superiority in the zone of an offensive was necessary for a successful attack. Indeed, this rule of thumb was consistent with statistical studies of the *overall* averages of Western Front battles. This bit of Great War doctrine was carried over into World War II and beyond. But how does this 3-to-1 doctrine look in the light of our data on surprise?

First, let's look at the World War I data. As Table 6.3 shows, of the 47 offensive

Table 6.3 Attacker-to-defender force ratio used to gain objectives in World War I

Achievement	*Surprise*		*No surprise*	
	N	*Force ratio*	*N*	*Force ratio*
Victory	7	2.1:1	0	–
About as planned	10	1.6:1	5	2.1:1
Below expectations	5	2.8:1	5	1.3:1
Defeat	3	5.9:1	12	1.8:1

Source: Whaley (1969), 199.

Table 6.4 Attacker-to-defender force ratio used to gain objectives, 1919–67

Achievement	*Surprise*		*No surprise*	
	N	*Force ratio*	*N*	*Force ratio*
Victory	18	1.2:1	1	2.5:1
About as planned	28	1.1:1	4	1.4:1
Below expectations	17	1.4:1	9	1.4:1
Defeat	4	1.0:1	20	0.9:1

Source: Whaley (1969), 199.

battles for which data were available, the average force ratio of attacker to defender was 2.1 to 1.

There is no obvious pattern here. But before comment, consider the comparable data in Table 6.4 for the half century following the Great War.

Tables 6.3 and 6.4, taken together, support two important conclusions. First, we see strong proof for the notion that a substantial superiority of force is needed, although my data yield force ratios of about 2 to 1 rather than the traditional 3 to 1. However, this general remark holds true only for the more usual type of offensive military operations, that is, those without surprise. Here, there is a fairly direct relationship between force and degree of success: the more the force, the greater the success. Second, and both more important and rather unexpected, is that surprise intervenes to shatter the direct and simple relationship between force and success. In other words, surprise – which is usually a result of deception – is a more important determinant of combat success than brute strength.

Casualty ratios reduced

It follows from deception theory that the more intense the surprise, the more the casualty ratios favor the surpriser. The empirical data available quite emphatically verify this, as Table 6.5 shows.

If deception is not only a main cause but also an enhancer of surprise, then I

Table 6.5 Effect of intensity of surprise on casualties, 1914–67

Intensity of surprise	*Mean casualty ratios*
0	1:1.1
1	1:1.7
2	1:4.5
3	1:5.4
4	1:4.1
5	1:11.5

Source: Whaley (1969), 194.

Table 6.6 Effects of deception and/or surprise on casualty ratios, 1914–67

	N	*Mean casualty ratios*
Surprise with deception	59	1:6.3
Surprise without deception	20	1:2.0
No surprise with deception	5	1:1.3
No surprise without deception	40	1:1.1
Total	122	

Source: Whaley (1969), 195.

predicted that the data must show (1) that casualty ratios are substantially greater in cases of surprise with deception than in those of surprise without deception, and (2) that casualty ratios between cases of no surprise and those without deception are similar. Both expectations are verified by the data in Table 6.6.

Cases illustrating costs and benefits

By any purely economic accounting, deception can be relatively cheap – indeed ridiculously cheap. The following case examples illustrate this point.

Case: Normandy, D-Day, 6 June 1944

In absolute economic terms, the most costly deception of all time was Operation Fortitude, the Allied deception operation designed to convince Hitler that the Allied invasion of Nazi-occupied Europe would arrive on the coast of Pas-de-Calais and not further West on the beaches of Normandy. The operation succeeded in keeping 19 of Hitler's divisions tied down in reserve for 66 days after the D-Day landings at Normandy. This benefit to the Allies was enormous, quite possibly making the difference between success and a disaster that would almost certainly have prolonged the war at least a year. But at what cost?

The total cost of all D-Day deception operations was well under 1 percent of that of the invasion force. The breakdown is as follows (I compiled these figures from various sources, beginning in 1972):

- The full-time services of some eighteen deception planners for about a year.
- The part-time attention of perhaps twenty senior staff planners and intelligence officers.
- Operational commitments of roughly 800 radio technicians (some 500 British and 296 Americans) who busily simulated the radio traffic of two phantom armies.
- Several camouflage and construction companies (about 2,000 troops drawn from the Royal Engineers and 379 from the US Corps of Engineers) building dummy installations.
- About 1,000 air force personnel to crew and service the ninety aircraft flying spoof missions on D-Day.
- About 200 sailors for the eighteen small launches involved in misleading the German coastal early warning system.
- Six Special Air Service (SAS) men to parachute on a diversionary mission.
- The ten members of Twenty ("Double-Cross") Committee directing the disinformation campaign of about eleven of the forty German double agents then in Britain.
- Perhaps 100 or so agents and underground members engaged in spreading false rumors on the Continent.
- And, finally, Lt. Clifton James, who was co-opted from the British Army Pay Corps to simulate General Montgomery.

The total diversion of personnel was about 4,500 soldiers, sailors, and airmen, of whom fewer than a quarter were combat personnel. This represents less than 0.2 percent of the available combat personnel, 0.5 percent of the naval craft, and only 0.6 percent of the aerial sorties flown on D-Day.

The total loss among the deception forces was one bomber (its crew safely parachuted), a few tons of aluminum strip air-dropped to deceive German radar, and four of the SAS paratroopers who were taken prisoner and executed (per Hitler's standing order regarding all behind-the-line paratroopers). In sum, the total diversion of well under 1 percent of all combat troops, aircraft, and naval craft at a cost of four personnel casualties, one aircraft, and zero naval craft – in short, the 1 percent solution that inspired the title of this chapter.

Case: RAF bomber offensive: opening the "Window," 1943

The Royal Air Force (RAF) Bomber Command introduced "Window" or "American Chaff" – strips of aluminized paper – into its bag of spoofing tricks on the night of 24/25 July 1943 (this account is drawn from Price 1967, 155, 158; and Webster and Frankland 1961, 156). The target was Hamburg. At 9:55 p.m., the entire Main Force bomber stream of 791 heavy bombers had assembled over the North Sea and set out due east at 225 mph in a typical stream twenty miles wide and 200 miles long. As was typical, 45 bombers aborted for various reasons and returned to base.

German early warning radar spotted the incoming stream at 12:15. At 12:19

the first air raid alarms were sounded and the German night fighters were alerted. Radar tracked the approach toward Hamburg, where the full air raid alarm sounded at 12:31. Simultaneously, each of the Main Force bombers, now twenty miles from the German coast, began dropping Window at the rate of one bundle a minute. To the watching German radar operators – upon whom the circling night fighters were entirely dependent – these 746 bombers suddenly looked like 11,000. The ground controllers were helpless. At 12:50 the bomber stream turned into its final attack bearing, straight to target, twenty-five miles away. At 12:57 the twenty Pathfinder bombers began their target marking and the attack was under way. Only then were the German ground controllers able to order the night fighters to Hamburg. But even then, the night fighters' radars were effectively jammed as they made attack after attack on empty drifting clouds of aluminum foil, while the equally jammed radar of the eighty anti-aircraft guns and twenty-two searchlight batteries sent their flak and candle-power on blind chases. The last of the 740 bombers that actually attacked withdrew at 1:50 after dropping 2,400 tons of bombs, which killed about 1,500 civilians. Only twelve RAF bombers were lost.

Window had been a stunning technological success. On this first use, forty tons of aluminized paper (92,000,000 separate strips) costing £770 subtracted only 1.7 percent of the raiders' bomb tonnage capacity. Yet it very likely saved about 35 British aircraft (with their own 110-ton bomb capacity) worth £1,400,000, plus their 245 aircrew, whose training alone was valued at £2,450,000 (figures calculated from data in Middlebrook 1974; Webster and Frankland [1961] incorrectly calculated a probable saving of seventy-eight bombers).

Case: RAF Window assessment, 1943–4

Window proved highly cost-effective. It was cheap to make and simple to use. The Allies dropped over 10,000 tons of this aluminum foil over Europe alone, simulating 12,500,000 heavy bombers. It cost only about £180,000 to make and subtracted less than 1 percent from the total bomb load of the Allied bombers that flew against Hitler's Fortress Europe under Window protection after mid-1943. Yet it enabled thousands of those bombers and their crews to survive and fight another day. Moreover, for a time, 90 percent of Germany's 7,800 invaluable high-frequency experts were diverted to the development of an effective electronic countermeasure.

The hypothetical savings achieved by Window cannot be calculated exactly, because the ever-changing factors (German countermeasures, new Allied tactics, other Allied deception methods, etc.) make overall before-and-after studies incomparable. Even rough estimates would require an exceedingly elaborate statistical operations research study. However, the available fragmentary evidence strongly suggests that the Germans were never able to overcome the effects of Window, as shown in Table 6.7.

Perhaps as good a summary statement on Window as any is Churchill's: "Its effects surpassed expectations.... For some months [after Hamburg] our bomber

Table 6.7 Percentages of bomber losses before and after Window raids began, 1943–4

Target city	*Dates*	*Percent lost with Window*	*Percent lost pre-Window*	*Percent saved*
Hamburg	24 July–3 August 1943	2.8	6.1	3.3
Essen	25/26 July 1943	3.4	5.4	2.0
Berlin	23 August–3 September 1943	7.2	*c.*10.0	*c.*2.8
Berlin	18 November–24 March 1944	5.4	*c.*10.0	*c.*4.6

Sources: Compiled and calculated from Webster and Frankland (1961) and Price (1967).

Table 6.8 RAF bomber command order-of-battle, 30 March 1944

Mission	*Squadrons*	
	N	%
Main force bombers	67	85.5
Secret ops	2	3.5
Ferrets	2	3.5
Diversionary	4	5.5
Jammers	1	1.5
Total	76	99.5

Source: Compiled from Middlebrook (1974), 322–5.

losses dropped to nearly half. And up to the end of the war, although the German fighter planes increased fourfold, our bomber losses never reached the same level that had been suffered before '*window*' was used" (Churchill 1950, 288–9).

Case: Assessing RAF Bomber Command deception cost, 1944

Another measure of RAF Bomber Command's deception costs is to calculate the percentages of its order-of-battle devoted to deception operations. As Table 6.8 shows, the total percentage cost in squadrons devoted to deception – diversionary plus jammers – was 7 percent of the total number of squadrons. However, as each of these deception squadrons cost only about half that of a Main Force squadron in aircraft and number of aircrew, the adjusted cost of the deception squadrons is only around 3.5 percent of Bomber Command's total strength.

Case: Maj. Ingersoll and the Battle of the Bulge, 1944

Urgent improvisation of a deception plan in its extreme form was suddenly needed during the critical Battle of the Bulge in December 1944. The German army had just broken through the Allies' weak point in a surprise attack. One measure of this surprise was that Col. William "Billy" Harris, who headed Gen.

Omar Bradley's 12th US Army Group's Special (i.e. deception) Plans Section, was away on other matters. Thus, at this crucial moment, command devolved upon his deputy, forty-two-year-old Maj. Ralph Ingersoll, a peacetime newspaper publisher. Ingersoll would be forced to improvise. His plan would prove to be a masterpiece of cost-beneficial deception. I call it the "Two Pattons Ruse." (This account is drawn from author's interview with Ralph Ingersoll, 12 May 1973; and Hoopes 1985, 304).

On 19 December, Gen. Eisenhower went to Bradley's rear headquarters at Verdun to meet Bradley and Patton and decide on a strategy to stop the German offensive. Ingersoll was present at the meeting. The battle-eager Lt. Gen. Patton assured Eisenhower and Bradley that he could rush his powerful Third Army north and stop the enemy advance. The two senior commanders were doubtful because of the immense logistical problems, but desperate measures were needed, and they unleashed Patton.

Urgency demanded that Patton's movement orders to his corps and division commanders would have to go out over his radio nets "in clear" (not coded), despite the certainty that the German signals intelligence teams would overhear them.

Everyone involved knew it was too late to mount any conventional deception operation to cover this counterattack. Nevertheless, Gen. Bradley turned to Maj. Ingersoll and asked, "Is there anything you and your people can think of to throw the Germans off – about where they [Patton's divisions] are going to strike?" Ingersoll replied, "Yes, sir. We can do ..." – he hesitated – "something." Bradley was momentarily silent before simply ordering, "Then do it" (Hoopes 1985, 304).

Ingersoll later said that the idea had come to him in a flash. Realizing during the meeting that time would be too short for any conventional deception, he'd been trying to think of some alternative. His mind lit on the recent example of the vast confusion to German military intelligence created on D-Day by the unintentional scattering of the two American and British airborne divisions. With this thought in mind, Ingersoll suddenly conceived – indeed invented – the unprecedented ruse of the Two Pattons.[1] He code-named the operation Kodak because he intended to confuse the enemy by giving them a "double-exposure" (*Official History of the 23rd Headquarters Special Troops*, 1945, 23).

The only resource available to Ingersoll was the Special Plans operational deception unit, the 23rd Headquarters Special Troops. It was too late to deploy their usual panoply of visual and aural spoofs. Radio deception alone would have to do the job. Accordingly, nineteen officers and twenty enlisted men were detached from the 23rd's Signal Company, rushed forward, and dispersed among advance units of Patton's Third Army. Another eleven officers and 205 EM provided support farther back. It was, as the official history called it, "a big (29 sets), loose radio show" (*Official History 23rd*, 1945, 23; also Enclosures I and V).

These radio operators normally spent their time simulating "notional" (fictitious) US Army units in a largely successful effort to dupe the enemy into wasting precious resources by either advancing or retreating from these phantoms. But now, for two days, 22 and 23 December, Ingersoll had them imitate two real units – the 80th Infantry Division and the Fourth Armored Division, which were spearheading Patton's advance. The 23rd's bogus radio show was timed to start when these division radio nets began broadcasting in clear.

The real 80th was getting ready to jump off due north from Luxembourg, and the real Fourth was preparing to roll up from Arlon to effect its historic relief of the 101st Airborne Division, which was trapped in Bastogne. "The 23rd mission was to show ... the presence of the 80th Infantry and 4th Armored Divisions slightly northeast of Luxembourg and in position to forestall any German plans of extending their counterattack southwest through Echternach" (*Official History 23rd*, 1945, 52).

In effect they had presented German army intelligence with two Gen. Pattons approaching the Bulge from slightly different directions. Of course, the Germans quickly understood the radio game being played against them, but they were still dazzled by the two fake divisions advancing eastward of the two real ones. In the event, the Germans were able to keep track of only one of Patton's advancing divisions. All others were either lost entirely on the German battle maps or, worse, mislocated. Consequently Patton was able to gain a substantial tactical surprise, which helped him break the back of the Wehrmacht's last offensive.

Long after the war, when I interviewed Ingersoll about his role in military deception, the Two Pattons story – unpublished at that time – emerged only as an afterthought. Ingersoll mentioned it, saying apologetically that, while proud of his improvisation, he didn't classify it as "deception," which he defined as making the victim certain but wrong. Instead he chose to make the enemy merely uncertain and confused and to hope for the best. Ingersoll's unique ruse does indeed fit one category of deception, the one called "dazzle" in the Bell-Whaley typology (Whaley 1982, 184; dazzle is the same as the "A type" [ambiguity-increasing] deception in Daniel and Herbig 1982, 5).

In sum, the benefits of Operation Kodak were a major defeat for the enemy's army and saving the Allied drive into Germany a delay of at least a couple of months. The cost was the employment of thirty officers, 225 enlisted men, and twenty-nine radio sets – under 0.1 percent of Bradley's 12th US Army Group's personnel and none of its combat personnel.

Case: Overall cost-benefit assessment of the US Army's First Tactical Deception Field Unit, 1944–5

This is a rare case where we have enough data on a specific military deception operations unit to assess its overall benefits and costs with some precision.

In January 1944, the 23rd Headquarters Special Troops was activated at Camp Forrest, Tennessee. This was the first self-contained US military organization specially and solely designed for tactical deception (this account is drawn from *Official History 23rd*, 1945; Mendelsohn 1989; Kneece 2001; Gawne 2002; Gerard 2002). The 23rd was a reinforced battalion-sized unit, commanded throughout its twenty-one-month existence by Col. Harry L. Reeder, who had been a regular officer since before World War I and had considerable tank experience and an unconcealed contempt for this noncombat assignment. At full strength, the 23rd mustered 1,106 troops (eighty officers, three warrant officers, and 1,023 enlisted men). It was structured as follows: Headquarters Company had 108 officers and men who exercised overall direction. The 603 Engineer Camouflage Battalion Special (with 379 officers and men, mainly New York and Philadelphia artists with an average IQ of 119), handled conventional camouflage and the inflatable rubber dummy tanks, trucks, and guns. The Signal Company Special (296 officers and men, of whom 196 were radio operators) handled the radio deceptions. The 3132 Signal Service Company Special (147 officers and men), the first such unit in the US Army, handled sonic deception. The 406 Engineer Combat Company Special (168 officers and men) handled security and rough jobs in general.

What was the 23rd's learning curve and what was its final score? Consider first the learning process, as outlined in Table 6.9. Here we see a pattern of ups

Table 6.9 Results of plans and operations of the 23rd Headquarters Special Troops in France and Germany, 1944–5

Date	*Operation or plan*	*Result*
1944		
6 June	No name	Aborted (situation found unsuitable)
7 June	Troutfly	Aborted (change of battle plan)
1–4 July	Elephant	Partial success
9–12 August	Brittany	Success
20–27 August	Brest	Partial success, partial backfire
1 September	No name	Aborted (too late)
15–22 September	Bettembourg	Partial success
4–10 October	Wiltz	Unknown
10 October	Vaseline	Aborted (change of battle plan)
2–10 November	Dallas	Probable success
3–12 November	Elsenborn	Partial success
4–9 November	Casanova	Success
6–14 December	Koblenz	Partial success
21 December	Koblenz II	Aborted (liaison officer lost trailer)
22–23 December	Kodak	Success exceeding expectations
28–31 December	Metz I	Unknown
1945		
6–9 January	Metz II	Unknown
10–13 January	L'Eglise	Unknown
17–18 January	Flaxweiler	Probable success
27–29 January	Steinsel	Probable success
28 January–2 February	Landonviller	Unknown
1–4 February	Whipsaw	Partial success
13–14 February	Merzig	Failed
1–11 March	Lochinvar	Partial success
11–13 March	Bouzonville	Partial success
18–24 March	Viersen	Success

Table 6.10 Overall results of the 23rd Headquarters Special Troops' deception plans, 1944–5

Result	*N*	*%*
Success	4	15.5
Partial or probable success	11	42.5
Failure	1	4.0
Unknown	5	19.0
Aborts	5	19.0
Total	26	100.0

Sources: Generated from *Official History 23rd*, 1945, Enclosure I; modified by details in pp. 8–31; and corrected by upgrading two cases from more recently available data, one from "unknown" to "successful."

and downs but a gradual overall improvement in performance. Aborted operations became less common because of closer liaison and coordination with the combat units being supported. Simulation of the army divisions being portrayed grew more realistic, as measured by the number and quality of their various signatures.

What was the final score? As Table 6.10 shows, the 23rd generated a total of twenty-six tactical deception plans in support of Gen. Bradley's twenty-nine US divisions during its ten operational months from Normandy in June 1944 until March 1945. Of these twenty-six plans, five were aborted and twenty-one became actual operations. Of the latter, four were entirely successful, eleven were at least partly successful, and only one failed completely. Of the five operations whose results are unknown, the very lack of evidence suggests failure or only negligible success. On that pessimistic assumption, we get fifteen "successes" against six "failures" plus five aborted operations – a cost-effective record given the small size of the unit and the nature of the German reaction to its deceptions.

We can't accurately calculate how many lives the 23rd saved during its eleven months in action. From the D-Day landings on 6 June 1944 until the German surrender on 8 May 1945, 135,000 Americans were killed and 450,000 wounded. One historian of the 23rd concluded, "It is not much of a stretch to say that its other twenty operations could have brought the total [saved] to about 40,000 Allied lives. Each time German divisions were frozen in place, it meant that they were not killing Americans elsewhere. It also meant they themselves suffered fewer casualties.... The Ghost Army was just a small unit of 1,100 men among 5,412,219 Allied troops under Eisenhower during the war in Europe. But they projected the work of thousands" (Kneece 2001, 280).

Defensive deception

Deception is cost-effective for the initiators and attackers. Less well appreciated is that it can also work well for the defender. Two spectacular examples are the 1915 evacuation of Suvla Bay and Anzac Beach and the 1916 evacuation of Cape Helles.

Case: Evacuation of Suvla Bay and Anzac Beach, 20 December 1915

This was a case where even the enemy could not help but admire the deception: "So long as wars exist, the British evacuation of the An Burnu [Anzac] and Anafarta [Suvla] fronts will stand before the eyes of all strategists of retreat as a hitherto unattained masterpiece" (*Vossische Zeitung*, 21 January 1916).

On 7 December 1915, the British Cabinet agreed to liquidate all of the Gallipoli enterprise except the tip position at Cape Helles (this account is drawn from Whaley 1969, A56–A62). This difficult and embarrassing decision had been made with the concurrence of the War Committee, Lord Kitchener (the minister for war), Lt. Gen. Sir Charles Monro (Ian Hamilton's successor as commander in the Aegean), all of Hamilton's former staff (now under Monro), and two of the three corps commanders on Gallipoli. The now isolated group favoring further escalation at Gallipoli was largely limited to Churchill, Commodore Keyes, Lord Curzon, the replaced Hamilton, one able Gallipoli corps commander (Birdwood), and the secretary of the War Committee (Colonel Hankey).

Moreover, the decision to evacuate was made in full concurrence with predictions of appalling numbers of casualties during the withdrawal phase. As early as 12 October, Hamilton had informed Kitchener (and through him, the War Committee) that an evacuation would entail 50 percent casualties and loss of all guns and stores. Although this was an exaggeration designed to forestall a decision to cancel Hamilton's pet operation, even he expected 35 to 45 percent casualties in such an event.

In late October, Hamilton's former staff, although generally biased in favor of withdrawal, presented the newly arrived Monro with a careful estimate that reckoned on a 50 percent loss in troops and 66 percent in guns. Monro notified Kitchener on 2 November of his own estimate of losses of 30 to 40 percent of both men (i.e. some 40,000) and material. On 15 November, Kitchener more optimistically reported to London that he expected about 25,000 casualties (i.e. 25 percent), which was also then the reduced estimate of the local staff. On 22 November, the British Army General Staff reported to the War Committee its estimate of 50,000 casualties (i.e. 50 percent).

These dire predictions were naturally based on the assumption that the evacuation would be contested. Lord Curzon, the Lord Privy Seal, in a memorandum circulated to the Cabinet on 25 November, depicted the operation in terms of doom:

> [The] evacuation and the final scenes will be enacted at night. Our guns will continue firing until the last moment … but the trenches will have been taken one by one, and a moment must come when a final *sauve qui peut* takes place, and when a disorganized crowd will press in despairing tumult on to the shore and into the boats. Shells will be falling and bullets ploughing their way into this mass of retreating humanity.… Conceive the crowding into the boats of thousands of half-crazy men, the swamping of craft, the nocturnal panic, the agony of the wounded, the hetacombs of slain.
>
> (Quoted in Ronaldshay, 1928, 130–1)

It is interesting that the differences among these various casualty predictions were directly associated with each predictor's bias for or against evacuation. Advocates of escalation pressed forward the more pessimistic figures, and advocates of evacuation advanced the optimistic estimates. Moreover, those individuals who changed their position on evacuation simultaneously adopted the psychologically appropriate statistics. After evacuation had been ordered by London, most of the local staff officers suddenly revised their private estimates downward to a comforting 15 percent. Thus, casualty estimates, far from being rational military calculations, were used by the military and political professionals alike as a political tool and a psychological crutch.

In planning the deception for the evacuation, Monro rejected the conventional feints or demonstrations as a supplement to the various ruses at the beachheads themselves; a feint could well arouse the Turks' suspicion, with disastrous results. Instead, Monro "came to the conclusion that our chances of success were infinitely more probable if we made no departure of any kind from the normal life which we were following both on sea and on land" (Monro despatch of 6 March 1916, as quoted in Callwell 1919, 276–7).

Portents of the decision to evacuate Gallipoli had been gradually accumulating before the public and enemy eye alike. Thus, the commander in chief of the expeditionary force, Hamilton, was relieved on 14 October. That same day, withdrawal was first openly advocated in the House of Lords by Milner. On 2 November, in the House of Commons, Prime Minister Asquith bluntly admitted the failure of the recent offensive. The announcement on 11 November of a reconstituted War Committee (to replace the Dardanelles Committee) revealed publicly that the two leading public proponents of the adventure, Churchill and Curzon, had been dumped.

Finally, on 18 November, Lord Ribblesdale, an outspoken advocate of withdrawal, asserted in the House of Lords that General Monro had "reported in favor of withdrawal from the Dardanelles, and adversely to the continuance of winter operations there." It is interesting that on learning of Milner's statement, the Turks discredited it as a deliberate effort at deception. They fully expected the Allies to remain on Gallipoli.

It was decided that the evacuation would take place at night, although some troops were embarked under cover of tarpaulin in daylight. Aviation patrols were depended on to keep the few German reconnaissance aircraft from getting too close a look at the denuded positions, particularly on the last day, when no fewer than five Allied aircraft maintained almost continuous air cover.

The monumentality of this task and its success are summarized in the timeline of the evacuation presented in Table 6.11. Yet it was completed without raising the suspicions of the Turks or the German commander, General (and Turkish field marshal) Liman von Sanders.

The evacuation on the last night was effected with no dead and only five wounded and with only ten out of 200 guns abandoned. Of some 5,000 draft animals, only fifty-six mules were left. Even most of the 2,000 carts were salvaged.

A measure of the effect of surprise can be dramatized by comparing predicted

Table 6.11 Diminishing strengths at Suvla and Anzac, 8–20 December 1915

Date and time	*Suvla*		*Anzac*		*Total*	
	Men	*Guns*	*Men*	*Guns*	*Men*	*Guns*
December						
8	43,000	90	40,000	110	83,000	200
10	?	?	?	?	80,000	?
17	?	?	?	?	44,000?	?
18/0600	16,000	?	22,000	?	38,000	?
2000	16,000	?	22,000	?	38,000	?
19/0530	9,000	16	11,000	10	20,000	26
2000	9,000	16	11,000	10	20,000	26
2200	2,000?	0	4,000?	10	6,000?	10
20/0100	1,322+	0	2,000?	10	3,500?	10
0400	550	0	800?	10	1,350?	10
0415	200	0	0	10	200	10
0515	0	0	0	10	0	10

Table 6.12 Predicted and actual losses in the Suvla and Anzac evacuation

		Predicted losses					
		Pessimistic		*Optimistic*		*Actual losses*	
Men and materials	*Total force*	N	%	N	%	N	%
Men	83,000	40,000	50.0	12,500	15.0	5	0.01
Guns	200	100	50.0	30	15.0	10	5.0
Horses and mules	5,000	2,500	50.0	750	15.0	56	1.1

and actual losses (Table 6.12). Even the "optimistic" predictions exceeded actual losses by a factor of 2,500 for troops, a factor of 3 for guns, and a factor of 14 for draft animals.

The cost-effectiveness of surprise was so little understood by many that one common but false rumor credited the success of this bloodless coup to the Turks' having been bribed to permit the evacuation without opposition.

Case: Evacuation of Cape Helles, January 1916

The success of the evacuation of the beachheads at Suvla Bay and Anzac Beach hurried the final decisions for the complete liquidation of the Gallipoli enterprise (this account is drawn from Whaley 1969, A63–A65). The only remaining beachhead was at Cape Helles, at the very tip of the peninsula. On 23 December, the British War Committee recommended withdrawal; the Cabinet agreed to it on the 27th, and General Monro, the commander in chief at Gallipoli, was notified the next day.

The situation on Gallipoli again seemed desperate. The remaining Allied force of four divisions (35,268 troops) were crowded into the twelve square mile beachhead at Cape Helles. Facing them was a vastly larger force of Turks directed by General von Sanders. With the British evacuation of Suvla and Anzac completed, von Sanders was concentrating his entire force of twenty-one divisions (120,000 troops) against the remaining Allied position. It was certain that a contested evacuation would be costly, and perhaps even disastrous.

Could a second surprise withdrawal be made? Again General Monro chose to depend on simulating normalcy and avoiding any feints (despatch of 16 March 1916, quoted in Callwell 1919, 310). The basic plan that had worked so well a fortnight before was used again, although this time with one new technique: the evacuation of Cape Helles was carried out under the cover of reinforcement and relief. This deception succeeded so well that even the troops on the beachhead did not fully recognize what was under way until some time later. As some relief and reinforcement had been in effect since early December anyway, "normalcy" at Helles already meant much coming-and-going. Now, however, beginning 29 December, the pattern of relief shifted slightly to bring in only units of IX Corps, which had experience in the special techniques of evacuation, having just extricated itself from Suvla. Published orders were given out that VIII Corps was being relieved by IX corps. Tables 6.13 and 6.14 summarize the evacuation and the casualties in the Cape Helles evacuation.

Table 6.13 Diminishing strength at Cape Helles, 29 December 1915–9 January 1916

Date and time	*Troops*	*Guns*
29 December	40,000	150
31 December	35,268	140
2 January	31,000	?
7 January	19,418	63
8 January 0500	17,118	54
2000	17,118	54
2330	9,918	17
9 January 0145	3,877	17
0300	200	17
0345	0	17

Table 6.14 Casualties at Cape Helles

Date	*British*		
	Men	*Guns*	*Turks*
7 January	164	0	"Heavy"
8 January	Few	0	Few
9 January	5	17	None

Camouflage and decoys

The hiding of certain targets and the deliberate show of others to work as decoys are common components of deception.

Case: Col. Turner's department: British airfield and port camouflage and decoys, 1939–45

During World War II the British built a total of 797 separate decoy sites on their island plus a few on the Normandy beachhead and in northern France. All of these sites were designed and installed by what, for security purposes, was called simply and officially "Colonel Turner's Department." Among these sites were 231 somewhat elaborate dummy airfields; 237 civil "Starfish" fire sites to simulate eighty-one cities; numerous decoys near twenty-nine ports; army decoys near forty-four posts; and eleven simple dummy buildings (Dobinson 2000, 241–89, and throughout his main text). Unfortunately, I've been unable to find specific statistics on the personnel and material costs of this British effort. However, the famously frugal-minded Col. Turner made extensive use of second-line military personnel, civilian volunteers, and co-opted motion-picture designers and carpenters from London's suburban Shepperton movie studios.

The British camouflage effort on the post-D-Day Normandy beaches was highly effective. Four sections of Gold Beach, where the British XXX Corps landed, were covered by dummy targets erected by a company of Royal Engineers that was supplemented by a small detachment of RAF camouflage and decoy personnel from the now redundant Starfish night decoys in Britain. These Normandy decoy sites, which operated from D-Night (6 June) until 15 September, were thought by local reporting officers to have diverted more than half the German bombs, although, admittedly, German bombing had been lighter than expected (Dobinson 2000, 202–3).

The overall British camouflage effort is summarized in Table 6.15. Given Col. Turner's meticulous accounting methods, these statistics are very likely underestimates (Dobinson 2000, 209–13, 290–8).

Table 6.15 Calculation of British decoy effectiveness in diverting German bombs, 1939–45

Target	*Decoy type*	*Number of attacks*	*Average tons diverted*	*Total tons diverted*
Airfields	Day decoys ("K sites")	47	2.5	118
Airfields	Night decoys ("Q sites")	521	1.7	860
Various	Night "QF" fires and "QL" lights)	173	1.5	265
Buildings	Day	9	1.3	12
Cities	Night decoys (Starfish)	119	5.6	968
Total		869		2,223

Source: Based on Dobinson 2000, appendix III, 298.

Table 6.16 Estimates of German bomb tonnage diverted by British decoys, 1939–45

Total German bomb tonnage	*Total tons diverted by decoys*	*% tons diverted by decoys*
56,000 by night	2,092	3.7
12,500 by day	129	1.0

Distilling these data, we obtain the figures in Table 6.16, which are conservative overall estimates of German bomb tonnage diverted for the entire war. The best adjusted overall proportion of tonnage diverted by decoys is probably 5 percent, and for crucial years and targets the proportions occasionally rise as high as 10 percent. If we accept the reasonable 5 percent estimate, this converts to a saving by moderately inexpensive decoys of more than 3,000 injured and 2,500 dead (Dobinson 2000, 210–13).

The bottom line is that for a modest cost in personnel and material, the British camoufleurs achieved a small but evidently cost-effective saving in casualties and damage to airfields, factories, cities, and port facilities.

Lures

Unlike a decoy, which draws attention away from a nearby real target, the lure stands alone. A lure is a worthless installation designed to draw attack by simulating high value or to provoke the opponent to rise to an unequal fight. Unlike the decoy, the lure is vigorously defended but on a killing ground of the defender's choice. Its purpose is to create a battle of attrition where the casualty ratios greatly favor the defender.

However, unless the lure is simultaneously an ambush, it is in effect no more than a form of attritional warfare and thus is profligate in material and personnel. In absolute losses, the lurer should expect to suffer about as much damage as he inflicts. This tactic of attrition is commonly justified on the grounds that it yields a relative advantage in mutual slaughter. This is a weak argument, for four reasons. First, advocates of attritional warfare often simply neglect taking into account the ability of their opponent to replace his losses; that is, time – and attrition is a matter of time – often works to the advantage of the opponent. Second, the time factor also gives the enemy a good opportunity to see through the intent and adjust his tactics accordingly. In other words, attritional warfare, being intrinsically slow paced, is seldom able to incorporate the most cost-effective factor – surprise. Third, the same result to the enemy at less cost to oneself can usually be achieved by more imaginative deception operations. And fourth, in practice, the attritional lure often backfires.

Lures apply across the range from minor tactics to grand strategy and from asymmetrical to conventional combat. The following is a successful operational-level case.

Case: The German ruse against Operation Crossbow, 1944

Operation Crossbow was Britain's effort in World War II to counter Hitler's super "Victory" weapons (this account is drawn from Irving 1964, 307–8). British intelligence, particularly its photo interpreters (PIs), took pride in their ability to locate the various launch sites before activation. This target information was used to plan 69,000 Allied aerial sorties that unloaded 122,000 tons of bombs at the high cost of 450 aircraft and 2,900 airmen. Yet, more than a quarter, and probably more than half, of this prodigious effort was wasted on obsolete targets. Indeed, not only was the entire 28,000-ton effort from January through May 1944 wasted, but at no time were delays imposed on German launch site preparations before they actually opened fire. This sorry record for the PIs was not only unsuspected at the time, it is still largely overlooked by historians and memoir writers.

The Germans' original "Flying Bomb" installations (which the British PIs called "ski sites," "large sites," and "supply sites") had become obsolete that January. They were henceforth superseded by the well-camouflaged "modified sites," which the PIs failed to detect until the sites began firing in June. Instead of simply abandoning the old sites, the Germans continued to simulate normal activity around them. For five months, the Allies expended their entire Crossbow attack on these now worthless targets, one of the larger of which had been appropriately renamed Concrete Lump. Even after the new "modified sites" revealed themselves in June through their activity, Crossbow still gave the old sites high target priority.

The PIs were adept at identifying *decoys*, that is, dummy installations simulating real ones at alternate locations, but they were quite unable to identify *lures*. A decoy simulates a real target but in an alternate place, and exaggerates value by simulating high payoff or value. A decoy is a pure case of dazzle in that it creates ambiguity of place. A fisherman's lure does not simulate another real object; it merely draws attention to itself in order to draw the fish to it. Even the hunter's "decoy duck" is actually a lure.

The RAF PIs were not only unable to identify lures, they were seemingly unaware of the possibility of lures. Lures are much more difficult to detect than decoys because they are real (not dummy) objects; the lure is the real thing – only its function is faked. Thus camera imagery alone is not enough to uncover the hoax. Here, human intelligence trumps imagery intelligence.

Lures that backfired

As noted earlier, the attritional lure often backfires, as was the case for the Germans at Verdun and the French at Dien Bien Phu. In both cases, the strategy was effective at the outset, but when the tide turned and the lurers' losses mounted, they were unable to relinquish the prized real estate they held, and in the end suffered enormous losses.

At Verdun, the strategy failed because General Erich von Falkenhayn had kept his plan secret from senior headquarters, from parallel headquarters, and from the Austrian allies – even his own artillery adviser was not informed until it was too late to appropriately modify the artillery program (von Falkenhayn 1920, 255, 264–5). Because his superiors and colleagues had no idea that his intention was a strategy of attrition, they soon joined the French in viewing Verdun as a symbol, a precious trophy that must be held at any cost; as a result, that blood-soaked patch of land become a killing ground for the Germans as well. When the battle ended ten months later, the total casualty rates only slightly favored the Germans – 337,000 Germans to 377,000 Frenchmen, or 1 to 1.1, compared with a ratio of 1 to 1.5 in other sectors of the Western Front during the same period.

General Henri Navarre very deliberately committed major elements of his French Army in Indochina to Dien Bien Phu. The French had taken the isolated mountaintop prize of Dien Bien Phu from the Viet Minh by airdrop in November 1953 and gradually increased its garrison to a peak strength of 13,000. Navarre's plan was to lure Giap into adopting a frontal attack or siege tactic; the bait had to appear temptingly weak while in fact being strong – a standard tactic among the French (see Fall 1967, 5, 49). Giap accepted the challenge and mounted a siege, bringing in more and more troops, just as Navarre had wanted. With complete command of the air, Navarre assumed he could reinforce and resupply as long as necessary. The early counterattacks were easily repulsed, with high casualty ratios (well over 3 to 1) for the Viet Minh.

All was going according to Navarre's plan. But then, as with Verdun, Dien Bien Phu started to become a household word not just in France but throughout the world; this trivial mountaintop was becoming transformed into a symbol of prestige and honor for France – and for Navarre. Meanwhile, everything was also going according to Giap's plan. Instead of trickling in one unit after another to be destroyed piecemeal, he gradually built up the besieging force to 50,000 combat troops and 55,000 support troops, winning the race of men and material. As the defense perimeter was squeezed tighter and tighter, the French helipads became vulnerable to artillery fire. Air resupply had become limited to airdrops, and reinforcement as well as evacuation had become impossible. The French were now caught in their own trap, and Dien Bien Phu was overrun six months after they took it, with losses nearly equal on both sides (18,916 French to 22,900 Viet Minh).

Diversionary attacks

The diversionary attack is one type of military deception operation that often proves quite costly, because it sends real combat units into real battle. At its worst, it is the most primitive form of deception. The diversionary attack exemplifies the relatively low level of deceptive thinking among most British, French, and German generals in World War I and the unprepared Russians and Americans early in World War II. For example, in 1917, in the three-day Battle of Scarpe and Vimy Ridge, twenty-two British and Canadian divisions were sent over the top, costing them 18,000 casualties and inflicting a slightly lower number of casualties on the German defenders. But this was merely a *limited diversionary attack* to cover the real Franco-British forty-six-division attack that followed elsewhere (Whaley 1969, A85–A87).

Although this strategy is common during war, such profligacy is not simply inelegant, it is seldom, if ever, necessary. Any number of other deceptions will very likely cost substantially less while providing a substantially higher payoff.

Comparative costs of operations with little or no deception

Most of the military cost-benefit statistics available can at best only compare data from before and after deception was instituted and hope that the class of all exceptional events were typical. A true scientific experiment would examine a stream of events and omit all deceptive measures in one or more randomly selected events. Of course, this normally is not an option. However, by pure chance, just such an "experiment" occurred once during the British bomber offensive against Germany during World War II. This instance was caused by an incredible break in standard operating procedures that produced the disastrous raid on Nuremberg on the night of 30/31 March 1944.

Case: RAF bomber raid on Nuremberg, 30/31 March 1944

That raid was typical of Main Force operations during the so-called Battle of Berlin, which comprised 35 raids on cities throughout Germany during the four and a half months from mid-November 1943 through March 1944 (this account is drawn from Middlebrook 1974; Webster and Frankland 1961, 192–3, 207–9; Price 1967, 196–8; and *The Royal Air Force* 1954, 28; see also Bekker 1968, 339–40; Irving 1964, 54–55; and Verrier 1968, 293). It was typical in all major particulars except one – deception. Strangely, Bomber Command omitted virtually all its best ruses. The Main Force of 795 heavy bombers flew an almost direct course toward their target, with only two minor turning points. None of the normal diversionary operations were undertaken. (Later, Bomber Command limply attributed this to "conditions over the North Sea" without explaining why other routine operations were able to proceed there.) A mine-laying raid by fifty Halifax bombers in the Heligoland Bight did not confuse the German ground controllers. Nor did the small-scale spoof raids on Kassel, Cologne, Frankfurt, and other minor targets, because the Germans immediately recognized the raiders as unthreatening Mosquito bombers.

An interceptor force of 246 night fighters was scrambled and waiting when the Main Force crossed into German airspace at midnight. In the clear night sky, lit by a half moon, the interceptors could see the bomber stream, so they no longer needed their ground controllers' instructions, which in any case were now being jammed. During the hour and a half the bomber stream was fighting its way the 250 miles into Germany, the night fighters shot down about fifty-eight bombers. Given the furious distraction of the aerial combat and the high tail winds, about twenty-seven other bombers aborted or lost their target, and the rest were so scattered that the ground controllers did not actually identify Nuremberg itself as the target until two minutes before the attack began by the 710 remaining bombers. This element of fortuitous surprise did catch seven squadrons of the single-engine "Wild Boar" interceptors hovering outside Berlin and Leipzig. However, twenty-one squadrons (75 percent of all scrambled interceptors) were clinging to the bomber stream and destroyed another twenty-two bombers over Nuremberg and a final fifteen on the homeward flight. In all, the British raiders suffered ninety-four bombers lost, an appalling rate of 11.8 percent. In addition, twelve bombers were damaged beyond repair and fifty-nine were moderately damaged. Nuremberg received only minor bomb damage.

If we merely take the "official history" statistics and recast them as in Table 6.17, the key point is made. During this same period, Mosquito light bombers flew 2,034 diversionary sorties, losing only 10 aircraft, or 0.4 percent (Webster and Frankland 1961, 199).

Three points are clear. First, the confusion and surprise produced by effectively applied deceptions yield average loss rates of only about 5 percent (and those other 34 raids included some where, despite good deception, the German ground controllers or some interceptors accidentally guessed the real target). Second, where deception was poor and surprise was slight, the loss rate soared to nearly 12 percent. Third, the costs of the electronic and spoofing operations themselves were small relative to their effect on the Main Force efficiency and loss rates.

This period (November–March) was one in which the 5.2 percent attrition rate in heavy bombers was "acceptable" to Bomber Command because replacement of planes and crews could keep up – indeed, the total daily average number of available heavies (with trained crews) increased from 864 to 974 over those five months. However, an attrition rate averaging 5.7 percent or more in that same period (i.e. costing an additional 110 planes) would have been "unacceptable." In other words, the already narrow margin of operational existence of Bomber Command in that period must be entirely credited to the deception planning, which was then "saving" about 7 percent (give or take a half percent) of the sorties, or at least 1,300 heavy bombers – more than the total force.[2]

Table 6.17 British heavy bomber losses in the Battle of Berlin, 18 November 1943–31 March 1944

Target	*Number of raids*	*Number of sorties*	*Bombers lost*	
			N	%
Nuremberg	1	795	84	11.8
Berlin	16	9,111	492	5.4
All other targets in Germany	18	10,318	461	4.5
Total	35	20,224	1,047	5.2

It is commonly argued that the strong needn't bother with deception. That may be so, but only in two circumstances – both illustrated by this case: first, when, in the short run, the stronger party isn't bothered by its losses; and second, in the long run, when it isn't concerned that the attrition rates from successive battles will exceed the rate of replacements.

Small-unit asymmetric combat: low-tech, high-deception units versus high-tech low-deception units

The accidental experiment described in the case example above and, more generally, in studies by Liddell Hart (1929, 1941, 1954), Jones (1957), Davies and Harris (1988), and Whaley (1969, 1973, 1979) have demonstrated the high benefit-to-cost ratio of military deception at the levels of grand strategy, strategy, and grand tactics (operational art) that cover entire wars and major battles. These studies suggest that the use of deception confers benefits across the entire strategic-to-tactical continuum. I've long suspected that the many studies that set up a rigid dichotomy between "strategic deception" and "tactical deception" perpetuate a largely false notion. But is this hypothesis on the right track? Since 1969 I've seen many small-unit tactical cases that support this conclusion, and I've found no counterexamples. But this is only anecdotal evidence. We need a robust, systematic operations research–type study focused on this question, on the model of Richardson (1960), Whaley (1969), Sherwin and Whaley (1982), and Dupuy (1985). The only effort in this direction of which I'm aware is a fairly sketchy one by Poole (2003). Poole's broad context is small-unit light-infantry tactics as practiced by the high-tech US Army and Marines against low-tech infantry, guerrillas, and other paramilitary units. Poole gives two statistical analyses – the Korean War and the Tet Campaign in Vietnam.

Case: Korean War, 1950–3

The best available statistics on battle casualties during the Korean War reveal a much less lopsided total than generally believed. If we combine the North Korean and the Chinese casualties, we get somewhere between 1.3 million and upwards of 1.5 million total communist battle casualties. If we combine the US and UN casualties and those of their South Korean allies, total casualties are somewhere between 1.13 million and 1.54 million. In other words, at close to 1.5 million each, the two sides achieved virtual parity. Given the fact that the US/UN forces had air-sea dominance and fielded a more high-tech heavy infantry, how did the Chinese and North Korean light infantry manage to match casualties on a nearly one-to-one basis? Poole's plausible explanation is that it was because the Chinese and North Korean light infantry were much better trained in deceptive tactics, right down to the squad level.

Case: Tet Offensive, Vietnam, 1968

The best available statistics on battle casualties during the sixty days of the Tet Offensive during the Vietnam War reveal a similarly much less lopsided total than generally assumed (this account is based on Poole 2003, 41, 42–3, 348). Poole remarks that, unlike in Korea, "this time, the Communists were not only badly outgunned, but also badly outnumbered." The combined Viet Cong and North Vietnamese Army (VC/NVA) attacked during Tet with only 67,000 combat troops against the combined 1,179,000 US and Army of the Republic of Vietnam (ARVN) defenders. Casualties for the VC/NVA ran from 19,000 to 40,000, against 45,000 defenders (24,000 US and 21,000 ARVN) – a surprisingly low casualty ratio for attackers. Again Poole, a Vietnam platoon commander, attributes this outcome to the enemy's greatly superior use of deception from division down to squad level.

Poole concludes that "[i]n effect, 'low-tech' Eastern armies have had to capitalize on the field skills of the individual soldier." These skills, practiced and honed to a degree found among American troops only in Special Forces, include all the arts of deception – camouflage, invisible movement, stealthy approach, night operations, decoys, lures, and so on (Poole 2003, 43–4).

Small-scale deception operations

In this section, I examine the costs and benefits of two types of small-scale deception operations – sniping and ambushes.

Sniping

Sniping is the smallest-scale and stealthiest military operation. Typically involving either a single sharpshooter or a two-person team, sniping is generally well worth the special efforts required in training, equipment, and administration. In addition to the opportunity for intelligence gathering, sniping operations lower enemy morale, slow their advancing patrols, and regularly yield the highest kill ratios of all types of infantry combat.

Sniping may be the single most cost-effective type of military operation, and it's certainly the most cost-effective infantry operation. This is demonstrated dramatically by the following statistics (Sasser and Roberts 1990, 152):

- At the end of World War I (1917–18) on the Western Front, American troops fired an average of about 7,000 bullets for each German soldier killed.
- In World War II (1939–45), Allied infantry fired approximately 25,000 bullets for each enemy soldier killed.
- During the Korean War (1950–3), UN troops fired 50,000 rounds for each enemy soldier killed.
- In Vietnam (1955–75) American combat troops (with M-14s and later with rapid-fire M-16s) fired at least 200,000 shots per kill.

- By contrast, the average trained sniper requires only 1.3 bullets per kill – just slightly more than the sniper's proud motto of "One shot – one kill" (Sasser and Roberts 1990, 2).

Ambushes

The ambush is another type of smaller-scale military deception operation that can be cost-effective. It has proved itself from ancient times to the Iraqi insurgency that began in 2003. However, the ambush has been grossly underrepresented in systematic studies (British Col. Malleson's *Ambushes and Surprises* [1885] notwithstanding). There is no excuse for this omission. The subject is both important and easily researched. I discovered this in 2003–4 when I collected over fifty personal-experience examples from as many of my American and foreign Special Ops students at the Naval Postgraduate School. My necessarily tentative conclusion from such anecdotal evidence is that ambushes are not only usually cost-beneficial but almost always so. And, while my evidence is skimpy, the same seems to be true of counterambushes.

Zero percent solutions

I refer to deception as the "1 percent solution," although admittedly, as some of the above cases show, it sometimes costs more. A 5 percent diversion of a commander's strength might not deter him from employing deception, but 10 percent would surely give him pause, and rightly so. And, except in extreme peril, planners would and should have difficulty in justifying a 20 percent or 25 percent diversion. But an unchallengeable case would be made if the cost was zero. Is that possible? Rarely, but yes, as the two following cases show.

Case: General Funston ends an insurrection, Philippines, 1901

One consequence of the American victory over Spain in 1898 in the Spanish-American War was that the United States annexed the Philippines (this account is drawn from Seelinger 2004; Editors of the *Army Times* 1963, 23–30; and Funston 1911). The American occupation led the 30-year-old Chinese-Tagalog leader *Presidente* Emilio Aguinaldo to launch a full-scale insurrection against the occupiers on 4 February 1899. Two years and 6,000 American soldier casualties later, there seemed no end to the affair. Moreover, with the first flush of victory past, the American press and public had turned against the US occupation.

Then, on 8 February 1901, thirty-five-year-old Medal of Honor recipient Brig. Gen. of Volunteers Frederick Funston, a district CO at San Isidro in central Luzon, read a dispatch intercepted with one of Aguinaldo's couriers. It ordered insurgent General Lacuna to send 200 soldiers from his brigade to Aguinaldo's headquarters. The courier revealed that this HQ was currently encamped at Palanan, a small riverside village in a remote region of northern Luzon. He further disclosed that Aguinaldo's HQ guard was only fifty strong. Funston knew that the only approach was overland up a long tortuous road through a dense forest inhabited by rebel-friendly villagers. He realized that any direct assault would only run a gauntlet of ambushes, ending at a deserted rebel HQ. So Funston proposed a bold and clever deception to his commander in chief, Gen. Arthur MacArthur, who approved.

Funston formed a party of four American officers and 85 soldiers from the Macabebe Regiment. He'd chosen the Macabebes because they physically resembled Tagalogs. Funston and his American officers (all disguised as privates) would pose as POWs of his Macabebes. The Macabebes would themselves simulate the advance half of the Tagalog reinforcements expected by Aguinaldo. Three weeks were spent in planning and preparations. The Macabebes were coached in Tagalog dialogue and outfitted with captured weapons and clothing. Previously seized letterhead of General Lacuna would provide forged documents to smooth their passage through enemy territory. On 6 March, Funston and his small party, totaling eighty-nine men, set out for Aguinaldo's HQ at Palanan.

The trip required several further ruses, each improvised to overcome unexpected problems, particularly during the final approach, a 110-mile march through jungle that began on 14 March. Nine days later, on 23 March, Funston's party reached Aguinaldo's camp.

On their arrival, Funston sent two of his trusted Filipino officers, Segovia and Placido, ahead, and they were personally greeted by Aguinaldo. Then, on Segovia's unobtrusive signal, Funston's main force opened fire on fifty of the surprised Tagalog guards, killing two and scattering the rest. Hearing the nearby rifle shots, Aguinaldo assumed it was his own guard firing in the air to celebrate the arrival of reinforcements and stepped to the window of his bungalow, where he was seized by Placido. Two Tagalog officers were killed and two surrendered before the rest of the insurgent officers and troops fled. Funston graciously accepted the *Presidente*'s surrender, along with that of two of his chief officers.

Funston's daring ploy effectively ended the Philippine insurrection. Aguinaldo's final act as President of the Philippine Republic was to order all his followers to cease resistance. Except for the separate insurgency by the Muslim Moros in the south, his order was fully obeyed within three weeks. That was the great strategic benefit of Funston's small-unit tactical deception operation. The cost had been, in effect, zero. And, had the expedition failed, it would have meant, at worst, the lives of 89 Americans and their allies. The only real cost of this successful deception was in reputation – the American press generally sneered at General Funston for having used such un-American tactics.

Case: Dr. Jones versus Hitler's flying bombs

The single most cost-beneficial deception ploy of World War II was perhaps the successful diversion of Hitler's rocket bomb attacks on London in 1944 (this account is drawn from Jones 1978, 420–4, and map opposite 301; and Irving

1964, 250–8). The Luftwaffe V-1 campaign aimed 2,340 of the terrifying "Flying Bombs" toward the heart of the British capital. V-1 bombs killed 5,500 Londoners and seriously injured 16,000. There would likely have been as many as 50 percent more casualties if it hadn't been for the Germans' persistent misperception that most of the hits were overshooting their intended aiming point, Tower Bridge, at the geographical center of the great metropolis. In fact, most bombs were falling short, onto the much less densely populated southern suburbs.

To adjust range, the Germans depended on individual time-and-place bomb reports from their spies in London. Unknown to them, *all* their agents were under British control. Taking full advantage of this fact, Dr. R. V. Jones, a world-class physicist at the Air Ministry, conceived and designed a scheme of plausible agent dispatches that reported real hits but faked their times of impact. This elegant ruse induced the Germans to steadily readjust their real aiming point ever further short of their intended one. By the end of the campaign, the Flying Bombs had been lured four miles south of their original bull's-eye.

There are two circumstances that could have foiled Dr. Jones' scheme. First, the balance of correct reporting on location of hits with misreporting of their times was designed to seem consistent with any enemy aerial photoreconnaissance evidence. In fact, but unknown to the British at the time, the Luftwaffe had managed (accidentally) to put only one photo-recon aircraft over London during the entire ten weeks of the V-1 bombing campaign. Second, and also not suspected by the British, the Germans had placed radios on several of their Flying Bombs precisely to get a radio direction-finding fix on the time and place of impact. German air intelligence correctly saw that the combined plots showed that they had been aiming low. However, they concluded that, because their eyewitness human intelligence was reliable, it must be the radio direction-finding equipment that was inexplicably faulty!

Costs and benefits of counterdeception

It is harder to judge the relative cost of counterdeception – the detection of deception. But it's fair to say that in the majority of cases, most of the overall costs of detecting deception are prepaid by already existing intelligence capabilities. Indeed, as I've pointed out elsewhere, consistently effective deception analysis doesn't require more analysts or a bigger budget but simply different recruitment criteria, teaching tools, training procedures, and operational protocols (Whaley, *The Maverick Detective, Or The Detection of Deception* [unpublished draft, 2005]).

Conclusion and recommendations

Military leaders are understandably reluctant to accept any additional demands on their budgets and manpower that might detract from the combat effectiveness of their units. Moreover, they will rightly resist institutional additions to their already cluttered tables of organization and equipment. For example, the innovation in World War II of special psychological warfare units met widespread and serious resistance from puzzled field commanders who (justifiably) failed to see precisely what concrete military benefits balanced the all too apparent expense and inconvenience of their leaflet, loudspeaker, and propaganda radio units. Thus, military commanders need assurance that the investment cost of deception is low – both absolutely and relatively – and that the combat benefits are high.

The main conclusion of this chapter is simple. Deception is worth the cost and effort in almost all situations. So how can we improve our ability to practice deception and make our deceptions more effective, with more consistent chances of success?

First, we can profit from further systematic research to verify the findings I've presented here and to extend this research into other types of deception/counterdeception situations as well as more recent examples. I strongly recommend cost-benefit studies of three particularly rich data sources:

1 The National Training Center at Fort Irwin, where the US Army collects extensive and detailed data on real-time war games in which deception strategies have a prominent role.
2 After-battle studies from Somalia, the Balkans, Afghanistan, Iraq, and Lebanon, with a particular focus on ambush and counterambush actions.
3 Detailed interviews and oral histories from Special Ops veterans of the above five combat zones. These warriors are a rich source of anecdote. For example, from nearly 100 American and foreign Special Ops officers in 2003–4 from whom I obtained details of ambush situations, all but one were able to give revealing details. These men had been personally involved in ambushes, either as ambushers or ambushees, in combat or during training exercises.

Data of this type should be collected and collated to support analyses of asymmetric insurgency and terrorist operations. While I believe the broad outlines of deception cost-benefit findings presented in this chapter will be confirmed by further study, new analyses will flesh out and fill in the details. This will help us to better anticipate and master the challenges posed by adaptable and deceptive opponents. However, one of the most revealing points that emerges from this survey of case studies is that commanders need not wait for their think tanks to come up with solutions. Most of the various commanders seen in the cases I presented proved themselves sufficiently imaginative to solve a difficult problem. They found what I call the "third option," an unexpected solution to a complex problem. But then these were exceptionally qualified staff officers like Maj. Ingersoll, commanders like Maj. Funston, and scientists like Dr. Jones – all of whom possessed more than the usual "prepared mind" plus the imagination to create appropriate solutions.

Notes

1 Elsewhere I dubbed this the "Five Patton's Ruse," an error later corrected by reading the then recently declassified official history of the 23rd Headquarters Special Troops. In my interview with Ingersoll, he had exaggerated – as his deputy, Capt. Wentworth Eldredge, had warned me Ingersoll tended to do. See Whaley and Bell 1982, 379–81.

2 Price (1967, 243) estimates that "more than one thousand [RAF] bombers and their crews" were saved from December 1942 to October 1944 by electronic countermeasures alone. During that period, 4,490 RAF bombers were lost on night raids. Baxter (1948, 169) reports an official American estimate that "radar counter-measures saved the United States Strategic Air Force based in England alone 450 planes and 4,500 casualties."

Annotated bibliography

Well over a thousand books and articles have been devoted to military deception. Of these, however, those below marked with an asterisk (*) are the very few I've found that discuss, explicitly or implicitly, both the benefits and the costs of deception or include relevant numerical or statistical data.

*Alcala, Robert W., *Effective Operational Deception: Learning the Lessons of Midway and Desert Storm*, Newport, RI: Naval War College, February 1995 (22 pp).

Atkinson, Rick, *Crusade: The Untold Story of the Persian Gulf War*, Boston: Houghton Mifflin, 1993.

*Barkas, Geoffrey, *The Camouflage Story (from Aintree to Alamein)*, London: Cassell and Company, 1952.

*Barkas, Geoffrey, and John Hutton, "Camouflage of Airfields in the Middle East, 1941–42," *RAF Quarterly*, vol. 5, no. 2 (April 1953): 112–20.

*Barton, S. W. [pseudonym of Barton Whaley], and Lawrence M. Martin [pseudonym of Ladislav Bittman], "Public Diplomacy and the Soviet Invasion of Czechoslovakia in 1968," in Gregory Henderson (ed.), *Public Diplomacy and Political Change: Four Case Studies*, New York: Praeger, 1973, 241–314. On the Soviet deception operations that largely succeeded in concealing the Soviet intent to invade Czechoslovakia.

Bekker, Cajus, *The Luftwaffe War Diaries*, Garden City, NY: Doubleday, 1968.

Bell, J. Bowyer and Barton Whaley, *Cheating and Deception*, New York: St. Martin's Press, 1982.

Bittman, Ladislav, *The Deception Game*, Syracuse, NY: Syracuse University Research Corporation, 1972.

*Brugioni, Dino A., "If You Can't See It, You Can't Hit It," *Air Power History*, vol. 45, no. 4 (Winter 1998): 18–25. On the use of smoke to conceal tactical military operations.

*Bryant, Melrose M., *Deception in Warfare: Selected References from Air University Library Collection*, rev. edn, Maxwell Air Force Base, AL: Air University Library, May 1986.

Callwell, Sir C. E. (Major-General), *The Dardanelles*, London: Constable, 1919.

Churchill, Winston, *The Second World War*, vol. 4, Boston: Houghton Mifflin, 1950.

*Cohen, Fred, Dave Lambert, Charles Preston, Nina Berry, Corbin Stewart, and Eric Thomas, *A Framework for Deception*, Fred Cohen and Associates, 13 July 2001. Available online at all.net/journal/deception/Framework/Framework.html. Finds deception to be cost-effective for protecting information from computer hackers.

Copeland, Miles, *Without Cloak or Dagger*, New York: Simon & Schuster, 1974.

*Daniel, Donald C. and Katherine L. Herbig (eds), *Strategic Military Deception*, New York: Pergamon Press, 1982.

Davies, Merton E., and William R. Harris, "RAND's Role in the Evolution of Balloon and Satellite Observation Systems and Related US Space Technology," Santa Monica: RAND, September 1988 (RAND Report R-3692-RC).

*Despres, John, Lilita Dzirkals, and Barton Whaley, *Timely Lessons of History: The Manchurian Model for Soviet Strategy*, Santa Monica: RAND, July 1976 (RAND Report R-1825-NA). A pioneer analysis of the Soviet surprise-through-deception invasion of Manchuria in August 1945. See particularly Chapter 2 ("Strategic Design for Surprise Offensive"), 13–24, which was Whaley's main contribution to this paper.

*Dobinson, Colin, *Fields of Deception: Britain's Bombing Decoys of World War II*, London: Methuen, 2000.

Dupuy, Trevor, *Numbers, Predictions, and War*, Fairfax, VA: Hero Books, 1985.

Editors of the Army Times, *The Tangled Web*, Washington, DC: Luce, 1963.

*Eldridge, Justin L. C. (Captain), "The Myth of Army Tactical Deception," *Military Review*, vol. 70, no. 8 (August 1990): 67–78. A devastating critique of US Army tactical deception practice and doctrine.

*Eller, Joseph M., *Estimate of Cost of False Targets as Air Defense Auxiliaries*, Baltimore: Operations Research Office, Johns Hopkins University, May 1961 ("Confidential," declassified in 1970; published in fifty copies; 11 pp). This is not really a useful product. It gives the specs and estimated costs of seventeen types of simulated decoys (ranging from tanks to trucks) and the percentage of some types of decoy equipment to the real as used in certain US Army units, with no comparison of these costs with those of the real items, no assessment of effectiveness, and no conclusions.

Fall, Bernard B., *Hell in a Very Small Place: The Siege of Dien Bien Phu*, Philadelphia: Lippincott, 1967.

*Feer, Fred[ric] S., *Thinking-Red-in-Wargaming Workshop: Opportunities for Deception and Counterdeception in the Red Planning Process*, Santa Monica, CA: RAND, May 1989 (published as paper P-7510; 12 pp.). See also Eldridge 1990.

**FM 90–2 (Tactical Deception)*, Washington, DC: Headquarters, Department of the Army, 2 August 1978. Particularly good in showing how high-quality operational deception can be had at little cost in diverting personnel and materiel.

**FM 90–2 (Battlefield Deception)*, Washington, DC: Headquarters, Department of Army, 3 October 1988.

Funston, Frederick, *Memories of Two Wars*, New York: Scribner's Sons, 1911.

*Gawne, Jonathan, *Ghosts of the ETO: American Tactical Deception Units in the European Theater, 1944–1945*, Havertown, PA: Casemate, 2002 (342 pp). On the US Army's 23rd Headquarters Special Troops. The most detailed book on the subject, comprehensive except that the author doesn't look on the other side of the hill to discover how all this flim-flam was being perceived by the enemy.

*Gerard, Philip, *Secret Soldiers: The Story of World War II's Heroic Army of Deception*, New York: Dutton, 2002. A detailed unit history of the US Navy's development of the Beach Jumpers under Lt. Cmdr. Doug Fairbanks Jr and the Army's 23rd Headquarters Special Troops and 3133rd Sonic Deception Company.

*Greenberg, Irwin, "The Role of Deception in Decision Theory," *Journal of Conflict Resolution*, vol. 26, no. 1 (March 1982): 139–56. A pioneering effort to factor deception into game-decision theory.

*Greenberg, Irwin, "The Effect of Deception on Optimal Decisions," *Operations Research Letters*, vol. 1, no. 4 (September 1982): 144–7.

*Hartcup, Guy, *Camouflage: A History of Concealment and Deception in War*, London: David and Charles, 1979. The subtitle is grossly misleading. This book is narrowly limited to British military camouflage history. The author's timid excursions into Russian and Japanese camouflage are either speculative or wrong.

*Hauser, Marc D., "Costs of deception: Cheaters Are Punished in Rhesus Monkeys (*Macaca mulatta*)," *Proceedings of the National Academy of Sciences of the United States of America*, vol. 89, no. 24 (15 December 1992): 12137–9. Direct observation and field experiments both indicate that the rhesus monkeys on Cayo Santiago, Puerto Rico, who attempt to conceal (by failing to call) their discovery of food from their group members and are discovered with food are subject to more aggression. "Such costs may constrain the frequency with which deception occurs in this and other populations" (p. 12137).

*Haveles, Paul A. (Captain), "Deception Operations in Reforger 88," *Military Review*, vol. 70, no. 8 (August 1990): 35–41.

*Hewish, Mark and Bill Sweetman, "Hide and Seek: Camouflage, Concealment, and Deception: You Can't Attack What You Can't Find," *Jane's International Defense Review* (April 1997): 26–32.

Hoopes, Roy, *Ralph Ingersoll: A Biography*, New York: Atheneum, 1985.

Irving, David, *The Mare's Nest*, London: William Kimber, 1964. A detailed account of the German V-1 "Flying Bomb" and V-2 rocket development and operations during World War II and the British countermeasures. A fine job, as Dr. R. V. Jones later commented, before Irving drifted off into Holocaust denial.

*Jones, R[eginald] V[ictor], *Most Secret War*, London: Hamish Hamilton, 1978 (556 pp). Republished in the United States with slight changes as *The Wizard War: British Scientific Intelligence, 1939–1945*, New York: Coward, McCann, and Geoghegan, 1978. I cite from the US edition.

Keegan, John, *The Iraq War*, New York: Knopf, 2004.

*Kneece, Jack, *Ghost Army of World War II*, Gretna, LA: Pelican, 2001. On the US Army 23rd Headquarters Special Troops. Rough cost-benefit estimates are given on pp. 79–93, 280.

Liddell Hart, B. H., *Great Captains Unveiled*, London: Faber & Faber, 1929.

Liddell Hart, B. H., *Greater Than Napoleon*, London: Faber & Faber, 1941.

*Liddell Hart, B. H., *Strategy: The Indirect Approach*, New York: Praeger, 1954. The classic study of military surprise through deception. It is an enlarged, updated, and rewritten version of the author's brilliant *The Decisive Wars of History: A Study in Strategy* (London: Bell, 1929).

*Malleson, G[eorge] B[ruce] (Colonel), *Ambushes and Surprises; Being a Description of Some of the Most Famous Instances of the Leading into Ambush and the Surprise of Armies, from the Time of Hannibal to the Period of the Indian Mutiny*, London: W. H. Allen & Co., 1885.

*Mendelsohn, John (general editor), *From Normandy into the Reich*, New York: Garland Publishing, 1989 (4 files). US Army tactical deception operations on the Continent, 1944–5, three by the American 23rd Headquarters Special Troops and one by US Sixth Army Group's 2nd Tactical Headquarters "A" Force.

Middlebrook, Martin, *The Nuremberg Raid*, New York: Morrow, 1974.

Moorehead, Alan, *Gallipoli*, New York: Harper, 1956.

**Official History 23rd Headquarters Special Troops*, 1945 (129 pp). On the US Army tactical deception unit in Europe, 1944–5. Particularly interesting in that, without access to German records, the 23rd greatly *underestimated* its success.

Photo Fakery: The History and Techniques of Photographic Deception and Manipulation, Dulles, VA: Brassey's, 1999.

*Poole, H. John, *The Tiger's Way: A US Private's Best Chance for Survival*, Emerald Isle, NC: Posterity Press, 2003. Adds Russian and German small unit tactics to his previous Asian model. Includes in "Appendix A: Casualty Comparisons" (pp. 347–8) some relevant statistics.

Possony, Stefan T., and J. E. Pournelle, *The Strategy of Technology: Winning the Decisive War*, Cambridge, MA: University Press of Cambridge, 1970.

*Price, Alfred, *Instruments of Darkness*, London: Kimber, 1967. There is a revised 1977 edition. On the aerial electronic hider-hunter game, where each new method of concealment provokes a new instrument for its detection and vice versa.

*Reed, George L. (Captain), "Voices in the Sand: Deception Operations at the NTC," *Armor*, vol. 97 (Sep.–Oct. 1988): 26–31. Detailed summary of the extensive but highly cost-effective use of deception by the OPFOR (then the 63rd Regiment of the First Armored Division) at the Fort Irwin (California) National Training Center. Capt. Reed was the 1–63 Armor S4 with the OPFOR at Fort Irwin at the time of writing.

Richardson, Lewis, *Statistics of Deadly Quarrels*, Pittsburgh, PA: Boxwood Press, 1960.

Ronaldshay, Earl of, *The Life of Lord Curzon*, vol. 3, London: Benn, 1928.

The Royal Air Force, vol. 3, London: Her Majesty's Stationery Office, 1954.

Sasser, Charles W. and Craig Roberts, *One Shot – One Kill*, New York: Pocket Books, 1990.

Seelinger, Mathew J., "A Desperate Undertaking: Funston Captures Aguinaldo," www.armyhistory.org, 2004.

*Sherwin, Ronald G. and Barton Whaley, "Understanding Strategic Deception: An Analysis of 93 Cases," in Donald C. Daniel and Katherine L. Herbig (eds), *Strategic Military Deception*, New York: Pergamon Press, 1982, 177–94. A secondary analysis of data on strategic-level deceptions in an updated version of Whaley (1969).

*Stanley, Roy M., II, *To Fool a Glass Eye: Camouflage versus Photoreconnaissance in World War II*, Washington, DC: Smithsonian Institution Press, 1998.

*Sykes, Steven, *Deceivers Ever: Memoirs of a Camouflage Officer, 1939–1945*, Tunbridge Sykes, Wells, Kent: Spellmount, 1990.

*Van Vleet, John Anton, *Tactical Military Deception*, Monterey, CA: Naval Postgraduate School, September 1985. Mainly a secondary analysis of the raw coded data on tactical cases in Whaley's *Stratagem*.

Verrier, Anthony, *The Bomber Offensive*, London: Bratsford, 1968.

Von Falkenhayn, [General] Erich, *The German General Staff and Its Decisions*, New York: Dodd, Mead, 1920.

Webster, Sir Charles and Noble Frankland, *The Strategic Air Offensive Against Germany, 1939–1945*, vol. 2, London: Her Majesty's Stationery Office, 1961.

*Whaley, Barton, *Stratagem: Deception and Surprise in War*, Cambridge, MA: Center for International Studies, MIT, 1969 (965 pp, multilithed, 150 copies).

Whaley, Barton, *Codeword Barbarossa*, Cambridge, MA: MIT Press, 1973.

*Whaley, Barton, "Toward a General Theory of Deception," *Journal of Strategic Studies* (London), vol. 5, no. 1 (March 1982): 179–93.

*Whaley, Barton, *Covert German Rearmament, 1919–1939: Deception and Misperception*, Frederick, MD: University Publications of America, 1984 (149 pp). A case study relevant to both deception and counterdeception, specifically arms control verification intelligence and order of battle-type intelligence. Written in 1979.

Whaley, Barton, *Detecting Deception: A Bibliography of Counterdeception across Time, Cultures, and Disciplines*, 2nd edn, Washington, DC: Foreign Denial and Deception

Section XVI:
When Your Deception Plan is Compromised: Plan B

CHAPTER 59

Battle of the Gothic Line, 25 Aug 1944: Operation OLIVE[1]

Barton Whaley

You know the opponent has bought your deception plan. Then circumstances change 180-degrees so that the obsolete deception plan has become the real plan. The opponent now expects your next move. But you can't abort or even postpone. What do you do?

Fortunately, this is rare; but here is a real situation and its largely successful solution. Note that this is also one way to counter if you discover that an opponent has detected your deception plan.

1. Barton Whaley, *Stratagem: Deception and Surprise in War,* Norwood, MA: Artech House, 2007, pp. 406–408.

Example B38. Battle of the Gothic Line, 25 Aug 1944: Operation OLIVE

For his decisive offensive to breech the German-held Gothic Line (*Gotenstellung*)[516] that reinforced the natural defenses of the Apennines, General Alexander, commanding the Allied 15th Army Group there, originally (in early June 1944) planned his main breakthrough to go through the mountainous center, with a feint along the Adriatic coast. Deception operations were set in motion that convinced his German counterpart, Field-Marshal Kesselring, that the Adriatic feint would be the main attack.[517] Then, on 5 July, Alexander was formally notified that the final decisions in Washington and London to proceed with the ANVIL/DRAGOON landings in southern France (see Case A48) had decreed he would lose many of his U.S. divisions and, particularly, all his French divisions on whose mountain warfare experience he was depending for his offensive.

Consequently, on 4 August, Lieutenant-General Oliver Leese, the Eighth Army commander, convinced Alexander to reverse his original plan, making the real attack up the Adriatic coast with the Eighth Army while the denuded Fifth conducted the demonstration attack at the center, above Florence–in other words–the very strategy that the current deception operations were communicating to the Germans.[518]

This reverse of strategy now required a plausibly adjusted stratagem. This I suppose is the most difficult trick to bring off. Moreover, with OLIVE D-day set for 25 August, Alexander's deception planners had only three weeks to reverse the perceptions of the enemy–a feat that history shows to be extremely rare.

The essence of the new deception plan was to work a double bluff by having the Germans now believe that the *old* "evidence" fed them had been and still was part of a deliberate bluff. This was attempted by a new and elaborate program of radio deception. This consisted, at least in part, of using radio to simulate the continued presence in the center of the departed divisions of V Corps.[519]

As far as I am aware, this is the only case where a deception plan was turned back upon itself–a stratagem to attempt to discredit earlier disinformation by exposing it for the stratagem it was.

The shift in the axis of attack from center to east involved transferring two Eighth Army corps totalling eight divisions from the center to the east coast behind the two divisions already manning that part of the line. This transfer was completed by D-minus-5 without the Germans quite realizing what had happened.[520] Headquarters of the opposing German Tenth Army had received

[516]Although the Germans had already renamed it the "Green Line," the earlier designation is more familiar.

[517]Orgill (67), 30. But compare Kesselring (54), 256.

[518]Jackson (67), 266.

[519]Orgill (67), 33.

[520]A vivid eyewitness report by the Collier's correspondent is Martha Gellhorn, *The Face of War* (New York: Simon and Schuster, 1959), pp. 116-175.

D+10	German 8th Army	British 10th Army
KIA/DOW		
WIA		
MIA		
POW		3,700
Totals:	8,000	3,700+
D+26		
KIA/DOW		
WIA		
MIA		
POW		8,000
Totals:	14,000	8,000+
Tanks:	250	
D+37		
Totals:	50,000	?

Table A.79: German 8th Army and British 10th Army Casualties and Losses at Gothic Line. References: Orgill (67), 112, 158, 160, 194; Jackson (67), 273, 276, 277.

reports of the noise and dust created by the large-scale Allied lateral shift. Army headquarters was interested but tended to accept the opinion of the commander of its 76th Panzer Corps, General Traugott Herr, that this movement was merely the *westward* shipment of war material that he assumed was now being unloaded in quantity at the recently captured Adriatic port of Ancona.[521] Allied air supremacy prevented the Germans from verifying this false hypothesis.

In any case, Kesselring was not entirely disabused about the *original* deception plan and had, accordingly, reinforced the Tenth Army with General Richard Heidrich's crack 1st Parachute Division. Nevertheless, he did perceive a center attack as sufficiently plausible that he insured himself against that alternative by holding two divisions in reserve near Bologna.[522]

When OLIVE began at one hour before midnight–with five divisions jumping the Metauro River and spearheading the three-corps attack–the usual preceding artillery preparation was dispensed with in order to insure maximum surprise. And surprise was indeed achieved, not only of place and strength but of timing as well–both the German field commander (Colonel-General Heinrich von Vietinghoff) and a key divisional commander (General Richard Heidrich) were still on leave.[523] Moreover, the assault caught one German division just during the period it was being relieved in the line.[524]

Surprise gave initial success, but the offensive was soon halted by the com-

[521] Orgill (67), 33-34.

[522] Orgill (67), 33.

[523] Orgill (67), 43.

[524] Kesselring (54), 256. Compare Orgill (67), 43, who, citing British sources, says *two* German divisions were being relieved.

bination of the difficult terrain, lack of reserve units, and the ability of the Germans to bring theirs up over a far better road system.

References

- **General**
  - Douglas Orgill, *The Gothic Line* (New York: Norton, 1967).
  - Jackson (67), 252-277.
- **Warnings**
  - Kesselring (54), 256.
- **Surprise**
  - Orgill (67), 43.
  - Kesselring (54), 256.
- **Deception**
  - Orgill (67), 30, 32, 33-34.
  - Jackson (67), 266, 268.
  - Lieut.-Col. G.W.L. Nicholson, *The Canadians in Italy, 1943-5* (Ottawa: Queen's Printer and Controller of Stationery, 1957), p. 491. (Not seen by me.)
  - Maj.-Gen. R.F.H. Nalder, *The Royal Corps of Signals* (London: Royal Signals Institution, 1958), p. 402. (Not seen by me.)

Section XVII: When Deception Backfires

CHAPTER 60

Polyakov's Run[1]

Raymond L. Garthoff

This article reveals two rare cases of a deception gone wrong. Specifically, the FBI's efforts through double agents (first U.S. Army Sgt. Joseph Cassidy in 1959, then Soviet Col. Dmitri Polyakov in 1967) to make the Soviets believe the U.S. had much greater capabilities in chemical and biological weapons (CBW). In his Cassidy's Run (2000) David Wise had speculated from weak evidence that Cassidy's efforts backfired when the Russian response was to expand and expedite its own chemical warfare (CW) program. Garthoff cites Wise and adds with better evidence that Polyakov's effort similarly backfired by provoking the Russians in 1969 to enlarge their biological warfare (BW) program.

When U.S. intelligence deceived the Soviets, were they being a little too clever?

To his Soviet handlers, SGT. Joseph Cassidy seemed like any other of the dozen or so well placed noncommissioned officers of the U.S. Army and Navy who had been recruited to be agents by Soviet Military Intelligence (the "GRU"). The U.S. Navy's Chief Warrant Officer John A. Walker, Jr. may have had the distinction of being paid more (well over $1 million) for the navy codes and other valuable tactical and operational secrets he passed to the Soviets—and Walker enjoyed a long "run" before his arrest in 1985.

But Cassidy had his own distinctions. First, he had the longest run of all—one that lasted more than 20 years, starting in 1959. The second was something the GRU did not learn about until much later: From the start, Sergeant Cassidy was a U.S. double agent.

His story was only recently revealed in *Cassidy's Run,* a fascinating account by veteran espionage observer David Wise, who evidently got a major assist from the FBI. Cassidy, who was recruited and run by the FBI and Army Intelligence, played the most important of his roles in a major deception operation in the 1960s, in the course of which he passed selected and real secret information—along with carefully prepared misinformation—about the U.S. Army chemical warfare program at the laboratory and test center at Edgewood Arsenal. It was a highly successful deception and disinformation operation.

1. Raymond L. Garthoff, "Polyakov's Run," *Bulletin of the Atomic Scientists,* Vol. 56, No. 5, September–October 2000, pp. 37–40.

Act One

Both the United States and the Soviet Union launched substantial chemical and biological weapons (CBW) development programs after World War II, largely building on German and Japanese wartime research. Both countries' programs were shrouded in secrecy. The purpose of the U.S. deception operation was to mislead the Soviet intelligence and military establishments into believing that the United States had a much more extensive and successful chemical weapons program than was known—and especially to convince Soviet leaders that the United States was developing a super nerve gas called GJ.

There had in fact been long and serious research on GJ, but it was finally decided that it would never be a practicable weapon. The objective of the disinformation program was to entice the Soviets into spending great efforts and resources on something American experts had concluded was a superficially promising prospect, but ultimately a dead end. The Soviets bought it.

Cassidy was used in the chemical weapons deception operation from 1966 through mid-1969, when the American managers of the game decided to shift him to other operations. Author Wise was apparently unaware that the United States had also been transmitting disinformation on chemical weapons through other channels, and that it continued the disinformation operation after Cassidy was transferred.

To all appearances at the time, the deception was a great success. The Soviet military did, indeed, expand and intensify its efforts to develop a super nerve gas like GJ. However, as time went on, there was increasing evidence that Soviet scientists had succeeded where their American counterparts had given up, and that the Soviets had successfully developed a usable, much more toxic, and highly effective nerve gas called Novichok. Whether Novichok is directly descended from GJ is not entirely clear, but the unexpected outcome raises an intriguing and important question.

Wise implies the answer: As a result of miscalculation by U.S. scientific and intelligence experts, the Soviets were encouraged to pursue a program that ultimately greatly increased their chemical warfare capabilities.

Because deception and disinformation operations like the nerve gas deception involve sensitive sources and methods, they are necessarily among the most tightly guarded secrets—even after many years have passed. Moreover, so long as these operations remain secret, they may continue to have effect. On the other hand, if they are blown, neither their unsuccessful originators nor the duped party have much incentive to disclose what has happened.

The possibility that such an operation, while successful in its own terms, might have wholly unexpected and undesired consequences is obviously one of the reasons for great care when authorizing such a measure. The need to keep the number of "witting" persons as limited as possible is understandable. But when the number of those apprised of a plan is kept to a minimum, the broader ramifications of the operation may not be adequately explored.

Intriguing, exciting, and perhaps as alarming as Wise's tale of Cassidy's Run is, it was in fact only the first act of a much more dramatic play. There was a second act, which might be called "Polyakov's Run."

"Top Hat"

On August 17, 1967, a top secret joint decree issued by the Central Committee-Council of Ministers of the Soviet Union reviewed the evidence for what was seen as an extensive and successful U.S. program in the field of chemical and biological warfare. The decree called for corresponding Soviet CBW preparations. Although my attempts to obtain this decree in the Russian archives have so far been unsuccessful, I was able to track down a reference to it in the index to the still-closed files of the Central Committee. This Soviet decision to continue to pursue CBW programs was heavily influenced by the U.S. deception operation—if not entirely prompted by it.

Because of the heightened Soviet concern about the U.S. CBW program, in 1967 Col. Dmitri Polyakov, one of the key Soviet military intelligence agents in the United States, was tasked with finding out all he could about the U.S. chemical weapons program.

Soviet intelligence, however, was unaware that since late 1961 Polyakov had been an FBI-run double agent code-named "Top Hat." (His CIA code name was "Bourbon.")

The report Polyakov sent back to Moscow was based on a continuation of the misinformation provided through Cassidy. In 1969 he told his American handlers that his report had "shocked" Soviet leaders and led to a decision to undertake a crash program to meet the U.S. CBW threat, especially in the area of nerve gas development. "Cassidy's run" on chemical weapons was coming to an end, but "Polyakov's run" was in full force.

The Soviet decision in 1969 to undertake the crash program was confirmed by other sources, notably by a second Soviet agent, a KGB officer, FBI code name "Fedora," who had also been recruited in the early 1960s. (His CIA code name was "Scotch.")

Red Herring?

Meanwhile, the United States conducted its last biological weapons test at Johnston Island in 1968, and was reassessing both its chemical and biological programs. On May 28, 1969, a national security study, Chemical and Biological Agents (NSSM-59), questioned the utility of those weapons—although the Defense Department did not want to rule out chemical weapons research.

Further studies and deliberations followed, and on November 25, 1969, President Richard Nixon renounced completely the use of biological weapons. He also announced that the United States would never be the first to use chemical weapons. He called for international agreements to ban both types of weapons, but declared that the restrictions he announced were not conditioned on reaching those agreements.

It was not until nearly a decade later that the first—and distorted—accounts of the role of Top Hat and Fedora were leaked to the press. In 1978, *New York Times* journalist David Binder reported that former FBI Deputy Director William Sullivan, along with several other FBI and CIA officials, had begun to suspect that both agents had been "turned" by Soviet counterintelligence, or perhaps had always been "triple agents." Sullivan claimed that Soviet double agents had provided falsely reassuring information on Soviet CBW programs in order to lead the United States to cut back its own programs. Incredibly, Sullivan attributed Nixon's renunciation of biological weapons and curbs on the use of chemical weapons to Soviet disinformation.

Somehow, Sullivan's leak did not result in the Soviet discovery of the identity of the two agents, nor did it compromise the U.S. deception operation. And, as Edward J.

Epstein made clear in his 1989 book *Deception,* Top Hat—and almost certainly Fedora as well-were working for the United States and not the Soviet Union. Top Hat, a.k.a. Dmitri Polyakov, by then a retired major general, paid with his life for his espionage activities when he was betrayed by CIA turncoat Aldrich Ames in 1985. (Fedora, who had earlier died a natural death, had never been revealed as an American agent.)

Nixon's renunciation of biological weapons was based on the judgment by American scientists that bioweapons would not be militarily useful, not on any presumptions about Soviet programs, much less on Soviet disinformation. Nixon's decision on biological weapons paralleled the earlier decision with respect to GJ and work on super nerve gas.

Act Two

The military intelligence chiefs and the FBI wanted to repeat their success, this time with a deception program on biological weapons. Polyakov's run was a direct sequel to Cassidy's run on nerve gas.

Accounts suggest that multiple channels, including Polyakov, were used to convey the misleading message that the United States was undertaking a clandestine biological weapons program, despite President Nixon's public announcement in November 1969 and the U.S. signature at the Biological Weapons Convention in April 1972. Soviet suspicions of U.S. perfidy in negotiating the biological weapons convention were seemingly confirmed by these clandestine disinformation channels.

In the late 1980s and early 1990s several high-ranking Russian scientists defected to the West. Because of their testimony, we now know that in January 1973—at the same time that agents were feeding misinformation about U.S. chemical and biological programs to the Soviets—the Central Committee-Council of Ministers decreed the establishment of a new, intensified Soviet program for research and development of biological weapons. A new cover organization known as Biopreparat, with both military and civilian research institutes and other facilities, including standby production capabilities, was created.

Throughout the 1970s and 1980s this extensive secret program continued—as did Soviet suspicions and concerns about the supposed clandestine U.S. biological weapons program. For example, I located in the archive of the Central Committee in Moscow a major study that the General Staff made in 1983 on foreign—and above all U.S.—biological and chemical weapons, with projections to the year 2000. No doubt there were other studies as well.

It is not clear when the U.S. disinformation operations on CBW ended—probably in the mid-1970s—but the operations appear not to have been compromised until 1985. In any case, it is evident that their effects continued long after.

Detailed information on the Soviet Biopreparat programs became available to Western intelligence only in the period 1989 to 1992, when three key scientists active in the biological weapons program defected to the United States and Britain. (Also, in 1991 a fourth scientist, still in the Soviet Union, publicly disclosed the secret super nerve gas program.) The most complete open account of the Soviet biological weapons program has appeared in a recent book by one of the three defectors, Kanatjan Alibekov, a Kazakh who rose to become deputy director of Biopreparat from 1988 to 1992, when he defected. Under his new name, Ken Alibek, he published *Biohazard* in 1999.

In addition to detailing the Soviet biological weapons program in the 1970s and 1980s, Alibek bears witness that Soviet scientists were constantly spurred on by being told that the United States had a huge biological weapons program. He writes that the precipitating factor in his own decision to defect was his shock on learning from a personal visit in late 1991 to Fort Detrick and Pine Bluff, the former U.S. biological weapons centers, that they had long before been closed down as active centers of research.

This former Soviet scientist—defector and other Russian scientists have made clear that leaders of the Soviet military-industrial complex were all too ready to use reports that the United States was working hard in the CBW field to support their efforts to get more resources for their own programs. U.S. disinformation not only influenced Soviet beliefs, it was also readily used by Soviet officials to serve their own purposes.

Act Three

The third act of this drama has been the legacy of the American CBW disinformation operation in the post-Cold War era. As Alibek and the other Soviet scientists have revealed, Biopreparat and the overall CBW program survived the end of the Cold War. These programs were curtailed by Mikhail Gorbachev in the late 1980s, and supposedly ended in 1990. Yet they continued on an attenuated clandestine basis, either because Gorbachev was persuaded that the United States also continued to carry out CBW research, or because the operation was so well covered even from him that he failed to cut it off.

Boris Yeltsin, in turn, by a decree on April 11, 1992, again officially terminated the program. But there remain strong suspicions that it continues covertly on a smaller scale.

An exchange of official inspection visits by American and British scientists to four key former Biopreparat laboratories was arranged under Gorbachev in early 1991. In turn, Soviet scientists—including Alibek—inspected former biological weapons facilities in the United States. Yet some Russian facilities and activities may remain concealed. It is very difficult to distinguish between research on epidemiology and defensive measures, which are permitted under the treaties, and research on actual chemical and biological weapons, which is banned.

In any case, the U.S. disinformation and deception operations of the 1960s and 1970s, designed to stimulate Soviet interest and investment in CBW, were only too successful, leading to the development of effective chemical and biological munitions and to a program that may still exist on a reduced scale a decade after the end of the Cold War.

This is a disastrous outcome. Although there is little danger of Russia (or the United States) using CBW, there is the new danger that remnants of the vast U.S.-encouraged Soviet programs may end up in rogue states or even in the hands of freelance terrorists. There have been charges that senior Russian officials in the program (including its former director, Lt. Gen. Anatoly Kuntsevich) supplied, or conspired to supply, advanced chemical weapons technology to Syria in the early 1990s.

Fortunately, the only known "brain drain" of top scientists from the Soviet program has been to the United States, Britain, and Israel. Nonetheless, in addition to advanced chemical and biological agents and technology another legacy of the U.S. deception operation is the existence of a large number of former Soviet CBW scientists who have had great difficulty finding new and acceptable roles in a shrunken Russian economy. In addition, Russia's remaining chemical munitions and CBW equipment and technology are inadequately protected, and the Russian government needs to institute a massive and costly program to dispose of its large stocks of chemical munitions.

There were, and to some extent remain, other less tangible but nonetheless real adverse consequences of the "successful" CBW deception operation. Not only did the U.S. disinformation program lead to unexpected and undesired Soviet achievements in developing chemical and biological weapons, it also contributed to mutual fear, suspicion, and tension.

The program undercut Soviet belief in the efficacy of arms control and in the integrity of American policy by misleading Soviet officials into believing that the United States was deliberately violating the Biological Weapons Convention, justifying their doing so as well. But were the potential benefits of the deception worth besmirching America's reputation?

These are some of the broader ramifications of strategic deception operations that were not in this instance—and generally are not assessed or even considered before decisions are made to forge ahead.

Section XVIII: Detecting Deception: When Detective Meets Deceiver

CHAPTER 61

Counterintelligence (CI) vs. Counterespionage (CE)[1]

Robin Winks

Robin Winks was a historian, diplomat, and author. A Fulbright scholar in New Zealand, Winks earned his Master's degree in Maori studies and went on to earn his Ph.D. from Johns Hopkins University. Between 1969 and 1971, he served as the U.S. Cultural Attaché to the American Embassy in London, and was a regular advisor to various government agencies. In addition to his diplomatic work, Winks wrote about detective fiction, publishing "Detective Fiction" (1980), "Modus Operandi" (1983), and "Colloquium on Crime" (1986), among others.

Intelligence is, of course, evaluated information, and there are hundreds of ways to assist a foreign evaluator to arrive at the wrong conclusions. While strategic deception was central in time of war, since it had immediate relevance to action, it is also important in time of peace, since it involves controlled, purposeful, and intimate contact with potential enemies. Counterintelligence and counterespionage work is, therefore, always in potential conflict with those in any intelligence organization whose primary concern is security. If security is the most important goal, the next logical step after the discovery of an opposition network is to make certain one has found all members of the network and then obliterate it. CE, on the other hand, would wish to exploit the network, to play it back to the enemy. This is highly dangerous if all members have not been uncovered, or if some only pretend to a willingness to be exploited and then pass information in such a way as to reveal that they have been doubled. Those who are security-minded will avoid foreign agents; CE will move in on them. Angleton knew which response he thought most effective, and he recognized that CE was not likely to be fully understood, could not even be fully explained since it depended upon obscuring intentions as well as operations. Security, operational branches, CI, and CE would often be at cross-purposes, and so there would always be suspicion and competition within any centralized intelligence agency.[2]

1. Robin Winks, "Counterintelligence (CI) vs Counterespionage (CE)," in *Cloak and Gown*, New Haven: Yale University Press, 1987, pp. 421–424.
2. Counterintelligence and counterespionage are two terms much abused, and the distinction between the two is discussed in several books. Buffs like to argue about whether hyphens are required or not and enjoy pointing out differences that arise when the words are seen as nouns, verbs, or adjectives. Norman Holmes Pearson's papers contain a document, stamped Secret, which sets out the definitions as used by the OSS. CI is "all efforts to neutralize, repress or eliminate the activity of enemy inimical or other persons or groups or governments or their representatives to secure intelligence the obtaining of which adversely affects the national security." CE is a subdivision of CI and involves human agents. It is "all efforts to neutralize, repress, or eliminate the practice of spying or employment

When counterintelligence actually works, it becomes a form of political action. As Lenin observed, every intelligence operation has a political object; CI helps to find what that objective is. Since no intelligence body in any democracy is supposed to engage in political activity, the success of any counterintelligence operation has to be hidden. Only failures generally become known. There can be no expectation of public praise, for anything deserving of praise must be kept secret. Counterintelligence functions are positive intelligence. Operations might be closed down, and yet, if penetration of the enemy has occurred, the operation can still run. CE is like putting a virus into *the* bloodstream of the enemy. Such an approach involves the longest-range views possible; intelligence can never be static, since the definition of the enemy may change, as postwar shifts had shown. One might never know who one's enemies will be, for revolution and subversion may turn even the staunchest allies into the bitterest of enemies decades down the road; one must hide long-range activities from friends as well. Inevitably this will seem, indeed for all practical purposes will be, paranoia. This is not to be helped. There are a few lines from Stephen Crane that elliptically sum up the delicate balance: "A man feared that he might find an assassin; another that he might find a victim. One was more wise than the other."

Counterintelligence, then, was not meant to apprehend enemy agents, or to operate against the enemy in a defensive way, despite the misleading implications of "counter." Rather, as an offensive operation, CE was meant to use the opposition's human initiatives against itself. This could scale the heights of the absurd if not in the hands of the most clever of men. In 1963, using the pseudonym Christopher Felix, James McCargar had written what is widely regarded as the best general book on the methods used in "the secret war," *The Spy and His Masters.* There he describes a scene from *Romanoff and Juliet,* as played by the actor Peter Ustinov. Ustinov is prime minister of a tiny, unaligned nation caught between the Russians and the Americans. He meets the American ambassador and learns that the Americans know of a secret operation which will give the Soviet Union dominance over Ustinov's country:

> After a bit of reflection, the Prime Minister . . . calls on the Soviet Ambassador. After some preliminaries, he says to the Soviet Ambassador, "They know." The Soviet Ambassador replies calmly, "We know they know." Back to the American Ambassador goes Ustinov. "They know you know," he says conspiratorially. The American Ambassador smiles confidently. "We know they know we know," he answers. Ustinov returns to the Soviet Ambassador. "They know you know they know," he says. To this the Soviet Ambassador replies, triumphantly, "We know they know we know they know." Once again the Prime Minister calls on the American Ambassador. "They know you know they know you know," he says, weary and curious at the same time. The American Ambassador repeats it after him, counting on his fingers. "What?" he suddenly cries in horror.

---

of secret agents by enemy inimical or other persons or groups or governments or their representatives to secure intelligence the obtaining of which adversely affects the interests of the United States." X-2 began as CI but in time was referred to as CE in much of the documentation. These definitions are found in typed documents on the theory of the running of Controlled Enemy Agents, on CI double agents in neutral countries, and similar subjects, in Pearson's wooden files, box 2, as noted in the previous chapter, and they appear to be lectures Pearson prepared for briefing newly arrived agents. One may surmise that Angleton was introduced to the terminology by Pearson.

The point was simply—not so simply—to "raise the tension one step higher": never to allow your ambassador to be the one who must count on his fingers. To some this was of the essence of successful spy craft; to others, it was madness, the essence of an escalating cold war which made no sense. The existentialists would tell us that to know what someone else knows is impossible in any case. Knowing what they know, one still would not know what action they might take on what you know, even with a deep understanding of national character. The whole thing was simply too dangerous, a game with such destructive potential it ought not to be played at all. Or so some felt; to others, CI and CE provided superior intelligence, were the very essence of the game, the ideal. Until Angleton ran into a director of central intelligence who simply could not or would not accept the value and methodology of counterintelligence (or, to be more charitable, took the first view, that the risks outweighed the possible gains of sending a Ustinov back and forth across the floor of the stage), he was able to pursue CE, and its most important technique, penetration, almost without hindrance.

Angleton was an ideal man for this game. He could keep track of the agents, doubled, tripled, turn and turn again. Christopher Felix describes the problem well: "The decisive question is precisely which turns are genuine and which are false. This is but one of the problems which makes CE an extraordinarily complex affair; the determination of which turns are genuine and which false is a painstaking exercise in the control of information, of who knows what when, that requires constant alertness and a simultaneous grasp of both large perspective and detail. It is obviously an intellectual exercise of almost mathematical complexity." Angleton saw everything in relation to everything else.

He saw questions in the round, immediately ticking off relationships, ramifications, subquestions which others did not think of. If a double agent were suspected, might one not learn about that duplicity by an examination of the precise time sequence and content of information, that is, by developing a "chron"? If there were a change in Israeli intelligence, how would this influence South Africa? If Communist China appeared to be breaking from the Soviet Union, might this not be merely a very long-range strategic deception?

CHAPTER 62

Counter-Deception Planning

William R. Harris

Harris's methodology most importantly identifies and analyzes three techniques for detecting deception and hedging against failures to detect deception:

1. *Reconstructive Inference: Attempts to reconstruct patterns of the "sprignals" (spurious signals).*
2. *Incongruity Testing: This is "the traditional form of intelligence analysis, which tests alternative hypotheses."*
3. *Vulnerability Assessment: Involves predicting "future vulnerabilities on the basis of Bayesian estimates of conditional probabilities derived from past cases of deception."*

Dr. Harris, a Harvard JD, had been a senior staff member in The RAND Corporation's Social Science Department, 1972–1991. He then became a freelance consultant whose efforts shifted away from the study of deception and counterdeception to matters of international law and, regrettably, vanished over the horizon.

Counter-Deception Planning: Methodologies

There is hardly an adequate theory of deception, much less a theory of counter-deception. In addition to Barton Whaley's *Stratagem,* which organizes conclusions from the empirical data in his appendixes, and which draws upon strategic concepts, of the late B. H. Liddell-Hart,[1] there are two stimulating pieces by R. V. Jones of the University of Aberdeen: "The Theory of Practical Joking—Its Relevance to Physics," *Bulletin of the Institute of Physics* (June 1957), pp. 193-201, and a lecture on "Irony as a phenomenon in natural science and human affairs," *Chemistry & Industry* (1968), pp. 470-477. I provide these references in the text because they are seminal pieces, printed in journals to which not everyone subscribes.

The Nature of a Counter-Deception System

Counter-deception planning, as I see it, involves three related concepts: (1) *detection* of adversary deceptions; (2) the adoption of countermeasures that reduce the likelihood and adverse consequences of these deceptions, and that may involve counter-deception deception operations, which for simplicity we will call *counter-stratagems;* and (3) the coordination of *detection* and *countermeasure* programs in a *counterdeception system.*

This section focuses upon the interrelated problems of *detection* and *countermeasures*; those who read between the lines may identify organizational implications for a *counter-deception system.*

It is worthwhile to distinguish between a *counter-deception system* and seat-of-the-pants counter-deception efforts. The latter generally consist of uncoordinated, perhaps sporadic efforts to detect and outwit foreign deception operations. The seat-of-the-pants method may involve overconfident intelligence analysts who know something about deception methods and who serve as self-appointed *detection* specialists. It may also involve the production of uncoordinated[2] analyses of particular components of a deception package—for example, a report on the design and apparent objective of various camouflage or "dummy" configurations; a content analysis of a channel of cryptologic intelligence, when one suspects one's adversary has identified the cipher insecurity; or a content analysis of data transmitted by enemy-controlled double agents for a particular circuit or period of time. The analysis of possible components in deception packages does not constitute a counter-deception system; such efforts are, however, ripe for organization into such a system.

Seat-of-the-Pants Counter-Deception Methods

Section II ["The Efficacy of Deception," above] is intended as an argument against the seat-of-the-pants approach to counter-deception planning, in which every analyst serves as his own counter-deception expert.[3] This method simply does not work. Those who are still unconvinced should read Barton Whaley's *Stratagem.* Even those who are already convinced could profit from *Stratagem,* for they would see that the very intelligence services and operations planners who excelled at strategic deception were also suckers at their own game without a counter-deception system.

The British were probably the outstanding experts at deception in World War II, yet they fell for the "threat" of a cross-channel German invasion (SEA LION) months after Hitler had terminated the program, leaving only enough traces to deceive the Soviets before Operation BARBAROSSA.[4] Similarly, although a frightfully clever British deception program consistently misled *Abwehr* intelligence in North Africa (1942–1943), the Germans controlled a British sabotage-intelligence network in Europe (Operation NORTH POLE). And when the British realized, after more than a year of being duped, that the entire NORTH POLE system was operating under German control, they in turn misled the Germans into believing that this was the only British sabotage network in Holland.[5]

As previously calculated (in Section II), patient development of deception capabilities can lead to the near-certainty of obtaining misestimates of intentions, given sufficient planning time and expenditure of substantial deception assets—without a counter-deception system. Those who still cling to the hope that the "well-rounded" intelligence professional can, by the seat-of-the-pants method, reverse the tide of deceptively induced misestimates should digest the following data from *Stratagem:*

Among 114 cases of deception or surprise, Whaley found 10 cases in which the victim of attack received detailed documentation respecting supposed enemy plans well before the attack itself. Of the 10 sets of enemy "plans," five were carefully designed disinformation packages and five were genuine breaches in security. All five of the false sets of enemy plans were accepted as genuine by the intended victim, and four of the five genuine plans were dismissed as unreliable.[6]

> Is there no justice in intelligence estimating? It would appear that amateur counter-deception efforts are counterproductive. Had these analysts chosen to flip coins instead of trusting to their judgment, in 99 percent of the cases they would have correctly identified more than one of 10 sets of documents.[7]

Detection of Deception: Three Theories

There are at least three techniques for the detection of deception operations and the identification of underlying verities. The three methods are interactive—one method may assist in the exploitation of another method, and vice versa. For convenience I label these methods with three shorthand designations: (1) reconstructive inference; (2) Incongruity testing; and (3) vulnerability assessment.

Reconstructive Inference

The reconstructive inference method involves attention to the *sprignal* [spurious signal] patterns transmitted by one's deception adversaries. Mrs. Wohlstetter's *Pearl Harbor* study demonstrates how difficult it is, without hindsight, to differentiate the signals from the noise. It is with some trepidation that one would venture to separate signals, sprignals, and noise, all at the same time. It is probably best first to concentrate upon the separation of sprignals from signals, recognizing that a bit of random noise or just plain junk will find its way into both categories.

To identify sprignals, one must guard against intelligence collectors who value their cleverness as analysts and who consequently throw out or block transmission of identified sprignals. It would be helpful to have a community-wide review of data collection—both intelligence and counter-intelligence. It may be that most of the agencies make at least a minor effort to collect what appear to be sprignals. Do all of the collectors and interim analysts know how important it is to collect sprignals as well as signals or to indicate all of the danger signs that may be a tip-off [that] a seeming signal is really a sprignal? Do the important personnel have much idea of the past track record of predictions? Have they read Whaley's *Stratagem* or a fuller equivalent based upon still-classified records? Do they know of an agency center for the transmission of sprignals, suspected sprignals, or danger signs, or do they just send along their suspect information with all of the rest, risking its loss in the massive flow of data?

When a cipher is broken too easily (perhaps with the assistance of a too-easily placed clandestine agent), when sophisticated camouflage is discovered, or when a double agent is found to be under the "control" of the opposition, are the proper steps taken to identify the spurious signals and to segregate those perhaps designed for our "witting" collection as opposed to those designed for our unwitting collection?[8] In short, is there a collecting and coordinating program to distinguish the sprignals from the signals, and the intentionally obvious sprignals from the others? One must be prepared to throw out obvious "cover plans" so as to concentrate upon more sensitive deceits, which may yield inferential clues.

To distinguish the patterns of deceit aimed at policy elites from those aimed at more diffused targets is not an easy task. The problem of differentiating strategic deception from propaganda comes immediately to mind. We can begin with F. M. Cornford's definition of propaganda: "that branch of the art of lying which consists in very nearly deceiving your friends without quite deceiving your enemies.[9]

> If the usual run of propaganda operations is somewhat dissociated from strategic deception programs, those who reconstruct foreign deception patterns may be led astray, particularly as to "black" propaganda, which is as likely as not to come through clandestine agent sources.

One [American] deception planner [H. Wentworth Eldredge] of World War II sheds light on this problem in a backhanded way, revealing the possibility that inconsistencies in propaganda and deception patterns might unveil genuine intentions, at the same time implying that "black" propaganda may be a poor index to the purposes of deception planners themselves:

> The biggest damn nuisance in our experience was the truly silly black and grey propaganda people on one's own side. They had a low security classification (so they never knew anything), they had high intelligence, enormous energy, quite a lot of resources, a great desire to win the war single-handed (thus they shot off in all sorts of private directions) which seemed to us to compromise what we were doing. They confused a target of the general enemy public (or his troops) with our target, the enemy commander through the intelligence system.[10]

The above confirms what we already expect: that there is noise in the collection of sprignals just as there is noise in the collection of signals. One should not neglect the noise from the intelligence of various profiteers, who foist their imagined data upon any willing buyer. Reconstructive inference can yield no helpful clues here, except perhaps as to the mental state of the involved entrepreneur. Another form of noise may involve low-level cover and deception plans, which scarcely lead to useful inferences. Again from [Eldredge's] World War II experience:

> Actually a lot of fairly obvious things [in a deception program] were really a cover for one's own "unwitting" people who were not to know that other more complex and more "classified" operations were going on to really get to the enemy command decisions. This is not to gainsay that routine cover and deception operations of this sort were not done to connect with all enemy intelligence systems.[11]

The foregoing suggests that it may be considerably easier to collect sprignals than to separate them from associated signals and noise.

Testing for Signals and Sprignals

Some sprignals may be self-identifying; if so, we may search for confirming data or channels that may yield already anticipated sprignals at some later date. A group that is sufficiently lucky—or successful in penetrating foreign deception systems—will learn of deception efforts from "agents in place," or from modes of collection "by technical means." A somewhat less well-established group will work out their incomplete jigsaw puzzles until some of the pieces fall into place inferentially. They will debrief defecting stratagematists with unusual care, reconstruct the sources of "bum" intelligence in past encounters, and seek to eliminate the circumstantial noise while labeling and watching the suspicious channels. They will emphasize the sprignal identification missions of the counterintelligence community. They will prod security experts for more pessimistic evaluations that identify those sources of insecurity that may have been turned against them. And they will watch those sources carefully.

The incidence of signals and sprignals may be gauged, in part, from analysis of past patterns of deception. A politician who deceives his compatriots (even in pursuit of "national" objectives abroad) may not enhance his esteem. It should not be surprising that most of the literal statements of leading statesmen are at least technically truthful, or at least so intended. The implications of tendentious remarks may be far less reliable—as with the implications of Khrushchev's boasts of growing Soviet missile production in 1957–1959.[12] And the literal or implied meaning of "background" statements not linked to a source whose credibility is highly valued may be of lessened reliability. So too with official press statements, as with the indirect *TASS* denial of September 11, 1962 that nuclear missiles would" be sited in Cuba.[13]

The careful analyst may wish to distinguish the past reliability of literal statements, of inferences from tendentious remarks ripe for the plucking, of "background" interpretations, and official but anonymous press releases. One analysis of perceived mendacity in and around Washington, D.C. is consistent with observations of foreign statesmen. Although infrequently finding "the flat lies ... and petty mistruths," Anthony Lake reports:

> There have been still more misleading statements—not quite lies, but partial revelations of the truth deliberately designed to fool the public into believing what the government wants it to believe. These must be considered the functional equivalent of lies. The public does not read the fine print in government statements, and when it is fooled the result is the same loss of confidence which follows a flat lie.[14]

The counter-deception analyst will learn to differentiate the literal from the implicit, the flat from the ambivalent lie, the deceived from the deceiving spokesman.[15] Further, by cross-checking deep penetrating technological sensors (by the "incongruity testing" method to be discussed shortly) it may be possible to establish with high confidence that a source or channel of data is indeed a stratagemic producer and not a source of signals or noise. If the exotic sources of the past have been the favorites of deception planners precisely because they were the ones that received top-level attention, these "special" sources will not be adopted as arbitrary standards against which potential sprignals will be tested.

It is equally important to guard against the temptation of resorting to a test standard derived from "traditionally reliable sources." Over extended contests between intelligence and counterintelligence adversaries, both sides tend to develop an understanding of what an adversary views as reliable, and deception planners target those channels. Data that appear "reliable" are likely to conform to preconceptions or shared expectations, and deception planners like to reinforce these vulnerabilities. Thus, traditionally reliable sources and data cannot be segregated as an automatic standard against which to test new sources or new data.

Separating truth from deception and noise involves more than a comparison of the unknown with a subset of the "traditionally reliable." It is more prudent to assume that all channels of information are unreliable in varying degrees and to concede the role of chance in strategic forecasting. By conceding the hazards of the intelligence business, one is more likely to embark upon an ongoing sifting and resifting effort: identifying tentative subsets of signals and sprignals, or inconsistent clusters of data, that demand critical and recurrent appraisal.

If the processes of stratagemic analysis are coordinated with the aid of computers, it may be possible to generate data banks and analytic specialists familiar with the styles of unseen adversaries. Having collected and identified a mass of sprignals but not all of those available, and having recognized that this collection is cluttered with residual noise, we will want to obtain a set of sprignal patterns or relationships, and by inference from this universe of data to fathom the ulterior purposes and strategic links that may unite; the strands.[16]

Reconstructive inference testing has more limitations than those resulting from the ambivalence associated with residual noise. It is important to recognize that the inferences drawn from perceived deception plans are sensitive to both time and context. If a pattern of sprignals is stale, it may provide clues to the last encounter, not the next one. One World War II operator, H. Wentworth Eldredge, reflects that "some of the best C. & D. operations (after an initial plan) were played by sensitive and daring men acting pragmatically." Detection of an initial stratagematic plan may not prepare sprignal analysts for the fluidity that follows. Similarly, as with women's fashions of recent years, one may be expected to see through more than what immediately meets the eye, and if one enjoys the phenomenon one may not worry about its design.

Stratagematists are at least as bright as lingerie designers. For example, in the Italian campaign of World War II General Alexander once changed his mind about the optimal place of attack only to discover that his deception staff had already encouraged the German command to concentrate resources at this newly chosen sector of the Front. A solution was found: intelligence was "leaked" to the Germans that the original deception target was just that, and information on the preliminary build-up at the now abandoned assault area convinced the Germans of the "real" target. A new deception plan reinforced the German smugness at their "discovery" of the prior stratagem, and General Alexander's forces attacked with surprise in the area of the initial deception target.[17]

Despite its limitations, reconstructive inference testing on the basis of sprignal configurations deserves a more systematic effort and higher priority than the public literature suggests it has received.

Upon occasion we may surprise ourselves with the success of our inferences. After all, most deception plans are conceived on the basis of fairly full knowledge of actual plans. What oversight or subconscious slip finds its way into the final stratagemic product may not be apparent until it is detected by the adversary. Such irony, which R. V. Jones defines, as "a contradictory outcome of events as if in mockery of the promise and fitness of things,"[18] is a faithful companion of history, and it should not be surprising if stratagems are uncovered that reveal far more than they need have done.

From the stratagematist's perspective, "We were simply scared to death *always* that the cover plan would be blown indicating the true plan with great loss of many real human beings; it was a soul-shattering experience to be the mirror image of a 'great captain.'"[19]

Once the importance of analyzing sprignals is more generally understood, a counter-deception group may be willing to pay a higher price to receive a good selection of such data, provided of course that it has high confidence in the tentative channel identifications. It may be willing to suffer the inconvenience and possible sense of injustice in retaining on its payrolls an unseemly collection of identified double agents. It may continue to digest cryptologic or electronic sources at substantial cost, although particular channels are likely to be carrying "chicken feed." I am not suggesting that responsible officials do not recognize that in sprignals there is some utility. I would expect that monitoring of the types mentioned already takes place. But are the relevant officials aware of the usefulness of collecting enough sprignals, sufficiently well coordinated and analyzed, to enhance predictive performance in the critical cases? Obviously there are limits as to what should be spent on the collection of misleading information.

As with any intelligence system, a sprignal collection and analysis network must be safeguarded against system "overload." According to the former Deputy Chief of the Czech "Disinformation Department," Ladislav Bittman, the Czech secret service alone mounts over 100 disinformation campaigns per year against Western targets. Of these, a good number are coordinated by the Soviet counterpart in Moscow (Department "D" of the First Chief Directorate, KGB) and orchestrated with supporting operations on the part of other East European services. It would be a mistake to add the number of annual operations of each of the East European secret services, since we would be recounting aspects of the same Moscow-coordinated operations in many cases—especially in the years since 1963–64 when most of the East European services established counterpart Department "D"s for liaison with the Soviet mother department. Nonetheless, we can assume that there are at least several hundred disinformation programs per year targeted against Western audiences.

This number of operations could completely occupy the efforts of a counter-deception system. Bittman's book, *The Deception Game,* indicates that most of these East European "special operations" are propaganda operations, mainly "black" operations and often involving exploitation of clandestinely obtained documents. Although influential Western audiences are the main targets, these operations are mainly disinformation programs rather than strategic deception operations seeking to have a direct influence on decisions of governments. A counter-deception system would do well to leave the analysis, detection, and mounting of counter operations to those segments of psychological warfare agencies specializing in these subjects.

Those who man the counter-deception system that guards governmental decisionmakers and their intelligence services from the strategic deceptions of their adversaries should protect their system from the deluge of relatively trivial problems associated with these disinformation programs. It would be more helpful, for example, to identify the channels than the contents of these disinformation programs. The same wire services' reporters, clandestinely influenced or owned newspapers, magazines, publishing houses, and the like that peddle disinformation packages are likely to be available for strategic deception in critical encounters. Detailed content analysis, refutation, and design of counter-operations should be left to psychological warriors. If for no other reason, counter-deception specialists should avoid detailed involvement so as to preserve their attention span and peace of mind for more critical contests.

There are three basic filtration barriers that can protect a counter-deception system from sprignal overload. The first is a feedback system that encourages the entire intelligence community (and its foreign liaison agencies) to restrict its collection of sprignals. Some reduction in the volume of incoming data and more rigorous quality control would ameliorate the problem at the outset. Even so, the collectors' judgment should not be trusted as to which sprignals are likely to provide the most helpful inferences; further, those agencies or segments of agencies that specialize in psychological operations will demand the collection of many sprignals that are of trivial consequence from the perspective of the counter-stratagematist. The counter-deception system must be equipped with a feedback and filtration system of its own, within the broader intelligence and counter-intelligence communities. This set of barriers should be complemented by the allocation of effort among the myriad clusters of sprignal packages available within the counter-deception system's data pool.

One of the reassuring aspects of Whaley's *Stratagem* is the disclosure that the same old bag of tricks that worked in World War I was tried with a few new wrinkles in World War II and afterward. Once counter-deception systems enter the picture, deception planners will adapt, but they too have their problems. If deception planners mount only small stratagems so as to avoid detection by the analysts of sprignals, they run the risk that their adversaries' penetrating intelligence will lead to accurate forewarning. If they mount moderate level stratagems along only a few channels, they run the risk that their adversaries' "incongruity testing" will unravel inconsistencies and unmask the underlying plans. If they mount "grand" stratagems they may reduce the risk of detection by "incongruity testing," but they enlarge the risks from reconstructive inference while jeopardizing many of their deception assets. If nothing else, a system for the detection of stratagems can increase the rate of depreciation of deception assets. One may still deceive, but less often, and with fewer undetected sprignal channels for play in later encounters.

Incongruity Testing

Incongruity testing involves alternative pattern matching and testing for internal and inter-pattern consistency.[20] It is no accident that many deception planners are great practical jokers as well, for both specialties thrive upon the same fertilizer, incongruity. Since incongruity testing is the guts of intelligence analysis, the Achilles' heel of deception, and the essence of joking, I cannot recommend too highly Jones' previously cited paper, "The Theory of Practical Joking."

Simple incongruities, direct or inverted, can be humorous enough, but the more advanced jokes usually involve a period of preparation and induction, sometimes elaborate, before the incongruity becomes apparent. They are then called hoaxes.[21]

The induction of incongruity in the perceptions of an adversary constitutes a stratagem or deception. At least theoretically, this induced incongruity could be discovered, given sufficient data and hypothesis testing. The contradictions inherent in the simultaneous presence of "reality" and a false image of that reality are amenable to both inductive and deductive testing. But as Klaus Knorr notes of "the inductive fallacy,"[22] one cannot assume, as some historians and intelligence analysts would like to believe, that "the facts speak for themselves." In his survey of predictive failures Benno Wasserman observed a tendency to postpone induction until "all the facts are in"—that is, until it is too late to provide useful warning.[23]

If an adversary provides the facts and perhaps a set of inferences sustaining one pattern of inductive reasoning, an analyst must overcome the natural inertia that prevents him from generating alternative hypotheses.[24] The human mind is not especially well suited to the task of generating alternative hypotheses and of juggling them all at the same time. This judgment leads me to a small difference with R. V. Jones with respect to the ideal intelligence organization. In a classic lecture on "Scientific Intelligence" in February 1947 Jones remarked:

> An Intelligence organization, despite the Encyclopedia Britannica dig about there being three kinds of Intelligence—human, animal and military, resembles in fact a human head very closely. The sources of Intelligence correspond to the sense organs of the head; the detailed resemblance here is in some cases remarkable, with photographic reconnaissance as the eyes and the radio listening service as the ears. The senses pass observations to the brain, where they are correlated, and a particular sound is associated with a particular visual object. In Intelligence, information from the sources is likewise fed to a collating centre, corresponding to the brain; and while the brain, to be successful, must have a good memory, an Intelligence organization must also have a good memory built up of the individual memories of its staff and its filed records. So far no machine has been found to perform these functions nearly so well as a good human mind, and the design of an Intelligence organization must be such as to make it resemble a single perfect human mind as closely as possible. It follows from this that the most successful Intelligence organization is likely to be that which employs the smallest number of individual minds each of the greatest possible, ability.[25]

In seeking to separate signals and sprignals by incongruity testing, a counter-deception system should compensate for the vulnerability of conservatism in human data evaluation. Carl Kaysen has run into the problem of cognitive rejection, when fresh data do not mesh with existing perceptual patterns:

> Many individual items of intelligence information must be rejected as unreliable, false, irrelevant, useless, and so on. The judgment on which rejection is based is often a consistency judgment. Rejection depends on whether a particular piece of information seems reasonable in the light of a large number of other pieces of information, or whether it contributes anything which is useful in the light of such other information. If the officers responsible for analysis were deprived of the opportunity of making such consistency judgments, extending, over the whole range of intelligence materials, then their judgments would necessarily be less good than they might be.[26]

This natural rejection cycle, the removal of the "odd" bits before cooking up a *puree* of consistent data, is pernicious. Analysts should *not* assume that non-access to "the whole range" of intelligence materials will "necessarily" impair intelligence judgments. On the contrary, computers that suppress segments of the data base from consideration may be beneficial. By using techniques of sequential analysis, analysts may be able to compensate for the weaknesses in human cognitive rigidity, screening out initial perceptual patterns and preconceptions.

Machines have one advantage over the human mind: the capability to forget on a temporary basis. In this respect, a computer-assisted intelligence system has a distinct advantage over the organization Professor Jones has commended, one using "the smallest number of individual minds each of the greatest possible ability."

Instead of "rejecting" data, a counter-deception system may seek to place such data in alternative perceptual patterns. It may turn out that in the last analysis (by which the humble mean not the best but the last analysis before events overtake them), analysts will reject the first consistent patterns in favor of another one. Since stratagematists often reinforce pre-existing or early patterns, they [the analysts] should resist the temptation to think that they were "right all along," unless the retesting of hypotheses provides frequent reassurance.

The human mind may repress valuable stored data, or dwell upon obsolete images. Fusing recalled experience and wishful thinking, the human mind assigns *a priori* probabilities and only reluctantly shifts these judgments. Small group experimentation suggests a tendency to underestimate low probability events and a tendency to overestimate high probability events, hence susceptibility to deceptive reinforcement of the "conventional wisdom," and undue disregard of rare events.[27] Vulnerability to prejudice, preconception, and disregard of low probability events is reinforced by the conservative memory system of man. Given new information, the conservative system of human cognition tends not to change *a priori* probability assessments as much as the new evidence should encourage. Having recognized his own weaknesses, man can delegate some analytic functions to nonconservative probabilistic information systems.[28]

By testing alternative hypotheses—both inductive and deductive—sometimes with the "assistance" of memory and sometimes without this burden, one can identify a set of plausible alternatives, eliminating enough of the hypotheses to provide sufficiently unequivocal warning of most critical events. There is a beneficial interaction between analytic method (1), reconstructive inference, and analytic method (2), incongruity testing. The set of possible sprignal patterns derived by method (1) can be tested against the broader mass of data analyzed by method (2). If the data are still ambiguous, "crash" intelligence collection (counter-deception intelligence) could resolve the doubts or at least lead to helpful probability refinements.

Moreover, conclusions from incongruity testing would help in discerning patterns and purposes of deception. Patterns of data may be identified as inconsistent with either a high-confidence expectation or a high-confidence post mortem. These patterns may be disaggregated and the component data and channels of data "tagged" with a conditional probability of being sprignal resources. Over time, various potential sprignal resources can be tested and the conditional probabilities adjusted upward or downward as necessary.[29] Clusters of high or low-probability sprignal data should assist in evaluating alternative theories.

Computer-aided incongruity testing offers substantial hope of predictive enhancement, but, like other analytic techniques, incongruity testing has its limitations. The two central limits of incongruity testing involve what might be called disjointed incongruities and false incongruities. Disjointed incongruities involve inconsistencies that have become, in the perceptions of the viewer if not in fact, separated or mis-matched. In simplest terms, a disjointed incongruity is one that is not recognized because the sets of inconsistent patterns are never paired. In contrast, a false incongruity involves the pairing of two or more apparently inconsistent patterns that represent a consistent underlying reality. Incongruity testing involves risks of both sorts of errors. False incongruities clutter our task of identifying and choosing among genuine incongruities in two ways: Some of these apparent but unreal incongruities are a matter of different perspectives and some are a consequence of the random distribution of noise in perceptual systems. The design of disjointed incongruities is, in fact, the design of stratagems. Induction of such incongruities is labeled in Jones' paper on "The Theory of Practical Joking" as the "Theory of Spoof." Since disjointed incongruities represent the primary headache in counterdeception planning, we should explore the two main varieties, clandestinely disjointed incongruities and mutually disjointed incongruities.

Clandestinely disjointed incongruities are discussed, but not labeled, by Amrom H. Katz:

> It is perhaps tautological to suggest that we have never found anything that the Soviets have successfully concealed. This is a self-contained proposition. But because we have indeed

> found many things that the Soviets have built and deployed, it is possible that a certain kind of self-congratulatory smugness has crept into our intelligence system.[30]

Some of the most dangerous stratagems in peacetime, and in arms control environments in particular, may involve such clandestinely disjointed incongruities. Until analysts are able to pair and test the inconsistent patterns, they are unlikely to identify even that an incongruity exists.

Incongruities whose preliminary foundations are being laid by stratagematists are a special case of clandestinely disjointed incongruities. Here the problem is more difficult than that of penetrating the security of clandestine behavior, assuming that there is no special access to the future intentions of foreign stratagematists. Where foreign stratagematists have been cultivating access to our intelligence systems and where they have been "chicken feeding" these systems with generally truthful data of a consistency similar to what we would obtain by our own efforts alone, there is only one slender present incongruity to detect—that the channels of reception are either controlled or known and targeted by our stratagemic opponents.

Since analysts are unlikely to detect such slender incongruities, they can at least take steps to accelerate the detection process once their adversaries begin to feed these channels with more broadly incongruous data. They may find that the background collection of "chicken feed" will drown out the substitution of sprignals designed to mislead them at critical junctures. Masterman insists upon "the prime necessity for truth whenever truth is possible. A lie when it is needed will only be believed if it rests on a firm foundation of previous truth."[31] Masterman also believes that "the force of this [misinformation] depends upon the reputation of the sender and that a long period of truthful reporting is usually a necessary preliminary for the passing over of the lie."[32]

If counter-deception analysts have some conception of the behavioral changes they should expect to encounter, through the use of sequential statistical analysis they may be able to reduce the detection time for a set of incongruities.[33] Thus, they may be able to identify sets of sprignals and noise in combination, which with sequential analysis will allow them to disaggregate the two over time. Then, more refined sets of sprignals may be subjected to incongruity testing until, it is hoped, a set of meaningful signals emerges.

Adversary deception planners are not attempting to allow the easy matching of these incongruities. Masterman notes "the obvious fact that cover schemes ought to be as near the 'real thing' as was safely possible."[34] This places the deception planner in a bind. On one hand, he wishes to establish a deception program sufficiently close to reality that detection of incongruities will be retarded. On the other hand, he wishes to establish a deception program in accord with the preconceptions of the intended victims. As an *Abwehr* officer explained, "On the German side we only attached importance to the reports of agents insofar as they fitted in with our own conceptions of the situation."[35]

The deception picture that is compatible with an adversary's preconceptions may be highly incompatible with what one intends to do. Consequently, the deception planner must choose between a deception plan that will be readily swallowed and one that is minimally incongruous with the planned operations.

So, the incongruity of two or more patterns may emerge sufficiently early that inconsistent patterns can be identified and matched.

The second form of unidentified incongruities involves a set of *mutually disjointed incongruities* (which R. V. Jones terms just "mutual incongruities").[36] Instead of seeing situation A as perception A, one observer, B, holds perception B and another, C, perception C. It may be that perceptions B and C are, from different perspectives, representations of the same situation, but this situation is *not* situation A. Both B and C share misleading conceptions of the true situation, A.

Jones describes a situation in which neither B nor C is likely to learn of his perceptual error:

> Induced incongruities mutual to two victims are possible. Each victim is led to believe in a false world-picture which is nevertheless consistent with and complementary to the false world-picture of the other victim. A simple example is the device of privately telling each of two people whom one is about to introduce that the other is a good fellow who has, however, been going through a severe nervous strain, and who is therefore apt to get both irritable and rude if contradicted. It is therefore advisable to humor him by agreeing with all that he says. The two victims then go to great lengths to agree with one another, and separate in the conviction that each has handled the other extremely well.[37]

Another set of mutually disjointed incongruities occurred during a "real" civil defense attack warning in the western United States, which lasted for some seven minutes on May 5, 1955. An Air Force warning through civil defense channels, a *warning yellow* ("enemy attack is probable"), resulted in 67 of 77 sampled California civil defense districts failing to start public sirens after verifying the validity of the warning. The failure of operators to follow instructions was in part the result of the mutual reinforcement of cognitive templates: civil defense operators did not start the sirens in part because radio stations had not adopted CONELRAD broadcasting on specified frequencies; radio broadcasters didn't hear the sirens. One siren operator explained that he does not want to be the first person to set off the siren switches—that if he heard sirens in other areas blowing then he wouldn't mind following them; and that even if he had received no alert signals he would set his sirens off if he heard others blowing.[38]

In this case, those who did not alert the public were right; there was no attack; but most were wrong in assuming they had not received a "real" Air Force warning and were affected by the mutual consistency of their perceptions and those of other warning system personnel. Mutual inaction bred reassurance. But the same result in an actual attack would result in the loss of many more lives.

In this same "attack," a microcosmic example of shared disjointed incongruity occurred when one civil defense operator heard the warning bell, saw the "yellow alert" light, and assumed that the telephone company had fouled up again. He called the repair service of the telephone company, and "reportedly was told by the repair service that a repairman would be sent to correct the [warning] device."[39] The warning device was working properly, but the imagined telephone system error provided mutual reassurance.

Sherman Kent, retired chairman of the Board of National Estimates, has often noted the disconcerting fact that many future intentions are not knowable. Indeed, even from within the mind of the decisionmaker it will be impossible to identify incongruities before an intention is formed, and it will be impossible to decide which of the incongruous patterns is correct so long as that intention may be switched.

The Whaley study demonstrates that all the major intelligence services have the utmost difficulty with such intentions estimates, and analysts should not expect their incongruity testing to reverse the abysmal predictive record. However, in conjunction with the other methods, they can expect incongruity testing to improve performance insofar as it deals with capabilities or partially executed intentions, which leave at least preliminary "trails" of data behind them. Such testing can at least identify ranges of uncertainty in the place of falsely held confidence in one's knowledge of future intentions.

These theoretical problems are exacerbated by the ecology of bureaucratic politics, some of whose characteristics do not escape the attention of sophisticated deception planners. Many intelligence services reflect the interests of their associated departments, however much they try to remain objective. Deception generally aims at the misdirection of policy decisions, not intelligence services. If intelligence services are "had" in the process, this is only incidental, a joke of passing pleasure to the deception planner. Deception planners not only feed information that confirms preconceptions, but where possible they feed sprignals that appear to reinforce self-interest.

Thus, the inertial bond between the sprignal and the bureaucrat is doubly strong: A preconception is reinforced at a time when the denial of this preconception would appear to be adverse to bureaucratic interests.

Part of the skill in deception work is to coax one's enemy into an assessment of intentions (perhaps by dangling information that confirms preconceptions) rather than an assessment of capabilities, in a situation where the adversary should know better than to expect much luck with an intentions estimate. In the words of a former deception planner [Eldredge], "all intelligence services seem seducible into an assessment of intentions rather than capabilities—which makes them such fall guys for C&D [cover and deception]."[40]

When cryptanalytic success is at the disposition of the deception planner, he has the kind of access to his adversary's decisionmaking system that enables him to spot the apparent self-interest and preconceptions that are most ripe for exploitation:

> No deception operation works against enemy intelligence but against the enemy command. It does little good to louse up X intelligence system, if the command believes Y system. Thus a basic assumption is that coordinated intelligence is a seldom thing and that commands (political and military) tend to listen to the intelligence operation that fits their preconceptions and even their personality set and natural cultural pattern... Hitler could be reached through the nitwit RSHA and tended to neglect his hard Army intelligence.[41]

Obviously there are organizational implications, if one wishes to do more than identify deceptive and underlying realities, if one wishes to be sure that decisionmakers accept patterns that are most sensible, not those that are most appealing. That topic will be discussed in Section IV. As for the substantive dangers of consensus formation, we should bear in mind the stratagemic vulnerability of all preconceptions and especially of those preconceptions that are generally shared. At this juncture we might consider the promising forecasting technique known as the DELPHI method and its vulnerability to stratagem.[42]

Through the exploration of expert opinion, obtained in anonymous responses and iterated with controlled feedback, group judgments can be elicited and refined. Although some of the later DELPHI techniques minimize centrist tendencies, there is the danger of generating too firm a consensus, or at least the risk that others will focus on group forecasting indicators that will enlarge vulnerabilities to deception. After all, foreign deception specialists are searching for the optimal preconceptions to reinforce, and they may be seeking their own tele-DELPHI estimates of others' expert consensus. On one hand, DELPHI forecasting indicators may provide above-average hypotheses to be tested for signal confirmation; on the other hand, DELPHI forecasting indicators may provide above-average targets for foreign stratagems. So DELPHI techniques may serve simultaneously to identify patterns of expectation and hypotheses for sprignal testing.

Guarding Against Expectation, Self-Deception, and Commitment

Thoughtful intelligence analysts may seek to challenge those stable and dominating assumptions they recognize as preconceptions, either theirs or those of other analysts or politicians. But in picking assumptions that deserve to be challenged by contrast with alternative data patterns, it is important to recognize that yesterday's preconception may lead to tomorrow's expectation. Deception planners work in a world of ongoing data flow. They cannot always reinforce a widely shared preconception but may be able to modulate the wave of subsequent expectations, preconceptions that have germinated, grown, and blossomed. The initial "recognition" of a pattern derived from a fresh set of data is "an act of imagination based on observation.[43] Thus, guarding against preconceptions also involves guarding against the failure to retest: the first apparently coherent pattern of newly received data.

We tend to find what we expect to find, whether or not it exists.[44] Deception planners who cannot reinforce our memories reinforce our expectations. A pseudohistory about World War II Soviet intelligence, *Shchit i Mech* [Shield and Sword], describes the phenomenon:

> The Soviet counterespionage had organized a tremendous catch on a wholesale scale and was draining out of the 'Vali Staff' [Abwehr, Eastern Front intelligence HQ] more and more people and funds.... Some of the Abwehr intelligence groups sent into the Soviet rear, from the time of their landing, had to engage in battle with the Soviet agents. The most sensible members of such groups began working under the control of the Soviet Chekists, [burdening the Abwehr staff departments with a variety of information which here was considered as absolutely reliable since it was likely. It is precisely because of its likelihood that it had a particularly destructive effect upon the work of the Wehrmacht staffs,] more even than the daring attacks of Soviet parachute units on such staffs.[45]

Conformist tendencies encourage minorities to yield in the deliberations of juries or in the small group demonstrations of the "Asch effect," where all but one unwitting person claim knowingly that the shorter of two rods is the longer one, and the unwitting subject agrees with the crowd. Conformity aside, human data processing may involve centrist tendencies, with overestimations of small magnitudes and underestimations of large magnitudes.[46] If analysts will recognize the human tendency to seek coherence and plausibility in data that may be incoherent,[47] to confirm expectations, and to dissolve minority resistance, then they should structure the incongruity testing process so as to protect against such weaknesses. Facilities for forgetting, for segregating and aggregating data sets, for identifying outmoded assumptions of past estimates, and for challenging the fashionable notions of the day will contribute to testing of inconsistent data against alternative assumptions.

Guarding against expectations is not enough. The tendency of the human psyche to deceive itself demands that we also guard against our fantasies, our unrealistic hopes, our fears that the future will repeat the particular historical disasters we have experienced. The Freudian conception of self-deception is found in the analysis of repression and "defense mechanisms."[48] This conception is reflected in Hannah Arendt's controversial essay on "Lying in Politics,"[49] in which the prevalence of deception in politics is ascribed primarily to the narcosis of self-deception, the deception of bystanders and adversaries being almost incidental to the freeing of imagination from the impediments to action.[50] As one confidence man puts it, "You've got to realize that the victim *wants* to believe you."[51]

Was the nonexistent "missile gap" of 1957–1961 the product of independent Soviet initiative or the product of anxieties within American society that Moscow harnessed for mutual satisfaction? Was the Suez invasion in 1956 or the 1968 invasion of Czechoslovakia a question of deception and surprise, or a combination of these, and a preference by many to be surprised in the event that an invasion was under way? Without quibbling over terminology, we may recognize that cognition involves a symbiotic relationship between transmitters and receivers of information.

Whatever the causes of self-deception, the result is vulnerability to foreign stratagematists. Since deception planners seek to influence decisions, not the track record of intelligence bureaucracies, they may not concentrate upon deception through the intelligence channels of an adversary. Confidants of the adversary leader, or known fantasies of key individuals, may be exploited through more direct channels.

One problem of the intelligence service guarding against the expectations and fantasies of decisionmakers is to identify cognitive vulnerabilities that are being independently reinforced. And those in power value the privacy of their thoughts, their hopes, and their fears. If they are to be protected against the blunders of estimation that have been so plentiful in the past, those in positions of power must recognize the abundant opportunities for their victimization. Only then will they share information on top-level negotiations. Only then will they encourage critical, heretical intelligence. Without such encouragement, established intelligence agencies may draw comfort from Goethe, who wrote, "Who destroys illusion in himself and in others, nature punishes tyrannically."[52]

Commitment, on the part of intelligence estimators or decisionmakers, is likely to increase resistance to attitude change. Prior agreement of a group may prolong conformity to erroneous judgments.[53] If significant actions have been taken, on the basis of this prior belief, resistance to attitude change may be more severe.[54] If past policies have appeared to be successful, there may be the further delusion that the quality of information has been improving, even when it has not.[55] Thus, attitude change is further discouraged. So, where action has been taken, decisionmakers may fail to request supplemental intelligence review. Those concerned with forecasts for better decisions should be mindful of the need for corrective reappraisal of forecasts that are the underpinnings of current policy and past decisions.

To assure that the horizons of decisionmakers are not unduly restricted, there is a role for structured "multiple advocacy" in the decision process, as suggested by Alexander George,[56] supported by the kind of management information systems that encourage use of intelligence hypotheses other than those conforming to fashionable expectations.[57] Incongruity testing is the primary method of overcoming preconceptions, expectations, self-deception, and prior commitment. When the relevant data are contradictory, even in small but important respects, alternative hypotheses to the most "plausible" deserve recurring scrutiny.

Before passing from the subject of incongruity testing, we should note two aspects of the bureaucratic milieu, pertaining to intelligence services and degrading their ability to detect their adversaries' stratagems by this method. First, there is the well-known tendency of the collecting agency to overvalue its own sources.[58] This is not a vast problem, except insofar as more-or-less untested intelligence is fed to the highest level decision-makers. In the case of communications intelligence intercepts, electronic transcripts, or deep penetration agent reports that arrive just before a necessary decision, there is a danger that clever sprignals will be funneled to just the people who should be sheltered from them. As Robert Amory puts it, policymakers like "the succulent taste of the hot poop," which often means the very latest in current intelligence, especially that from exotic sources at the expense of the more pedestrian formal estimates, which involve a greater effort to eliminate sprignals.[59] Those channels not subjected to rigorous predistribution testing are those most tempting to the peddlers of sprignals.

Second, there is the less well-known tendency of intelligence agencies to overcollect data, at the expense of quality control.[60] This can have seriously adverse effects upon intelligence performance. Over the last two million years, the size of the human brain has been expanding at a relatively steady but slow rate.[61] If only the appetites of intelligence collectors grew at such a leisurely pace indigestion would be less troublesome. The sheer bulk of incoming data encourages only cursory examination of an overwhelming volume with a higher percent of both noise and sprignals than would be found in a more select collection program. If one had a small core of carefully scrutinized and cross-checked data that could serve as a basis for incongruity testing, one would have a head start in the predictive contest over those who are flooded by reconnaissance photos or signal intercepts but have neither the manpower nor the understanding of the importance of incongruity testing and retesting.

If we are puzzled as to why a modest collection of deceptively induced sprignals can overwhelm genuine signals arriving in larger volume, and from a wider array of sources, we may derive some understanding from studies of signal detection. J. A. Swets observes: "The accuracy of knowledge about signal characteristics is less critical for strong signals, since strong signals carry more information about these characteristics with themselves."[62]

If deception planners work to supplant an array of weak but genuine signals with strong sprignals, the seeming significance of perceived sprignals is likely to obscure the relative ignorance respecting their sources, characteristics, and channels of transmittal. The inundation of intelligence systems with increasing masses of data encourages only minimal attention to signal (and sprignal) characteristics. If an overloaded intelligence system contains sprignals with stronger impulses than those in background signals and noise, we should not be shocked if the sprignals dominate resulting perceptual patterns.

Those who think that the budgeting of another squadron of recce aircraft or the launching of another set of satellites will solve our intelligence problems should contemplate the possibility that more massive but indiscriminate data collection will have the following result: to inundate the signals with the high volume of noise, while adversary deception planners work feverishly and successfully to bombard the system at critical junctures with enough of the sprignal data to control the final pattern. If deception planners fear that their critical sprignals will be buried in the backlog of unprocessed data, they can arrange to "leak" some last-minute gem, assuring the hurdling of filtration barriers. There are too many bureaucratic and budgetary pressures encouraging overinvestment in intelligence collection systems. With fewer data and more care, with computer assistance and a counterdeception system, we can obtain better estimative results.

Vulnerability Assessment

Various statistical approaches may be adopted to predict future vulnerabilities through knowledge of those in the past. With Bayes' theorem and multivariate statistical analysis we can arrive at conditional probabilities of deception in various modes under various conditions. Game theory, simulation methods, and historical studies may yield conclusions about the likelihood and usefulness of deception in various modes, and about the payoffs through inducing wrongful rejection of accurate warnings (by labeling them sprignals). With estimates of the risks and costs of making either a Type I (lack of warning), or Type II (false positive warning) error, a reconstructive inference system and an incongruity testing system can be designed so as to optimize system outputs.[63] Studies such as Barton Whaley's *Stratagem* provide at least a tentative set of *a priori* probabilities. A set of classified studies could further refine some of these probabilities.

Aside from historical studies, war gaming, and simulation exercises, a set of field exercises could generate close-to-real-life probability ranges that would help assess the risks of various arms control postures. Although suggested from time to time since 1961, Amrom Katz's "Hiders and Finders" proposal has never been tested in the field.[64] The Arms Control and Disarmament Agency's efforts in Project CLOUD GAP (renamed FIRST LOOK) did not constitute a full-fledged field test of stratagematic vulnerabilities. Even more disconcerting are the results of one of the few off-site verification experiments as yet conducted: "Aerial and ground observations in a study by independent research teams failed to distinguish from the exterior which of three American manufacturing plants was producing the deadly VX nerve gas."[65] Although it was known that one of three plants was engaged in clandestine manufacture, off-site verification efforts failed without any complications introduced by deception specialists. With the "assistance" of "tips" as to yet other sites said to be clandestine producers, a team with even a small quota of on-site inspections might well expend that quota inspecting perfectly innocent plants.

If the United States wishes to enter into arms control arrangements that are not safe unless protected by accurate intelligence, planners should run field tests in which the "hiders'" resources include not only passive security but strategic deception as well. If they are wary of deceptions mounted by particular states or clusters of states (as, for example, the Warsaw Pact) they should seek to understand past stratagemic styles and practices.

Past national styles and practices may assist in assessing vulnerabilities to prospective stratagems and in the detection of these practices by the reconstruction of sprignal patterns. No rigorous studies of national styles of stratagem have been identified, thus the research base does not support firm conclusions. Impressionistic writings do exist and are the basis of the following tentative observations of national styles of deception.

NOTE: Here follows several informative pages on Soviet and Chinese styles of stratagem. Harris called in 1973 for a "research base" that would "support firm conclusions." Thirty-four years later and I still don't know of one. A starting point might be the thin preliminary effort in Whaley, The Prevalence of Guile: Deception through Time and across Cultures and Disciplines *(Washington, DC: FD&DC, Office of the Director of National Intelligence, 2007), 11-57, 60-64.]*

Countermeasures

Even if deception analysts have only partially succeeded in detecting those stratagems of which they are the target, they can undertake three sets of countermeasures. The first is a program to make their intelligence systems less vulnerable to stratagem; some of the systems and techniques have been previously discussed. As preventive medicine they should try to ensure that their intelligence system is designed to provide secure, variable, corroborative data. In general the U.S. intelligence system provides such data. Perhaps the greatest weakness in the system is with variability. The United States tends to collect data in accustomed ways; further, we have bought expensive hardware and associated analytic equipment, and it is costly to vary collecting patterns. It is important to collect intelligence "sources and methods" with whatever security we can provide since the greater the uncertainty about our collection parameters, the more difficult for our adversary to custom tailor sprignal packages. It is important to corroborate data in "all available lights," and where necessary to undertake "crash" collection programs (counter-deception intelligence), that authenticate or contradict sets of suspected signals and sprignals.

The second set of counter-measures involves the allocation of resources, the design of weapon systems and of arms control postures that reduce the costs of intelligence misjudgments, taking account of stratagemic vulnerabilities. When we make intentions estimates about matters that have not left a trail of evidence, we should take into account the high *a priori* probabilities of vulnerability to adversary deception. We may attempt to prolong our options or to withhold a more-substantial component of our reserves until we have greater confidence in our judgments. The withholding of resources may, in some cases, be more costly than a choice in the face of uncertainty. Where we have detected or anticipated a threat accompanied by stratagemic feints or decoys that we cannot expect to segregate, we might prefer to hold our reserves for specific parries, but we will find it irrational to do so if the "harder" indicators we seek are unlikely to materialize.

In the extreme case we may face the threat of a randomly generated choice. No matter how clever an intelligence system, it will be unable to identify the genuine threat before the choice is made. Further, if the random choice among N possible options is accompanied by N stratagems, one associated with each option, and if after the choice is made the probability of timely detection actions of a stratagem that does not accompany the actual choice is pD, then the probability of identifying the actual choice before the choice is made = 1/N, and the probability of the timely elimination of all of the misleading stratagems (of which there are N#M1) = the combination of detecting (N#Ml) events when taken (N#Ml) events at a time (#n#m1 #n#m#1) with the probability of detection = pD, or (pD) #n#m#l. This net probability will be quite low when pD is low or N is large. Further, if the probability of incorrectly eliminating (as a stratagem) the chosen option is much above zero, the expectation of reaching a correct and timely identification will be further degraded.

Where a random choice has been made and where post-choice detection of stratagems would not allow timely countermeasures (pD = 0), if one mounts P parries against

P of the N options, the probability of actually mounting a parry against the correct target is P/N. There is the risk of both spreading countermeasures too thinly to be effective and of concentrating them against the wrong targets. Depending upon the utilities and costs of various levels of counteraction, the rational game theorist will seek to minimize his maximum vulnerabilities (a *minimax* strategy) or to maximize his expected payoffs in counter-attack (a mixed strategy). In either case, dispositions are made that do not depend upon the detection of adversary stratagems.

For example, in designing an active defense and retaliatory system in the nuclear missile age, one may calculate that among other penetration aids one's adversary is likely to mount decoys that cannot be distinguished from incoming ICBMs until after the decision must be made to launch ABM interceptors. A more discriminating detection system would be preferable; perhaps the system can discriminate among most of the incoming decoys, but one cannot count upon this event. At the least, dummy ABM interceptors would increase uncertainty levels on the part of the would-be attacker. One can raise the "entry cost" to any particular target by forcing the adversary to assign a greater range of uncertainty to his estimate of the distribution of "real" interceptors (even if the adversary knows that a certain percentage of the total apparent interceptors is only a set of dummies). Similarly, if one fears a counterforce attack one can increase the misdirection of the adversary's initial attack by building ICBM launcher dummies and by camouflaging actual sites. In such cases, the value of intelligence that uncovers these deceits may be quite high; thus the designer of such "positive" and "negative" camouflage should assume that the adversary will mount expensive and sophisticated detection systems against these intelligence targets.

The point is not to deny the value of sprignal intelligence but to recognize that in some circumstances optimal allocations must be made without waiting for such intelligence. Where threats are randomly generated or where sprignal packages cannot be pierced, waiting for counter-deception indicators may be like waiting for Godot.

The third set of counter-measures involves the "playing back" of deception operations against their sponsors: counter-deception deceptions, or what I have termed "counter-stratagems." Where a set of deception patterns are identified and their strategic nexus uncovered, it is sometimes advisable to "feed" falsely confirming indicators back to the deception planners so as to lull them into the belief that they achieved the misallocations they were seeking.

An example of a partly successful counter-stratagem from World War II experience involved limited German perceptions that the Pas de Calais sector was too obviously only a deception target for the OVERLORD landings:

> I [H. W. Eldredge] talked briefly to von Blumentritt (Rundstedt's Chief of Staff) after V-E Day and he said that they had the Pas de Calais only lightly guarded (he used the word "crust") and that if we broke through and went on to Paris and cut off their main forces they would have been in a hell of a mess. He implied that they had been misleading us on the Pas de Calais' real strength (no reserves too). Thus, both he and Rundstedt had a Pas de Calais sensitivity. Despite this, as I recall, ... OKW and the mad Fuehrer, who saw the "big picture," i.e. the deception operation, bit and hung on. (Possibly, again, it's hard to admit you've been had.)

An effective counter-deception system should support the mounting of more than partly successful tactical counter-stratagems. These can be effected in combination, too, so as to provide strategic misallocations on the part of the stratagemic initiator.

A somewhat more mischievous variation involves a set of counterstratagems that do more than falsely confirm the stratagemic success of opposing deception planners. Instead of returning sprignal packages indicating total acceptance of the original deceits, one can send back counter-stratagems suggesting modified acceptance of sprignal inferences with such modifications designed to cause opposing planners to modify their own commitments. If done skillfully, these induced modifications may uncover new vulnerabilities that further counter-stratagems can perpetuate.

Chess players—of whom the Russians are among the best—will appreciate the endless layers that can be added: counter-countermeasures, and beyond. But the sheer complexity and bureaucratic inefficiency should cause most of these additional layers to collapse of their own weight.

Those who view stratagem as preposterous in the first instance will wonder what inanities I really recommend. Once more, I cite the abysmal predictive records of the past and note that I am deadly serious.

But we are not above making fun of ourselves; and I close the discussion of counter-deception strategy with a quotation from Peter Ustinov's *Romanoff and Juliet:*

> GENERAL: Incidentally, they know your code.
> AMERICAN AMBASSADOR (beaming): We know they know our code .... We only give them things we want them to know.
> GENERAL: Incidentally, they know you know their code.
> SOVIET AMBASSADOR (smiling): We have known for some time that they knew we knew their code. We have acted accordingly—by pretending to be duped.
> GENERAL: Incidentally, you know-—they know you know they know you know
> AMERICAN AMBASSADOR (genuinely alarmed): What? Are you sure?

Notes

1. B. H. Liddell-Hart, *Strategy: The Indirect Approach.* (New York: Praeger, 1954). Liddell-Hart discusses British utilization of deception in the Mareth Line campaign of February-March 1943 and in Tunisia, 6 May 1943, at pp. 282, 286-286, 289-290, in addition to his general discussion of a multiple option strategy.
2. By "uncoordinated" I mean that the analytic inferences as to the underlying intentions were not the product of all-channels intelligence analysis. One assumes that such "uncoordinated" analyses would be widely distributed among various Intelligence consumers, but this constitutes distribution, not "coordination." On the evils of disjointed, compartmented analysis, see Carl Kaysen, *Notes on Strategic Air Intelligence in World War II* (ETO) R-165 (Santa Monica, Calif.: The Rand Corporation, October 1949), pp. 24-26.
3. Professor HI. Wentworth Eldredge conceived of some of the more spontaneous stratagems as "seat-of-the-pants" deceptions. I have converted his usage in the cause of counter-deception analysts, who seem to rest more firmly on their seats.
4. See Barton Whaley, *Codeword BARBAROSSA* (Cambridge, Mass.: M.I.T. Press, 1973). See also *Stratagem* (1969 ed.), p. 146.
5. See sources cited in-Section II, Note 6. These examples are somewhat complicated by the remarkable success of the British counter-espionage services (both the Security Service, MI-5, and Section V of the Intelligence Service, MI-6) in penetrating their German counterparts. Consequently, British deceptions operated on the German services from the outside, admittedly with cryptologic insights, and through their contacts on the inside.

It is thus not surprising that Admiral Canaris, often pictured as a "hero" of the German resistance, should be portrayed by Hugh Trevor-Roper, who ran the German section of MI-6's Section V, as rather incompetent. See Trevor-Roper's portrait of the *Abwehr* chief in *The Philby Affair.*

6. Whaley, *Stratagem* (1969 ed.), p. 230. The five sets of fake plans swallowed whole involved Whaley's cases A6 (Gaza), A8 (St.-Mihiel), A38 (North Africa, TORCH), A53 (Bavarian Redoubt), and B27 (Alam Haifa). The four sets of genuine plans that were disregarded involved Whaley's cases A10 (Warsaw), A20 (Belgium), A21 (France), and A28 (USSR, BARBAROSSA). The only genuine plan accepted as such involved Whaley's case A34 (German summer offensive, Russia, 1942, BLAU). One Soviet account of an NKGB penetration of German intelligence in World War II, Vadim Kozhevnikov' s *Shchit i Mech* [Shield and Sword] (Moskva: Sovetskiy Pisatel, 1965), translated in JPRS 56046 (19 May 1972), V. II, p. 14, claims that the German war plan captured at Hechelen-sur-Meuse in January 1940 was a deception plan, not the genuine plan that Whaley claims it was. The German attack was postponed after compromise of this document and temporary military alerts in the West, but the attack on May 10, 1940, achieved substantial surprise.

   NOTE: Whaley's 1969 data was slightly corrected and added to in Barton Whaley, *Meinertzhagen's Haversack Exposed: The Consequences for Counterdeception Analysis* (Washington, DC: FD&DC, Office of the Director of National Intelligence, 2007), 25-30.

7. Calculating the binomial distribution of 10 choices at a time, with the probability of success, p=0.5, yields the following random distribution:

R, Number of Correct (p=0.5) Predictions	*pR*	*Successful Predictions*				
0	of	10	1/1,	024	=	.001
1	of	10	10/1,	024	=	.010
2	of	10	45/1,	024	=	.044
3	of	10	120/1,	024	=	.117
4	of	10	210/1,	024	=	.205
5	of	10	252/1,	024	=	.246
6	of	10	210/1,	024	=	.205
7	of	10	120/1,	024	=	.117
8	of	10	45/1,	024	=	.044
9	of	10	10/1,	024	=	.010
10	of	10	1/1,	024	=	.001
1000						

   Probability of correctly predicting 0 or 1 of 10 = 0.011 Probability of correctly predicting 2 or more of .10 = 0.989 [Source:] Hubert M. Blalock, Jr., *Social Statistics* (New York: McGraw- Hill, I960), pp. 17-118.

8. Carl Kaysen has noted "the vital importance of not attributing to an intelligence source credibility in respect to information which the source cannot reasonably be expected to possess. This is, of course, an obvious remark, but its importance was not often appreciated in World War II." *Notes on Strategic Air Intelligence in World War II (ETO)*, p. 20.

9. F. M. Cornford, in *Microcosmographia Academica* (London: Bowes & Bowes, 1953).

10. Personal Communication from H. Wentworth Eldre[d]ge.

11. Ibid.

12. Arnold L. Horelick and Myron Rush, *Strategic Power and Soviet Foreign Policy* (Chicago: University of Chicago Press, 1966), Pt. 2m, "The Politics of Soviet Missile Deception, 1957-1961," pp. 5-102.

13. "The Government of the Soviet Union authorized TASS to state that there is no need for the Soviet Union to shift its weapons ... to any other country, for instance Cuba. Our nuclear weapons are so powerful in their explosive force and the Soviet Union has such powerful rockets to carry these nuclear warheads, that there is no need to search for sites for them beyond the boundaries of the Soviet Union." *TASS*, September 11, 1962, in Graham T. Allison, *Essence of Decision: Explaining the Cuban Missile Crisis* (Boston: Little, Brown, 1971), p. 40.

14. Anthony Lake, "Lying Around Washington: The Foreign Policy Bureaucracy (2)," *Foreign Policy,* n. 2 (Spring 1971), pp. 92-93. See also David Wise, *The Politics of Lying* (New York: Random House, 1973).
15. Allison, *The Essence of Decision,* pp. 40-41, 135, discusses the apparent ignorance of ambassadors whose protestations of innocence further deception plans. In the Cuban missile crisis, assurances of Foreign Minister Gromyko were not interpreted as acts of an ignorant official. Those of Ambassador Dobrynin were. However, in the "standstill ceasefire" understanding of 1970, it was reported that "U.S. officials ... [branded Ambassador Dobrynin's] assurances to [Department of] State Secretary Rogers that the Russians would not move their missiles in the Suez as a 'deliberate lie.'" *Newsweek* (September 28, 1970), p. 17.
16. Whaley, *Stratagem* (1969), pp. 147-149, advocates the collection and analysis' of the "signals of stratagem." See also Harris, *Intelligence and National Security.*
17. The best general account of Allied deception operations in the Italian campaign is W.G.F. Jackson, [The Battle for Italy] (London and New York: Harper & Row, 1967), esp. pp. 25-26, 44, 46-48, 51 (Sicily); 110-111 (Salerno); 133 (Volturno); 146-147 (Sangro); 155 (Monte Camino); 173, 182-183 (Anzio); 177, 181 (Garigliano); 204207, 222, 225-229, 230, 236-237 (DIADEM); 266, 268 (Gothic Line); 299-3C0, 304-305 (Po Valley). See also Whaley's *Stratagem,* Cases A38, A41,A44, B31, B33, B38, and B41, for a fuller account of the stratagems but less detail of their context.
18. R. V. Jones, "Irony as a Phenomenon in Natural Science and Human Affairs," *Chemistry & Industry* (1968), p.470. See also pp. 472-474 for unanticipated revelations.
19. Personal Communication of H. Wentworth Eldredge.
20. See 0. T. Campbell, "Pattern Matching as an Essential in Distal Knowing,"' in K. R. Hammond, ed. , *The Psychology of Egon Brunswick* (New York: Holt, 1966), pp. 81-106.
21. R. V. Jones, "The Theory of Practical Joking—Its Relevance to Physics," p. 95.
22. Klaus Knorr, *Foreign Intelligence and the Social Sciences,* Research Monograph No. 17 (Princeton, N.J.: Center for International Studies, June 1, 1964), p. 23.
23. Benno Wasserman, "The Failure of Intelligence Prediction,"*Political Studies,* Vol. 8 (June 1960), 156-169. This study is marred by its failure to probe for details of the cases cited and glibly analyzed in footnotes. Wasserman misses the presence of stratagem in most of his cases, many of which are better handled in Whaley's [Stratagem].
24. See the sources on cognitive dissonance and attitude change cited in Section II, Note 55.
25. R. V. Jones, "Scientific Intelligence," *Journal of the Royal United Services Institution,* Vol. 92 (1947), p. 354.
26. Kaysen, *Notes on Strategic Air Intelligence in World War II (ETO),* p. 25.
27. M. Al[l]port and H. Raiffa, "A Progress Report on the Training of Probability Assessors," unpublished note, August 28, 1969; N. C. Dalkey and B. Brown, *Comparison of Group Judgment Techniques with Short-Range Predictions and Almanac Questions,* R-67 8-ARPA, The Rand Corporation, May 1971; Thomas A. Brown, *An Experiment in Probabilistic Forecasting,* R-944-ARPA, The Rand Corporation, forthcoming.
28. Ward Edwards, *Nonconservative Probabilistic Information Processing Systems,* Report ESD-Tk-66-404 (Ann Arbor, Mich.: Institute of Science and Technology, University of Michigan, December 1966). Andrew W. Marshall brought this and related studies to my attention.
29. The shifting conditional probability evaluation, (or "tagging") of intelligence sources and intelligence contents would probably require computer facilities, if only to keep track of both source and contents evaluations and to adjust confidence levels in the light of post mortems or special intelligence considered highly reliable. This approach should be contrasted with three other static evaluation systems: the traditional Al to E6 source-content rating system, the abandonment of quantitative evaluation in favor of a fixed verbal judgment, and the weighted "magnitude estimation scaling" approach. On the last-mentioned method see T. Meeland and R. F. Rhyne, A *Confidence Scale for Intelligence Reports: An Application of Magnitude Estimation Scaling,* SRI Technical Note RSSC-TN 4923-31 (Menlo Park, Calif.: Stanford Research Institute, June 1967). The computer-aided Bayesian approach has the advantage of freeing the analytic data base from the vulnerabilities of earlier preconceptions.
30. Amrom H. Katz, personal communication.
31. Masterman, *The Double-Cross System,* p. 46.
32. *Ibid.,* p. 34.
33. This technique is suggested by Andrew W. Marshall. Sequential analysis can both identify shifting channel exploitation and flag perceptual patterns that might not otherwise appear suspicious if analyzed in small pieces over an

extended period. In this connection, note Robert Jervis' conclusions derived from the literature on attitudinal change:

"Actors can more easily assimilate into their established image of another actor information contradicting that image if the information is transmitted and considered bit by bit than if it all comes at once...When the information arrives in a block, the contradiction between it and the prevailing view is apt to be much clearer." Robert Jervis, "Hypotheses on Misperception," *World Politics,* v. 20 (April 1968), pp. 4 65-4 66.

34. Masterman, *The Double-Cross System,* p. 164.
35. Delmer, *The Counterfeit Spy,* p. 25.
36. Jones, "The Theory of Practical Joking," p. 195.
37. Ibid., pp. 195-196.
38. Richard A. Blum, *Emergency Operations Procedures in a Civil Defense Situation* (Menlo Park, California: Stanford Research Institute, December 1955), p. 23.
39. *Ibid.,* p. 40.
40. Personal Communication of H. Wentworth Eldredge. See also Amrom H. Katz's warning to "count on nothing, and especially count not on intentions which can change faster than capabilities can be developed." Letter to the Editor, *Air Force/Space Digest* (September 1963).
41. Personal Communication of H. Wentworth Eldredge. On the vulnerability of intelligence to emotion when fused with political interest, see Irving, *The Mare's Nest;* and R. V. Jones, "Emotion, Science, and the Bomber Offensive," *The Listener* (November 30, 1961).
42. On DELPHI technique and its development, see The Rand-Corporation, *A Bibliography of Selected RAND Publications: Long-Range Forecasting and Future Computer Technology,* SB-1019 (Santa Monica, Calif.: The Rand Corporation, February 1972), and especially Norman C. Dalkey, *The DELPHI Method: An Experimental Study of Group Opinion,* RM-5 888-PR (Santa Monica, Calif.: The Rand Corporation, June 1969).
43. Lee J. Cronbach, [Essentials of Psychological Testing]. (New York: Harper, 2d. ed., 1960), quoted in Eugene J. Webb, "Individual and Organizational Forces Influencing the Interpretation of Indicators," MS, April 1969, p. 12.
44. Donald T. Campbell,. "Systematic Error on the Part of Human Links in Communication Systems," *Information and Control,* I (195-S), pp. 349-350; J. B. Juhasz and T. R. Sarbin, "On the false alarm metaphor in psychophysics," *Psychological Record,* Vol.16 (1966), pp. 322-327; M. D. Vernon, ed., *Experiments in Visual Perception* (Harmondsworth, Penguin, 1966), ch. 5. Early "cues" may trigger a series of seemingly independent expectations; when 22 witnesses identified the wrong individual in an English criminal case, the real criminal being discovered years later, or where in a U.S. case involving forgeries attributed to one person, 30 witnesses attributed the forgeries to a person who was acquitted upon proof that he was in prison at the time of at least one forgery. See Patrick Wall, *Eye-Witness Identification* (Springfield, Illinois: Thomas, 1965), p. 12. In another case, the police replicated a hunting accident in which a hunter was mistaken for a deer. The police, at the same distance, could tell that the figure was that of a man, not a deer. But "the hunters, expecting to see a deer, 'saw' a deer; the police expected to see a man and therefore 'saw' a man." D. S. Greer, "Anything but the Truth? The Reliability of Testimony in Criminal Trials, *British Journal of Criminology,* Vol. 11 (April 1971), p. 43. Expectation of an unidentified "sane" applicant, at a mental hospital led to substantial reductions in abnormal admission diagnoses, and unidentified "normal" admittees were treated as if they were abnormal. D. L. Rosenhan, "On Being Sane in Insane Places," *Science,* Vol. 179 (January 19, 1973), pp. 250-257.
45. Vadim Kozhevnikov, *Shchit i Mech* [Shield and Sword], Moskva: Sovetskiy Pisatel, 1965, translated as JPRS 56046, 19 May 1972, pp. 102-103 (emphasis added).
46. HI. L. Hlollingsworth, "The inaccuracy of movement with special reference to constant errors," *Arch. Psychology,* n.13 (1909), and citations in Campbell, "Systematic Error on the Part of Human Links in Communications Systems," pp. 345-346.
47. "The output of the human transmission and memory unit, no matter what degree of information loss, is apt to appear to a later, human unit, as intelligible and usable, as a base of action. This appearance of plausibility and comprehensibility in the output can accompany a total loss of the input message. Human beings as transmission units have this characteristic of rationalizing, of filling gaps, of providing outputs that lead to action rather than paralysis." Campbell, "Systematic Error on the Part of Human Links in Communications Systems;" p. 341.
48. Anna Freud, *The Ego and the Mechanisms of Defence* (1936), translated by Cecil Baines (New York: International Universities Press, 1946).

49. Hannah Arendt, "Lying in Politics: Reflections on the Pentagon Papers," *New York Review of Books,* Vol. 17, No. 8, November 18, 1971, pp. 30-39; in Hannah Arendt, *Lying in Politics, Civil Disobedience, on Violence, Thoughts on Politics and Revolution* (New York: Harcourt Brace, Jovanovich, 1972).
50. Also see Campbell, note 47 above.
51. Richard H. Blum, [Deceivers and Deceived] (Springfield, Illinois: Thomas, 1972), p. 42.
52. J. W. von Goethe, "Die Nature," (1782), in *Schriften uber die Natur* (Leipzig, A. Kroner Verlag, n.d.).
53. James W. Julian, C. Robert Regula, and Edwin P. Hollander, "Effects of Prior Agreement by Others on Task Confidence and Conformity," *Journal of Personality & Social Psychology 9* (1968), pp. 171-178.
54. Charles A. Kiesler et al., *The Psychology of Commitment:Experiments Linking Behavior to Belief* (New York & London: Academic Press, 1971), chs. 4 and 5.
55. Siegfried Streufert and Carl H. Castore, "Effects of Increasing Success and Failure on Perceived Information Quality," *Psychonomic Science 11* (1968), pp. 63-64.
56. Alexander L. George, "The Case for" Multiple Advocacy in Making Foreign Policy," *The American Political Science Review,* Vol. 66, No. 3 (September 1972), pp. 751-785.
57. See Seyom Brown, Paul Y. Hammond, William M. Jones, and Robert L. Patrick, *An Information System for the National Security Community,* RM-6054 (The Rand Corporation, Santa Monica), August 1969; and Bruce. F. Goeller, Paul Y. Hammond, John E. Koehler, and William B. Quandt, *Information System Applications for a High Level Staff,* R-S40 (The Rand Corporation, Santa Monica), August 1972.
58. This tendency is especially noteworthy with established espionage agents, whose case officers defend their reliability. Masterman remarks on the difficulty of "blowing" a well-established agent: "The fact that we have so many facts and details under review makes it obvious to [us] [deception controllers] that a certain message if sent ought to 'blow' an agent. But in truth it probably or almost certainly will not." *The Double-Cross System,* p. 57. "In short, it was extremely, almost fantastically difficult to 'blow' a well-established agent." *Ibid.,* p. 58. On departmental perspectives, see Dearborn and Simon, "Selective Perception: A Note on the Departmental Identification of Executives," *Sociometry,* Vol.21 (June 1958), pp. 140-144.
59. In this connection, note Carl Kaysen's judgment, derived from his World War II experience: "If intelligence from certain sources or intelligence derived by certain methods is considered supersecret and superior to the general run of intelligence, it tends to penetrate upward to command levels immediately without passing through the machinery of shirt-sleeve intelligence analysis. Intelligence items of this character can be falsely interpreted and wrongly evaluated by commanders and high-ranking staff officers who do not themselves possess the necessary background possessed by the whole intelligence staff as a unit." *Notes on Strategic Air Intelligence in World War II (ETO),* R-165 (The Rand Corporation, Santa Monica, California), October 1949, p. 22.
60. Any information processing system involves selectivity choices, whether recognized or not. See "The 'Gate Keeper': A Case Study in the Selection of News," *Journalism Quarterly,* Vol. 27 (Fall 1950), pp. 383-390. On overcollection see Amrom H. Katz, *Some Ramblings and Musings on Tactical Reconnaissance,* P-2722, The Rand Corporation, March 1963, in *Air Force/Space Digest* (August 1963) ; and Patrick J. McGarvey, *CIA: The Myth and the Madness* (New York: Saturday Review Press, 1972), pp. 4, 14, 20, 23-24, 27, 115, 213-214, and 225-227. McGarvey quotes Gilbert Fitzhugh, Chairman of a Blue Ribbon panel on the Department of Defense in July 1970: "I believe that the Pentagon suffers from too much intelligence. They can't use what they get because there is so much collected" (p. 24). The availability of digital computers postpones the necessity of remedying excess collection by providing increased processing capabilities. Edwin W. Paxson observes, "Like scientific journals, command and control requirements are exponentiating. And neither scholars nor generals can encompass the plethora of information. We shouldn't forget that a plethora is a morbid condition." *Computers and National Security,* P-4728, The Rand Corporation, January 1972, p. 14. The data processing requirements of intelligence are second only to those of logistics, according to Paxson. Special surveillance systems may generate exponential increases in data processing requirements. Major General G. T. Gould, Jr., referred to annual USAF data traffic increases of about percent, noting, "This figure ... [does] not include the datalike requirements related to imagery or special sensor-generated systems. Systems of these types require separate analysis. ...On a fulltime basis, one of these systems, for example, could generate ten times more data than all data systems are generating today." "Computers and Communications in the Information Age," *Air University Review,* Vol. 21 (May-June 1970), p. 1. Ambassador John W. Tuthill writes that recruitment of intelligence agents "can become a kind of infection in the intelligence services, "Operation TOPSY," *Foreign Policy,* n. 8 (Fall 1972), p. 74.

61. Phillip V. Tobias, *The Brain in Hominid Evolution* (New York: Columbia. University Press, 1971).
62. J. A. Swets, "Detection Theory and Psychophysics: A Review." *Psychornetrika,* Vol. 26 (1961), pp. 49-63.
63. A Bayesian model for the assessment of pattern verification capabilities on the basis of *a priori* probabilities is presented by R. C. Dixon and P. E. Boudreau, in "Mathematical Model for Pattern Verification," *IBM Journal of Research and Development,* Vol. 13 (November 1969), pp. 717-721.
64. Amrom H. Katz, *Hiders and Finders: An Approach to Evasion and Inspection Technology,* P-2432 (Santa Monica, Calif.: The Rand Corporation), April 26, 1961.
65. "Verification Task Cited at Arms Talk," *The New York Times* (July 1970), p. 3. Midwest Research Institute's summary report, *Verification Aspects of a Chemical and Biological Weapons Arms Control Agreement,* ACDA/ST-150 (Kansas City, Mo.: Midwest Research Institute, May 29, 1969), Vol. 1, p. 20, observes: "Regardless of the type of evasive tactics which may be practiced in either CW or BW activity, the estimated inspection effectiveness is high for onsite access, medium for plant-perimeter access, and low to negligible for extra-territorial access levels. This indicates that inspections conducted solely at locations, outside the border of a treaty nation will not provide the necessary assurance of compliance with a CW arms-control agreement." But if deceptive measures degrade the supposedly "high" effectiveness of on-site inspections, physical access may not be a panacea.

Conclusion

The art of deception reached a high point during the Second World War. Still, there was reluctance among Americans to accept deception as a part of modern warfare, despite the fact that the Japanese attack on Pearl Harbor was a successful deception operation of great magnitude. The slow evolution from reluctance to acceptance was encouraged and assisted by the British. But, the evolution was not an easy one, and most U.S. military leaders remained skeptical about the benefits of high-level deception operations. The success of the Fortitude plans changed American attitudes.[1] The obvious benefits of successful deception operations could not be ignored. However, American deception staffs were never as organized as the British, and the descent from the high point reached in World War II was rapid. The pendulum has swung almost completely.[2] Why is deception so difficult for the U.S. military?

A Cultural Handicap?

Officially sanctioned deception does not come easy to Americans. The remarkable Frenchman Alexis de Tocqueville observed in the 1840s that Americans are an open, honest lot. Americans have a general respect for honesty mostly based on the practical notion that honesty is necessary for efficient, normal transactions among people. Another Frenchman, lesser known, but equally remarkable, Marquis de Custine, chronicled his journey to Russia about the same time that Tocqueville was visiting America. Custine observed that Russians are secretive, deceptive, and careless with the truth both individually and as representatives of the state. Concealment, deception and lying are deep-seated in Russian official and interpersonal relations.[3]

Tocqueville and Custine were not saying that all Americans are honest and all Russians are liars. However, the premium placed on the truth by Americans is contrary to the style of relations standard among Russians. Perhaps the authoritarian nature of the Russian government, then and now, which demands loyalty to the state and enforces this norm through a system of secret police who report any deviant behavior, explains Russian dishonesty. After all, no reasonable person can truly be loyal to such a cruel state. Deception and lying become necessary for survival.[4]

Characteristics of the Russian counterintelligence state are present in many of the other state and nonstate actors who now threaten or challenge the interests of the United States. These new actors use deception and lying for the same reasons as the Russians.

1. Charles Cruickshank, *Deception in World War II,* New York: Oxford University Press, 1979, pp. 213–216.
2. Many will argue that effective deception operations would remain secret and therefore unknown to scholars. However, more than four decades of very close association with the U.S. military directly supporting operations leads me to believe that deception is a lost art in the U.S. military.
3. Brian D. Dailey and Patrick J. Parker, editors, *Soviet Strategic Deception,* Hoover Institution Press, 1987, p. 511
4. Ibid., 512.

As a result, Americans are apt to suffer in war and in international politics from a lack of cunning and a vulnerability to fall too easily into the enemy's traps. General Sir Archibald Wavell, British field marshal and a commander of British Army forces in the Middle East during the Second World War, was a master of deception. During a lecture, he referred to a time-honored definition of an ideal infantryman as an "athlete, stalker and marksman." Wavell went on to say that *his* ideal infantryman was a "cat burglar, poacher and gunman." Wavell's point was that his ideal infantryman risked his life and liberty in the exercise of his profession and success was directly related to wits, while the "athlete, stalker and marksman" was not. In fact, the former is a peacetime definition of a good infantryman, and the latter a wartime definition. Wavell's objective was to stimulate interest to consider methods of deceiving the enemy based on the historical record.[5]

Not unlike Wavell's observation of Englishmen and reflecting on his wartime experience, President Eisenhower recognized the fragile place deception holds in American military thinking as he felt compelled to advise the War Department to "take those steps that are necessary to keep alive the arts of ... cover and deception."[6] What Eisenhower couldn't foresee was the expansion of impediments that undermine developing and implementing stratagem as a tool to further U.S. interests. Regrettably, these impediments are mostly internally generated.

The Erosion of Strategy and Stratagem

The terms *strategy* and *stratagem* are derived from the Greek word strategos, which means: military leader, general, he who would decide strategy. This distinct origin for both terms is nontrivial. Strategy and stratagem must be connected, and strategy almost always precedes stratagem. Shallow strategic thinking can only deliver inept stratagems.

The complexities of the modern world have been surpassed by the complexities of the organizations whose leaders are charged with developing strategy. Overly complex organizations induce an internal focus at the expense of an external one that is necessary to more fully understand the threat and to develop strategies to defeat it.

The great labyrinth of joint, interagency, and combined or international headquarters, centers, and task forces require a style of military planning, coordination, and control that brutally subordinates the need for strategic thinking to the convenience of bureaucratic harmony. Commanders are compelled to focus more on their own internal administration and operations. As a result, they are less responsive to the external structural elements of strategy consisting of the enemy, the terrain, and other specific phenomena particular to the conflict. In other words, these organizations are so complex on the inside that they cannot possibly be responsive to the varied and often exotic phenomena on the outside.[7] Consequently, strategy and stratagem suffer.

An example of the triumph of bureaucratic harmony over strategic thought is evident in the recent awakening in the Department of Defense to revitalize military deception (MILDEC). Several years ago, a well-resourced program to educate selected

5. Archibald Wavell, *Speaking Generally—Broadcasts, Orders and Addresses in Time of War (1939–43)*, London: MacMillan & Co. LTD, 1946, p. 80.
6. Dwight D. Eisenhower, *The Papers of Dwight David Eisenhower: The Chief of Staff*, Louis Galambos, editor, Vol. VIII, Baltimore: Johns Hopkins Press, 1978, p. 1763.
7. Edward N. Luttwak, "Notes on Low Intensity Conflict," *Parameters*, December 1983, pp. 335–338.

personnel from high-level military commands on the art and science of MILDEC was established. A knowledgeable and experienced cadre of mostly contractors was assembled to travel to the various commands and conduct multiday, on-site MILDEC seminars. Unfortunately, the seminars focused exclusively on the formidable administrative hurdles required to secure approval to conduct deception and never addressed the fundamentals of how to deceive an opponent in war or peace. The seminar may have had the chilling effect of deterring units from planning deception because of the heavy burden associated with getting a deception plan approved.[8]

Competent Commanders are Good Deceivers

The National Training Center (NTC) is a relatively straightforward, realistic, and unconstrained simulation of war for brigade and battalion commanders. It is perhaps the only dynamic field environment that provides opportunities and penalties tied to the practice of deception. An unpublished RAND Corporation study on deception concluded that well-planned and executed operations contain within themselves the potential for deception. In fact, successful commanders at the NTC usually attempted deception in battle. In contrast, commanders who demonstrated low competence in battle used deception much less often. However, the link between the use of deception and success seems less clear than the link between competently executing plans and success.[9]

The point is that, given the opportunity and incentives, competent commanders will choose to deceive. Defeating the experienced, proficient NTC opposing force (OPFOR) requires using every trick in the book. Good strategy and execution are necessary for defeating the OPFOR, but usually insufficient without a supporting stratagem. It is important to note that the simplicity of the NTC environment does not replicate today's jumble of joint, interagency, and combined or international headquarters, centers, and task forces that unfortunately subordinates strategic thinking to the convenience of bureaucratic harmony. We should avoid making war more complex than it naturally is.

More Deception is Coming

The stated purpose of this book is to set in motion a renaissance for using deception as an instrument of statecraft. Deception is likely to be used more in war and in political and diplomatic interactions in the future. It is fair to assume that the chances of small wars will increase as the likelihood of major war diminishes. In fact, small and irregular wars have dominated history. Political and diplomatic stratagems can be important tools for those best prepared to use them. Current and future political and military leaders will need to be at least as clever, if not more so, than those of the past. The incentives to use deception by weaker actors are great. With the world being a trickier place, detecting deception is very important and carrying out a few of our own will almost certainly be prudent.

8. I attended one of these seminars in the summer of 2012 as this book was being put together. I was shocked to find that not once during the seminar was anything about the art and science of deception discussed. I reported this to the appropriate office in the Pentagon. HSR

9. F. Feer, "Tactical Deception at the National Training Center," The Rand Corporation, working draft, 1989, p. vii.

Glossary of Terms Relating to Deception

"A" Type Deception "Ambiguity Deception" geared toward creating general confusion.

"M" Type Deception "Misleading Deception" designed to mislead an adversary into a specific and preconceived direction.

Active Camouflage The artificial creation of the image or impression that you have a force or capability that does not actually exist.

Active Deception Any attempt to create the impression of intentions and capabilities which you do not, in fact, possess.

Conditioning The repetition of what could be preparations for a hostile action without conducting hostilities thereby lulling the victim into a false sense of security. This is a variation of the "familiarity breeds contempt" theme.

Cover The use of an apparently nonthreatening activity to disguise preparation for or initiation of a hostile act. A common example is the use of a training exercise to hide preparations for an attack.*Note:* Conditioning and cover may occur in combination with one another and they can be mutually supportive. A common example is a military training exercise.

Deception (as used in the manipulation of information and perceptions combination "Denial and to induce the target of that deception to take or not Deception") to take an action, thereby benefiting the deceiver. *Note:* "Denial and deception are interrelated. Denial is the basis for a successful deception. One cannot manipulate or blur the truth or lie convincingly unless the truth is first concealed." John Yurechko, Defense Intelligence Agency, "DoD Briefing on Iraqi Denial and Deception," Tuesday, October 8, 2002, 12:58 p.m. EDT.

Demonstration The deployment of forces to distract an enemy, but such a deployment does not usually include actual contact or combat. The purpose of a diversion is simply to mislead an enemy away from your real operations and objectives.

Denial Methods used to conceal state and military secrets, particularly from foreign intelligence collection.

Dezinformatsia The dissemination of false or misleading information intended to confuse, discredit, or embarrass the enemy. (Marshals of the Soviet Union A. A. Grechko and N. V. Ogarkov [successive Chairmen of the Main Editorial Commission], *The Soviet Military Encyclopedia; English Language Edition*, Vol. 1, William C. Green and W. Robert Reeves, ed. and trans., Boulder, CO Westview Press, 1993, pp. 345–346.)

Diversion The intentional distraction of an enemy's attention away from the area of interest or attack. Two basic types: feint and demonstration.

Fabrication The *creation* of false information or images to mislead an adversary as to your intentions and/or capabilities. This is deception *via* manufactured data (e.g., forgeries).

Feint An attack by friendly forces to distract enemy attention from your main area of interest or attack.

Manipulation The use of true or factual data in such a way as to create a false impression. The information is not false, but through using it out of context, leaving out some of the details, or providing a false balance of emphasis, the impression is skewed (e.g., being quoted out of context).

Maskirovka "A means of securing the combat operations and daily activity of forces; a complex of measures designed to mislead the enemy as to the presence and disposition of forces and various military objects, their condition, combat readiness and operations and also the plans of the commander . . . *Maskirovka* contributes to the achievement of surprise for the actions of forces, the preservation of combat readiness and the increased survivability of objects" (Grechko and Ogarkov, pp. 277–280).

Operational Deception Deception that confuses or diverts an adversary in regard to a *specific* operation or action you are preparing to conduct.

Passive Camouflage The disguise or cloaking of forces and/or facilities to prevent their detection by an enemy.

Passive Deception Efforts designed to prevent detection of your actual capabilities and intentions.

Strategic Deception Deception that disguises your *basic* objectives, intentions, strategies, and capabilities.

Tactical Deception Deception that misleads others while they are *actively involved* in competition with you, your interests, or your forces.

About the Editors

Hy Rothstein teaches in the Defense Analysis Department at the Naval Postgraduate School in Monterey, CA. He served in the U.S. Army as a Special Forces officer for more than 26 years prior to embarking on a teaching career. He is a graduate of the United States Military Academy at West Point and has a Ph.D. in international relations from the Fletcher School at Tufts University. Dr. Rothstein has written and edited books about Afghanistan (*Afghanistan and the Troubled Future of Unconventional Warfare* (2006) and *Afghan Endgames Strategy and Policy Choices for America's Longest War* (Feb 2012)), Iraq (*The Three Circles of War* (2010)), and an anthology that explores the similarities between insurgency and gang violence (*Gangs & Guerrillas* (2011)). He teaches courses and conducts research on the strategic utility of special operations, military deception, and psychological warfare.

Barton Whaley, a specialist in deception analysis, earned a Ph.D. in political science from M.I.T. in 1969. His dissertation, later published as *Codeword Barbarossa,* was the first application of the analysis of competing hypotheses (ACH) method to unmask the secret deception operation underlying a major surprise attack. He has been called one of the world's preeminent military-political deception theorists since 1969. Dr. Whaley is also a student of magic as deception and has published an award-winning reference book and biographies on that subject. Dr. Whaley has been a visiting professor in the Defense Analysis Department at the Naval Postgraduate School in Monterey, CA, teaching deception, the detection of deception, and intelligence analysis. His book, *Stratagem: Deception and Surprise in War* (reprinted recently by Artech House), is arguably the most cited work on the subject of military deception.

www.ingramcontent.com/pod-product-compliance
Lightning Source LLC
LaVergne TN
LVHW080646210926
848485LV00003B/333